Mathematics

Phase 3

NOTION PRESS

NOTION PRESS

India. Singapore. Malaysia.

ISBN xxx-x-xxxxx-xx-x

Dedicated to our Beloved Prime Minister Shree Narendra Modi ji, whose life gives the Inspiration to every Individual

*If a Man **Decides** to achieve something in his life nothing is Impossible*

CHAPTERS

1. SEQUENCES AND SERIES

1. SEQUENCE

1.1 Introduction

A sequence can be defined as an ordered collection of things (usually numbers) or a set of numbers arranged one after another. Sometimes, sequence is also referred as progression. The numbers a_1, a_2, a_3 .....a_n are known as terms or elements of the sequence. The subscript is the set of positive integers 1, 2, 3..... that indicates the position of the term in the sequence. T_n is used to denote the nth term.

Some examples of a sequence are as follows:

0, 7, 26......................, 1, 4, 7, 10......................, 2, 4, 6, 8......................

Note: The minimum number terms in a sequence should be 3.

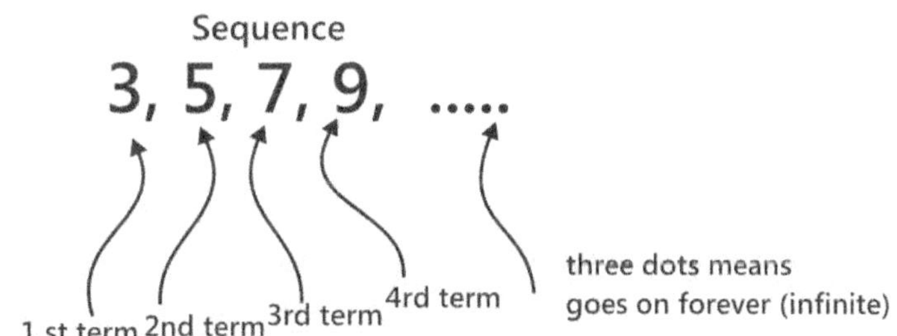

("term"", element" or " member"" mean the same thing)

Figure 3.1

1.2 Finite and Infinite Sequences

A sequence containing a finite number of terms is called a finite sequence. If the sequence contains a infinite number of terms, it is known as an infinite sequence. It is infinite in the sense that it never ends. Examples of infinite and finite sequences are as follows:

{1, 2, 3, 4......} is an infinite sequence

{20, 25, 30, 35....} is an infinite sequence

{1, 3, 5, 7} is the sequence of the first 4 odd numbers, which is a finite sequence

1.3 Rule

A sequence usually has a rule, on the basis of which the terms in the sequence are built up. With the help of this rule, we can find any term involved in the sequence. For example, the sequence {3, 5, 7, 9} starts at the number 3 and jumps 2 every time.

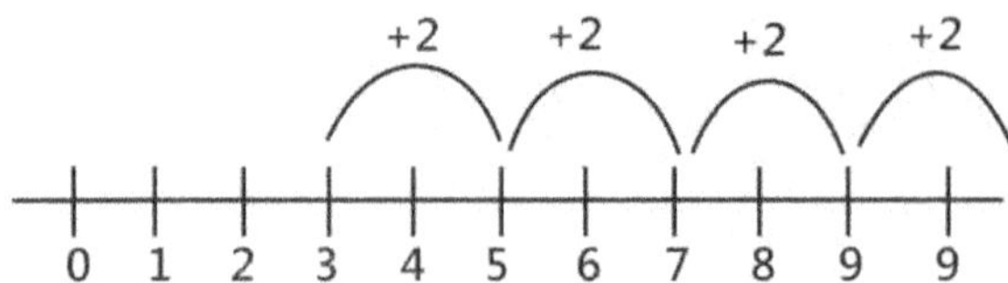

Figure 3.2

As a Formula:

Saying 'start at the number 3 and jump 2 every time' is fine, but it does not help to calculate the 10th term or 100th term or n^{th} term. Hence, we want a formula for the sequence with "n" in it (where n is any term number). What would the rule for {3, 5, 7, 9.......} be? First, we can see the sequence goes up 2 every time; hence, we can guess that the rule will be something like '2 times n' (where 'n' is the term number). Let us test it out.

n	Test Rule	Term
1	$2n = 2 \times 1 = 2$	3
2	$2n = 2 \times 2 = 4$	5
3	$2n = 2 \times 3 = 6$	7

That nearly worked! But it is less by 1 every time. Let us try changing it to 2n+1.

n	Test Rule	Term
1	$2n + 1 = 2 \times 1 + 1 = 3$	3
2	$2n + 1 = 2 \times 2 + 1 = 5$	5
3	$2n + 1 = 2 \times 3 + 1 = 7$	7

That Works: Therefore, instead of saying 'starts at the number 3 and jumps 2 every time,' we write the expression $2n + 1$. We can now calculate, e.g. the 100th term as $2 \times 100 + 1 = 201$.

1.4 Notation

The notation T_n is used to represent the general term of the sequence. Here, the position of the term in the sequence is represented by n. To mention for the '$5'^{th}$ term, just write T_5.

Thus, the rule for {3, 5, 7, 9...} can be written as the following equation: $T_n = 2n + 1$.

To calculate the 10th term, we can write $T_{10} = 2n + 1 = 2 \times 10 + 1 = 21$

Illustration 1: Find out the first 4 terms of the sequence, $\{T_n\} = \{-1/n\}^n$. **(JEE MAIN)**

Sol: By substituting n = 1, 2, 3 and 4 in $\{T_n\} = \{-1/n\}^n$, we will get the first 4 terms of given sequence.

$$T_1 = (-1/1)^1 = -1$$
$$T_2 = (-1/2)^2 = 1/4$$
$$T_3 = (-1/3)^3 = -1/27$$
$$T_4 = (-1/4)^4 = 1/256$$
$$\Rightarrow \quad \{T_n\} = \{-1, \tfrac{1}{4}, -1/27, 1/256 \ldots\}$$

Illustration 2: Write the sequence whose n^{th} term is (i) 2^n and (ii) $\log(nx)$. **(JEE MAIN)**

Sol: By substituting $n = 1, 2, 3$........, we will get the sequence.

(i) n^{th} term $= 2^n$

      $a_1 = 2^1, a_2 = 2^2$.......

      Sequence $\Rightarrow 2^1, 2^2, ...2^n$

(ii) n^{th} term $= a_n = \log(nx)$

      $a_1 = \log(x)$

      $a_2 = \log(2x), a_n = \log(nx)$

      Sequence $\Rightarrow \log(x), \log(2x), ...\log(nx)$

2. SERIES

Series is something that we get from a given sequence by adding all the terms. If we have a sequence as $T_1, T_2, ... , T_n$, then the series that we get from this sequence is $T_1 + T_2 + ... + T_n$. S_n is used to represent the sum of n terms. Hence, $S_n = T_1 + T_2 + ... + T_n$

3. SIGMA AND PI NOTATIONS

3.1 Sigma Notation

The meaning of the symbol Σ (sigma) is summation. To find the sum of any sequence, the symbol Σ (sigma) is used before its nth term. For example:

(i) $\displaystyle\sum_{n=1}^{9} n = 1+2+3+.........+9$

(ii) $\displaystyle\sum_{r=1}^{n} r^a$ or $\displaystyle\sum n^a = 1^a + 2^a + 3^a + ...+ n^a$

(iii) $\displaystyle\sum_{i=1}^{5} \frac{i+1}{2i+4} = \frac{1+1}{2\times1+4} + \frac{2+1}{2\times2+4} + \frac{3+1}{2\times3+4} + \frac{4+1}{2\times4+4} + \frac{5+1}{2\times5+4}$

Properties of Σ (Sigma)

(i) $\displaystyle\sum_{i=1}^{k} a = a + a + a... \text{ (k times)} = ka$, where a is a constant.

(ii) $\displaystyle\sum_{i=1}^{k} ai = a\sum_{i=1}^{k} i$, where a is a constant.

(iii) $\displaystyle\sum_{r=1}^{n} (a_r \pm b_r) = \sum_{r=1}^{n} a_r \pm \sum_{r=1}^{n} b_r$

(iv) $\displaystyle\sum_{i=i_0}^{i_n} \sum_{j=j_0}^{j_n} a_i a_j = \sum_{j=j_0}^{j_n} \sum_{i=i_0}^{i_n} a_i a_j$

3.2 Pi Notation

The symbol Π denotes the product of similar terms. For example:

(i) $\displaystyle\prod_{n=1}^{6} n = 1 \times 2 \times 3 \times 4 \times 5 \times 6$

(ii) $\displaystyle\prod_{n=1}^{k} n^m = 1^m \times 2^m \times 3^m \times 4^m \times .... \times k^m$

(iii) $\displaystyle\prod_{n=1}^{k} n = 1 \times 2 \times 3 \times .... \times k = k!$

4. ARITHMETIC PROGRESSION

The sequence in which the successive terms maintain a constant difference is known as an arithmetic progression (AP). Consider the following sequences:

      $a, \quad a + d, \quad a + 2d, \quad a + 3d$

      $T_1, \quad\quad T_2, \quad\quad T_3, \quad\quad T_4$

$\because T_2 - T_1 = T_3 - T_2 = T_4 - T_3$ = constant (common difference)

The given sequence is an example of AP. The set of natural numbers is also an example of AP.

4.1 General Term

General term (nth term) of an AP is given by $T_n = a + (n - 1)\,d$, where a is the first term of the sequence and d is the common difference of the sequence.

Note:

(i) General term is also denoted by ℓ (last term).

(ii) n (number of terms) always belongs to the set of natural numbers.

(iii) Common difference can be zero, + ve or – ve.

If $d > 0 \Rightarrow$ increasing AP and the sequence tends to $+\infty$

If $d < 0 \Rightarrow$ decreasing AP and the sequence tends to $-\infty$

If $d = 0 \Rightarrow$ constant AP (all the terms remain same)

(iv) The nth term from end is $(m - n + 1)$ term from the beginning, where m is the total number of terms and is given by the following expression:

$T_{m-n+1} = T_m - (n-1)\,d$

NOMORECLASS CONCEPTS

- If the mth term is n and the nth term is m, then the $(m + n)^{th}$ term is 0.
- If m times the mth term is equal to n times the nth term, then the $(m + n)^{th}$ term is 0.

Illustration 3: If the 5th term of an AP is 17 and its 7th term is 15, then find the 22th term. **(JEE MAIN)**

Sol: Using the formula $T_n = a + (n - 1)\,d$, we can solve above problem.

Given $a + 4d = 17$ and $a + 6d = 15$

$\Rightarrow \quad 2d = -2 \Rightarrow d = -1,\ a = 21$

$\therefore \quad T_{22} = 21 - 21 = 0$

Illustration 4: If 11 times the 11th term of an AP is equal to 9 times the 9th term, then find the 20th term. **(JEE MAIN)**

Sol: By solving $11\,(a + 10d) = 9\,(a + 8d)$, we will get the value of a and d.

$\therefore\ 2a = -38\,d \Rightarrow a = -19d$

$\therefore$ 20th term $= a + 19d = 0$

Illustration 5: Check whether the sequences given below are AP or not. **(JEE MAIN)**

(i) $T_n = n^2$ (ii) $T_n = an + b$

Sol: By taking the difference of two consecutive terms, we can check whether the sequences are in AP or not.

(i) $T_n = n^2;\ T_{n-1} = (n - 1)^2$

Difference = $T_n - T_{n-1} = n^2 - (n-1)^2 = n^2 - (n^2 - 2n + 1) = 2n - 1$

This difference varies with respect to the term. Hence, the sequence is not an AP.

(ii) $T_n = an + b;\ T_{n-1} = a(n-1) + b$

Difference = $(an + b) - (a(n-1) + b) = a$ (constant)

Hence, the sequence is an AP.

Illustration 6: The 2nd, 31st and the last term of an AP are given as $7\dfrac{3}{4}, \dfrac{1}{2}$ and $-6\dfrac{1}{2}$, respectively. Find the first term and the number of terms. **(JEE MAIN)**

Sol: Using $T_n = a + (n-1)\, d$, we can get the first term and common difference. Suppose a be the first term and d be the common difference of the AP.

Given, $T_2 = 7\dfrac{3}{4} \Rightarrow a + d = \dfrac{31}{4}$ .... (i)

$T_{31} = \dfrac{1}{2} \Rightarrow a + 30d = \dfrac{1}{2}$ .... (ii)

Subtracting (i) from (ii), we get $29d = \dfrac{1}{2} - \dfrac{31}{4} = -\dfrac{29}{4} \Rightarrow d = \dfrac{-1}{4}$

Putting the value of d in (i), we get $a - \dfrac{1}{4} = \dfrac{31}{4} \Rightarrow a = \dfrac{31}{4} + \dfrac{1}{4} = \dfrac{32}{4} = 8$

Suppose the number of terms be n, so that $T_n = -\dfrac{13}{2}$

i.e. $a + (n-1)\, d = -\dfrac{13}{2} \Rightarrow 8 + (n-1)\left(-\dfrac{1}{4}\right) = -\dfrac{13}{2}$

$\Rightarrow 32 - n + 1 = -26 \Rightarrow n = 59$

Hence, the first term = 8 and the number of terms = 59.

Illustration 7: Prove that the square roots of three unequal prime numbers cannot be three terms of an AP. **(JEE ADVANCED)**

Sol: Here by Considering $\sqrt{p}$, $\sqrt{q}$, $\sqrt{r}$ to be the λth, μ^{th} and v^{th} terms of an AP and solving them using $T_n = a + (n-1)\, d$, we prove the problem.

If possible let $\sqrt{p}$, $\sqrt{q}$, $\sqrt{r}$ be the three terms of an AP. $a, a + d, a + 2d$........., where $p \neq q \neq r$ and they are prime numbers.

Let them be the λth, μ^{th} and v^{th} terms, respectively.

$\therefore\ \sqrt{p} = a + (\lambda - 1)\, d$

$\sqrt{q} = a + (\mu - 1)\, d$

$\sqrt{r} = a + (v - 1)\, d$

$\therefore\ \sqrt{p} - \sqrt{r} = (\lambda - \mu)\, d$

Also, $\sqrt{q} - \sqrt{r} = (\mu - v)\, d$

$\therefore\ \dfrac{\sqrt{p} - \sqrt{q}}{\sqrt{q} - \sqrt{r}} = \dfrac{\lambda - \mu}{\mu - v}$ or $\dfrac{\left(\sqrt{p} - \sqrt{q}\right)\left(\sqrt{q} + \sqrt{r}\right)}{\left(\sqrt{q} - \sqrt{r}\right)\left(\sqrt{q} + \sqrt{r}\right)} = \dfrac{\lambda - \mu}{\mu - v}$

or $\sqrt{pq} + \sqrt{pr} - q - \sqrt{qr} = \dfrac{\lambda - \mu}{\mu - v}(q - r)$ or $\sqrt{pq} + \sqrt{pr} - \sqrt{qr} = q + \dfrac{\lambda - \mu}{\mu - v}(q - r)$ = rational number

Since p, q, r are unequal primes, $\sqrt{pq}, \sqrt{pr}$ and $\sqrt{qr}$ are unequal pure irrational numbers. Thus, LHS is irrational, but irrational $\neq$ rational.

Hence, the problem is proved.

Illustration 8: If x, y and z are real numbers satisfying the equation $25\,(9x^2 + y^2) + 9z^2 - 15\,(5xy + yz + 3zx) = 0$, then prove that x, y and z are in AP.   **(JEE ADVANCED)**

Sol: By solving the equation $25\,(9x^2 + y^2) + 9z^2 - 15\,(5xy + yz + 3zx) = 0$, we can prove that x, y and z are in AP.

We have

$$(15x)^2 + (5y)^2 + (3z)^2 - (15x)(5y) - (5y)(3z) - (3z)(15x) = 0$$

$$\Rightarrow \quad (15x - 5y)^2 + (5y - 3z)^2 + (3z - 15x)^2 = 0$$

$$\Rightarrow \quad 15x - 5y = 0,\; 5y - 3z = 0,\; 3z - 15x = 0$$

$$\Rightarrow \quad 15x = 5y = 3z \Rightarrow \frac{x}{1} = \frac{y}{3} = \frac{y}{5} \;(= k \text{ say})$$

$$\therefore \quad x = k,\; y = 3k,\; z = 5k$$

Thus, x, y and z are in AP.

Illustration 9: Let $a_1, a_2, a_3, \ldots, a_n$ be in AP, where $a_1 = 0$ and the common difference $\neq 0$. Show that

$$\frac{a_3}{a_2} + \frac{a_4}{a_3} + \frac{a_5}{a_4} + \ldots \ldots \frac{a_n}{a_{n-1}} - a_2 \left(\frac{1}{a_2} + \frac{1}{a_3} + \ldots \ldots + \frac{1}{a_{n-2}} \right) = \frac{a_{n-1}}{a_2} + \frac{a_2}{a_{n-1}}$$

(JEE ADVANCED)

Sol: Given $a_1 = 0$ and $d = a_2 - a_1 = a_2 - 0 = a_2$

By solving LHS and RHS separately, we can solve the problem.

$$\text{LHS} = \frac{a_3 - a_2}{a_2} + \frac{a_4 - a_2}{a_3} + \frac{a_5 - a_2}{a_4} + \ldots + \frac{a_{n-1} - a_2}{a_{n-2}} + \frac{a_n}{a_{n-1}}$$

$$= \frac{a_2}{a_2} + \frac{a_3}{a_3} + \frac{a_4}{a_4} + \ldots \ldots + \frac{a_{n-2}}{a_{n-2}} + \frac{a_n}{a_{n-1}}$$

$$= (n-3) + \frac{a_n}{a_{n-1}} = (n-3) + \frac{a_1 + (n-1)d}{a_1 + (n-2)d} = (n-3) + \frac{n-1}{n-2} \;\{\because a_1 = 0\}$$

$$\text{RHS} = \frac{a_1 + (n-2)d}{a_2} + \frac{a_2}{a_1 + (n-2)d} = (n-2) + \frac{1}{n-2} = (n-3) + 1 + \frac{1}{n-2} = (n-3) + \frac{n-1}{n-2}$$

$$\therefore \text{LHS} = \text{RHS}$$

NOMORECLASS CONCEPTS

A sequence obtained by multiplication or division of corresponding terms of two APs may not be in AP

For example, let the first AP be 2, 4, 6, 8,……………… and the second AP be 1, 2, 3, 4, 5…….

Multiplying these two, we get 2,8,18,32, ………, which is clearly not an AP

4.2 Series of an AP

Series of an AP can be obtained as

S_n = a + (a + d) + (a + 2d) ...... [a+(n – 1)d]

S_n = [a + (n – 1)d] + [a+(n – 2)d] ...... + a (writing in the reverse order)

$\therefore$ $2S_n$ = n(2a + (n –1) d)

$\therefore$ Sum to n terms, $S_n = \dfrac{n}{2}(2a + (n-1)d) = \dfrac{n}{2}\left(T_1 + T_n\right)$

Illustration 10: Find the sum of the first 19 terms of an AP when $a_4 + a_8 + a_{12} + a_{16} = 224$. **(JEE MAIN)**

Sol: We need to find out the sum of the first 19 terms of an AP, i.e. $\dfrac{19}{2}$ (2a + 18d), and we can represent the given equation as (a + 3d) + (a + 7d) + (a + 11d) + (a + 15d) = 224.

Given (a + 3d) + (a + 7d) + (a + 11d) + (a +15d) = 224

$\Rightarrow$ 4a + 36d = 224 $\Rightarrow$ a + 9d = 56

Sum of the first 19 terms $\Rightarrow$ S = $\dfrac{19}{2}$ (2a + 18d) = $\dfrac{19}{2} \times 2 \times 56 = 1064$

Illustration 11: The sum of n terms of two arithmetic progressions is in the ratio of $\dfrac{7n+1}{4n+27}$. Find ratio of the 11th terms? **(JEE MAIN)**

Sol: Since we know the sum of n terms, i.e. $S_n = \dfrac{n}{2}\left[2a+(n-1)d\right]$, we can write the equation as

$\dfrac{7n+1}{4n+27} = \dfrac{\dfrac{n}{2}(2a_1 + (n-1)d_1)}{\dfrac{n}{2}(2a_2 + (n-1)d_2)}$. Hence, by putting n = 11 in this equation, we can obtain the ratio of the 11th terms.

$$\dfrac{7n+1}{4n+27} = \dfrac{a_1 + \left(\dfrac{n-1}{2}\right)d_1}{a_1 + \left(\dfrac{n-1}{2}\right)d_2}$$

We want the ratio of $\dfrac{a_1 + 10d_1}{a_2 + 10d_2}$. Hence, $\dfrac{n-1}{2} = 10 \Rightarrow n = 21$

$\Rightarrow \dfrac{a_1 + 10d_2}{a_2 + 10d_2} = \dfrac{148}{111}$

Illustration 12: In an AP of n terms, prove that the sum of the kth term from the beginning and the kth term from the end is independent of k and equal to the sum of the first and last terms. **(JEE MAIN)**

Sol: Using the formula $T_k = a + (k-1)d$ and $T_{n-k+1} = [a + (n-k)d]$, we can obtain the kth term from the beginning and end, respectively, and after that by adding these values we can prove the given problem.

Suppose a be the first term and d be the common difference of the AP.

$\therefore$ kth term from the beginning = $T_k = a + (k-1)d$ ... (i)

Let l be the last term of the AP and l = a + (n – 1) d

The kth term from the end of the given AP is the $(n - k + 1)^{th}$ term from the beginning.

$$\therefore \quad T_{n-k+1} = [a + (n-k)d] \qquad \qquad \text{...(ii)}$$

Adding (i) and (ii), we get

$$\therefore \text{ The required sum } = T_k + T_{n-k+1} = [a+(k-1)d] + \Big[a+(n-k)d\Big]\, 2a + (k-1+n-k)\,d$$

$$= 2a + (n-1)\,d \qquad \qquad \text{.... (iii), which is independent of } k$$

Moreover, the sum of the first and last terms $= a + l = a + [a + (n-1)\,d] = 2a + (n-1)\,d.$ ...(iv)

Thus, the sum of the first and last terms is independent of k and (3) = (4). Hence proved.

Illustration 13: If $a_1, a_2, a_3, ..., a_n$ is an AP of non-zero terms , then prove that   **(JEE ADVANCED)**

$$\frac{1}{a_1 a_2} + \frac{1}{a_2 a_3} + ... + \frac{1}{a_{n-1} a_n} = \frac{n-1}{a_1 a_n}$$

Sol: By considering a as the first term and d as the common difference, we can write a_n as a + (n − 1) d, where n = 1, 2, 3,... n.

$$\frac{1}{a_1 a_2} + \frac{1}{a_2 a_3} + ... + \frac{1}{a_{n-1} a_n} = \frac{1}{a(a+d)} + \frac{1}{(a+d)(a+2d)} + ... \frac{1}{[a+(n-2)d][a+(n-1)d]}$$

$$= \frac{1}{d}\left(\frac{1}{a} - \frac{1}{a+d}\right) + \frac{1}{d}\left(\frac{1}{a+d} - \frac{1}{a+2d}\right) + ... + \frac{1}{d}\left(\frac{1}{a+(n-2)d} - \frac{1}{a+(n-1)d}\right)$$

$$= \frac{1}{d}\left(\frac{1}{a} - \frac{1}{a+(n-1)d}\right) = \frac{a+(n-1)d-a}{ad(a+(n-1)d)} = \frac{(n-1)d}{ad(a+(n-1)d)} = \frac{n-1}{a(a+(n-1)d)} = \frac{n-1}{a_1 a_n}$$

NOMORECLASS CONCEPTS

Facts:

- If each term of an AP is increased, decreased, multiplied or divided by the same non-zero number, the resulting sequence is also an AP.

- The sum of the two terms of an AP equidistant from the beginning and end is constant and is equal to the sum of the first and last terms.

$$a_1 + a_n = a_2 + a_{n-1} = a_3 + a_{n-2} = ...$$

Illustration 14: Split 69 into three parts such that they are in AP and the product of the two smaller parts is 483.

    (JEE MAIN)

Sol: By considering the three parts as a − d, a, and a + d and using the given conditions, we can solve the given problem.

Sum of the three terms $= 69 \quad \Rightarrow \quad (a-d) + a + (a+d) = 69$

$\Rightarrow \qquad 3a = 69 \qquad \qquad \Rightarrow \quad a = 23$ ..... (i)

Product of the two smaller parts = 483 $\Rightarrow$ a (a – d) = 483

$\Rightarrow$ 23(23 – d) = 483 $\qquad\Rightarrow$ 23 – d = 21 $\quad\Rightarrow$ d = 23 – 21 = 2

Hence, the three parts are 21, 23 and 25.

Illustration 15: Divide 32 into four parts that are in AP such that the ratio of the product of extremes to the product of mean is 7: 15. **(JEE MAIN)**

Sol: We can consider the four parts as (a – 3d), (a – d), (a + d) and (a + 3d).

Sum of the four parts = 32

$\Rightarrow$ (a – 3d) + (a – d) + (a + d) + (a + 3d) = 32 $\qquad\Rightarrow$ 4a = 32 $\Rightarrow$ a = 8

And $\dfrac{(a-3d)(a+3d)}{(a-d)(a+d)} = \dfrac{7}{15} \Rightarrow \dfrac{a^2-9d^2}{a^2-d^2} = \dfrac{7}{15} \Rightarrow \dfrac{64-9d^2}{64-d^2} = \dfrac{7}{15}$

$\Rightarrow 128\,d^2 = 512 \Rightarrow d^2 = 4 \Rightarrow d = \pm\,2$

Thus, the required parts are 2, 6, 10 and 14.

Illustration 16: If a + b + c $\neq$ 0 and $\dfrac{b+c}{a}, \dfrac{c+a}{b}, \dfrac{a+b}{c}$ are in AP, then prove that $\dfrac{1}{a}, \dfrac{1}{b}, \dfrac{1}{c}$ are also in AP.

(JEE ADVANCED)

Sol: Here $\dfrac{b+c}{a}, \dfrac{c+a}{b}, \dfrac{a+b}{c}$ are in AP; therefore, by adding 1 to each term and then by dividing each term by

a + b +c, we will get the required result $\dfrac{b+c}{a}, \dfrac{c+a}{b}, \dfrac{a+b}{c}$ are in AP

Adding 1 to each term, find that $\left(\dfrac{b+c}{a}+1\right), \left(\dfrac{c+a}{b}+1\right), \left(\dfrac{a+b}{c}+1\right)$ are in AP

i.e. $\dfrac{a+b+c}{a}, \dfrac{c+a+b}{b}, \dfrac{a+b+c}{c}$ are in AP

Dividing each term by a + b + c, we find that $\dfrac{1}{a}, \dfrac{1}{b}, \dfrac{1}{c}$ are in AP

Illustration 17: If $a\left(\dfrac{1}{b}+\dfrac{1}{c}\right), b\left(\dfrac{1}{a}+\dfrac{1}{c}\right), c\left(\dfrac{1}{a}+\dfrac{1}{b}\right)$ are in AP, then prove that a, b, c are in AP. **(JEE ADVANCED)**

Sol: By adding 1 and then multiplying by $\left(\dfrac{abc}{ab+bc+ac}\right)$ to each term, we will get the result.

$a\left(\dfrac{1}{b}+\dfrac{1}{c}\right), b\left(\dfrac{1}{a}+\dfrac{1}{c}\right), c\left(\dfrac{1}{a}+\dfrac{1}{b}\right)$ are in AP $\Rightarrow$ $a\left(\dfrac{b+c}{bc}\right), b\left(\dfrac{a+c}{ac}\right), c\left(\dfrac{b+a}{ab}\right)$ are in AP

Adding 1, we find that $\dfrac{ab+ac}{bc}+1, \dfrac{ab+bc}{ac}+1, \dfrac{bc+ac}{ab}+1$ are in AP

$\Rightarrow$ $\dfrac{ab+ac+bc}{bc}, \dfrac{ab+bc+ac}{ac}, \dfrac{bc+ac+ab}{ab}$ are in AP

Multiplying by $\left(\dfrac{abc}{ab+bc+ac}\right)$ to all the terms, we find that a, b, c are in AP

4.3 Arithmetic Mean

The arithmetic mean (AM) A of any two numbers a and b is given by the equation (a + b)/2. Please note that the sequence a, A, b is in AP. If a_1, a_2, ..., a_n are n numbers, the (AM) A, of these numbers is given by:

$$A = \frac{1}{n}(a_1 + a_2 + ... + a_n)$$

Inserting 'n' AMs between 'a' and 'b'

Suppose A_1, A_2, A_3, ..., A_n be the n means between a and b. Thus, a, A_1, A_2 ,...., A_n, b is an AP and b is the $(n + 2)^{th}$ term.

Thus, $b = a + (n + 1)d$ $\Rightarrow$ $d = \dfrac{b-a}{n+1}$

Now,

$A_1 = a + d$

$A_2 = a + 2d$

$\vdots$

$\underline{A_n = a + nd}$

$$\sum_{i=1}^{n} A_i = na + (1 + 2 + 3 + ... + n)d = na + \left(\frac{n(n+1)}{2}\right)d = na + \left(\frac{n(n+1)}{2}\right)\left(\frac{b-a}{n+1}\right)$$

$$= \frac{n}{2}[2a + b - a] = nA \text{ where, } A = \frac{a+b}{2}$$

Note: The sum of the n AMs inserted between a and b is equal to n times A.M. between them.

Illustration 18: Insert 20 AMs between the numbers 4 and 67. **(JEE MAIN)**

Sol: Given, a = 4 and b = 67; therefore by using the formula $d = \dfrac{b-a}{n+1}$, we can solve it.

$$d = \frac{67 - 4}{20 + 1} = 3$$

$A_1 = a + d$ $\Rightarrow$ $A_1 = 7$

$A_2 = a + 2d$ $\Rightarrow$ $A_2 = 10$

$A_3 = a + 3d$ $\Rightarrow$ $A_3 = 13$

$A_{20} = a + 20 d$ $\Rightarrow$ $A_{20} = 63$

Thus, between 4 and 67, 20 AMs are 7, 10, 13, 16, ..., 63.

Illustration 19: If $\dfrac{a^n + b^n}{a^{n-1} + b^{n-1}}$ is the A.M. between a and b, find the value of n. **(JEE ADVANCED)**

Sol: Since the A.M. between a and b = $\dfrac{a+b}{2}$, we can obtain the value of n by equating this to $\dfrac{a^n + b^n}{a^{n-1} + b^{n-1}}$.

$\Rightarrow \dfrac{a^n + b^n}{a^{n-1} + b^{n-1}} = \dfrac{a+b}{2}$ [Given] $\Rightarrow 2a^n + 2b^n = a^n + ab^{n-1} + a^{n-1}b + b^n$

$\Rightarrow a^n - a^{n-1}b = ab^{n-1} - b^n$ $\Rightarrow a^{n-1}(a - b) = b^{n-1}(a - b)$

$\Rightarrow a^{n-1} = b^{n-1}$ [$\because a \neq b$] $\Rightarrow \left(\dfrac{a}{b}\right)^{n-1} = 1 = \left(\dfrac{a}{b}\right)^0 \left(\because \left(\dfrac{a}{b}\right)^0 = 1\right)$

$\Rightarrow n - 1 = 0$ $\Rightarrow n = 1$

Illustration 20: Between 1 and 31, m arithmetic means are inserted in such a way that the ratio of the 7th and $(m - 1)$th means is 5: 9. Calculate the value of m. **(JEE MAIN)**

Sol: AMs inserted between 1 and 31 are in AP. Thus, by considering d to be the common difference of AP and obtaining the 7th and $(m - 1)$th means we can solve the problem.

Suppose $A_1, A_2, A_3, A_4 \ldots\ldots A_m$ be the m AMs between 1 and 31.

Thus, $1, A_1, A_2, \ldots A_m, 31$ are in AP

The total number of terms is $m + 2$ and $T_{m+2} = 31$

$$1 + (m + 2 - 1) d = 31 \quad \Rightarrow (m + 1) d = 30 \Rightarrow d = \frac{30}{m+1}$$

$$A_7 = T_8 = a + 7d = 1 + 7 \times \frac{30}{m+1} = \frac{m+1+210}{m+1} = \frac{m+211}{m+1}$$

$$A_{m-1} = T_m = 1 + (m-1) d = 1 + (m-1) \times \frac{30}{m+1} = \frac{m+1+30m-30}{m+1} = \frac{31m-29}{m+1}$$

$$\frac{A_7}{A_{m-1}} = \frac{(m+211)/(m+1)}{(31m-29)/(m+1)} = \frac{m+211}{31m-29} = \frac{5}{9} \qquad \text{[Given]}$$

$$\Rightarrow \quad \frac{m+211}{31m-29} = \frac{5}{9} \quad \Rightarrow \quad 9m + 1899 = 155m - 145$$

$$\Rightarrow \quad 146m = 2044 \quad \Rightarrow \quad m = \frac{2044}{146} = 14 \ ; \text{Thus, m = 14}$$

Illustration 21: Gate receipts at the show of "Baghbaan" amounted to Rs 9500 on the first night and showed a drop of Rs 250 every succeeding night. If the operational expenses of the show are Rs 2000 a day, find out on which night the show ceases to be profitable? **(JEE MAIN)**

Sol: Here, $a = 9500$ and $d = -250$. The show ceases to be profitable on the night when the receipts are just Rs 2000. Thus, by considering that it will happen at nth night and using $T_n = a + (n - 1) d$, we can solve this problem.

We have the cost of gate receipt on the first night $(a) = 9500$

Common difference $(d) = -250$

Suppose, it happens on the nth night, then

$$2000 = 9500 + (n - 1) (-250) \quad \Rightarrow \quad 2000 - 9500 = -250n + 250$$

$$\Rightarrow \quad -7500 - 250 = -250n \qquad \Rightarrow \quad -7750 = -250n \qquad \Rightarrow \quad n = \frac{7750}{250} = 31$$

NOMORECLASS CONCEPTS

(a) If the sum of n terms S_n is given, then the general term $T_n = S_n - S_{n-1}$, where S_{n-1} is sum of $(n-1)$ terms of AP.

(b) In a series, if S_n is a quadratic function of n or T_n is a linear function of n, the series is an AP.

   (i) If $T_n = an + b$, the series so formed is an AP and its common difference is a.

   (ii) If $S_n = an^2 + bn + c$, the series so formed is an AP and its common difference is 2a.

(c) If in a finite AP, the number of terms is odd, then its middle term is the A.M. between the first term and last term and its sum is equal to the product of the middle term and number of terms.

(d) It is found that the sum of infinite terms of an AP is ∞, if $d > 0$ and $-\infty$, if $d < 0$.

(e) If for an AP, the pth term is q and the qth term is p, then the mth term is $= p + q - m$.

(f) If for an AP, the sum of p terms is q and sum of q terms is p, then the sum of $(p + q)$ terms is $-(p + q)$.

(g) If for an AP, the sum of p terms is equal to the sum of q terms, then the sum of $(p + q)$ terms is zero.

(h) If for different APs, $\dfrac{S_n}{S'_n} = \dfrac{f_n}{\phi_n}$, then $\dfrac{T_n}{T'_n} = \dfrac{f(2n-1)}{\phi(2n-1)}$.

(i) If for two APs, $\dfrac{T_n}{T'_n} = \dfrac{An+B}{Cn+D}$, then we find that $\dfrac{S_n}{S'_n} = \dfrac{A\left(\dfrac{n+1}{2}\right)+B}{C\left(\dfrac{n+1}{2}\right)+D}$.

An Important Property of AP: A sequence is said to be an AP if the sum of its n terms is of the form $An^2 + Bn$, where A and B are constants. Thus, the common difference of the AP is 2A.

Proof: Suppose, a and d be the first term and common difference of AP, respectively, and S_n be the sum of n terms.

$$S_n = \frac{n}{2}[2a + (n-1)d]$$

$$\Rightarrow \quad S_n = an + \frac{n^2}{2}d - \frac{n}{2}d = \left(\frac{d}{2}\right)n^2 + \left(a - \frac{d}{2}\right)n$$

$$\Rightarrow \quad S_n = An^2 + Bn, \text{ where } A = \frac{d}{2} \text{ and } B = a - \frac{d}{2}$$

Hence, the sum of n terms of an AP is of the form $An^2 + Bn$.

Conversely, suppose the sum S_n of n terms of a sequence $a_1, a_2, a_3, \ldots\ldots a_n\ldots$ is of the form $An^2 + Bn$. Then, we have to prove that the sequence is an AP.

We have $\quad S_n = An^2 - Bn$

$\Rightarrow \qquad S_{n-1} = A(n-1)^2 + B(n-1) \qquad$ [On replacing n by n = 1]

Now, $\qquad A_n = S_n - S_{n-1}$

$\Rightarrow \qquad A_n = \{An^2 + Bn\} - \{A(n-1)^2 - B(n-1)\} = 2An + (B - A)$

$\Rightarrow \qquad A_{n+1} = 2A(n+1) + (B - A) \qquad$ [On replacing n by (n + 1)]

$\therefore \qquad A_{n+1} - A_n = \{2A(n + 1) + B - A\} - \{2An + (B - A)\} = 2A$

Since $A_{n+1} - A_n = 2A$ for all $n \in N$, the sequence is an AP with a common difference 2A.

For example, if $S_n = 3n^2 + 2n$, we can say that it is the sum of the n terms of an AP with a common difference of 6.

5. GEOMETRIC PROGRESSION

A sequence of non-zero numbers is called a geometric progression (GP) if the ratio of successive terms is constant.

In general, G.P. is written in the following form: $a, ar, ar^2, \ldots, ar^{n-1}, \ldots$

where a is the first term and r is the common ratio

5.1 General Term

If a is the first term and r is the common ratio, then $T_n = ar^{n-1}$.

Illustration 22: The 5th, 8th and 11th terms of a G.P. are given as p, q and, s respectively. Prove that $q^2 = ps$.

(JEE MAIN)

Sol: By using $T_n = ar^{n-1}$ and solving it, we can prove the problem.

Given, $\quad T_5 = p, T_8 = q, T_{11} = s$ $\hfill$ ....(i)

Now, $\quad T_5 = ar^{5-1} = ar^4 \implies ar^4 = p$ $\hfill$ .... (ii) [Using (i)]

$\quad\quad\quad T_8 = ar^{8-1} = ar^7 \implies ar^7 = q$ $\hfill$ ... (iii) [Using (i)]

$\quad\quad\quad T_{11} = ar^{11-1} = ar^{10} \implies ar^{10} = s$ $\hfill$ ..... (iv)

On squaring (iii), we get

$$q^2 = a^2 r^{14} = a \cdot a \cdot r^4 \cdot r^{10} = (ar^4)(ar^{10})$$

$\implies \quad q^2 = ps$ $\hfill$ [Using (ii) and (iv)] proved.

5.2 Series of GP

Let us suppose $S_n = a + ar + ar^2 + \ldots + ar^{n-1}$ $\hfill$ ... (i)

Multiplying 'r' on both the sides of (i) and shifting the RHS terms by one place, we get

$$S_n r = 0 + ar + ar^2 + \ldots + ar^n$$ $\hfill$ ... (ii)

By subtracting (ii) from (i), we get

$$S_n(1-r) = a - ar^n = a(1-r^n)$$

$$S_n = \frac{a(1-r^n)}{1-r}, \text{ where } r \neq 1$$

Thus, the sum of the first n terms of a G.P. is given by $\implies S_n = \left(\frac{a(r^n-1)}{r-1}\right) = \left(\frac{T_{n+1} - a}{r-1}\right)$

And $S_n = na$, when $r = 1$

Note: If $r = 1$, then the sequence is of both AP and GP, and its sum is equal to na, i.e. $S_n = na$.

If $|r| < 1$, the nth term of G.P. converges to zero and the sum becomes finite.

The sum to infinite terms of G.P. $= \lim_{n \to \infty} S_n = \lim_{n \to \infty} \frac{a(r^n-1)}{r-1}$

As $|r| < 1$ $r^n \to 0$ as $n \to \infty$ $\quad \therefore \quad S_\infty = \frac{a}{1-r}$

5.3 Geometric Mean

If a, b and c are three positive numbers in GP, then b is called the geometrical mean (GM) between a and c, and $b^2 = ac$. If a and b are two real numbers of the same sign and G is the G.M. between them, $G^2 = ab$.

"

Note: If a and b are two number of opposite signs, then the G.M. between them does not exist.

To Insert 'n' GMs Between a and b: If a and b are two positive numbers and we have to insert n GMs, G_1, G_2, ..., G_n, between the two numbers 'a' and 'b' then a, G_1, G_2, ..., G_n, b will be in GP. The series consists of (n + 2) terms and the last term is b and the first term is a.

Thus, $b = ar^{n+2-1} \Rightarrow b = ar^{n+1} \Rightarrow r = \left(\dfrac{b}{a}\right)^{\frac{1}{n+1}}$

$\Rightarrow \quad G_1 = ar,\ G_2 = ar^2\ ....G_n = ar^n$ or $G_n = a^{yn+1} \cdot b^{yn+1} = (ab)^{yn+1}$

Note: The product of n GMs inserted between 'a' and 'b' is equal to the nth power of the single G.M. between 'a' and 'b,' i.e.

$\displaystyle\prod_{r=1}^{n} G_r = (G)^n$, where $G = \sqrt{ab}$ (GM between a and b)

5.4 Relation between A.M. and GM

For any two non-negative number A.M. $\geq$ G.M.

Proof: Let two non-negative numbers be $\sqrt{a}$ and $\sqrt{b}$.

Now, we can write $\left(\sqrt{a} - \sqrt{b}\right)^2 \geq 0 \Rightarrow a - 2\sqrt{ab} + b \geq 0 \quad \Rightarrow a + b \geq 2\sqrt{ab} \Rightarrow \dfrac{a+b}{2} \geq \sqrt{ab} \quad \Rightarrow \text{A.M.} \geq \text{GM}$

Note: (i) Equality for AM, G.M. (i.e. A.M. = GM) exists when a = b.

(ii) Since A.M. $\geq$ GM; $(AM)_{min} = $ GM; $(GM)_{max} = $ AM

Illustration 23: If x, y and z have the same sign, then prove that $\dfrac{x}{y} + \dfrac{y}{z} + \dfrac{z}{x} \geq 3$. **(JEE ADVANCED)**

Sol: As we know that A.M. $\geq$ G.M., therefore by obtaining A.M. and G.M. of $\dfrac{x}{y}$, $\dfrac{y}{z}$ and $\dfrac{z}{x}$ we can prove the problem.

Let $\dfrac{x}{y} = x_1$; $\dfrac{y}{z} = x_2$; $\dfrac{z}{x} = x_3$

$\therefore \dfrac{x_1 + x_2 + x_3}{3} \geq (x_1 x_2 x_3)^{1/3} \Rightarrow \dfrac{\dfrac{x}{y} + \dfrac{y}{z} + \dfrac{z}{x}}{3} \geq 1$

Hence proved.

Illustration 24: Calculate the values of n so that $\dfrac{a^{n+1} + b^{n+1}}{a^n + b^n}$ may be the G.M. between a and b. **(JEE ADVANCED)**

Sol: We know that the G.M. between a and b = $\sqrt{ab}$, but here G.M. between a and b is $\dfrac{a^{n+1} + b^{n+1}}{a^n + b^n}$.

$\Rightarrow \dfrac{a^{n+1} + b^{n+1}}{a^n + b^n} = \sqrt{ab}$ $\qquad \Rightarrow a^{n+1} + b^{n+1} = (a^n + b^n)(ab)^{1/2}$

$\Rightarrow a^{n+1} + b^{n+1} = a^{n+\frac{1}{2}} \cdot b^{\frac{1}{2}} + a^{\frac{1}{2}} \cdot b^{n+\frac{1}{2}}$ $\qquad \Rightarrow a^{n+1} - a^{n+\frac{1}{2}} \cdot b^{\frac{1}{2}} = a^{\frac{1}{2}} \cdot b^{n+\frac{1}{2}} - b^{n+1}$

$\Rightarrow a^{n+\frac{1}{2}}\left(a^{\frac{1}{2}} - b^{\frac{1}{2}}\right) = b^{n+\frac{1}{2}}\left(a^{\frac{1}{2}} - b^{\frac{1}{2}}\right)$ $\qquad \Rightarrow a^{n+\frac{1}{2}} = b^{n+\frac{1}{2}}$

$\Rightarrow \left(\dfrac{a}{b}\right)^{n+\frac{1}{2}} = 1 = \left(\dfrac{a}{b}\right)^{0}$ $\qquad \Rightarrow n + \dfrac{1}{2} = 0 \Rightarrow n = -\dfrac{1}{2}$

Illustration 25: Find the sum to n terms for the series 9 + 99 + 999 ...... n. **(JEE ADVANCED)**

Sol: The given series can be written as $S = (10 - 1) + (10^2 - 1) + (10^3 - 1) ..... + (10^n - 1)$.

Thus, by using $S_n = \dfrac{a(1 - r^n)}{1 - r}$, we can find out the required sum.

$\therefore \quad S = (10 + 10^2 + 10^3 + ... + 10^n) - n; \quad S = \dfrac{10(1 - 10^n)}{1 - 10} - n$

Illustration 26: If a_1, a_2 and a_3 are in G.P. with a common ratio r (r > 0 and a > 0), then values of r for which inequality $9a_1 + 5a_3 > 14a_2$ hold good are? **(JEE ADVANCED)**

Sol: Since $a_1 = \dfrac{a}{r}$, $a_2 = a$, $a_3 = ar$, by substituting these values to the given inequality we will get the result.

$a_1 = \dfrac{a}{r}$, $a_2 = a$, $a_3 = ar$ $\qquad\qquad$ Now, $\dfrac{9a}{r} + 5ar > 14a$

$\Rightarrow \quad 5ar^2 - 14ar + 9a > 0 \qquad\qquad \Rightarrow \quad 5r^2 - 14r + 9 > 0$

$\Rightarrow \quad 5r^2 - 5r - 9r + 9 > 0 \qquad\qquad \Rightarrow \quad 5r(r - 1) - 9(r - 1) > 0$

$\Rightarrow \quad (5r - 9)(r - 1) > 0 \qquad\qquad \Rightarrow \quad r \in R - \left(1, \dfrac{5}{9}\right)$

NOMORECLASS CONCEPTS

- The product of n geometric means between a and (1/a) is 1.

- Let the first term of a G.P. be negative; if r > 1, then it is a decreasing G.P. and if 0 < r < 1, then it is an increasing GP.

- If a_1, a_2, a_3,...., a_n are in AP, $a^{a_1}, a^{a_2}, a^{a_3}, ..., a^{a_n}$ will be in G.P. whose common ratio is a^d.

Illustration 27: On a certain date, the height of a plant is 1.6 m. If the height increases by 5 cm in the following year and if the increase in each year is half of that in the preceding years, show that the height of the plant will never be 1.7 m. **(JEE MAIN)**

Sol: Here, the sum of the increases in the height of the plant in the first, second, third, ... year is equal to (1.7 − 1.6) m = 0.1 m = 10 cm.

According to the question, increases in the height of the plant in the first, second, third, ... year are 5, $\dfrac{5}{2}, \dfrac{5}{4}$, ... cm, respectively.

Let it reach the height of 1.7 m (i.e. increases [1.7 − 1.6] m = 0.1 m = 10 cm).

Therefore, the sum of 5, $\dfrac{5}{2}, \dfrac{5}{4}$, ...to n terms = 10

$\Rightarrow \quad \dfrac{5\left(1 - \dfrac{1}{2^n}\right)}{1 - \dfrac{1}{2}} = 10 \qquad\qquad \left[\because a = 5, r = \tfrac{1}{2}, S_n = \dfrac{a(1 - r^n)}{1 - r}\right]$

$\Rightarrow \quad 10\left(1 - \dfrac{1}{2^n}\right) = 10 \Rightarrow 1 - \dfrac{1}{2^n} = 1 \Rightarrow \dfrac{1}{2^n} = 0$, which does not hold for any n. Thus, the plant will never reach the height of 1.7 m.

Illustration 28: A manufacturer reckons that the value of a machine (price = Rs 15,625) will depreciate each year by 20%. Calculate the estimated value at the end of 5 years. **(JEE MAIN)**

Sol: Here the value of the machine after 5 years= ar^5, where a= 15,625. We will obtain the value of r using the given condition.

The present value of the machine = Rs 15,625

The value of the machine in the next year = Rs $15,625 \times \dfrac{80}{100}$

The value of the machine after 2 years = Rs $15,625 \times \dfrac{80}{100} \times \dfrac{80}{100}$

The values of the machine in the present year, after 1 year and after 2 years are

Rs 15,625, Rs $15,625 \times \dfrac{80}{100}$ and Rs $15,625 \times \dfrac{80}{100} \times \dfrac{80}{100}$, respectively

These values form a GP.

Here, the first term is Rs 15,625 and the common ratio is $\dfrac{80}{100}$,i.e. $\dfrac{4}{5}$.

Thus, thee value of the machine after 5 years = ar^5 = Rs $15,625 \times \left(\dfrac{4}{5}\right)^5 = \dfrac{15625 \times 1024}{625 \times 5} = 1024 \times 5 =$ Rs 5120

5.5 Properties of GP

(a) If each term of a G.P. is multiplied or divided by the same non-zero quantity, then the resulting sequence is also a GP.

(b) If in a finite GP, the number of terms is odd, then its middle term is the G.M. of the first and last terms.

(c) If a, b and c are in GP, then $\dfrac{b}{a} = \dfrac{c}{b} \Rightarrow b^2 = ac$ (which is the condition of GP).

(d) The reciprocals of the terms of a given G.P. also give a G.P. with a common ratio of $\dfrac{1}{r}$.

Proof: Let $a_1, a_2, a_3, a_4, ..., a_n, ...$ be the terms of a G.P. with the common ratio r.

Then, $\dfrac{a_{n+1}}{a_n} = r$ for all $n \in N$ ... (i)

The sequence formed by the reciprocals of the terms of the above G.P. is given by

$$\dfrac{1}{a_1}, \dfrac{1}{a_2}, \dfrac{1}{a_3}, ..., \dfrac{1}{a_n}, ...$$

Now, $\dfrac{1/a_{n+1}}{1/a_n} = \dfrac{a_n}{a_{n+1}} = \dfrac{1}{r}$ [Using (i)]

Hence, the new sequence is also a G.P. with the common ratio 1/r.

(e) If each term of a G.P. is raised to the same power (say k), then the resulting sequence also forms a G.P. with the common ratio as r^k.

Proof: Let $a_1, a_2, a_3, a_4, ..., a_n....$ be the terms of a G.P. with the common ratio r.

Then, $\dfrac{a_{n+1}}{a_n} = r$ for all $n \in N$ (i)

Let k be a non-zero real number. Consider the sequence. $a_1^k, a_2^k, a_3^k, ..., a_n^k, ...$

Here, $\dfrac{a_{n+1}^k}{a_n^k} = \left(\dfrac{a_{n+1}}{a_n}\right)^k = r^k$ for all $n \in d$ [Using (i)]

Thus, $a_1^k, a_2^k, a_3^k, \ldots, a_n^k, \ldots$ is a G.P. with a common ratio r^k.

(f) In a GP, the product of the terms equidistant from the beginning and the end is always the same and it is equal to the product of the first and last terms (only for finite GP).

Proof: Let $a_1,\ a_2,\ a_3,\ \ldots,\ a_n$ be a finite G.P. with the common ratio r. Then,

k^{th} term from the beginning $= a_k = a_1 r^{k-1}$

k^{th} term from the end $= (n - k + 1)^{th}$ term from the beginning

$\quad\quad = a_{n-k+1} = a_1 r^{n-k}$

$\therefore$ (k^{th} term from the beginning) (k^{th} term from the end) $= a_k a_{n-k+1}$

$\quad = a_1 r^{k-1} a_1 r^{n-k} = a_1^2\, r^{n-1} = a_1 . a_1 r^{n-1} = a_1 a_n$ for all $k = 2,3,\ldots,n-1$

Thus, the product of the terms equidistant from the beginning and the end is always the same and is equal to the product of the first and last terms.

(g) If the terms of a G.P. are chosen at regular intervals, the new sequence so formed also forms a G.P. with the common ratio as r^p, where p is the size of interval.

For example:

2, 4, 8, 16, 32, 64, 128,..... (GP, where r = 2)

4, 16, 64 ..... (also a GP, where r = 4)

(h) If $a_1,\ a_2,\ a_3,\ldots,\ a_r\ldots$ is a G.P. of non-zero, non-negative terms, then $\log a_1,\ \log a_2,\ldots,\ \log a_n,\ldots$ is an AP and vice versa.

(i) If $T_1,\ T_2,\ T_3\ldots$ and $t_1,\ t_2,\ t_3$ are two GPs, $T_1 t_1,\ T_2 t_2,\ T_3 t_3,\ldots$ is also in GP.

Proof: Let the two GPs be $T_1,\ T_2,\ \ldots,\ T_n,\ldots$ with the common ratio R

$$\Rightarrow \dfrac{T_{n+1}}{T_n} = R \qquad\qquad\qquad \text{... (i)}$$

and $t_1,\ t_2,\ldots,t_n,\ldots$ with common the ratio r

$$\Rightarrow \dfrac{t_{n+1}}{t_n} = r \qquad\qquad\qquad \text{.... (ii)}$$

Multiplying each term of the sequence (i) by the corresponding term of (ii), we get $\left(\dfrac{T_{n+1}}{T_n}\right)\left(\dfrac{t_{n-1}}{t_n}\right) = Rr$

Thus, the resulting sequence is also in G.P. with the common ratio Rr.

(j) The resulting sequence thus formed by dividing the terms of a G.P. by the corresponding terms of another G.P. is also a GP.

Proof: Let the two GPs be $T_1, T_2, \ldots, T_n, \ldots$ with the common ratio R

$$\Rightarrow \dfrac{T_{n+1}}{T_n} = R \qquad\qquad\qquad \text{... (i)}$$

and $t_1, t_2, \ldots, t_n, \ldots$ with the common ratio r

$$\Rightarrow \dfrac{t_{n+1}}{t_n} = r \qquad\qquad\qquad \text{...(ii)}$$

Dividing each term of the sequence (i) by the corresponding term of (ii), we get

$$\frac{\dfrac{T_{n+1}}{t_{n+1}}}{\dfrac{T_n}{t_n}} = \left(\frac{R}{r}\right) \qquad\qquad\qquad \text{[Using (i) and (ii)]}$$

Thus, the resulting sequence is also in G.P. with the common ratio $\left(\dfrac{R}{r}\right)$.

Illustration 29: If sum of infinite terms of G.P. is 15 and sum of squares of infinite terms of G.P. is 45, then find GP.

(JEE MAIN)

Sol: As the sum of infinite terms $S_\infty = \dfrac{a}{1-r}$, therefore by using this formula we can obtain the value of the common ratio.

$$\frac{a}{1-r} = 15$$

Now, $a^2,\ a^2r^2,\ a^2r^4,\dots;$ $\qquad \dfrac{a^2}{1-r^2} = 45 \qquad \therefore \quad \dfrac{225(1-r)(1-r)}{(1-r)(1+r)} = 45$

$225 - 225r = 45 + 45r; \qquad\quad 180 = 270\,r \qquad \therefore \quad r = 2/3$

Illustration 30: If $x = 1 + a + a^2 + \dots \infty$, $y = 1 + b + b^2 + \dots \infty$ and $|a| < 1$, $|b| < 1$, then prove that
$1 + ab + a^2b^2 + \dots = \dfrac{xy}{x+y-1}$

(JEE MAIN)

Sol: By using the formula $S_\infty = \dfrac{a}{1-r}$, we can solve problem.

$$x = 1 + a + a^2 + \dots \text{ to } \infty = \frac{1}{1-a} \qquad\qquad (\because |a| < 1)$$

$$\Rightarrow \quad 1 - a = \frac{1}{x} \Rightarrow a = 1 - \frac{1}{x} \quad \Rightarrow a = \frac{x-1}{x} \qquad\qquad\qquad \dots \text{(i)}$$

$$\text{Also, } y = 1 + b + b^2 + \dots \text{ to } \infty = \frac{1}{1-b} \qquad\qquad (\because |b| < 1)$$

$$\Rightarrow \quad 1 - b = \frac{1}{y} \Rightarrow b = 1 - \frac{1}{y} \quad \Rightarrow b = \frac{y-1}{y} \qquad\qquad\qquad \dots \text{(ii)}$$

$$\therefore \quad 1 + ab + a^2b^2 + \dots \text{ to } \infty = \frac{1}{1-ab} \quad (\because |a| < 1, |b| < 1 \Rightarrow |ab| < 1)$$

$$= \frac{1}{1 - \dfrac{x-1}{x}\cdot\dfrac{y-1}{y}} \text{ [Using (i) and (ii)]} = \frac{xy}{xy - xy + x + y - 1} = \frac{xy}{x+y-1}$$

Hence proved.

Illustration 31: If $S_1, S_2, S_3, \dots, S_p$ denote the sum of an infinite G.P. whose first terms are 1, 2, 3, ..., p, respectively and whose common ratios are $\dfrac{1}{2}, \dfrac{1}{3}, \dfrac{1}{4}, \dots, \dfrac{1}{(p+1)}$, respectively, show that $S_1 + S_2 + S_3 + \dots + S_p = \dfrac{p(p+3)}{2}$.

(JEE ADVANCED)

Sol: By using $S_\infty = \dfrac{a}{1-r}$ we can obtain $S_1, S_2, S_3, \dots, S_p$ and after that by adding them we can prove the given equation.

For S_1, we have a = 1, $\qquad$ $r = \dfrac{1}{2}$ $\qquad$ $\therefore$ $\quad S_1 = \dfrac{1}{1-\dfrac{1}{2}} = 2$

For S_2, we have $\quad$ a = 2, $\qquad$ $r = \dfrac{1}{3}$ $\qquad$ $\therefore$ $\quad S_2 = \dfrac{2}{1-\dfrac{1}{3}} = 3$

For S_3, we have $\quad$ a = 3, $\qquad$ $r = \dfrac{1}{4}$ $\qquad$ $\therefore$ $\quad S_3 = \dfrac{3}{1-\dfrac{1}{4}} = 4$

For S_p, we have $\quad$ a = p, $\qquad$ $r = \dfrac{1}{p+1}$ $\qquad$ $\therefore$ $\quad S_p = \dfrac{p}{1-\dfrac{1}{p+1}} = p + 1$

Adding all these, we get $\quad S_1 + S_2 + S_3 + \ldots\ldots + S_p = 2 + 3 + 4 + \ldots\ldots + (p + 1)$

$$= \dfrac{p}{2}\,[2 + (p + 1)] = \dfrac{p}{2}\,[p + 3] = \dfrac{p(p+3)}{2}$$

Hence proved.

6. ARITHMETIC GEOMETRIC PROGRESSION

A series formed by multiplying the corresponding terms of AP and G.P. is called arithmetic geometric progression (AGP).

Let a = first term of AP, b = first term of GP, d = common difference and r = common ratio of GP, then

AP: a, a + d, a + 2d, a+ 3d, ..., a + (n − 1) d

GP: b, br, br^2, br^3,....., br^{n-1}

AGP: ab, (a + d) br, (a + 2d) br^2 .... (a + (n − 1) d) br^{n-1} (Standard appearance of AGP)

The general term (n^{th} term) of an AGP is given as $T_n = [a + (n − 1)d]\,br^{n-1}$.

6.1 Series of AGP

To find the sum of n terms of an AGP, we suppose that its sum as S_n and then multiply both the sides by the common ratio of the corresponding G.P. and then subtract as in the following way. Thus, we get a G.P. whose sum can be easily obtained.

$S_n = ab + (a + d)\,br + (a + 2d)\,br^2 + \ldots + (a + (n− 1)d)\,br^{n-1}$ $\hfill$ ...(i)

$rS_n = 0 + abr + (a + d)\,br^2 + \ldots\ldots + (a + (n − 1)d)\,br^n$ $\hfill$ ...(ii)

After subtraction, we get

$S_n(1 − r) = ab + [dbr + dbr^2 + \ldots\ldots + \text{up to } (n − 1) \text{ terms}] − [\,(a + (n − 1)d)br^n\,]$

$S_n(1 − r) = ab + \dfrac{dbr(1 − r^{n-1})}{1 − r} − (a + (n − 1)d)br^n$

$S_n = \dfrac{ab}{1 − r} + \dfrac{dbr(1 − r^{n-1})}{(1 − r)^2} − \dfrac{(a + (n − 1)d)br^n}{1 − r}$. This is the sum of n terms of AGP

For an infinite AGP, as $n \to \infty$, then $r^n \to 0$ $(\because |r| < 1)$

$\Rightarrow \quad S_\infty = \dfrac{ab}{1 − r} + \dfrac{dbr}{(1 − r)^2}$

Illustration 32: If $|x| < 1$, then find the sum $S = 1 + 2x + 3x^2 + 4x^3 \, ... + \infty$. **(JEE MAIN)**

Sol: The sum can be found out by calculating the value of $Sx - S$.

$$Sx = x + 2x^2 + 3x^3 + 4x^3 + ...\infty$$

$$S(1-x) = 1 + x + x^2 + x^3 + ... + \infty; \quad S(1-x) = \frac{1}{(1-x)} \Rightarrow S = \frac{1}{(1-x)^2}$$

Illustration 33: If $|x| < 1$, then find the sum $S = 1 + 3x + 6x^2 + 10x^3 + ...\infty$. **(JEE ADVANCED)**

Sol: Similar to above illustration.

$$S = 1 + 3x + 6x^2 + 10x^3 + ...\infty$$

$$Sx = x + 3x^2 + 6x^3 ....\infty$$

$$S(1-x) = 1 + 2x + 3x^2 + 4x^3 ....\infty$$

$$S(x)(1-x) = x + 2x^2 + 3x^3 ....\infty$$

$$S(1-x)^2 = 1 + x + x^2 + ...\infty$$

$$S(1-x)^2 = \frac{1}{1-x} \quad S = \frac{1}{(1-x)^3}$$

7. MISCELLANEOUS SEQUENCES

Type 1: Some Standard Results

(i) Sum of the first n natural numbers $\quad = \quad \sum_{r=1}^{n} r = \frac{n(n+1)}{2}$

(ii) Sum of the first n odd natural numbers $\quad = \quad \sum_{r=1}^{n}(2r-1) = n^2$

(iii) Sum of the first n even natural numbers $\quad = \quad \sum_{r=1}^{n} 2r = n(n+1)$

(iv) Sum of the squares of the first n natural numbers $= \quad \sum_{r=1}^{n} r^2 = \frac{n(n+1)(2n+1)}{6}$

Proof: $\sum_{n=1}^{n} n^2 = \frac{n(n+1)(2n+1)}{6}$

Consider $(x + 1)^3 = x^3 + 1 + 3x^2 + 3x$

$$(x + 1)^3 - x^3 = 3x^2 + 3x + 1$$

Put $x = 1, 2, 3.... n$

$$2^3 - 1^3 = 3.1^2 + 3.1 + 1$$

$$3^3 - 2^3 = 3.2^2 + 3.2 + 1$$

$$(n + 1)^3 - n^3 = 3n^2 + 3.n + 1$$

Adding all, we get

$$\Rightarrow \quad (n + 1)^3 - 1 = 3(1^2 + 2^2 + 3^2 + ... + n^2) + 3(1 + 2 + ... + n) + n$$

$$\Rightarrow \quad (n + 1)^3 - 1 = 3\Sigma n^2 + 3\frac{3n(n+1)}{2} + n \Rightarrow \quad 3\Sigma n^2 = (n + 1)^3 - 1 - \frac{3n(n+1)}{2} - n$$

$$\Rightarrow \quad 3\Sigma n^2 = \quad n^3 + 1 + 3n^2 + 3n - 1 - \frac{3n(n+1)}{2} - n$$

$$\Rightarrow \quad 3\Sigma n^2 = n^3 + 3n^2 + 2n - \frac{3n(n+1)}{2}$$

$$\Rightarrow \quad 3\Sigma n^2 = \frac{2n^3 + 6n^2 + 4n - 3n^2 - 3n}{2} \quad \Rightarrow 3\Sigma n^2 = \frac{2n^3 + 3n^2 + n}{2}$$

$$\Rightarrow \quad 3\Sigma n^2 = \frac{2n^3 + 3n^2 + n^2 + n}{2} \quad \Rightarrow 3\Sigma n^2 = \frac{2n^2(n+1) + n(n+1)}{2}$$

$$\Rightarrow \quad 3\Sigma n^2 = \frac{n(n+1)\times(2n+1)}{2} \quad \Rightarrow \Sigma n^2 = \frac{n(n+1)\times(2n+1)}{6}$$

(v) Sum of the cubes of first n natural numbers $\displaystyle\sum_{r=1}^{n} r^3 = \left[\frac{n(n+1)}{2}\right]^2$

Proof: Consider $(x + 1)^4 - x^4 = 4x^3 + 6x^2 + 4x + 1$

Put $x = 1, 2, 3 \ldots\ldots n$

$$2^4 - 1^4 = 4\cdot1^3 + 6\cdot1^2 + 4\cdot1 + 1$$

$$3^4 - 2^4 = 4\cdot2^3 + 6\cdot2^2 + 4\cdot2 + 1$$

$$4^4 - 3^4 = 4\cdot3^2 + 6\cdot2^2 + 4\cdot3 + 1$$

$$\vdots$$

$$(n + 1)^4 - n^4 = 4\cdot n^3 + 6\cdot n^2 + 4\cdot n + 1$$

Adding all, we get

$$(n + 1)^4 - 1^4 = 4(1^3 + 2^3 + \ldots + n^3) + 6(1^2 + 2^2 + \ldots + n^2) + 4(1 + 2 + 3 \ldots + n) + n$$

$$= 4\Sigma n^3 + 6\left(\frac{n(n-1)(2n+1)}{6}\right) + 4\left(n\left(\frac{n+1}{2}\right)\right) + n$$

On simplification, we get

$$\Sigma n^3 = \left(n\left(\frac{n+1}{2}\right)\right)^2$$

(vi) Sum of the fourth powers of the first n natural numbers (Σn^4)

$$\Sigma n^4 = 1^4 + 2^4 + \ldots + n^4 \quad ; \quad \Sigma n^4 = \frac{n(n+1)(2n+1)(3n^2 + 3n - 1)}{30}$$

[The result can be proved in the same manner as done for Σn^3]

Illustration 34: Find the value of $\displaystyle\sum_{i=1}^{n}\sum_{j=1}^{i}\sum_{k=1}^{j}(1)$. (JEE MAIN)

Sol: Using the formula $\displaystyle\sum_{r=1}^{n} r = \frac{n(n+1)}{2}$ and $\displaystyle\sum_{r=1}^{n} r^2 = \frac{n(n+1)(2n+1)}{6}$, we can solve the problem.

Let $S = \displaystyle\sum_{i=1}^{n}\sum_{j=1}^{i}\sum_{k=1}^{j}(1)$

$$S = \sum_{i=1}^{n}\sum_{j=1}^{i}(j) = \sum_{i=1}^{n}\frac{i(i+1)}{2} = \frac{1}{2}[\Sigma n^2 + \Sigma n] = \frac{1}{2}\left[\frac{n(n+1)(2n+1)}{6} + \frac{n(n+1)}{2}\right] = \frac{n(n+1)(2n+4)}{12} = \frac{n(n+1)(n+2)}{6}$$

Illustration 35: Find the sum of 1.2.3 + 2.3.4 + 3.4.5...... n terms. $\hfill$ **(JEE MAIN)**

Sol: The given series is in the form of $T_n = n(n + 1)(n + 2) = n^3 + 3n^2 + 2n$.

Therefore, by using $\displaystyle\sum_{r=1}^{n} r^3 = n^2 \left(\frac{n+1}{2}\right)^2$ $\displaystyle\sum_{r=1}^{n} r^2 = \frac{n(n+1)(2n+1)}{6}$ and $\displaystyle\sum_{r=1}^{n} r = \frac{n(n+1)}{2}$, we can solve the problem.

$$T_n = n(n + 1)(n + 2) = n(n^2 + 3n + 2)$$

$$T_n = n^3 + 3n^2 + 2n$$

$$\Sigma T_n = \Sigma n^3 + 3\Sigma n^2 + 2\Sigma n = n^2\left(\frac{n+1}{2}\right)^2 + \frac{3n(n+1)(2n+1)}{6} + n(n+1)$$

Type 2: Using Method of Difference: If $T_1, T_2, T_3, T_2, T_4, T_5....$ is a sequence whose terms are sometimes in AP and sometimes in GP, then for such series we first compute their nth term and then compute the sum to n terms using sigma notation.

Illustration 36: Find $S_n = 6 + 13 + 22 .... + T_n$. $\hfill$ **(JEE ADVANCED)**

Sol: By calculating $[S_n + (- S_n)]$, we will get T_n. After that we will obtain ΣT_n and thus we will get the result.

$$\Sigma_n = 6 + 13 + 22 .... T_n$$

$$-\Sigma_n = -6 - 13 .... - T_{n-1} - T_n$$

$$\Rightarrow \quad 0 = 6 + (7 + 9 + 11 ..... (T_n - T_{n-1})) - T_n$$

$$\Rightarrow \quad T_n = 6 + (7 + 9 + 11.... (T_n - T_{n-1})) = 6 + (n - 1)(7 + n - 2) = 6 + (n - 1)(n + 5)$$

$$\Rightarrow \quad T_n = 6 + n^2 + 4n - 5 = n^2 + 4n + 1$$

$$\Sigma_n = \Sigma n^2 + 4\Sigma n + n = \frac{n(n+1)(n+1)}{6} + 2n(n+1) + n$$

Illustration 37: Find $S = 1 + \left(1 + \dfrac{1}{3}\right) + \left(1 + \dfrac{1}{3} + \dfrac{1}{3^2}\right) + ... $ n terms. $\hfill$ **(JEE ADVANCED)**

Sol: Given, $T_n = 1 + \dfrac{1}{3} + \dfrac{1}{3^2} + ... + \dfrac{1}{3^{n-1}}$; therefore by obtaining ΣT_n we will get the result.

$$S = 1 + \left(1 + \frac{1}{3}\right) + \left(1 + \frac{1}{3} + \frac{1}{9}\right).......$$

$$T_n = 1 + \frac{1}{3} + \frac{1}{3^2} + ... + \frac{1}{3^{n-1}} = \frac{3\left(1 - \dfrac{1}{3^n}\right)}{2}$$

$$\Sigma T_n = \frac{3n}{2} - \frac{3}{2}\Sigma\frac{1}{3^n} = \frac{3n}{2} - \frac{3}{2}\left(\frac{1}{3} + \frac{1}{3^2} ...... \frac{1}{3^n}\right) = \frac{3}{2}\left(n - \frac{3}{2}\left(1 - \frac{1}{3^n}\right)\right)$$

Type 3: Splitting the nth term as a difference of two: Here, S is a series in which each term is composed of the reciprocal of the product of r factors in an AP.

Illustration 38: Find the sum of n terms of the series $\dfrac{1}{1\cdot2\cdot3\cdot4}+\dfrac{1}{2\cdot3\cdot4\cdot5}+\dfrac{1}{3\cdot4\cdot5\cdot6}+.....$ **(JEE ADVANCED)**

Sol: Here n^{th} term of the series will be $T_n = \dfrac{1}{n(n+1)(n+2)(n+3)}$.

By considering $S_n = c - \lambda$, where $\lambda = \dfrac{1}{3(n+1)(n+2)(n+3)}$, we will get the result.

First calculate the nth term, $T_n = \dfrac{1}{n(n+1)(n+2)(n+3)}$

Now, let the sum of the above series be given by:

$S_n = c - \lambda$, where λ is obtained by replacing the first factor by (last factor − first factor) ... (i)

Hence,

$$\lambda = \dfrac{1}{3(n+1)(n+2)(n+3)} \quad \begin{bmatrix} \text{First}\ \text{factor}=n \\ \text{Last}\ \text{factor}=n+3 \end{bmatrix} \qquad \text{... (ii)}$$

Using (ii) $\Rightarrow \quad S_n = c - \dfrac{1}{3(n+1)(n+2)(n+3)}$ ... (iii)

To calculate 'c', put n = 1 in (iii)

$$S_1 = c - \dfrac{1}{1\cdot2\cdot3\cdot4} \Rightarrow \dfrac{1}{1\cdot2\cdot3\cdot4} = c - \dfrac{1}{3\cdot2\cdot3\cdot4} \qquad \Rightarrow \quad c = 1/18$$

Put the value of 'c' in (iii)

$$S_n = \dfrac{1}{18} - \dfrac{1}{3(n+1)(n+2)(n+3)},\ S_n = \dfrac{1}{3}\left\{\dfrac{1}{6} - \dfrac{1}{(n+1)(n+2)(n+3)}\right\}$$

Remark $\Rightarrow$ If we want to calculate S_∞, then $n \to \infty$, $\dfrac{1}{(n+1)(n+2)(n+3)} \to 0 \Rightarrow S_\infty = \dfrac{1}{18}$

Note: The above method is applicable only when the series looks like as follows:

$$\dfrac{1}{a(a+d)(a+2d)}+\dfrac{1}{(a+d)(a+2d)(a+3d)}+\dfrac{1}{(a+2d)(a+3d)(a+4d)}+....$$

Type 4: Vn Method: This is method of resolving the nth term into partial fraction and summation by telescopic cancellation. First, find the nth term of the series and try to create a denominator part in the numerator by using partial fraction whenever the series is in the form of fraction or T_n is in the form of fraction.

For example, let us suppose a summation where the nth term is like the following:

$$T_n = \dfrac{2}{n^2 - 1}$$

Using the partial fraction, we can write the nth term as $T_n = \dfrac{1}{n-1} - \dfrac{1}{n+1}$

Now, when we find the summation, there will be telescopic cancellation and thus we will get the sum of the given series.

Type 5: Dealing with Sn⁴: This technique is valid for Σn^2 and Σn^3. In this type, there is a series in which each term is composed of factors in an AP, i.e. factors of several terms being in AP.

$$T_n = \frac{1}{5}[(n+1)(n+2)(n+3)(n+4)[n-(n-1)]] = \frac{1}{5}(n(n+1)(n+2)(n+3)(n+4) - (n-1)(n+1)(n+2)(n+3)(n+4))$$

$$T_1 = \frac{1}{5}(1\cdot2\cdot3\cdot4\cdot5 - 0)$$

$$T_2 = \frac{1}{5}(2\cdot3\cdot4\cdot5\cdot6 - 1\cdot2\cdot3\cdot4\cdot5)$$

$$T_3 = \frac{1}{5}(3\cdot4\cdot5\cdot6\cdot7 - 2\cdot3\cdot4\cdot5\cdot6)$$

$$T_n = = \frac{1}{5}(n(n+1)(n+2)(n+3)(n+4) - (n-1)(n+1)(n+2)(n+3)(n+4))$$

Adding all, we have

$$S_n = \frac{1}{5}\Big(n(n+1)(n+2)(n+3)(n+4)\Big)$$

Note: This method will be applicable only when the series looks like the following:

a(a + d) (a+ 2d) + (a + d) (a + 2d) (a + 3d) + (a + 2d) (a + 3d) (a + 4d)+….+ up to n term, where a = first term and d = common difference

NOMORECLASS CONCEPTS

- $$\frac{1}{1.2} + \frac{1}{2.3} + \frac{1}{3.4} + \dots + \frac{1}{n(n+1)} = \frac{n}{n+1}$$

$$\frac{1}{1.2.3} + \frac{1}{2.3.4} + \dots + \frac{1}{n(n+1)(n+2)} = \frac{1}{4} - \frac{1}{2(n+1)(n+2)}$$

$$\frac{1}{a_1 a_2 \dots a_r} + \frac{1}{a_2 a_3 \dots a_{r+1}} + \dots + \frac{1}{a_n a_{n+1} \dots a_{n+r-1}} = \frac{1}{(r-1)(a_2 - a_1)}\left[\frac{1}{a_1 a_2 \dots a_{r-1}} - \frac{1}{a_{n+1} a_{n+2} \dots a_{n+r-1}}\right]$$

- $$a_1 a_2 \dots a_r + a_2 a_3 \dots a_{r+1} + \dots + a_n a_{n+1} \dots a_{n+r-1} = \frac{1}{(r+1)(a_2 - a_1)}\left[a_n a_{n+1} \dots a_{n+r} - a_0 a_1 a_2 \dots a_n\right]$$

    Where $a_1 a_2 \dots a_n$ are in AP and $a_0 = a_1 - d$

8. HARMONIC PROGRESSION

A sequence will be in harmonic progression (HP) if the reciprocals of its terms are in AP, e.g. if $a_1,\ a_2,\ a_3,\ \dots$ are in HP, then $\dfrac{1}{a_1}, \dfrac{1}{a_2}, \dfrac{1}{a_3} \dots$ are in AP. For every AP, there will be a corresponding HP, and the standard H.P. is

$$\frac{1}{a},\ \frac{1}{a+d},\ \frac{1}{a+2d} + \dots\dots + \frac{1}{a+(n-1)d}.$$

The terms of a harmonic series are the outcomes of an AP.

Note:

(i) 0 cannot be a term of H.P. because ∞ is not a term of AP, but ∞ can be a term of HP.

(ii) There is no general formula for finding the sum to n terms of HP.

(iii) If a, b and c are in HP, then $\dfrac{1}{a}, \dfrac{1}{b}, \dfrac{1}{c}$ are in AP.

$$\therefore \frac{2}{b} = \frac{1}{a} + \frac{1}{c} \qquad \Rightarrow \qquad b = \frac{2ac}{a+c}$$

$\Rightarrow$ a, b and c are in HP

Moreover, $\dfrac{1}{b} - \dfrac{1}{a} = \dfrac{1}{c} - \dfrac{1}{b}$ $\qquad$ i.e. $\dfrac{a-b}{ab} = \dfrac{b-c}{bc}$; $\qquad$ i.e. $\dfrac{a}{c} = \dfrac{a-b}{b-c}$

Illustration 39: If the 3rd, 6th and last terms of a H.P. are $\dfrac{1}{3}, \dfrac{1}{5}, \dfrac{3}{203}$, then find the number of terms. $\qquad$ **(JEE MAIN)**

Sol: If n^{th} term of a H.P. is $\dfrac{1}{a}$, then the nth term of the corresponding AP will be a. Thus, by using

$T_n = a + (n - 1)\, d$, we will get the result.

Let a be the first term and d be the common difference of the corresponding AP.

If the 3rd term of H.P. $= \dfrac{1}{3}$; then the 3rd term of the corresponding AP = 3

$\Rightarrow \quad a + 2d = 3$ $\hfill$ ..... (i)

If the 6th term of H.P. $= \dfrac{1}{5}$; then the 6th term of the corresponding AP = 5

$\Rightarrow \quad a + 5d = 5$ $\hfill$ ..... (ii)

From (i) and (ii), we get $\quad d = \dfrac{2}{3} \Rightarrow a = \dfrac{5}{3}$

If the nth term of H.P. $= \dfrac{3}{203}$; then n^{th} term of AP $= \dfrac{203}{3}$

$$a + (n - 1)\, d = \frac{203}{3}; \qquad \frac{5}{3} + (n - 1)\,\frac{2}{3} = \frac{203}{3}$$

$5 + 2n - 2 = 203; \quad n = 100$

Illustration 40: If $a_1, a_2, \ldots\ldots a_n$ are in H.P. then the expression $a_1 a_2 + a_2 a_3 + \ldots\ldots + a_{n-1} a_n$ is equal to.

$\hfill$ **(JEE ADVANCED)**

Sol: As $\dfrac{1}{a_1}, \dfrac{1}{a_2}, \ldots\ldots \dfrac{1}{a_n}$ are in AP, taking $\dfrac{1}{a_2} - \dfrac{1}{a_1} = \dfrac{1}{a_3} - \dfrac{1}{a_2} = \ldots \dfrac{1}{a_n} - \dfrac{1}{a_{n-1}} = d$, we can obtain the values of $a_1 a_2, a_2 a_3$ and so on.

$a_1, a_2, \ldots\ldots a_n$ are in HP

$\dfrac{1}{a_1}, \dfrac{1}{a_2}, \ldots\ldots \dfrac{1}{a_n}$ are in AP

$\Rightarrow \quad \dfrac{1}{a_2} - \dfrac{1}{a_1} = \dfrac{1}{a_3} - \dfrac{1}{a_2} = \ldots \dfrac{1}{a_n} - \dfrac{1}{a_{n-1}} = d$ (say)

$\Rightarrow \quad a_1 a_2 = \dfrac{1}{d}(a_1 - a_2),\ a_2 a_3 = \dfrac{1}{d}(a_2 - a_3), \ldots\ldots, a_{n-1} a_n = \dfrac{1}{d}(a_{n-1} - a_n)$

Hence, $\quad a_1a_2 + a_2a_3 + ... + a_{n-1}a_n = \dfrac{1}{d}[a_1 - a_2 + a_2 - a_3 + ... + a_{n-1} - a_n] = \dfrac{1}{d}(a_1 - a_n)$

But $\quad \dfrac{1}{a_n} = \dfrac{1}{a_1} + (n-1)\,d \;\Rightarrow\; \dfrac{a_1 - a_n}{a_n a_1} = (n-1)\,d$

$\therefore \quad a_1a_2 + a_2a_3 + ... + a_{n-1}a_n = (n-1)a_1 a_n$

8.1 Harmonic Mean

If a, b and c are in HP, then the middle term is called the harmonic mean (HM) between them. If H is the HM between a and b, then a, H, c are in H.P. and $H = \dfrac{2ac}{a+c}$.

To Insert n HMs Between a and b

Let $H_1, H_2, ..., H_n$ be the n HMs between a and b.

Thus, a, $H_1, H_2, ..., H_n$ b are in HP.

$\dfrac{1}{a}, \dfrac{1}{H_1}, \dfrac{1}{H_2} .... \dfrac{1}{H_n}, \dfrac{1}{b}$ are in AP.

$\Rightarrow \dfrac{1}{b} = \dfrac{1}{a} + (n+1)d; \dfrac{1}{b} - \dfrac{1}{a} = (n+1)\,d;\; d = \dfrac{a-b}{ab(n+1)}$

$\Rightarrow \dfrac{1}{H_1} = \dfrac{1}{a} + d \qquad \Rightarrow \quad \dfrac{1}{H_2} = \dfrac{1}{a} + 2d$

$\Rightarrow \dfrac{1}{H_3} = \dfrac{1}{a} + 3d \qquad \Rightarrow \quad \dfrac{1}{H_n} = \dfrac{1}{a} + nd$

Adding all, we get

$$\sum_{i=1}^{n} \dfrac{1}{H_i} = \dfrac{n}{a} + \dfrac{d(n)(n+1)}{2} = \dfrac{n}{a} + \dfrac{n(n+1)}{2}\dfrac{(a-b)}{ab(n+1)} = n\left[\dfrac{1}{a} + \dfrac{a-b}{2ab}\right] = \dfrac{n}{2ab}[2b + a - b] = \dfrac{n(a+b)}{2ab} = n\dfrac{1}{H}$$

Note: The sum of the reciprocals of all the n HMs between a and b is equal to n times the reciprocal of the single HM between a and b.

For example, between 1 and $\dfrac{1}{100}$ if 100 HMs are inserted, then $\displaystyle\sum_{i=1}^{100} \dfrac{1}{H_i} = 5050$.

8.2 Sum of the Reciprocal of 'n' Harmonic Means

The sum of reciprocal of n harmonic means $= \dfrac{n(a+b)}{2ab}$

To Insert n Harmonic Means Between a and b

a, $H_1, H_2, H_3 ....H_n$, b $\to$ H.P.

$\dfrac{1}{a}, \dfrac{1}{H_1}, \dfrac{1}{H_2}, \dfrac{1}{H_3} .... \dfrac{1}{H_n}, \dfrac{1}{b} \to$ A.P.

$\dfrac{1}{b} = \dfrac{1}{a} + (n+1)\,d \;\Rightarrow\; (n+1)\,d = \dfrac{a-b}{ab}$

$$d = \frac{a-b}{(n+1)ab}$$

Illustration 41: Find the sum of $\dfrac{1}{H_1} + \dfrac{1}{H_2} + \dfrac{1}{H_3} \ldots\ldots\ldots \dfrac{1}{H_n}$.

Sol: Using $\dfrac{1}{H_n} = \dfrac{1}{a} + nd$, we can obtain the values of $\dfrac{1}{H_1}$, $\dfrac{1}{H_2}$ and so on. Then, by obtaining the value of $\displaystyle\sum_{n=1}^{n}\dfrac{1}{H_n}$, we will get the result.

$$\frac{1}{H_1} = \frac{1}{a} + d \qquad\qquad \frac{1}{H_2} = \frac{1}{a} + 2d$$

$$\frac{1}{H_n} = \frac{1}{a} + nd \qquad \Rightarrow \qquad \sum_{n=1}^{n}\frac{1}{H_i} = \frac{n}{a} + \frac{n(n+1)}{2}d$$

$$= \frac{n}{a} + \frac{n(n+1)\times(a-b)}{2(n+1)ab} = \frac{n}{a}\frac{(2b+a-b)}{2b} = \frac{n}{2ab}(a+b)$$

(i) For 3 numbers a, b and c, HM is defined as the reciprocals of the mean of the reciprocals of a, b and c, i.e. means

of reciprocal $= \dfrac{1}{3}\left(\dfrac{1}{a} + \dfrac{1}{b} + \dfrac{1}{c}\right)$; $\quad$ HM $= \dfrac{3}{\left(\dfrac{1}{a} + \dfrac{1}{b} + \dfrac{1}{c}\right)}$

(ii) If a_1, a_2, a_3, ..., a_n are n numbers, then

$$\text{AM} = \left(\left|\frac{a_1 + a_2 + a_3 + a_3 + \ldots\ldots + a_n}{n}\right|\right)$$

$$\text{GM} = (a_1\, a_2\, a_3 \ldots\ldots a_n)^{1/n}$$

$$\text{HM} = \left(\frac{n}{\dfrac{1}{a_1} + \dfrac{1}{a_2} + \dfrac{1}{a_2} + \ldots\ldots \dfrac{1}{a_n}}\right)$$

Illustration 42: If a^2, b^2, c^2 are in AP, then show that b + c, c + a, a + b are in HP.

Sol: Given that a^2, b^2 and c^2 are in AP. Thus, by adding ab + ac + bc to each term and then dividing each term by (a + b)(b + c)(c + a), we will get the result.

By adding ab + ac + bc to each term, we find that $a^2 + ab + ac + bc$, $b^2 + ba + bc + ac$, $c^2 + ca + cb + ab$ are in AP, i.e.

(a + b)(a + c), (b + c)(b + a), (c + a)(c + b) are in AP

$\therefore$ Dividing each terms by (a + b)(b + c)(c + a), we find that

$\dfrac{1}{b+c}, \dfrac{1}{c+a}, \dfrac{1}{a+b}$ are in AP, i.e.

b + c, c + a, a + b are in HP

Illustration 43: If H_1, H_2, ..., H_n are n harmonic means between a and b ($\neq a$), then find the value of $\dfrac{H_1 + a}{H_1 - a} + \dfrac{H_n + b}{H_n - b}$.

Sol: As a, H_1, H_2, ..., H_n, b are in HP, $\dfrac{1}{a}, \dfrac{1}{H_1}, \dfrac{1}{H_2} \ldots\ldots \dfrac{1}{H_n}, \dfrac{1}{b}$ are in AP. By considering d as the common difference of this

AP and using $T_n = a + (n - 1)d$ we can solve this problem.

$$\frac{1}{b} = \frac{1}{a} + (n+1)d \quad \text{and} \quad \frac{1}{H_n} - \frac{1}{H_1} = (n-1)d$$

Now, $\quad \dfrac{H_1 + a}{H_1 - a} = \dfrac{1/a + 1/H_1}{1/a - 1/H_1} = \dfrac{1/a + 1/H_1}{-d}$

and $\quad \dfrac{H_n + b}{H_n - b} = \dfrac{1/b + 1/H_n}{1/b - 1/H_n} = \dfrac{1/b + 1/H_n}{d}$

$\therefore \quad \dfrac{H_1 + a}{H_1 - a} + \dfrac{H_n + b}{H_n - b} = \dfrac{1/a + 1/H_1}{-d} + \dfrac{1/b + 1/H_n}{d} = \dfrac{1}{d}\left[\left(\dfrac{1}{b} - \dfrac{1}{a}\right) + \left(\dfrac{1}{H_n} - \dfrac{1}{H_1}\right)\right] = 2n$

9. RELATION BETWEEN AM, G.M. AND HM

If a and b are two positive numbers, then it can be shown that $A \geq G \geq H$ and A, G, H are in GP, i.e. $G^2 = AH$.

Proof: Given that, $A = \dfrac{a+b}{2}$, $G = \sqrt{ab}$ and $H = \dfrac{2ab}{a+b}$

$\therefore \quad A - G = \dfrac{a+b}{2} - \sqrt{ab}$

$\Rightarrow \quad A - G = \dfrac{(\sqrt{a} - \sqrt{b})^2}{2} \geq 0$

$\Rightarrow \quad A \geq G \qquad\qquad\qquad\qquad\qquad\qquad\qquad\qquad\qquad\qquad ......(i)$

$G - H = \sqrt{ab} - \dfrac{2ab}{a+b}$

$\Rightarrow \quad G - H = \sqrt{ab}\left(\dfrac{a+b-2\sqrt{ab}}{a+b}\right) \quad \Rightarrow \quad G - H = \dfrac{\sqrt{ab}}{a+b}(\sqrt{a} - \sqrt{b})^2 \geq 0$

$\Rightarrow \quad G \geq H \qquad\qquad\qquad\qquad\qquad\qquad\qquad\qquad\qquad\qquad ......(ii)$

Using (i) and (ii), we find that

$$A \geq G \geq H$$

Please note that the equality holds only when a = b.

Proof of $G^2 = AH$

Proof: $A = \dfrac{a+b}{2}$, $G = \sqrt{ab}$ and $H = \dfrac{2ab}{a+b}$

Now, $AH = ab = G^2 \qquad \Rightarrow \quad$ A, G & H are in G.P.

Moreover, $\dfrac{A}{G} = \dfrac{G}{H}$; $\qquad \therefore \quad A \geq G \quad \Rightarrow \quad G \geq H$

Therefore, $A \geq G \geq H$; in fact, $RMS \geq A.M. \geq G.M. \geq HM$ (where RMS is root mean square).

- If a and b are two positive quantities, then AM, G.M. and HM are always in GP, i.e. only for two numbers.
- If there are three numbers. then AM, G.M. and HM are in G.P. only when the three numbers. are in GP.

  For example, 2, 4, 8 $\rightarrow$ GP

  $GM = 4$; A.M. $= \dfrac{14}{3}$; HM $= \dfrac{24}{7}$

- For two positive numbers, it has been shown that $A \geq G \geq H$, equality holding for equal numbers.
- For n non-zero positive numbers, it has been shown that $A \geq G \geq H$, equality holding when all the numbers are equal.

Illustration 44: If a, b and c are unequal positive numbers in HP, then prove that

$$\frac{a+b}{2a-b}+\frac{c+b}{2c-b}>4.$$

(JEE ADVANCED)

Sol: As a, b and c are in HP, therefore $\dfrac{2}{b}=\dfrac{1}{a}+\dfrac{1}{c}$. Thus, by substituting this to LHS, we can prove the given problem.

$$\text{LHS} = \frac{\dfrac{1}{b}+\dfrac{1}{a}}{\dfrac{2}{b}-\dfrac{1}{a}}+\frac{\dfrac{1}{b}+\dfrac{1}{c}}{\dfrac{2}{b}-\dfrac{1}{c}} = \frac{\dfrac{1}{b}+\dfrac{1}{a}}{\dfrac{1}{c}}+\frac{\dfrac{1}{b}+\dfrac{1}{c}}{\dfrac{1}{a}}, \text{ using (i)}$$

$$= \frac{c}{b}+\frac{c}{a}+\frac{a}{b}+\frac{a}{c} = \frac{a+c}{b}+\frac{a}{c}+\frac{c}{a}.$$

Now, A.M. > G.M. $\Rightarrow \dfrac{\dfrac{a}{c}+\dfrac{c}{a}}{2} > \sqrt{\dfrac{a}{c}\cdot\dfrac{c}{a}}$ or $\dfrac{a}{c}+\dfrac{c}{a} > 2.$

$$\frac{a+c}{b} = \frac{a+c}{\dfrac{2ac}{a+c}} = \frac{(a+c)^2}{2ac} = \frac{(a-c)^2}{2ac}+2$$

$$\therefore \text{LHS} = \frac{a+c}{b}\left(\frac{a}{c}+\frac{c}{a}\right) > 2 + 2 = 4$$

10. PROPERTIES OF AM, G.M. AND HM

(i) The equation with a and b as its roots is $x^2 - 2Ax + G^2 = 0$

Proof: The equation with a and b as its roots is $x^2 - (a + b) x + ab = 0$

$$\Rightarrow \quad x^2 - 2Ax + G^2 = 0 \qquad\qquad (\because A = \frac{a+b}{2}, G = \sqrt{ab})$$

(ii) If A, G and H are the arithmetic, geometric and harmonic means, respectively, between three given numbers a, b and c, then the equation having a, b, c as its roots is $x^3 - 3Ax^2 + \dfrac{3G^3}{H}x - G^3 = 0$

Proof: As given, $A = \dfrac{a+b+c}{3}$, $G = (abc)^{1/3}$ and $\dfrac{1}{H} = \dfrac{\frac{1}{a}+\frac{1}{b}+\frac{1}{c}}{3}$

$\Rightarrow \quad a + b + c = 3A$, $abc = G^3$ and $\dfrac{3G^3}{H} = ab + bc + ca$

The equation having a, b and c as its roots is $x^3 - (a + b + c)\, x^2 + (ab + bc + ca)\, x - abc = 0$

$\Rightarrow \quad x^3 - 3Ax^2 + \dfrac{3G^3}{H}x - G^3 = 0$

Illustration 45: The harmonic means between two numbers is given as 4, their A.M. is A, and G.M. is G, satisfy the relation $2A + G^2 = 27$. Determine the two numbers. **(JEE ADVANCED)**

Sol: Let a and b be the two numbers and H = 4 be the harmonic mean between them. Therefore, by using A.M. = $\dfrac{a+b}{2}$ and G.M. = $\sqrt{ab}$, we can obtain the values of a and b.

H = 4 (given)

As A, G and H are in GP, therefore $\quad G^2 = AH \Rightarrow \quad G^2 = 4A$

Also, $2A + G^2 = 27 \qquad$ (given; $\therefore G^2 = 4A$)

$\therefore \quad 6A = 27$

$\Rightarrow \quad A = \dfrac{9}{2} \qquad\qquad \Rightarrow \quad \dfrac{a+b}{2} = \dfrac{9}{2} \Rightarrow \quad a + b = 9$

We have, $G^2 = 4A$ and $A = 9/2 \Rightarrow \quad G^2 = 18 \quad \Rightarrow \quad ab = 18$

The quadratic equation having a and b as its roots is $x^2 - (a + b) x + ab = 0$ or, $x^2 - 9x + 18 = 0$

$\Rightarrow x = 3, 6$

Thus, the two numbers are 3 and 6.

Illustration 46: If $2a + b + 3c = 1$ and $a > 0$, $b > 0$, $c > 0$, then find the greatest value of $a^4b^2c^2$ and obtain the corresponding values of a, b and c. **(JEE ADVANCED)**

Sol: Since there is a^4,. take four equal parts of 2a; as there is b^2, take two equal parts of b; as there is c^2, take two equal parts of 3c. Since A.M. $\geq$ G.M., obtaining A.M. and G.M. of these numbers will help in solving this illustration.

Let us consider the positive numbers $\dfrac{2a}{4}, \dfrac{2a}{4}, \dfrac{2a}{4}, \dfrac{2a}{4}, \dfrac{b}{2}, \dfrac{b}{2}, \dfrac{3c}{2}, \dfrac{3c}{2}$.

For the numbers, $A = \dfrac{\frac{2a}{4}+\frac{2a}{4}+\frac{2a}{4}+\frac{2a}{4}+\frac{b}{2}+\frac{b}{2}+\frac{3c}{2}+\frac{3c}{2}}{4+2+2} = \dfrac{2a+b+3c}{8} = \dfrac{1}{8}$

$(\therefore 2a + b + 3c = 1)$

$G = \left(\dfrac{2a}{4}\cdot\dfrac{2a}{4}\cdot\dfrac{2a}{4}\cdot\dfrac{2a}{4}\cdot\dfrac{b}{2}\cdot\dfrac{b}{2}\cdot\dfrac{3c}{2}\cdot\dfrac{3c}{2}\right)^{\frac{1}{8}} = \left(\dfrac{1}{2^4}\cdot\dfrac{1}{2^2}\cdot\dfrac{1}{2^2}\cdot 3^2 a^4 b^2 c^2\right)^{\frac{1}{8}}$

$\therefore \quad A \geq G \qquad \Rightarrow \quad \dfrac{1}{8} \geq \left(\dfrac{3^2}{2^8}a^4 b^2 c^2\right)^{\frac{1}{8}}$

or $\quad \dfrac{1}{8^8} \geq \dfrac{3^2}{2^8}a^4 b^2 c^2$ or $\quad \dfrac{2^8}{3^2.8^8} \geq a^4 b^2 c^2$ or $\quad \dfrac{1}{9.4^8} \geq a^4 b^2 c^2$.

Hence, the greatest value of $a^4b^2c^2 = \dfrac{1}{9.4^8}$

It has been found that when the equality holds, the greatest value takes place.

We know that A = G when all the numbers are equal, i.e.

$$\dfrac{2a}{4} = \dfrac{b}{2} = \dfrac{3c}{2} \qquad \Rightarrow \quad a = b = 3c$$

$\therefore \quad \dfrac{a}{3} = \dfrac{b}{3} = \dfrac{c}{1} = k \qquad \therefore \quad a = 3^k, b = 3k, c = k$

$\therefore \quad 2a + b + 3c = 1 \qquad \Rightarrow \quad 6k + 3k + 3k = 1$

$\therefore \quad k = \dfrac{1}{12} \qquad \therefore \quad a = \dfrac{3}{12}, b = \dfrac{3}{12}, c = \dfrac{1}{12},$ i.e. $a = \dfrac{1}{4}, b = \dfrac{1}{4}, c = \dfrac{1}{12}$

Arithmetic Mean of the mth power

Suppose $a_1, a_2, \ldots, a_n$ be n positive real numbers (not all equal) and let m be a real number, then

$$\dfrac{a_1^{\,m} + a_2^{\,m} + \ldots a_n^{\,m}}{n} > \left(\dfrac{a_1 + a_2 + \ldots a_n}{n} \right)^m , \text{ if } m \in R -[0, 1]$$

If $m \in (0, 1)$, then $\dfrac{a_1^{\,m} + a_2^{\,m} + \ldots a_n^{\,m}}{n} < \left(\dfrac{a_1 + a_2 + \ldots a_n}{n} \right)^m$

Thus, if $m \in \{0, 1\}$, then $\dfrac{a_1^{\,m} + a_2^{\,m} + \ldots a_n^{\,m}}{n} = \left(\dfrac{a_1 + a_2 + \ldots a_n}{n} \right)^m$

PROBLEM-SOLVING TACTICS

(a) When looking for a pattern in a sequence or series, writing out several terms will help you see the pattern, do not simplify directly. If you do this way, it is often easier to spot the pattern (if you leave terms as products, sums, etc.).

(b) If each term of an AP is multiplied by (or divided by a non-zero) fixed constant C, the resulting sequence is also an AP, with a common difference C times $\left(\text{or } \dfrac{1}{c} \text{times} \right)$ the previous.

(c) **Tips for AP problems**

    (i) When the number of terms are three, then we take the terms as a – d, a, a + d;

        Five terms as a - 2d, a - d, a, a + d, a + 2d

        Here, we take the middle term as 'a' and common difference as 'd'.

    (ii) When the number of terms is even, then we take:

        Four terms as a – 3d, a – d, a + d, a + 3d;

        Six terms as a – 5d, a – 3d, a – d, a + d, a + 3d, a + 5d

        Here, we take 'a – d' and 'a + d' as the middle terms and common difference as '2d'.

    (iii) If the number of terms in an AP is even, then take the number of terms as 2n and if odd then take it as (2n + 1).

(d) Tips for G.P. problems

(i) When the number of terms is odd, then we take three terms as a/r, a, ar; five terms as $\dfrac{a}{r^2}, \dfrac{a}{r}, a, ar, ar^2$.

Here, we take the middle term as 'a' and common ratio as 'r'.

(ii) When the number of terms is even, then we take four terms as $\dfrac{a}{r^3}, \dfrac{a}{r}, ar, ar^3$; six terms as $\dfrac{a}{r^5}, \dfrac{a}{r^3}, \dfrac{a}{r}, ar, ar^3, ar^5$.

Here, we take $\dfrac{'a'}{r}$ and 'ar' as the middle terms and common ratio as 'r^2'.

(e) Tips for H.P. problems

For three terms, we take as $\dfrac{1}{a-d}, \dfrac{1}{a}, \dfrac{1}{a+d}$

For four terms, we take as $\dfrac{1}{a-3d}, \dfrac{1}{a-d}, \dfrac{1}{a+d}, \dfrac{1}{a+3d}$

For five terms, we take as $\dfrac{1}{a-2d}, \dfrac{1}{a-d}, \dfrac{1}{a}, \dfrac{1}{a+d}, \dfrac{1}{a+2d}$

FORMULAE SHEET

Arithmetic Progression: Here, a, d, A and S_n represent the first term, common difference, A.M. and sum of the numbers, respectively, and T_n stands for the nth term.

1.	$T_n = a + (n-1)\, d$	4.	$S_n = \dfrac{n}{2}\left[2a + (n-1)d\right]$
2.	$T_n = \dfrac{T_{n-1} + T_{n+1}}{2}$	5.	$A = \dfrac{\left(a_1 + a_2 + \; ..... \; + a_n\right)}{n}$
3.	$S_n = \dfrac{n}{2}\left(a + T_n\right)$	6.	Insertion of n arithmetic means between a and b is A_n $= a + \dfrac{n(b-a)}{n+1}$

Geometric Progression: Here, a, r, S_n and G represent the first term, common ratio, sum of the terms and G.M., respectively, and T_n stands for the nth term.

1.	$T_n = a.r^{n-1}$	4.	$S_n = \dfrac{a(r^n - 1)}{r-1}$
2.	$T_n = \sqrt{T_{n-1} \cdot T_{n+1}}$	5.	$S_\infty = \dfrac{a}{1-r}$ (for $-1 < r < 1$)
3.	$S_n = \dfrac{T_{n+1} - a}{r-1}$	6.	Insertion of n geometric means between a and b is $G_1 = ar,\ G_2 = ar^2\ ....G_n = ar^n$ or $G_n = b/r$, where $r = \left(\dfrac{b}{a}\right)^{\frac{1}{n+1}}$

Arithmetic Geometric Progression: Here, a = the first term of AP, b = the first term of GP, d = common difference and r = common ratio of GP.

1.	$S_n = ab + (a + d)br + (a + 2d)br^2 + (a + 3d)br^3 + \dots$
2.	$S_n = \dfrac{ab}{1-r} + \dfrac{dbr(1-r^{n-1})}{(1-r)^2} - \dfrac{\left[a+(n-1)d\right]br^n}{1-r}$
3.	$S_\infty = \dfrac{ab}{1-r} + \dfrac{dbr}{(1-r)^2} \quad (\text{for} -1 < r < 1)$

Harmonic Progression

1. $a_n = \dfrac{1}{a+(n-1)d}$, where $a = \dfrac{1}{a_1}$ and $d = \dfrac{1}{a_2} - \dfrac{1}{a_1}$

2. $\dfrac{1}{H} = \dfrac{1}{n}\left(\dfrac{1}{a_1} + \dfrac{1}{a_2} + \dots + \dfrac{1}{a_n}\right)$

3. Insertion of n harmonic means between a and b

$$\dfrac{1}{H_1} = \dfrac{1}{a} + \dfrac{a-b}{(n+1)ab}$$

$$\dfrac{1}{H_2} = \dfrac{1}{a} + \dfrac{2(a-b)}{(n+1)ab} \text{ and so on} \Rightarrow \left[\dfrac{1}{H_n} = \dfrac{1}{a} + \dfrac{n(a-b)}{(n+1)ab}\right]$$

1.	The sum of n natural numbers	$\displaystyle\sum_{r=1}^{n} r = \dfrac{n(n+1)}{2}$
2.	The sum of n odd natural numbers	$\displaystyle\sum_{r=1}^{n} (2r-1) = n^2$
3.	The sum of n even natural numbers	$\displaystyle\sum_{r=1}^{n} 2r = n(n+1)$
4.	The sum of squares of n natural numbers	$\displaystyle\sum_{r=1}^{n} r^2 = \dfrac{n(n+1)(2n+1)}{6}$
5.	The sum of cubes of n natural numbers	$\displaystyle\sum_{r=1}^{n} r^3 = \left[\dfrac{n(n+1)}{2}\right]^2$

Solved Examples

JEE Main/Boards

Example 1: Find the rth term if the pth term of an AP is q and the qth term is p .

Sol: Using $T_n = a + (n - 1)\,d$, we can obtain the pth, q^{th} and r^{th} terms.

Let the initial term and common difference of the given AP be a and d, respectively.

As given,

$q = a + (p - 1)\,d$ ... (i)

$p = a + (q - 1)\,d$ ... (ii)

Subtracting (i) by (ii), we find that

$q - p = (p - q)\,d$

$\therefore d = -1$

Putting $d = -1$ in (i), we get

$a = q + p - 1$

$\therefore t_r = a + (r - 1)d$

$= (q + p - 1) - r + 1 = p + q - r$

Example 2: Find out the number of terms in a given AP 20, 25, 30, 35, ... 100.

Sol: We know that $T_n = a + (n - 1)\,d$.

Given, $a = 20$, $d = 5$ and $T_n = 100$. Therefore, by solving the equation, we will get the number of terms.

Let the number of terms be n.

Given, $T_n = 100$, $a = 20$, $d = 5$

$T_n = a + (n - 1)\,d$

$\Rightarrow 100 = 20 + (n - 1)\,5 \Rightarrow 80 = (n - 1)\,5$

$\Rightarrow 16 = (n - 1) \Rightarrow n = 17$

Example 3: Solve the following series:

99 + 95 + 91 + 87 + ... to 20 terms

Sol: Using $S_n = \dfrac{n}{2}\left[2a + (n - 1)d\right]$, we can solve the given problem.

We know that the terms of the given series are in AP. Given,

$D = -4$, $a = 99$ and $n = 20$

$\because S_n = \dfrac{n\left(2a + (n - 1)d\right)}{2}$

$S_{20} = \dfrac{20}{2}[2 \times 99 + (20 - 1)(-4)]$

$= 10[198 + 19 \times (-4)] = 10(198 - 76) = 1220$

Example 4: Find out the G.P. if the fifth and second terms of a G.P. are 81 and 24, respectively.

Sol: We know that in GP, the nth term is given by $T_n = a \cdot r^{n-1}$. Thus, by using this formula, we can find the GP.

Given, $T_5 = 81$ and $T_2 = 24$

$\therefore \quad 81 = ar^4$ ... (i)

and $\quad 24 = ar$ ... (ii)

Dividing (i) by (ii), we get

$\dfrac{81}{24} = r^3 \Rightarrow r^3 = \dfrac{27}{8} \Rightarrow r^3 = \left(\dfrac{3}{2}\right)^3 \Rightarrow r = \dfrac{3}{2}$

Substituting the value of r in (ii), we get, $a = 16$

Thus, the required G.P. is 16, 24, 36, 54 ,....

Example 5: If the sum of four numbers in AP is 50 and the greatest of them is four times the least, then find the numbers.

Sol: Let the four numbers in AP be a, a + d, a + 2d, a + 3d with $d > 0$.

As given, sum of the numbers is 50.

$a + (a + d) + (a + 2d) + (a + 3d) = 50$

$\therefore 4a + 6d = 50$

$\Rightarrow 2a + 3d = 25$ ...(i) and

$a + 3d = 4a$

$\Rightarrow 3d = 3a$

$\therefore d = a$

$\therefore$ Equation (i) becomes $5a = 25$

Thus, $a = 5 = d$

Therefore, the four number are 5, 10, 15 and 20.

Example 6: If S_1, S_2, S_3,....., S_p are the sum of p infinite geometric progression whose first terms are 1, 2, 3,...., p and whose common ratios are $\dfrac{1}{2}, \dfrac{1}{3}, \dfrac{1}{4}, ... \dfrac{1}{p+1}$, respectively, then prove that $S_1 + S_2 + ... + S_p = \dfrac{p(p+3)}{2}$.

Sol: As we know $S_\infty = \dfrac{a}{1-r}$, therefore by using this formula we can obtain the value of $S_1, S_2, ...S_p$.

We know that $S_\infty = \dfrac{a}{1-r}$

$\therefore \quad S_1 = \dfrac{1}{1-\dfrac{1}{2}} = 2 ; \ S_2 = \dfrac{2}{1-\dfrac{1}{3}} = 3$

$S_p = \dfrac{p}{1-\dfrac{1}{p+1}} = p+1$

$S_1 + S_2 + ...+ S_p = \dfrac{p}{2}[2 \times 2 + (p-1)1] = \dfrac{p}{2}[p+3]$

Example 7: Solve the series $1 + 2 \cdot 2 + 3 \cdot 2^2 + 4 \cdot 2^3 + ... + 100 \cdot 2^{99}$.

Sol: Let $S = 1 + 2 \cdot 2 + 3 \cdot 2^2 + 4 \cdot 2^3 + ... + 100 \cdot 2^{99}$. Therefore, by multiplying 2 on both the sides and then taking the difference, we can solve the given problem.

Given,

$S = 1 + 2 \cdot 2 + 3 \cdot 2^2 + 4 \cdot 2^3 + ...+ 100 \cdot 2^{99}$ ... (i)

Multiplying 2 on both the sides,

$2S = 1 \cdot 2 + 2 \cdot 2^2 + 3 \cdot 2^3 + ... + 99 \cdot 2^{99} + 100 \cdot 2^{100}$ ... (ii)

Subtracting (ii) from (i), we get

$-S = 1 + 1 \cdot 2 + 1 \cdot 2^2 + 1 \cdot 2^3 + ... + 1 \cdot 2^{99} - 100 \cdot 2^{100}$

$-S = \dfrac{1 - 2^{100}}{1-2} - 100 \cdot 2^{100} ;$

$\Rightarrow S = 99 \cdot 2^{100} + 1$

Example 8: If $(5n + 4) : (9n + 6)$ is the ratio of the sums of the nth terms of two APs, then find out the ratio of their 13th terms.

Sol: Let a_1 and a_2 be the first terms of the two APs and d_1 and d_2 be their respective common difference.

Applying $S_n = \dfrac{n}{2}\left[2a + (n-1)d\right]$, we can solve the given problem.

Given,

$\dfrac{\dfrac{n}{2}[2a_1 + (n-1)d_1]}{\dfrac{n}{2}[2a_2 + (n-1)d_2]} = \dfrac{5n+4}{9n+6}$

$\Rightarrow \dfrac{a_1 + \dfrac{n-1}{2}d_1}{a_2 + \dfrac{n-1}{2}d_2} = \dfrac{5n+4}{9n+6}$ ... (i)

The ratio of the 13th terms is $\dfrac{a_1 + 12d_1}{a_2 + 12d_2}$ [which is obtained from (i) with n = 25]

$\therefore \quad \dfrac{a_1 + 12d_1}{a_2 + 12d_2} = \dfrac{129}{231}$

Example 9: If the 7th and 8th terms of an H.P. are 8 and 7, respectively, then find its 15th term.

Sol: We know that $t_n = \dfrac{1}{a + (n-1)d}$. Therefore, by using this formula we can solve the given problem.

Given, $T_7 = 8$ =and $T_8 = 7$

$\therefore \dfrac{1}{a + 6d} = 8 \Rightarrow 8a + 48d - 1 = 0$ ..(i)

$\dfrac{1}{a + 7d} = 7 \Rightarrow 7a + 49d - 1 = 0$ ..(ii)

By solving these two equations, we find that $d = a$

$\therefore$ From eq. (i), we get $\dfrac{1}{7a} = 8$

$\Rightarrow a = d = \dfrac{1}{56}$

$\therefore \quad T_{15} = \dfrac{1}{a + 14d} = \dfrac{56}{15}$

Example 10: Suppose x y and z are positive real numbers, which are different from 1.

If $x^{18} = y^{21} = z^{28}$, then show that 3, $3\log_y(x)$. $3\log_z(y)$ and $7 \log_x(z)$ are in AP.

Sol: By applying log on $x^{18} = y^{21} = z^{28}$, we can find the values of $\log_y x$, $\log_z y$ and $\log_x z$.

Given, $x^{18} = y^{21} = z^{28}$

Taking log, we find that

$18 \log x = 21 \log y = 28 \log z$

$\log_y x = \dfrac{\log x}{\log y} = \dfrac{7}{6}$

$\Rightarrow 3\log_y x = \dfrac{7}{2}$ ... (i)

$\log_z y = \dfrac{\log y}{\log z} = \dfrac{4}{3} ; \quad 3\log_z y = 4$ ... (ii)

$$\log_x z = \frac{\log z}{\log x} = \frac{9}{14}$$

$$\Rightarrow 7\log_x z = \frac{9}{2} \qquad \text{... (ii)}$$

The numbers 3, $\frac{7}{2}$, 4 and $\frac{9}{2}$ are in AP with common difference = $\frac{1}{2}$.

$\therefore$ 3, $\log_y x$, $3\log_z y$ and $7\log_x z$ are in AP.

Example 11: If $\sqrt[x]{a} = \sqrt[y]{b} = \sqrt[z]{c}$ and if a, b and c are positive and in GP, then prove that x, y and z are in AP.

Sol: This problem can be solved by taking log on $\sqrt[x]{a} = \sqrt[y]{b} = \sqrt[z]{c}$.

$$a^{1/x} = b^{1/y} = c^{1/z}$$

$$\Rightarrow \frac{\log a}{x} = \frac{\log b}{y} = \frac{\log c}{z} = k$$

$$\Rightarrow \quad \log a = kx, \ \log b = ky, \ \log c = kz \qquad \text{... (i)}$$

a, b and c are in GP

$$\Rightarrow \quad b^2 = ac$$

$$\therefore \quad 2\log b = \log a + \log c$$

$$\Rightarrow \quad 2ky = kx + kz \text{ by (i)}$$

$$\Rightarrow \quad 2y = x + z$$

$$\Rightarrow \quad x, \ y \text{ and } z \text{ are in AP.}$$

Example 12: Determine the relation between x, y and z if 1, $\log_y x$, $\log_z y$, $-15\log_x z$ are in AP.

Sol: By considering the common difference as d and obtaining its value by $\log_y x = 1 + d$ and $\log_z y = 1 + 2d$, we can determine the required relation.

Suppose d be the common difference of the given AP, then

$$\log_y x = 1 + d \Rightarrow x = y^{1+d} \qquad \text{... (i)}$$

$$\log_z y = 1 + 2d \Rightarrow y = z^{1+2d} \qquad \text{... (ii)}$$

$$15\log_x z = -(1 + 3d)$$

$$\Rightarrow \quad z = x^{\frac{1+3d}{-15}} \qquad \text{... (iii)}$$

Elimination y and z from equations (i), (ii) and (iii), we get

$$x = x^{\frac{(1-d)(1+2d)(1+3d)}{-15}}$$

$$\therefore \quad 1 = \frac{(1+d)(1+2d)(1+3d)}{-15}$$

or (1 + d) (1 + 2d) (1 + 3d) + 15 = 0

or (d + 2) (6d^2 + 5d + 8) = 0

$$\Rightarrow \quad d = -2$$

The other factors do not give any real solution.

$\therefore$ $x = y^{-1}, \ y = z^{-3}, \ z = x^{1/3}$ or $\quad x = y^{-1} = z^3$

Example 13: There are four numbers of which the first three are in G.P. and the last there are in AP, with a common difference of 6. If the first number and the last number are equal, then find the numbers.

Sol: Let the four numbers be a, a − 2d, a − d, a, where d = 6

a, a − 12, a − 6 are in GP.

$$\Rightarrow \quad a(a - 6) = (a - 12)^2$$

$$\Rightarrow \quad a^2 - 6a = a^2 - 24a + 144$$

$$\Rightarrow \quad 18a = 144$$

$$\Rightarrow \quad a = 8$$

The numbers are 8, − 4, 2 and 8.

Example 14: a, b and c are the pth, qth and rth terms of both an AP and a GP, respectively, then prove that $a^{b-c} \cdot b^{c-a} \cdot c^{a-b} = 1$ (both progressions have the same first term.)

Sol: By using formula $T_n = a + (n - 1) d$ and $T_n = a.r^{n-1}$, we can obtain the pth, qth and rth terms of both an AP and a GP.

$$T_p = a = a_1 + (p - 1) d_1 = a_1(r_1)^{p-1} \qquad \text{... (i)}$$

$$T_q = b = a_1 + (q - 1)d_1 = a_1(r_1)^{q-1} \qquad \text{... (ii)}$$

$$T_r = c = a_1 + (r - 1)d_1 = a_1(r_1)^{r-1} \qquad \text{... (iii)}$$

From (i), (ii), (iii)

$$a - b = (p - q) d_1$$

$$b - c = (q - r) d_1$$

$$c - a = (r - p) d_1$$

Therefore, $a^{b-c} \cdot b^{c-a} \cdot c^{a-b}$

$$= (a_1 r_1^{p-1})^{b-c} (a_1 r_1^{q-1})^{c-a} (a_1 r_1^{r-1})^{a-b}$$

$$= a_1^{b-c+c-a+a-b} r_1^{(p-1)(b-c)+(q-1)(c-a)+(r-1)(a-b)}$$

$$= a_1^0 \cdot r_1^{(p-1)(q-r)d_1 +(q-1)(r-p)d_1 +(r-1)(p-q)d_1}$$

$$= a_1^0 \cdot r_1^0 = 1$$

JEE Advanced/Boards

Example 1: If the sum of first n terms of three arithmetic progressions are S_1, S_2 and S_3, the first term of each being 1 and the common differences being 1, 2 and 3, respectively, then prove that $S_1 + S_3 = 2S_2$

Sol: Using $S_n = \dfrac{n}{2}\left[2a + (n-1)d\right]$, we can get the values of S_1, S_2 and S_3.

Given, $a = 1$, $d_1 = 1$, $d_2 = 2$, $d_3 = 3$

$S_1 = \dfrac{n}{2}[2a + (n-1)\,d_1]$

$= \dfrac{n}{2}[2 \times 1 + (n-1)\,1] \quad = \dfrac{n}{2}[1 + n]$

$S_2 = \dfrac{n}{2}[2a + (n-1)d_2] = \dfrac{n}{2}[2 \times 1 + (n-1)\,2] = n^2$

$S_3 = \dfrac{n}{2}[2a + (n-1)\,d_3] = \dfrac{n}{2}[2 \times 1 + (n-1)\,3]$

$= \dfrac{n}{2}[3n - 1]$

$S_1 + S_3 = \dfrac{n}{2}[1 + n + 3n - 1] = 2n^2 = 2S$

Example 2: Calculate the sum to n terms of the series:
$8 + 88 + 888 + \ldots$

Sol: We can solve this problem by taking 8 as common from given series and applying various operations.

Let $S_n = 8 + 88 + 888 + \ldots$ to n terms $= 8\,[1 + 11 + 111 + \ldots]$

$= \dfrac{8}{9}\,[9 + 99 + 999 + \ldots] = \dfrac{8}{9}\,[(10 - 1) + (100 - 1) + (1000 - 1) + \ldots$ to n terms$]$

$= \dfrac{8}{9}\,[10 + 100 + 1000 + \ldots + \text{to n terms}] - \dfrac{8}{9}\,n = \dfrac{8}{9}\dfrac{(10^n - 1)}{9} - \dfrac{8n}{9}$

$= \dfrac{8}{81}\,[10^{n+1} - 9n - 10]$

Example 3: If $a_1, a_2, a_3, \ldots, a_n$ are in AP, where $a_i > 0$ for all I, then show that

$\dfrac{1}{\sqrt{a_1} + \sqrt{a_2}} + \dfrac{1}{\sqrt{a_2} + \sqrt{a_3}} + \ldots +$

$\dfrac{1}{\sqrt{a_{n-1}} + \sqrt{a_n}} = \dfrac{n-1}{\sqrt{a_1} + \sqrt{a_n}}$

Sol: We can write $x - y = p$ as

$\left(\sqrt{x} + \sqrt{y}\right)\left(\sqrt{x} - \sqrt{y}\right) = p$.

Thus, by following this method we can represent difference of $a_1, a_2, a_3, \ldots, a_n$.

$a_1, a_2, a_3, \ldots, a_n$ are in AP.

$a_2 - a_1 = a_3 - a_2 = \ldots = a_n - a_{n-1} = d$ (say)

$\Rightarrow \quad \left(\sqrt{a_2} + \sqrt{a_1}\right)\left(\sqrt{a_2} - \sqrt{a_1}\right)$

$= \left(\sqrt{a_3} + \sqrt{a_2}\right)\left(\sqrt{a_3} - \sqrt{a_2}\right) \ldots$

$= \left(\sqrt{a_n} + \sqrt{a_{n-1}}\right)\left(\sqrt{a_n} - \sqrt{a_{n-1}}\right) = d$

$\Rightarrow \dfrac{1}{\sqrt{a_2} + \sqrt{a_1}} = \dfrac{1}{d}\left(\sqrt{a_2} - \sqrt{a_1}\right), \dfrac{1}{\sqrt{a_3} + \sqrt{a_2}}$

$= \dfrac{\sqrt{a_3} - \sqrt{a_2}}{d}, \ldots$

$\dfrac{1}{\sqrt{a_n} + \sqrt{a_{n-1}}} = \dfrac{\sqrt{a_n} - \sqrt{a_{n-1}}}{d}$

$\text{LHS} = \dfrac{1}{\sqrt{a_1} + \sqrt{a_2}} + \dfrac{1}{\sqrt{a_2} + \sqrt{a_3}} + \ldots +$

$= \dfrac{1}{d}\left(\sqrt{a_n} - \sqrt{a_1}\right)$

$\because d = \dfrac{a_n - a_1}{n - 1}$

$\Rightarrow \text{LHS} = \dfrac{\sqrt{a_n} - \sqrt{a_1}}{a_n - a_1}\,(n-1) = \dfrac{n-1}{\sqrt{a_n} + \sqrt{a_1}} = \text{RHS}$

Example 4: A series of natural numbers is divided into groups: (1); (2, 3, 4); (5, 6, 7, 8, 9) and so on. Prove that the sum of the numbers in the nth group is $(n-1)^3 + n^3$.

Sol: In this problem, the last term of each group is the square of the corresponding number of the group. Thus, the first term of the nth group is $(n-1)^2 + 1 = n^2 - 2n + 2$. Hence, by using $S_n = \dfrac{n}{2}\left[2a + (n-1)d\right]$, we can solve the problem.

The number of terms in the first group = 1

The number of terms in the second group = 3

The number of terms in the third group = 5

$\therefore$ The number of terms in the nth group = $2n-1$

The common difference of the numbers in the nth group
$= 1$

The required sum $= \dfrac{2n-1}{2} \, [2(n^2 - 2n + 2) + (2n - 2) \, 1]$

$= \dfrac{2n-1}{2} \, [2n^2 - 2n + 2] = (2n - 1)[n^2 - n + 1]$

$= 2n^3 - 3n^2 + 3n - 1 = n^3 + (n - 1)^3$

Example 5: Find the sum of the series

$1 + \dfrac{1}{5} + \dfrac{3}{5^2} + \dfrac{5}{5^3} + \ldots$ to ∞.

Sol: We can write given series as $S_\infty = 1 + S_1$, where $S_1 = \dfrac{1}{5} + \dfrac{3}{5^2} + \dfrac{5}{5^3} + \ldots$ to ∞ Thus, by multiplying $\dfrac{1}{5}$ on both the sides and subtracting, we can obtain the required sum.

Let $S_\infty = 1 + \dfrac{1}{5} + \dfrac{3}{5^2} + \dfrac{5}{5^3} + \ldots$ to ∞

$\therefore S_\infty = 1 + S_1$ ....(i)

Where

$\therefore \quad \dfrac{1}{5} S_1 = \dfrac{1}{5^2} + \dfrac{3}{5^3} + \dfrac{5}{5^4} + \ldots \ldots$ to ∞ ....(ii)

Subtracting, (ii) from (i)

$\therefore \quad \dfrac{4}{5} S_1 = \dfrac{1}{5} + 2 \left(\dfrac{1}{5^2} + \dfrac{1}{5^3} + \ldots\ldots \text{to} \quad \infty \right)$

$= \dfrac{1}{5} + 2 \dfrac{\frac{1}{5^2}}{1 - \frac{1}{5}} = \dfrac{1}{5} + \dfrac{2}{4}\left(\dfrac{1}{5}\right) = \dfrac{3}{10},$

$S_1 = \dfrac{3}{8}$

From (i), $S_\infty = 1 + \dfrac{3}{8} = \dfrac{11}{8}$

Example 6: If $n \in N$ and $n > 1$, then prove that

(a) $n^n \geq 1.3.5 \ldots\ldots(2n - 1)$ (b) $2^n \geq 1 + n \, 2^{\frac{n-1}{2}}$

Sol: (a) Use the inequality A.M. $\geq$ GM.

(b) By solving $\dfrac{1 + 2 + 2^2 + \ldots + 2^{n-1}}{n} \geq (1.2.2^2 \ldots\ldots 2_{n-1})^{1/n}$,
we can prove the given equation.

(a) $\therefore \dfrac{1 + 3 + 5 + \ldots\ldots + (2n - 1)}{n} \geq \left[1.3.5 \ldots\left(2n - 1\right) \right]^{1/n}$

$\Rightarrow \dfrac{\frac{n}{2}\left[1 + (2n - 1)\right]}{n} \geq \left[1.3.5 \ldots\left(2n - 1\right) \right]^{1/n}$

$\Rightarrow n = \left[1.3.5 \ldots\left(2n - 1\right) \right]^{1/n}$

$\therefore n^n = 1.3.5 \ldots\left(2n - 1\right)$

(b) $\dfrac{1 + 2 + 2^2 + \ldots + 2^{n-1}}{n} \geq (1.2.2^2 \ldots\ldots 2_{n-1})^{1/n}$

$\Rightarrow \dfrac{2^n - 1}{2 - 1} \times \dfrac{1}{n} \geq \left(2^{\frac{n(n-1)}{2}} \right)^{1/n}$

$\Rightarrow \dfrac{2^n - 1}{n} \geq 2^{n - 1/2}$

$\Rightarrow 2^n \geq n\sqrt{2^{n-1}} + 1$

Exercise 1

Q.1 In a G.P. sum of n terms is 364. First term is 1 and common ratio is 3. Find n.

Q.2 The sum of an infinite geometric progression is 2 and the sum of the geometric progression made from the cubes of this infinite series in 24. Then find the series.

Q.3 Sum of n terms of the series,

(i) $0.7 + 0.77 + 0.777 + \ldots$ (ii) $6 + 66 + 666 + \ldots$

Q.4 If a, b, c are in A.P., prove that

(i) b + c, c + a, a + b are also in A.P.

(ii) $\dfrac{1}{bc}, \dfrac{1}{ca}, \dfrac{1}{ab}$ are also in A.P.

(iii) $a^2(b + c), b^2(c + a), c^2(a + b)$ are also in A.P.

(iv) $a\left(\dfrac{1}{b}+\dfrac{1}{c}\right), b\left(\dfrac{1}{c}+\dfrac{1}{a}\right), c\left(\dfrac{1}{a}+\dfrac{1}{b}\right)$ are also in A.P.

Q.5 The sum of three numbers in A.P. is 12 and the sum of their cubes is 288. Find the numbers.

Q.6 Find the sum of the integers between 1 and 200 which are

(i) Multiple of 3   (ii) Multiple of 7

(iii) Multiple of 3 and 7

Q.7 The sum of first n terms of two A.P.'s are in the ratio $(3n - 3): (5n + 21)$. Find the ratio of their 24th terms.

Q.8 If the pth term of an A.P. is x and q^{th} term is y, show that the sum first $(p + q)$ terms is $\dfrac{p+q}{2}\left\{x+y+\dfrac{x-y}{p-q}\right\}$

Q.9 If a, b, c are in H.P. prove that $\dfrac{a}{b+c}, \dfrac{b}{c+a}, \dfrac{c}{a+b}$ are in H.P.

Q.10 Find the sum of n terms of the series $1+\dfrac{4}{5}+\dfrac{7}{5^2}+\dfrac{10}{5^3}+\dots\dots.$

Q.11 Let a, b, c, d, e be five real numbers such that a, b, c are in A.P.; b, c, d are in G.P.; c, d, e are in H.P. If a = 2 and e = 18, find all possible values of b, c and d.

Q.12 Find the sum of first n terms of the series: $1\cdot2\cdot3 + 2\cdot3\cdot4 + 3\cdot4\cdot5 + \dots\dots$

Q.13 Find the sum of first 2n terms of the series: $1^2+ 2 + 3^2 + 4 + 5^2 + 6 + \dots.$

Q.14 The H.M of two numbers is 4 and their A.M. (A) and G.M. (G) satisfy the relation $2A + G^2 = 27$. Find the numbers.

Q.15 Find the sum of first 10 terms of the series: $(3^3 - 2^3) + (5^3 - 4^3) + (7^3 - 6^3) + \dots$

Q.16 Find the sum of first 20 terms of the series: $1\cdot3^2 + 2\cdot5^2 + 3\cdot7^2 + \dots.$

Q.17 Find three numbers a, b, c between 2 and 18 such that:

(i) Their sum is 25.

(ii) The numbers 2, a, b are consecutive terms of an A.P. and

(iii) The numbers b, c, 18 are consecutive terms of a G.P.

Q.18 If a > 0, b > 0 and c > 0, prove that:

$(a + b + c)\left(\dfrac{1}{a}+\dfrac{1}{b}+\dfrac{1}{c}\right)\geq 9$

Q.19 If $A_1, A_2, G_1, G_2,$; and H_1, H_2 be two A.M.'s, G.M.'s and H.M's between two numbers, then prove that:

$$\dfrac{G_1 G_2}{H_1 H_2}=\dfrac{A_1 + A_2}{H_1 + H_2}$$

Q.20 Find the coefficient of x^{99} and x^{98} in the polynomial: $(x - 1) (x - 2) (x - 3) \dots (x - 100)$.

Q.21 The interior angles of a polygon are in A.P. The smallest angle is $120°$ and the common difference is $5°$. Find the number of sides of the polygon

Q.22 A number consists of three digits in G.P.. The sum of the digits at units and hundreds place exceeds twice the digit at tens place by 1 and the sum of the digits at tens and hundreds place is two third of the sum of the digits at tens and units place. Find the number.

Q.23 25 trees are planted in a straight line at intervals of 5 meters. To water them the gardener must bring water for each tree separately from a well 10 meters from the first tree in line with the trees. How far he will have to cover in order to water all the tree beginning with the first if he starts from the well.

Q.24 Natural numbers have been grouped in the following way 1 ; (2, 3) ; (4, 5, 6); (7, 8, 9, 10) ; $\dots$ Show that the sum of the numbers in the nth group is $\dfrac{n(n^2 +1)}{2}$.

Q.25 In three series of GP's, the corresponding numbers in G.P. are subtracted and the difference of the numbers are also found to be in G.P. Prove that the three sequences have the same common ratio.

Q.26 If $a_1, a_2, a_3, \dots$ Are in A.P such that $a_i \neq 0$, show that

$$S = \dfrac{1}{a_1 a_2}+\dfrac{1}{a_2 a_3}+\dots+\dfrac{1}{a_n a_{n+1}}=\dfrac{n}{a_1 a_{n+1}}$$

Also evaluate $\lim\limits_{a \to \infty} S$.

Q.27 If 9 arithmetic means and 9 harmonic means be inserted between 2 and 3, prove that $A + \dfrac{6}{H} = 5$, where A is any arithmetic mean and H the corresponding harmonic mean.

Q.28 If $x + y + z = 1$ and x, y, z are positive numbers, show that $(1 - x)(1 - y)(1 - z) \geq 8xyz$.

Q.29 Show that any positive integral power (greater than 1) of a positive integer m, is the sum of m consecutive odd positive integers. Find the first odd integer for $m^r (r > 1)$.

Exercise 2

Single Correct Choice Type

Q.1 If a, b, c are distinct positive real in H.P., then the value of the expression, $\dfrac{b+a}{b-a} + \dfrac{b+c}{b-c}$ is equal to

(A) 1   (B) 2   (C) 3   (D) 4

Q.2 The sum of infinity of the series

$\dfrac{1}{1} + \dfrac{1}{1+2} + \dfrac{1}{1+2+3} + \ldots\ldots$ is equal to

(A) 2   (B) 5/2   (C) 3   (D) None

Q.3 Along a road lies an odd number of stones placed at intervals of 10 m. These stones have to be assembled around the middle stone. A person can carry only one stone at a time. A man carried out the job starting with the stone in the middle, carrying stones in succession, thereby covering a distance of 4.8 km. Then the number of stones is

(A) 15   (B) 29   (C) 31   (D) 35

Q.4 If $S = 1^2 + 3^2 + 5^2 + \ldots + (99)^2$ then the value of the sum $2^2 + 4^2 + 6^2 + \ldots + (100)^2$ is

(A) S + 2550   (B) 2S   (C) 4S   (D) S + 5050

Q.5 In an A.P. with first term and the common difference d (a, d $\neq$ 0), the ratio 'S' of the sum of the first n terms to sum of n terms succeeding them does not depend on n. Then the ratio "a/d" and the ratio 'ρ', respectively are

(A) $\dfrac{1}{2}, \dfrac{1}{4}$ (B) $2, \dfrac{1}{3}$ (C) $\dfrac{1}{2}, \dfrac{1}{3}$ (D) $\dfrac{1}{2}, 2$

Q.6 If $x \in R$, the numbers $(5^{1+x} + 5^{1-x})$, $a/2 \, (25^x + 25^{-x})$ form an A.P. then 'a' must lie in the interval

(A) [1, 5]   (B) [2, 5]   (C) [5, 12]   (D) [12, ∞]

Q.7 If the sum of the first 11 terms of an arithmetical progression equals that of the first 19 terms, then the sum of its first 30 terms, is

(A) Equal to 0   (B) Equal to – 1

(C) Equal to 1   (D) Non unique

Q.8 Let $s_1, s_2, s_3 \ldots$ and $t_1, t_2, t_3 \ldots$ are two arithmetic sequence such that $s_1 = t_1 \neq 0$; $s_2 = 2t_2$ and $\displaystyle\sum_{i=1}^{10} s_i = \sum_{i=1}^{15} t_i$. Then the value of $\dfrac{s_2 - s_1}{t_2 - t_1}$ is

(A) 8/3   (B) 3/2   (C) 19/8   (D) 2

Q.9 Let a_n, $n \in I$ be the nth term an A.P. with common difference 'd' and all whose terms are non-zero. If n approaches infinity, then the sum

$\dfrac{1}{a_1 a_2} + \dfrac{1}{a_2 a_3} + \ldots + \dfrac{1}{a_n a_{n+1}}$ will approach

(A) $\dfrac{1}{a_1 d}$ (B) $\dfrac{2}{a_1 d}$ (C) $\dfrac{1}{2a_1 d}$ (D) $a_1 d$

Q.10 The sum of the first three terms of an increasing G.P. is 21 and the sum of their squares is 189. Then the sum of its first n term is

(A) $3(2^n - 1)$ (B) $12\left(1 - \dfrac{1}{2^n}\right)$

(C) $6\left(1 - \dfrac{1}{2^n}\right)$ (D) $6(2^n - 1)$

Q.11 The sum $\displaystyle\sum_{n=1}^{\infty}\left(\dfrac{n}{n^4 + 4}\right)$ is equal to

(A) 1/4   (B) 1/3   (C) 3/8   (D) 1/2

Q.12 If $a \neq 1$ and $(\ln a^2) + (\ln a^2)^2 + (\ln a^2)^3 + \ldots = 3 [\ln a + (\ln a)^2 + (\ln a)^3 + (\ln a)^4 + \ldots]$, then 'a' is equal to

(A) $e^{1/5}$ (B) $e^{1/2}$ (C) $3e^{1/2}$ (D) $e^{1/4}$

Previous Years' Questions

Q.1 If a, b, c d and p are distinct real numbers such that $(a^2 + b^2 + c^2)p^2 - 2(ab + bc + cd)p + (b^2 + c^2 + d^2) \le 0$, then a, b, c, d **(1987)**

(A) Are in A.P.

(B) Are in G.P.

(C) Are in H.P.

(D) Satisfy ab = cd

Q.2 Sum of the first n terms of the series $\frac{1}{2} + \frac{3}{4} + \frac{7}{8} + \frac{15}{16} + \ldots$ is equal to

(A) $2^n - n - 1$

(B) $1 - 2^{-n}$

(C) $n + 2^{-n} - 1$

(D) $2^n + 1$

Q.3 If $x > 1$, $y > 1$, $z > 1$ are in G.P. then $\frac{1}{1 + \ln x}, \frac{1}{1 + \ln y}, \frac{1}{1 + \ln z}$ are in **(1998)**

(A) AP

(B) H.Π.

(C) G.Π.

(D) None

Q.4 If a, b, c, d are positive real number such that $a + b + c + d = 2$, then $M = (a + b)(c + d)$ satisfies the relation **(2000)**

(A) $0 < M \le 1$

(B) $1 \le M \le 2$

(C) $2 \le M \le 3$

(D) $3 \le M \le 4$

Q.5 Let the positive numbers a, b, c, d be in A.P. then abc, abd, acd, bcd are **(2001)**

(A) not in AP/GP/HP

(B) in AP

(C) in GP

(D) in HP

Q.6 Suppose a, b, c are in AP and a^2, b^2, c^2 are in G.P. If $a < b < c$ and $a + b + c = \frac{3}{2}$, then the value of a is **(2002)**

(A) $\frac{1}{2\sqrt{2}}$

(B) $\frac{1}{2\sqrt{3}}$

(C) $\frac{1}{2} - \frac{1}{\sqrt{3}}$

(D) $\frac{1}{2} - \frac{1}{\sqrt{2}}$

Q.7 An infinite G.P. has first term x and sum 5, then x belongs to **(2004)**

(A) $x < -10$

(B) $-10 < x < 0$

(C) $0 < x < 10$

(D) $x > 10$

Q.8 If the sum of first n terms of an AP is cn^2, then the sum of squares of these n terms is **(2009)**

(A) $\dfrac{n(4n^2 - 1)c^2}{6}$

(B) $\dfrac{n(4n^2 + 1)c^2}{3}$

(C) $\dfrac{n(4n^2 - 1)c^2}{3}$

(D) $\dfrac{n(4n^2 + 1)c^2}{6}$

Q.9 The first two terms of a geometric progression add up to 12. The sum of the third and the fourth terms is 48. If the terms of the geometric progression are alternately positive and negative, then the first term is **(2008)**

(A) –4

(B) –12

(C) 12

(D) 4

Q.10 The sum to the infinity of the series is $1 + \frac{2}{3} + \frac{6}{3^2} + \frac{10}{3^3} + \frac{14}{3^4} + \ldots$ **(2009)**

(A) 2

(B) 3

(C) 4

(D) 6

Q.11 If m is the A. M. of two distinct real numbers l and n(l, n > 1) and G1, G2 and G3 are three geometric means between l and n, then $G_1^4 + 2G_2^4 + G_3^4$ equals: **(2015)**

(A) $4\, l^2\, mm$

(B) $4\, lm^2\, n$

(C) $4\, lmn^2$

(D) $4\, l^2 m^2 n^2$

Q.12 The sum of first 9 terms of the series is $\frac{1^3}{1} + \frac{1^3 + 2^3}{1 + 3} + \frac{1^3 + 2^3 + 3^3}{1 + 3 + 5} + \ldots$ **(2015)**

(A) 71

(B) 96

(C) 142

(D) 192

Q.13 If the 2nd, 5th and 9th terms of a non-constant A.P. are in G.P., then the common ratio of this G.P. is: **(2016)**

(A) $\frac{4}{3}$

(B) 1

(C) $\frac{7}{3}$

(D) $\frac{8}{5}$

Q.14 If the sum of the first ten terms of the series is $\frac{16}{5}m$ then m is equal to $\left(1\frac{3}{5}\right)^2 + \left(2\frac{2}{5}\right)^2 + \left(3\frac{1}{5}\right)^2 + 4^2 + \left(4\frac{4}{5}\right)^2 + \ldots$ **(2016)**

(A) 101

(B) 100

(C) 99

(D) 102

Q.15 Three positive numbers form an increasing G.P. If the middle term in this G.P. is doubled, the new numbers are in A.P. Then the common ratio of the G.P. is **(2014)**

(A) $2 - \sqrt{3}$

(B) $2 + \sqrt{3}$

(C) $\sqrt{2} + \sqrt{3}$

(D) $3 + 2$

Q.16 Statement-I: The sum of the series $1 + (1 + 2 + 4) + (4 + 6 + 9) + (9 + 12 + 16) + \ldots\ldots + (361 + 380 + 400)$ is 8000.

Statement-II: $\displaystyle\sum_{k=1}^{n}(k^3 - (k-1)^3) = n^3$ for any natural number n. **(2012)**

(A) Statement-I is false, statement-II is true

(B) Statement-I is true, statement-II is true; statement-II is a correct explanation for statement-I

(C) Statement-I is true, statement-II is true; statement-II is not a correct explanation for statement-I

(D) Statement-I is true, statement-II is false

Q.17 Statement-I: The sum of the series $1 + (1 + 2 + 4) + (4 + 6 + 9) + (9 + 12 + 16) + \ldots\ldots + (361 + 380 + 400)$ is 8000.

Statement-II: $\displaystyle\sum_{k=1}^{n}(k^3 - (k-1)^3) = n^3$ for any natural number n.

Q.18 If 100 times the 100th term of an AP with non zero common difference equals the 50 times its 50th term, then the 150th term of this AP is **(2012)**

(A) –150   (B) 150 times its 50th term

(C) 150   (D) Zero

Q.19 The real number k for which the equation, $2x3 + 3x + k = 0$ has two distinct real roots in $[0, 1]$ **(2013)**

(A) Lies between 1 and 2

(B) Lies between 2 and 3

(C) Lies between -1 and 0

(D) Does not exist.

Q.20 If the equations $x^2 + 2x + 3 = 0$ and $ax^2 + bx + c = 0$, $a,b,c \in R$, have a common root, then $a : b : c$ is **(2013)**

(A) $1 : 2 : 3$ (B) $3 : 2 : 1$ (C) $1 : 3 : 2$ (D) $3 : 1 : 2$

Q.21 The sum of first 20 terms of the sequence 0.7, 0.77, 0.777,……, is **(2013)**

(A) $\dfrac{7}{81}(179 - 10^{-20})$ (B) $\dfrac{7}{9}(99 - 10^{-20})$

(C) $\dfrac{7}{81}(179 + 10^{-20})$ (D) $\dfrac{7}{9}(99 + 10^{-20})$

Q.22 If x, y, z are in A.P. and tan-1x, tan-1y and tan-1z are also in A.P., then **(2013)**

(A) $x = y = z$ (B) $2x = 3y = 6z$

(C) $6x = 3y = 2z$ (D) $6x = 4y = 3z$

JEE Advanced/Boards

Exercise 1

Q.1 (i) The harmonic mean of two numbers is 4. The arithmetic mean A & the geometric mean G satisfy the relation $2A + G^2 = 27$. Find the two numbers.

(ii) The A.M. of two numbers exceeds their G.M. by 15 and HM by 27. Find the numbers.

Q.2 If the 10th term of an H.P. is 21 and 21st term of the same H.P. is 10, then find the 210th term.

Q.3 If sinx, $\sin^2 2x$ and cosx.sin4x form an increasing geometric sequence, then find the numerical value of cos2x. Also find the common ratio of geometric sequence.

Q.4 If a, b, c, d, e be 5 numbers such that a, b, c are in AP; b, c, d are in G.P. & c, d, e are in H.P. then,

(i) Prove that a, c, e are in GP

(ii) Prove that $e = (2b - a)^2/a$

(iii) If a = 2 & e = 18, find all possible values of b, c, d.

Q.5 Let a_1 and a_2 be two real values of α for which the numbers $2\alpha^2$, α^4, 24 taken in that order form an arithmetic progression. If β_1 and β_2 are two real values of β for which the numbers 1, β^2, $6 - \beta^2$ taken in that order form a geometric progression, then find the value of $(\alpha_1^2 + \alpha_2^2 + \beta_1^2 + \beta_2^2)$.

Q.6 Two distinct, real infinite geometric series each have a sum of 1 and have the same second term. The third term of one of the series is 18. If the second term of both the series can be written in the form $\dfrac{m-n}{p}$, where m, n and p are positive integers, and m is not divisible by the square of any prime, find the value of $100m + 10n + p$.

Q.7 Let $S = \sum\limits_{n=1}^{99} \dfrac{5^{100}}{(25)^n + 5^{100}}$. Find [s].

Where [y] denotes largest integer less than or equal to y.

Q.8 Given that the cubic $ax^3 - ax^2 + 9bx - b = 0$ $(a \neq 0)$ has all three positive roots. Find the harmonic mean of the roots independent of a and b, hence deduce that the root are all equal. Find also the minimum value of $(a + b)$, if a and $b \in N$.

Q.9 A computer solved several problems in succession. The time it took the computer to solve each successive problem was the same number of times smaller than the time it took to solve the preceding problem. How many problems were suggested to the computer if it spent 63.5 min to solve all the problems except for the first, 127 min to solve all the problems except for the last one, an 31.5 min to solve all the problems except for the first two?

Q.10 The sequence $a_1, a_2, a_3, \ldots\ldots a_{98}$ satisfies the relation $a_{n+1} = a_n + 1$ for $1,2,3,\ldots.\ 97$ and has the sum equal to 4949. Evaluate $\sum\limits_{k=1}^{49} a_{2k}$.

Q.11 Let a and b be positive integers. The value of xyz is 55 or $\dfrac{343}{55}$, according as a, x, y, z, b are in arithmetic progression or harmonic progression resp.. Find the value of $(a^2 + b^2)$.

Q.12 If the roots of $10x^3 - cx^2 - 54x - 27 = 0$ are in harmonic progression, then find c and all the roots.

Q.13 If a, b, c be in G.P. & $\log_c a$, $\log_b c$, $\log_a b$ be in AP, then find the common difference of the AP if $\log_a c = 4$.

Q.14 The first term of a geometric progression is equal to $b - 2$, then third term is $b + 6$, and the arithmetic mean of the first and third term to the second term is in the ratio 5: 3. Find the positive integral value of b.

Q.15 In a G.P. the ratio of the sum of the first eleven terms to the sum of the last eleven terms is 1/8 and the ratio of the sum of all the terms without the first nine to the sum of all the terms with out the last nine is 2. Find the number of terms in the GP.

Q.16 If sum of first n terms of an AP (having positive terms) is given by $S_n = (1 + 2T_n)(1 - T_n)$

where T_n is the nth term of series then $T_2^2 = \dfrac{\sqrt{a} - \sqrt{b}}{4}$ (a, $b \in N$). Find the value of $(a + b)$.

Q.17 Given a three digit number whose digits are three successive terms of a G.P. If we subtract 792 form it, we get a number written by the same digits in the reverse order. Now if we subtract four from the hundred's digit of the initial number and leave the other digits unchanged, we get a number whose digits are successive terms of an A.P. Find the number.

Q.18 For $0 < \theta < \dfrac{\pi}{4}$, let $S(\theta) = 1 + (1 + \sin\theta)\cos\theta +$ $(1 + \sin\theta + \sin^2\theta)\cos^2\theta + \ldots\ldots \infty$.

Then find the value of $\left(\dfrac{\sqrt{2} - 1}{\sqrt{2}}\right) S\left(\dfrac{\pi}{4}\right)$.

Q.19 If $\tan\left(\dfrac{\pi}{12} - x\right), \tan\dfrac{\pi}{12}, \tan\left(\dfrac{x}{12} + x\right)$ in order are three consecutive terms of a G.P., then sum of all the solutions in [0, 314] is $k\pi$. Find the value of k.

Exercise 2

Single Correct Choice Type

Q.1 The arithmetic mean of the nine numbers in the give set {9, 99,999,……999999999} is a 9 digit number N, all whose digits are distinct. The number N does not contain the digit

(A) 0   (B) 2   (C) 5   (D) 9

Q.2 $\sum\limits_{k=1}^{360}\left(\dfrac{1}{k\sqrt{k+1} + (k+1)\sqrt{k}}\right)$ is the ratio of two relative prime positive integers m and n. The value of $(m + n)$ is equal to

(A) 43   (B) 41   (C) 39   (D) 37

Q.3 The sum $\sum\limits_{k=1}^{100} \dfrac{k}{k^4 + k^2 + 1}$ is equal to

(A) $\dfrac{4950}{10101}$ (B) $\dfrac{5050}{10101}$

(C) $\dfrac{5151}{10101}$ (D) None of these

Q.4 A circle of radius r is inscribed in a square. The mid point of sides of the square have been connected by line segment and a new square resulted. The sides of these square were also connected by segments so that a new square was obtained and so on, then the radius of the circle inscribed in the nth square is

(A) $\left[2^{\frac{1-n}{2}}\right]r$

(B) $\left[2^{\frac{3-3n}{2}}\right]r$

(C) $\left[2^{\frac{n}{2}}\right]r$

(D) $\left[2^{\frac{5-3n}{2}}\right]r$

Assertion Reasoning Type

(A) Statement-I is true, statement-II is true and statment-II is correct explanation for statment-I.

(B) Statement-I is true, statement-II is true and statment-II is NOT the correct explanation for statment-I.

(C) Statement-I is true, statement -II is false.

(D) Statement-I is false, statement-II is true

Q.5 Statement-I: If $27\ abc \geq (a + b + c)^3$ and $3a + 4b + 5c = 12$ then $\dfrac{1}{a^2} + \dfrac{1}{b^3} + \dfrac{1}{c^5} = 10$, where a, b, c are positive real numbers.

Statement-II: For positive real numbers A.M. $\geq$ G.M.

Multiple Correct Choice Type

Q.6 Let a_1, a_2, a_3 .... and b_1, b_2, b_3 .... be arithmetic progressions such that $a_1 = 25$, $b = 75$ and $a_{100} + b_{100} = 100$. Then,

(A) The difference between successive terms in progression 'a' is opposite of the difference in progression 'b'.

(B) $a_n + b_n = 100$, for any n.

(C) $(a_1 + b_1)$, $(a_2 + b_2)$, $(a_3 + b_3)$,... are in AP

(D) $\displaystyle\sum_{r=1}^{100}(a_r + b_r) = 10000$

Q.7 If $\sin (x - y)$, $\sin x$ and $\sin (x + y)$ are in H.P. then the value of $\sin x.\sec \dfrac{y}{2} =$

(A) 2   (B) $2^{1/2}$ (C) $- 2$ (D) $-2^{1/2}$

Q.8 The sum of the first three terms of the G.P. in which the difference between the second and the first term is 6 and the difference between the fourth and the third term 54, is

(A) 39   (B) –10.5   (C) 27   (D) –27

Q.9 If the roots of the equation $x^3 + px^2 + qx - 1 = 0$ form an increasing GP, where p and q are real, then

(A) $p + q = 0$

(B) $p \in (-3, \infty)$

(C) One of the roots is unity

(D) One root is smaller than 1

Q.10 If the triplets log a, log b, log c and (log a – log 2b), (log 2b – log3c), (log 3c – loga) are in arithmetic progression then

(A) $18 (a + b + c)^2 = 18(a^2 + b^2 + c^2) + ab$

(B) a, b, c are in GP

(C) a, 2b, 2c are in HP

(D) a, b, c can be the lengths of the sides of a triangle.

(Assume all logarithmic terms to be defined)

Q.11 x_1, x_2 are the roots of the equation $x^2 - 3x + A = 0$; x_3, x_4 are roots of the equation $x^2 - 12x + B = 0$, such that x_1, x_2, x_3, x_4 form an increasing G.P. then

(A) $A = 2$ (B) $B = 32$

(C) $x_1 + x_3 = 5$ (D) $x_2 + x_4 = 10$

Previous Years' Question

Q.1 If the first and the $(2n - 1)^{th}$ term of an AP, G.P. and H.P. are equal and their n^{th} terms are a, b, and c respectively, then   *(1988)*

(A) $a = b = c$ (B) $a \geq b \geq c$

(C) $a + c = b$ (D) $ac - b^2 = 0$

Q.2 Let S_1, S_2.... be squares such that for each $n \geq 1$ the length of a side of S_n equals the length of a diagonal of S_{n+1}. If the length of a side of S_1 is 10 cm, then for which of the following values of n is the area of S_n less than 1 sq. cm ?   *(1999)*

(A) 7   (B) 8   (C) 9   (D) 10

Q.3 Let S_k, $k = 1,2,....., 100$, denotes the sum of the infinite geometric series whose first term is $\dfrac{k-1}{k!}$ and the common ratio is $\dfrac{1}{k}$. Then the value of $\dfrac{100^2}{100!} + \displaystyle\sum_{k=1}^{100} |(k^2 - 3k + 1)S_k|$ is   *(2010)*

Q.4 Let a_1, a_2, a_3,....., a_{11} be real numbers satisfying $a_1 = 15$, $27 - 2a_2 > 0$ and $a_k = 2a_{k-1} - a_{(k-2)}$, for k = 3, 4,...., 11. If $\dfrac{a_1^2 + a_2^2 + ... + a_{11}^2}{11} = 90$, then the value of $\dfrac{a_1 + a_2 + ...a_{11}}{11}$ is equal to *(2010)*

Q.5 Le a_1, a_2, a_3,.... a_{100} be an arithmetic progression with $a_1 = 3$ and $S_p = \sum_{i=1}^{p} a_i$, $1 \le p \le 100$. For any integer n with $1 \le n \le 20$, let m = 5n. If $\dfrac{S_m}{S_n}$ does not depend on n, then a_2 is .... *(2011)*

Paragraph 1: Let A_1, G_1, H_1 denote the arithmetic, geometric and harmonic means, respectively, of two distinct positive numbers. For n > 2, let A_{n-1} and H_{n-1} have arithmetic, geometric and harmonic means as A_n, G_n, H_n respectively.

Q.6 Which one of the following statements is correct?

(A) $G_1 > G_2 > G_3 > ......$

(B) $G_1 < G_2 < G_3 < ......$

(C) $G_1 = G_2 = G_3 = ...$

(D) $G_1 < G_3 < G_5 < ......$ and $G_2 > G_4 > G_6 > ......$

Q.7 Which one of the following statements is correct?

(A) $A_1 > A_2 > A_3 > ......$

(B) $A_1 < A_2 < A_3 < ......$

(C) $A_1 > A_3 > A_5 > ......$ and $A_2 < A_4 < A_6 < ......$

(D) $A_1 < A_3 < A_5 < ......$ and $A_2 > A_4 > A_6 > ......$

Q.8 Which one of the following statements is correct?

(A) $H_1 > H_2 > H_3 > ......$

(B) $H_1 < H_2 < H_3 < ......$

(C) $H_1 > H_3 > H_5 > ......$ and $H_2 < H_4 < H_6 < ......$

(D) $H_1 < H_3 < H_5 < ......$ and $H_2 > H_4 > H_6 > ......$

Paragraph 2: Let V_r denote the sum of the first 'r' terms of an arithmetic progression (A.P.), whose first term is 'r' and the common difference is $(2r - 1)$. Let $T_r = V_{r+1} - V_{r-2}$ and $Q_r = T_{r+1} - T_r$ for r = 1, 2, ... *(2007)*

Q.9 The sum $V_1 + V_2 + ... + V_n$ is

(A) $\dfrac{1}{12} n(n + 1)(3n^2 - n + 1)$

(B) $\dfrac{1}{12} n(n + 1)(3n^2 + n + 2)$

(C) $\dfrac{1}{2} n(2n^2 - n + 1)$

(D) $\dfrac{1}{3} (2n^3 - 2n + 3)$

Q.10 T_r is always

(A) An odd number   (B) An even number

(C) A prime number   (D) A composite number

Q.11 Which one of the following is a correct statement?

(A) Q_1, Q_2, Q_3....... are in A.P. with common difference 5.

(B) Q_1, Q_2, Q_3....... are in A.P. with common difference 6.

(C) Q_1, Q_2, Q_3....... are in A.P. with common difference 11.

(D) $Q_1 = Q_2 = Q_3 = ...$

Q.12 Let $S_n = \displaystyle\sum_{k=1}^{n} \dfrac{n}{n^2 + kn + k^2}$ and $T_n = \displaystyle\sum_{k=0}^{n-1} \dfrac{n}{n^2 + kn + k^2}$

for n = 1, 2, 3, ... Then, *(2008)*

(A) $S_n < \dfrac{\pi}{3\sqrt{3}}$ (B) $S_n > \dfrac{\pi}{3\sqrt{3}}$

(C) $T_n < \dfrac{\pi}{3\sqrt{3}}$ (D) $T_n > \dfrac{\pi}{3\sqrt{3}}$

Q.13 Suppose four distinct positive numbers a_1, a_2, a_3, a_4 are in G.P. Let $b1 = a_1$, $b_2 = b_1 + a_2$, $b_3 = b_2 + a_3$ and $b_4 = b_3 + a_4$. *(2008)*

Statement-I: The numbers b_1, b_2, b_3, b_4 are neither in A.P. Nor in G.P.

Statement-II: The numbers b_1, b_2, b_3, b_4 are in H.P.

(A) Statement-I is true, statement-II is true; statement-II is a correct explanation for statement-I

(B) Statement-I is true, statement-II is true; statement-II is not a correct explanation for statement-I.

(C) Statement-I is true, statement-II is false

(D) Statement-I is false, statement-II is true

Q.14 If the sum of first n terms of an A.P. is cn^2, then the sum of squares of these n terms is *(2009)*

(A) $\dfrac{n(4n^2-1)c^2}{6}$

(B) $\dfrac{n(4n^2+1)c^2}{3}$

(C) $\dfrac{n(4n^2-1)c^2}{3}$

(D) $\dfrac{n(4n^2+1)c^2}{6}$

Q.15 Let S_k, $k = 1, 2,...., 100$, denote the sum of the infinite geometric series whose first term is $\dfrac{k-1}{k!}$ and the common ratio is $\dfrac{1}{k}$. Then the value of $\dfrac{100^2}{100!} + \sum_{k=1}^{100} \left|(k^2 - 3k + 1)S_k\right|$ is *(2010)*

Q.16 Let $a_1, a_2, a_3, ..., a11$ be real numbers satisfying $a1 = 15$, $27 - 2a_2 > 0$ and $a_k = 2a_k-1 - a_k-2$ for k $= 3, 4, ...,11$. If $\dfrac{a_1^2 + a_2^2 + ... + a_{11}^2}{11} = 90$, then the value of $\dfrac{a_1 + a_2 + ... + a_{11}}{11}$ is equal to *(2010)*

Q.17 Let b = 6, with a and c satisfying (E). If α and β are the roots of the quadratic equation $ax^2 + bx + c = 0$, then $\sum_{n=0}^{\infty} \left(\dfrac{1}{\alpha} + \dfrac{1}{\beta}\right)^n$ is *(2011)*

(A) 6 (B) 7 (C) $\dfrac{6}{7}$ (D) ∞

Q.18 Let $a_1, a_2, a_3, ...$ be in harmonic progression with $a_1 = 5$ and $a_{20} = 25$. The least positive integer n for which an< 0 is *(2012)*

(A) 22 (B) 23 (C) 24 (D) 25

Q.19 Let $S_n = \sum_{k=1}^{4n} (-1)^{\frac{k(k+1)}{2}} k^2$. Then S_n can take value(s) *(2013)*

(A) 1056 (B) 1088 (C) 1120 (D) 1332

Q.20 Let a, b, c be positive integers such that $\dfrac{b}{a}$ is an integer. If a, b, c are in geometric progression and the arithmetic mean of a, b, c is b + 2, then the value of $\dfrac{a^2 + a - 14}{a + 1}$ is _______ *(2014)*

Q.21 Suppose that all the terms of an arithmetic progression (A.P.) are natural numbers. If the ratio of the sum of the first seven terms to the sum of the first eleven terms is 6 : 11 and the seventh term lies in between 130 and 140, then the common difference of this A.P. is *(2015)*

Q.22 Let $b_i > 1$ for i = 1, 2, ..., 101. Suppose loge b1 loge b_2, ..., loge b_{101} are in Arithmetic Progression (A.P.) with the common difference loge 2. Suppose $a_1, a_2, ..., a_{101}$ are in A.P. such that $a_1 = b_1$ and $a_{51} = b_{51}$. If $t = b_1 + b_2 + ...+ b_{51}$ and $s = a_1 + a_2 + ... + a_{51}$, then *(2016)*

(A) s > t and $a_{101} > b_{101}$

(B) s > t and $a_{101} < b_{101}$

(C) s < t and $a_{101} > b_{101}$

(D) s < t and $a_{101} < b_{101}$

Important Questions

JEE Main/Boards

Exercise 1

Q.3	Q.11	Q.14
Q.17	Q.21	Q.25
Q.27		

Exercise 2

Q.2	Q.4	Q.10
Q.13		

Previous Years' Questions

Q.2	Q.5	Q.8

JEE Advanced/Boards

Exercise 1

Q.6	Q.9	Q.12
Q.15	Q.17	

Exercise 2

Q.1	Q.4	Q.5	Q.12

Previous Years' Questions

Q.1	Q.3	Q.4
Q.6	Q.7	Q.8

Answer Key

JEE Main/Boards

Exercise 1

Q.1 $n = 6$

Q.2 $3, -\dfrac{3}{2}, \dfrac{3}{4}, -\dfrac{3}{8}, \ldots$ is the series.

Q.3 (a) $\dfrac{7n}{9} - \dfrac{7}{81}\left(1 - \left(\dfrac{1}{10}\right)^n\right)$; (b) $\dfrac{2}{27}[10^{n+1} - 9n - 10]$

Q.5 $(2, 4, 6)$ or $(6, 4, 2)$

Q.6 (i) 6633 (ii) 2842 (iii) 945

Q.7 69 : 128

Q.10 $\dfrac{35}{16} - \dfrac{3}{16(5^{n-2})} - \dfrac{(3n-2)}{4(5^{n-1})}$

Q.11 $c = 6, b = 4, d = 9$; $b = -2, c = -6, d = -18$

Q.12 $\dfrac{1}{4}n(n+1)(n+2)(n+3)$

Q.13 $\dfrac{1}{3}n(4n^2 + 3n + 2)$

Q.14 6, 3

Q.15 4960

Q.16 188090

Q.17 $a = 5, b = 8, c = 12$.

Q.20 $-5050, \dfrac{1}{2}[(5050)^2 - 338350]$

Q.21 9

Q.22 469

Q.23 3370 m

Q.26 $\dfrac{1}{a_1(a_2 - a_1)}$

Exercise 2

Single Correct Choice Type

Q.1 B	**Q.2** A	**Q.3** C	**Q.4** D	**Q.5** C	**Q.6** D
Q.7 A	**Q.8** C	**Q.9** A	**Q.10** A	**Q.11** C	**Q.12** D

Previous Years' Questions

Q.1 B	**Q.2** C	**Q.3** B	**Q.4** A	**Q.5** D	**Q.6** D
Q.7 C	**Q.8** C	**Q.9** B	**Q.10** B	**Q.11** B	**Q.12** B
Q.13 A	**Q.14** A	**Q.15** B	**Q.16** B	**Q.17** D	**Q.18** D
Q.19 A	**Q.20** C	**Q.21** A			

JEE Advanced/Boards

Exercise 1

Q.1 (i) 6, 3 ; (ii) 120, 30   **Q.2** 1   **Q.3** $\dfrac{\sqrt{5}-1}{2} ; \sqrt{2}$

Q.4 (iii) b = 4, c = 6, d = 9 or b = – 2, c = – 6, d = – 18   **Q.5** 12   **Q.6** 518   **Q.7** 49

Q.8 28   **Q.9** 8 problems, 127.5 minutes   **Q.10** 2499   **Q.11** 50

Q.12 C = 9; (3, –3/2, – 3/5)   **Q.13** 13/4   **Q.14** 3   **Q.15** n = 38

Q.16 6   **Q.17** 931   **Q.18** 2   **Q.19** 4950

Exercise 2

Single Correct Choice Type

Q.1 A	**Q.2** D	**Q.3** B	**Q.4** A

Assertion Reasoning Type

Q.5 D

Multiple Correct Choice Type

Q.6 A, B, C, D	**Q.7** B, C	**Q.8** A, B	**Q.9** A, C, D	**Q.10** B, D	**Q.11** A, B, C, D

Previous Years' Questions

Q.1 A, B, D	**Q.2** B, C, D	**Q.3** 4	**Q.4** 0	**Q.5** 3 or 9	**Q.6** C
Q.7 A	**Q.8** B	**Q.9** B	**Q.10** D	**Q.11** B	**Q.12** A, D
Q.13 C	**Q.14** C	**Q.15** 3	**Q.16** 0	**Q.17** B	**Q.18** D
Q.19 A, D	**Q.20** 6	**Q.21** 9	**Q.22** B		

JEE Main/Boards

Exercise 1

Sol 1: Sum of n terms is 364

$a + ar + ar^2 \ldots\ldots\ldots ar^{n-1} = 364$

$\dfrac{a(r^n - 1)}{r - 1} = 364$

given $r = 3$, $a = 1$

$\Rightarrow \dfrac{(3^n - 1)}{(3 - 1)} = 364 \Rightarrow 3^n - 1 = 728$

$\Rightarrow 3^n = 729$

$\Rightarrow \boxed{n = 6}$

Sol 2: Sum of infinite G.P. is $2 \Rightarrow \dfrac{a}{-r + 1} = 2$

$\Rightarrow a = -2(r - 1)$

Series is $a, ar, ar^2 \ldots\ldots (|r| < 1)$

$\Rightarrow a^3, (ar)^3, (ar^2)^3 \ldots\ldots\ldots\ldots \ldots(2)$

First term of this infinite series is a^3 and ratio is r^3

Hence sum of this infinite series is $\dfrac{a^3}{-r^3 + 1}$

Given $\dfrac{a^3}{-r^3 + 1} = 24$

$\dfrac{8(r-1)^3}{\left(-r^3 + 1\right)} = 24 \Rightarrow \dfrac{(r-1)^2}{(r^2 + r + 1)} = 3$

$\Rightarrow r^2 + 1 - 2r = 3r^2 + 3r + 3$

$\Rightarrow 2r^2 + 5r + 2 = 0 \Rightarrow 2r^2 + 4r + r + 2 = 0$

$\Rightarrow (2r + 1)(r + 2) = 0$

$\Rightarrow r = -2, r = -\dfrac{1}{2}$

$|r| < 1 \Rightarrow r = -\dfrac{1}{2} \Rightarrow a = +3$

Series is $3, -\dfrac{3}{2}, \dfrac{3}{4}, -\dfrac{3}{8}, \ldots..$

Sol 3: (a) Sum upto n terms

$S_n = 0.7 + 0.77 + 0.777 \ldots\ldots\ldots\ldots$ n terms

$= 7[0.1 + 0.11 + 0.111 + \ldots] = \dfrac{7}{9}[0.9 + 0.99 + 0.999 \ldots]$

$= \dfrac{7}{9}[1 - 0.1 + 1 - 0.01 + 1 - 0.001 \ldots]$

$= \dfrac{7}{9}[n - (0.1 + 0.01 + 0.001 \ldots)]$

$= \dfrac{7}{9}\left[n - \dfrac{0.1(1 - (0.1)^n)}{1 - 0.1}\right] = \dfrac{7}{9}\left[n - \dfrac{1}{9}(1 - (0.1)^n)\right]$

$= \dfrac{7n}{9} - \dfrac{7}{81}\left(1 - \left(\dfrac{1}{10}\right)^n\right)$

(b) $6 + 66 + 666 = 6[1 + 11 + 111 \ldots]$

$= \dfrac{6}{9}[9 + 99 + 999 + \ldots..] = \dfrac{2}{3}[10 - 1 + 100 - 1 + 1000 - 1 \ldots]$

$= \dfrac{2}{3}\left[\dfrac{10(10^n - 1)}{10 - 1} - n\right] = \dfrac{2}{3} \times \dfrac{10}{9}(10^n - 1) - \dfrac{2n}{3}$

$= \dfrac{2}{27}\left(10^{n+1} - 10\right) - \dfrac{2n \times 9}{3 \times 9} = \dfrac{2}{27}\left[10^{n+1} - 9n - 10\right]$

Sol 4: a, b, c are in AP

(i) $b + c, c + a, a + b$ are also in AP

a, b, c are in AP $\Rightarrow 2b = a + c$

$\Rightarrow b - a = c - b$

$\Rightarrow \boxed{a - b = b - c}$

Difference between term of given AP $= a - b, b - c$ which are equal by equation (i)

Hence $b + c, c + a, a + b$ is an AP

(ii) $\dfrac{1}{bc}, \dfrac{1}{ac}, \dfrac{1}{ab}$ are also in AP

Common difference $= \dfrac{(b - a)}{cab}, \dfrac{(c - b)}{abc}$

By equation (i) $b - a = c - b$ ie difference between terms is same

Hence the given series is in AP

(iii) $a^2(b + c), b^2(c + a), c^2(a + b)$

Difference $= \underbrace{b^2c + b^2a - a^2b - a^2c}_{d_1}, \underbrace{c^2a + c^2b - b^2c - b^2a}_{d_2}$

$d_1 = c(b^2 - a^2) + ab(b - a) = (ca + ab + cb)(b - a)$

$= (ca + ab + bc)(c - b)$ [from eq.(i)

$d_2 = a(c^2 - b^2) + bc(c - b) = (ac + ab + bc)(c - b)$

$d_1 = d_2$

Hence given series is an AP

(iv) $a\left(\dfrac{1}{b} + \dfrac{1}{c}\right),\ b\left(\dfrac{1}{c} + \dfrac{1}{a}\right),\ c\left(\dfrac{1}{a} + \dfrac{1}{b}\right)$

$\Rightarrow d_1 = \dfrac{1}{c}(b - a) + \dfrac{b}{a} - \dfrac{a}{b} \Rightarrow d_2 = \dfrac{1}{a}(c - b) + \dfrac{c}{b} - \dfrac{b}{c}$

$d_1 = \dfrac{b - a}{c} + \dfrac{(b - a)(b + a)}{ab}$

$d_2 = \dfrac{c - b}{a} + \dfrac{(c - b)(c + b)}{bc} = (b - a)\left[\dfrac{1}{c} + \dfrac{1}{a} + \dfrac{1}{b}\right]$

$= (b - a)\left[\dfrac{1}{c} + \dfrac{1}{a} + \dfrac{1}{b}\right] = (c - b)\left[\dfrac{1}{c} + \dfrac{1}{a} + \dfrac{1}{b}\right]$

From eq^n (i) $\Rightarrow d_1 = d_2$

Hence given series is also an AP

Sol 5: Sum of first 3 numbers in AP is 12

Let $a - r,\ a,\ a + r$ be the first 3 numbers

$3a = 12 \Rightarrow a = 4$

$\Rightarrow (a - r)^3 + a^3 + (a + r)^3 = 288$

$\Rightarrow 2a^3 + 6ar^2 + a^3 = 288 \Rightarrow 3a^3 + 6ar^2 = 288$

$\Rightarrow 6 \times 4 \times r^2 = 288 - 3(4^3) \Rightarrow 24r^2 = 96$

$\Rightarrow r = \pm 2$

So numbers are, $4 - 2, 4, 4 + 2 = (2, 4, 6)$ (for $r = 2$)

$(4 + 2, 4, 4 - 2)$ for $(r = -2)$

$\Rightarrow (6, 4, 2)$

Sol 6: (i) Sum of integers between 1 & 200 which are multiple of 3

$3, 6, 9, \dots 198 \Rightarrow n = 66$

Hence sum $= \dfrac{n}{2}[a + l]$

$= \dfrac{(66)}{2}[3 + 198] = 33\,[201] = 6633$

(ii) Multiple of 7

$7, 14, 21 \dots 196 \Rightarrow n = 28$

Now sum $= \dfrac{n}{2}[a + l] = \dfrac{28}{2}[7 + 196] = 14[203] = 2842$

(iii) Multiple of 3 and 7

$21, 42, 63 \dots 189 \Rightarrow n = 9$

Sum $= \dfrac{9}{2}[21 + 189] = \dfrac{9}{2} \cdot 210 = 945$

Sol 7: Sum of first n terms of 2 AP's are in ratio

$= \dfrac{3n - 3}{5n + 21}$

$\Rightarrow$ Let the AP be $a_1,\ a_1 + d_1 \dots$

2nd AP be $a_2,\ a_2 + d_2 \dots$

$\dfrac{\dfrac{n}{2}[2a_1 + (n - 1)d_1]}{\dfrac{n}{2}[2a_2 + (n - 1)d_1]} = \dfrac{3n - 3}{5n + 21}$

$\Rightarrow \dfrac{a_1 + \dfrac{(n - 1)}{2}d_1}{a_2 + \dfrac{(n - 1)d_2}{2}} = \dfrac{3n - 3}{5n + 21}$ .....(i)

Ratio of 24th term well be $\dfrac{a_1 + 23d_1}{a_2 + 23d_2}$

Putting $n = 47$ in equation (i), we well get desired ratio

$\dfrac{a_1 + 23d_1}{a_2 + 23d_2} = \dfrac{3(47) - 3}{5(47) + 21} = \dfrac{138}{256} = \dfrac{69}{128}$

Sol 8: given

$a + (p - 1)d = x$ ...(i)

$a + (q - 1)d = y$ ...(ii)

sum of first $(p + q)$ terms

$= \dfrac{p + q}{2}[2a + (p + q - 1)d]$ ...(iii)

Subtracting (ii) from (i)

$(p - q)d = x - y$

$\Rightarrow d = \dfrac{x - y}{p - q}$ and putting this value in equation (i)

$a + \dfrac{(p - 1)(x - y)}{p - q} = x$

$\Rightarrow a = x - \dfrac{(px - py - x + y)}{p - q} = \dfrac{-qx + py + x - y}{p - q}$

Putting values of a and d in equation (iii)

$S_{p+q} = \dfrac{p + q}{2}\left[\dfrac{2x - 2y - 2qx + 2py + (p + q - 1)(x - y)}{p - q}\right]$

$$= \frac{p+q}{2}\left[\frac{x-y-qx+py+px-qy}{p-q}\right]$$

$$= \frac{p+q}{2}\left[(x+y)\frac{(p-q)}{(p-q)}+\frac{(x-y)}{p-q}\right] = \frac{p+q}{2}\left[x+y+\frac{x-y}{p-q}\right]$$

Sol 9: a, b, c are in HP

i.e $\frac{2}{b} = \frac{1}{a} + \frac{1}{c} \Rightarrow b = \frac{2ac}{a+c}$ ...(i)

$\frac{2a_1 c_1}{a_1 + c_1}$ [where a_1 & c_1 are 1st & 3rd terms of given series]

$$= \frac{\dfrac{2ac}{(a+b)(b+c)}}{\dfrac{a}{b+c}+\dfrac{c}{a+b}} = \frac{2ac}{a^2+ab+c^2+bc} = \frac{2ac}{a^2+c^2+b(a+c)}$$

$$= \frac{2ac}{a^2+c^2+(2ac)} \quad \text{(from equation (i))}$$

$$= \frac{2ac}{(a+c)(a+c)} = \frac{b}{a+c} = b_1$$

Middle term of given series, hence $\frac{2a_1 c_1}{a_1 + c_1} = b_1$ ie given Series is H.P

Sol 10: $1 + \frac{4}{5} + \frac{7}{5^2} + \frac{10}{5^3}$

Sum of first n terms

$$S_n = \frac{1}{5^0} + \frac{4}{5} + \frac{7}{5^2} + \frac{10}{5^3} \dots \frac{1+(n-1)3}{5^{n-1}} \quad \text{...(i)}$$

$$\frac{S_n}{5} = \frac{1}{5} + \frac{4}{5.5} + \frac{7}{5.5^2} + \frac{10}{5.5^3} \dots \frac{1+(n-1)3}{5^n} \quad \text{...(ii)}$$

Subtracting (ii) from (i)

$$= 1 + \frac{3}{5} + \frac{3}{5^2} + \frac{3}{5^3} \dots - \left(\frac{3n-2}{5^n}\right)$$

$$= 1 - \frac{(3n-2)}{5^n} + 3\left[\frac{1}{5} + \frac{1}{5^2} \dots \frac{1}{5^{n-2}}\right]$$

$$= 1 - \frac{(3n-2)}{5^n} + 3.\frac{1}{5}\frac{\left(1-\dfrac{1}{5^{n-1}}\right)}{\left(1-\dfrac{1}{5}\right)}$$

$$= 1 - \left(\frac{3n-2}{5^n}\right) + \frac{3}{5}\frac{(5^{n-1}-1)\times 5}{5^{n-1}\times 4}$$

$$\frac{4s_n}{5} = 1 - \left(\frac{3n-2}{5^n}\right) + \frac{15}{4}\frac{(5^{n-1}-1)}{5^n}$$

$$S_n = \frac{5}{4} - \frac{(3n-2)}{4.5^{n-1}} + \frac{75}{16}\left(\frac{1}{5}-\frac{1}{5^n}\right)$$

$$= \frac{5\times 4}{4\times 4} + \frac{15}{16} - \frac{3}{16\times 5^{n-2}}\frac{-(3n-2)}{4\times 5^{n-1}}$$

$$= \frac{35}{16} - \frac{3}{16(5^{n-2})}\frac{-(3n-2)}{4(5^{n-1})}$$

Sol 11: a, b, c are in AP

$\Rightarrow 2b = a + c$ ...(i)

b, c, d in GP

$c^2 = bd$ ...(ii)

c, d, e are in HP

$d = \frac{2ce}{c+e}$ ...(ii)

Given that a = 2, e = 18

We have $(2b - 2) = c$ from (i) and $\frac{(2b-2)^2}{b} = d$ from (ii)

and also $\frac{(2b-2)^2}{b} = \frac{2\times(2b-2)18}{(2b-2)+18}$ from (iii)

$\Rightarrow (2b-2) = \frac{36b}{2b+16} \Rightarrow (b-1)(b+8) = 9b$

$\Rightarrow b^2 + 7b - 8 = 9b \Rightarrow b^2 - 2b - 8 = 0$

$\Rightarrow b = 4, -2$

$\Rightarrow c = 6, -6$

$\Rightarrow d = 9, -18$

b, c, d = [4, 6, 9] and [-2, -6, -18]

Sol 12: $S_n = 1.2.3 + 2.3.4 + 3.4.5 \dots$

$T_n = n(n+1)(n+2) = n(n^2+3n+2)$

$= n^3 + 3n^2 + 2n = n^3 + 3n^2 + 2n$

$\Sigma_n = \Sigma T_n = \Sigma n^3 + 3\Sigma n^2 + 2\Sigma n$

$$= \left[\frac{n(n+1)}{2}\right]^2 + 3\frac{(n)(n+1)(2n+1)}{6} + 2\frac{n(n+1)}{2}$$

$$= \frac{n(n+1)}{2}\left[\frac{n(n+1)}{2} + \frac{3(2n+1)}{3} + 2\right]$$

$$= \frac{n(n+1)}{2}\left[\frac{n^2+n+4n+2+4}{2}\right]$$

$$= \frac{n(n+1)}{4}(n^2+5n+6) = \frac{n(n+1)(n+2)(n+3)}{4}$$

Sol 13: $S_n = 1^2 + 2 + 3^2 + 4 + 5^2 + 6$

(first 2n numbers)

$1^2 + 3^2 + 5^2 \dots$ n terms $+ 2 + 4 + 6 \dots$ n terms

$= 2(1 + 2 + 3 \dots + n) + 1^2 + 3^2 + 5^2 \dots$

$= 2n\dfrac{(n+1)}{2} + 1^2 + 2^2 + 3^2 \dots (2n-1)^2 - 2^2$

$\quad - 4^2 - 6^2 \dots (2n-2)^2$

$= n(n+1) + \dfrac{(2n-1)(2n-1+1)(4n-2+1)}{6}$

$\quad - 2^2\{1^2 + 2^2 + \dots (n-1)^2\}$

$= n(n+1) + \dfrac{(2n-1)n(4n-1)}{3}$

$\quad - 2^2\dfrac{(n-1)n(2n-1)}{6}$

$= n(n+1) + \dfrac{(2n-1)(n)(4n-1)}{3}$

$\quad - \dfrac{2^2(n-1)n(2n-1)}{6}$

$= n(n+1) + \dfrac{n}{3}\left[(2n-1)(4n-1) - 2(n-1)(2n-1)\right]$

$= n(n+1) + \dfrac{n}{3}(2n-1)\,[2n+1]$

$= n(n+1) + \dfrac{\left(4n^2-1\right)n}{3} = \dfrac{n}{3}[4n^2+3n+2]$

Sol 14: HM of 2 numbers is 4 ie $\dfrac{2ab}{a+b} = 4$

$$\boxed{ab = 2a + 2b} \qquad \dots(i)$$

$AM = \dfrac{a+b}{2}$ and $G.M. = \sqrt{ab}$

We have $2A + G^2 = 27 \Rightarrow a + b + ab = 27$

$a + b + 2a + 2b = 27$ [from (i)]

$\Rightarrow a + b = 9$ and $ab = 18$

$[a = 6, b = 3]$; $[a = 3, b = 6]$

Sol 15: $(3^3 - 2^3) + (5^3 - 4^3) + (7^3 - 6^3) + \dots$

$= 3^3 + 5^3 + 7^3 + \dots + 21^3$

$\quad - 2^3(1 + 2^3 + 3^3 \dots 10^3) + 1 - 1$

$= 1^3 + 2^3 + 3^3 + 4^3 \dots 21^3 - 2^4(1 + 2^3 \dots 10^3) - 1$

$$= \left[\frac{21\times(21+1)}{2}\right]^2 - 2^4\left[\frac{(10)(10+1)}{2}\right]^2 - 1$$

$= (21 . 11)^2 - 16(5 . 11)^2 - 1$

$= 11^2(21 . 21 - 16 . 25) - 1$

$= 121 \times 41 - 1 = 4961 - 1 = 4960$

Sol 16: $S_n = 1.3^2 + 2.5^2 + 3.7^2 + \dots$

$T_n = n(2n+1)^2 = 4n^3 + n + 4n^2$

$\Sigma_n = \Sigma T_n = 4\Sigma n^3 + \Sigma n + 4\Sigma n^2$

$$= 4\left[\frac{n(n+1)}{2}\right]^2 + \frac{n(n+1)}{2} + 4\frac{n(n+1)(2n+1)}{6}$$

$= n^2(n+1)^2 + \dfrac{n(n+1)}{2} + \dfrac{2}{3}n(n+1)(2n+1)$

$= n(n+1)\left[n^2 + n + \dfrac{1}{2} + \dfrac{4n}{3} + \dfrac{2}{3}\right]$

$= n(n+1)\left[n^2 + \dfrac{7n}{3} + \dfrac{7}{6}\right]$

$S_{20} = 20 \times 21\left[20^2 + \dfrac{7}{3}\times 20 + \dfrac{7}{6}\right]$

$= \dfrac{420}{6}\left[2400 + 280 + 7\right] = 70(2687) = 188090$

Sol 17: $a, b, c \in (2, 18)$

$a + b + c = 25 \qquad \dots(i)$

$2a = 2 + b \qquad \dots(ii)$

$c^2 = 18 b \qquad \dots(iii)$

$b = 2a - 2$

$c = 25 - a - 2a + 2 = 27 - 3a$

$\Rightarrow (27 - 3a)^2 = 18(2a - 2)$

$\Rightarrow (9 - a)^2 = 4(a - 1) \Rightarrow a^2 + 81 - 18a = 4a - 4$

$\Rightarrow a^2 - 22a + 85 = 0$

$a = 17, 5$

$b = 32, 8$

$c = 24, 12$

Numbers are (5, 8, 12)

Sol 18: $a > 0$, $b > 0$, $c > 0$

To prove $(a + b + c)\left(\dfrac{1}{a} + \dfrac{1}{b} + \dfrac{1}{c}\right) \geq 9$

We know that A.M. $\geq$ G.M. $\geq$ HM

Therefore, A.M. $\geq$ HM

For 3 numbers a, b, c

$AM = \dfrac{a+b+c}{3}$, $HM = \dfrac{3}{\dfrac{1}{a} + \dfrac{1}{b} + \dfrac{1}{c}}$

$\Rightarrow \dfrac{a+b+c}{3} \geq \dfrac{3}{\dfrac{1}{a} + \dfrac{1}{b} + \dfrac{1}{c}}$

$\Rightarrow (a + b + c)\left(\dfrac{1}{a} + \dfrac{1}{b} + \dfrac{1}{c}\right) \geq 9$

Sol 19: Let the 2 number be a, b

a, A_1, A_2, b are in AP, (a, G_1, G_2, b) are in GP,

(a, H_1, H_2, b) are in HP

$A_1 + A_2 = a + b$ ...(i)

$G_1 G_2 = ab$ ...(ii)

By properties of respective series

$\dfrac{H_1 H_2}{H_1 + H_2} = \dfrac{ab}{a+b}$

$\dfrac{H_1 H_2}{H_1 + H_2} = \dfrac{G_1 G_2}{A_1 + A_2}$

$\dfrac{G_1 G_2}{H_1 H_2} = \dfrac{A_1 + A_2}{H_1 + H_2}$ Hence proved

Sol 20: $(x - 1)(x - 2)(x - 3) \ldots (x - 100)$

Coefficient of $x^{99} = \dfrac{-b}{a}$

We can see that $a = 1$

and $b = 1 + 2 + 3 \ldots 100 = \dfrac{100 \times 101}{2} = 5050$

$\therefore$ Coefficient of $x^{99} = -5050$

Coefficient of x^{98}

$= 1 \times 2 + 1 \times 3 + \ldots + 1 \times 100 + 2 \times 3 + 2 \times 4 + \ldots + 99 \times 100$

$= \dfrac{1}{2}\left\{(1 + 2 + \ldots + 100)^2 - \left(1^2 + 2^2 + 100^2\right)\right\}$

$= \dfrac{1}{2}\left\{(5050)^2 - 338350\right\}$

Sol 21: Given that, for a polygon of "n" sides, we have

$\alpha = 120°$; $d = 5$

Sum of interior angle

$(n - 2)\,180° = a + (n - 1)d$

$\qquad = \dfrac{n}{2}\left[2(120) + (n - 1)5\right] = \dfrac{n}{2}\left[5n + 235\right]$

$5n^2 + 235n = 360n - 720$

$\Rightarrow 5n^2 - 125n + 720 = 0$

$\Rightarrow n^2 - 25n + 144 = 0 \Rightarrow n = 16, 9$

If $n = 16$, then interior angle will be greater than $180°$.

Hence the answer is 9.

Sol 22: Let the number be a, b, c

$b^2 = ac$, $a + c = 2b + 1$, $b + a = \dfrac{2}{3}(b + c)$

$\Rightarrow a = 2b + 1 - c$

$\Rightarrow 3b + 1 - c = \dfrac{2}{3}(b + c)$

$\Rightarrow 2b + 2c = 9b + 3 - 3c$

$\Rightarrow 7b = 5c - 3$

$c = \dfrac{7b + 3}{5} \Rightarrow a = 2b + 1 - \dfrac{(7b + 3)}{5}$

$a = \dfrac{3b + 2}{5} \Rightarrow b^2 = \dfrac{(3b + 2)(7b + 3)}{25}$

$\Rightarrow 25b^2 = 21b^2 + 23b + 6 \Rightarrow 4b^2 - 23b - 6 = 0$

$\Rightarrow 4b^2 - 24b + b - 6 = 0$

$\Rightarrow b = 6, c = 9$

$\Rightarrow a = 4$ Number is 469

Sol 23:

$S_n = 10 + 10 + 15 + 15 + 20 + 20$

$\ldots 85 + 85 + \ldots + 125 + 125 + 130$

$= 2[10 + 15 + 20 \ldots 130] - 130$

$= 10[2 + 3 + 4 \ldots 26] - 130$

$= 10\left[\dfrac{26 \times 27}{2} - 1\right] - 130 = 10 \times 350 - 130$

$= 3500 - 130 = 3370$ m

Sol 24: Number of elements in nth group = n

First number in the group will be $\dfrac{n(n-1)+2}{2}$

$S_n = \dfrac{n}{2} \left[n(n-1) + 2 + (n-1)1 \right]$

$= \dfrac{n}{2} \left[n^2 - n + 2 + n - 1 \right] = \dfrac{n}{2} \left[n^2 + 1 \right]$

Sol 25: Let the 3 number in G.P. be a, ar, ar^2 & other 3 numbers be $a_1, a_1 r_1, a_1 r_1^2$

$(a_1 r_1 - ar)^2 = (a_1 r_1^2 - ar^2)(a_1 - a)$

$a_1^2 r_1^2 + a^2 r^2 - 2aa_1 r r_1 = a_1^2 r_1^2 - aa_1 r^2 - aa_1 r^2 + a^2 r^2$

$2rr_1 = r^2 + r_1^2$

$\Rightarrow (r_1 - r)^2 = 0$

$\Rightarrow r = r_1$

Ratio for third G.P. $= \dfrac{a_1 r_1 - ar}{a_1 - a} = \dfrac{(a_1 - a)r}{(a_1 - a)} = r$

Hence ratio of the three G.P. is same

Sol 26: $S = \dfrac{1}{a_1 a_2} + \dfrac{1}{a_2 a_3} \cdots \dfrac{1}{a_n a_{n+1}}$

$= \dfrac{1}{a_2 - a_1} \left[\dfrac{a_2 - a_1}{a_1 a_2} + \dfrac{a_3 - a_2}{a_2 a_3} \cdots \right]$

As $a_2 - a_1 = a_3 - a_2 = \dots a_{n+1} - a_n$

$= \dfrac{1}{a_2 - a_1} \left[\dfrac{1}{a_1} - \dfrac{1}{a_2} + \dfrac{1}{a_2} - \dfrac{1}{a_3} \cdots + \dfrac{1}{a_n} - \dfrac{1}{a_{n+1}} \right]$

$= \dfrac{1}{a_2 - a_1} \left[\dfrac{1}{a_1} - \dfrac{1}{a_{n+1}} \right] = \dfrac{a_{n+1} - a_1}{(a_2 - a_1)a_1 a_{n+1}}$

$= \dfrac{nd}{d(a_1 a_{n+1})} = \dfrac{n}{a_1 a_{n+1}}$

$S = \dfrac{n}{a_1 a_{n+1}} = \dfrac{n}{a_1(a_1 + nd)}$

$\lim_{n \to \infty} S = \lim_{n \to \infty} \dfrac{n}{a_1(a_1 + nd)} = \lim_{n \to \infty} \dfrac{1}{a_1 \left(d + \dfrac{a_1}{n} \right)} = \dfrac{1}{a_1 d}$

Sol 27: 2, AM_1, AM_2 ... AM_9, 3

2, HM_1, HM_2 ... HM_9, 3

Suppose we take AM_n and HM_n

$\Rightarrow 2 + 10d = 3 \Rightarrow d = \dfrac{1}{10}$

$AM_n = 2 + nd = 2 + \dfrac{n}{10}$

$\dfrac{1}{2}, \dfrac{1}{HM_1} \cdots \dfrac{1}{HM_9}, \dfrac{1}{3}$ in AP

$\Rightarrow \dfrac{1}{2} + 10d = \dfrac{1}{3} \Rightarrow d = \dfrac{-1}{60}$

$HM_n = \dfrac{60}{30 - n}$

$A + \dfrac{6}{H} = 2 + \dfrac{n}{10} + \dfrac{6}{60}(30 - n)$

$= 2 + \dfrac{n}{10} + 3 - \dfrac{n}{10} = 5$

Hence proved

Sol 28: $x + y + z = 1$

For x, y, z. A.M. $\geq$ HM

$\Rightarrow \dfrac{x + y + z}{3} \geq \dfrac{3}{\dfrac{1}{x} + \dfrac{1}{y} + \dfrac{1}{z}}$

$\Rightarrow \dfrac{1}{x} + \dfrac{1}{y} + \dfrac{1}{z} \geq 9$

Therefore, $xy + yz + zx - 9xyz \geq 0$

Hence proved.

Sol 29: We must prove that for some m and p;

$M^p = \dfrac{m}{2} \left[2a + (m-1)2 \right]$, for some odd a

$= m[a + m - 1]$

Let us prove this by induction

Taking, $P = 2$

$m^2 = m[a + m - 1] \Rightarrow a = 1$ is the required a.

$m^{p+1} = m^p . m$

$= m[a + m - 1]\, m = m[ma + m^2 - m]$

$= m[ma + m^2 - 2m + 1 + m - 1]$

$= m\left[ma + (m - 1)^2 + m - 1 \right]$

$= m\left[A + m - 1 \right]$

We must prove that A is odd.

A is odd

For even m, ma is even and $(m - 1)^2$ is odd $\Rightarrow$ A is odd
for odd m, ma is odd and $(m - 1)^2$ is even $\Rightarrow$ A is odd

$\therefore$ By induction hypothesis,

$M^p = \dfrac{m}{2} \left[2a + (m-1)2 \right]$, with odd a.

Hence proved

Exercise 2

Single Correct Choice Type

Sol 1: (B) $b = \dfrac{2ac}{a+c}$ [a, b, c are in HP] ...(i)

$$= \frac{b+a}{b-a} + \frac{b+c}{b-c} = \frac{\frac{2ac}{a+c}+a}{\frac{2ac}{a+c}-c} + \frac{\frac{2ac}{a+c}+c}{\frac{2ac}{a+c}-c}$$

$$= \frac{3ac+a^2}{ac-a^2} + \frac{3ac+c^2}{ac-c^2} = \frac{3c+a}{c-a} + \frac{(3a+c)}{a-c}$$

$$= \frac{3c+a}{c-a} - \frac{(3a+c)}{c-a} = \frac{2c-2a}{c-a} = 2$$

Sol 2: (A) Given summation is, $\dfrac{1}{1} + \dfrac{1}{1+2} + \dfrac{1}{1+2+3}$

$$\therefore T_n = \frac{2}{n(n+1)} \Rightarrow T_n = 2\left[\frac{1}{n} - \frac{1}{n+1}\right]$$

$$S_n = S T_n = 2\left[\frac{1}{1} - \frac{1}{2} + \frac{1}{2} - \frac{1}{3} + \frac{1}{3} ... \frac{1}{\infty}\right] = 2$$

Sol 3: (C)

·   ·   ·   ·   ·   ·   ·

←10m→   ←middle→   n stones

     n stones

$\Rightarrow 2[20 + 40 + 60 ... n \text{ terms}] = 4800$

$\Rightarrow 120 = 1 + 2 ... n \text{ terms}$

$\Rightarrow 120 = \dfrac{n(n+1)}{2} \Rightarrow n^2 + n = 240$

$\Rightarrow n = 15$

Total no of stone = 2n+1 = 31 [C]

Sol 4: (D) $S = 1^2 + 3^2 + 5^2 ... 99^2$

$S_1 = 2^2 + 4^2 + 6^2 ... (100)^2$

$S_1 - S = 2^2 - 1^2 + 4^2 - 3^2 + 6^2 - 5^2 ...$

$= 2+1+4+3+6+5+ ... +100+99$

$= \dfrac{100}{2}[1+100] = 50 [101] = 5050$

Sol 5: (C) $a, a + d, ... a + (n-1)d, a + nd, ... a + (2n-1)d$

$S_1 = \dfrac{n}{2}[2a + (n-1)d]$

$S_2 = \dfrac{n}{2}[2a + 2nd + (n-1)d]$

$\dfrac{s_1}{s_2} = \dfrac{2a+(n-1)d}{2a+2nd+(n-1)d}$

$$\Rightarrow \frac{2+(n-1)\frac{d}{a}}{2+(3n-1)\frac{d}{a}} = s$$

$\Rightarrow 2 + (n-1)\dfrac{d}{a} = 2s + (3n-1)\dfrac{d}{a}s$

$\Rightarrow 2s - 2 = \dfrac{d}{a}[n - 3ns + s - 1]$

$(2s - 2)\dfrac{a}{d} = n(1 - 3s) + s - 1$

This is independent of n ie coefficient of n will be zero

$\Rightarrow s = \dfrac{1}{3}; \Rightarrow \dfrac{a}{d} = \dfrac{1}{2}$

Sol 6: (D) $a = 5^{1+x} + 5^{1-x} + 25^x + 25^{-x}$

$\Rightarrow a = 5.5^x + \dfrac{5}{5^x} + (5^x)^2 + (5^{-x})^2$

Let $5^x = t \Rightarrow a = 5t + \dfrac{5}{t} + t^2 + \dfrac{1}{t^2}$

$\dfrac{5t + \frac{5}{t}}{2} \geq 5; \dfrac{t^2 + \frac{1}{t^2}}{2} \geq 1$ hence $a \geq 10 + 2$

$\therefore \quad a \geq 12$

Sol 7: (A) $S_{11} = S_{19}$

$\dfrac{11}{2}[2a + 10d] = \dfrac{19}{2}[2a + 18d]$

$16a = -232 d \Rightarrow \dfrac{a}{d} = \dfrac{-29}{2}$

$S_{30} = \dfrac{30}{2}[2a + 29d] = 30\left[a + \dfrac{29d}{2}\right] = 0$

Sol 8: (C) $S_2 = 2t_2$

$S_1 + d_s = 2(t_1 + d_t)$

$d_s = t_1 + 2d_t$

$\Rightarrow \dfrac{10}{2}[2s_1 + 9d_s] = \dfrac{15}{2}[2t_1 + 14d_t]$

$\Rightarrow 18d_s = 2t_1 + 42d_t \Rightarrow 18d_s = 2d_s - 4d_t + 42d_t$

$\Rightarrow 16d_s = 38d_t$

$\dfrac{d_s}{d_t} = \dfrac{19}{8}$

Sol 9: (A) The given expression is equal to

$$\frac{1}{d}\left(\frac{1}{a_1} - \frac{1}{a_2} + \frac{1}{a_2} - \frac{1}{a_3} + ...\right) = \frac{1}{a_1 d}$$

Sol 10: (A) $a + ar + ar^2 = 21$ ...(i)

$a^2 + a^2 r^2 + a^2 r^4 = 189$ ...(ii)

Squaring equation (i) & then dividing by (ii)

$$\Rightarrow \frac{a^2(1+r+r^2)^2}{a^2(1+r^2+r^4)} = \frac{441}{189}$$

$$\Rightarrow \frac{(1+r+r^2)(1+r+r^2)}{(1+r+r^2)(1-r+r^2)} = \frac{441}{189}$$

$$\Rightarrow 2r^2 - 5r + 2 = 0$$

$$\Rightarrow r = 2, \frac{1}{2} \Rightarrow a = 3, 12$$

GP is 3, 6, 12 ... $S_n = 3\dfrac{(2^n - 1)}{2-1} = 3(2^n - 1)$ Hence, (A) is the correct choice

Sol 11: (C) $\displaystyle\sum_{n=1}^{\infty} \frac{n}{n^4 + 4} = \frac{1}{n\left(n^2 + \dfrac{4}{n^2} + 4 - 4\right)}$

$$= \sum \frac{1}{n\left[\left(n+\dfrac{2}{n}\right)^2 - 4\right]} = \sum \frac{1}{n\left[n+\dfrac{2}{n}-2\right]\left[n+\dfrac{2}{n}+2\right]}$$

$$= \sum \frac{n}{(n^2 + 2 - 2n)(n^2 + 2n + 2)}$$

$$= \frac{1}{4}\sum\left[\frac{1}{n^2 - 2n + 2} - \frac{1}{n^2 + 2n + 2}\right]$$

$$= \frac{1}{4}\sum_{n=1}^{\infty}\left[\frac{1}{(n-1)^2 + 1} - \frac{1}{(n+1)^2 + 1}\right]$$

$$= \frac{1}{4}\left(1 + \frac{1}{2} + \sum_{n=1}^{\infty}\left[\frac{1}{(n+1)^2 + 1} - \frac{1}{(n+1)^2 + 1}\right]\right) = \frac{3}{8}$$

Sol 12: (D) $\ln a^2 + (\ln a^2)^2 + (\ln a^2)^3 .....$

$= 3\{\ln a + (\ln a)^2 + (\ln a)^3 + ... \}$

$$\Rightarrow \frac{2\ln a}{1 - 2\ln a} = \frac{3\ln a}{1 - \ln a}$$

$\Rightarrow 2 - 2 \ln a = 3 - 6 \ln a$

$\Rightarrow 1 = 4 \ln a$

$\Rightarrow a = e^{1/4}$

Sol 1: (B) Here, $(a^2 + b^2 + c^2) p^2 - 2(ab + bc + cd)p + (b^2 + c^2 + d^2) \le 0$

$\Rightarrow (a^2 p^2 - 2abp + b^2) + (b^2 p^2 - 2bcp + c^2) + (c^2 p^2 - 2cdp + d^2) \le 0$

$\Rightarrow (ap - b)^2 + (bp - c)^2 + (cp - d)^2 \le 0$

(Since, sum of squares is never less than zero)

$\Rightarrow$ Each of the squares is zero

$\therefore (ap - b)^2 = (bp - c)^2 = (cp - d)^2 = 0$

$$\Rightarrow p = \frac{b}{a} = \frac{c}{b} = \frac{d}{c}$$

$\therefore$ a, b, c are in G.P.

Sol 2: (C) Sum of the n terms of the series $\dfrac{1}{2} + \dfrac{3}{4} + \dfrac{7}{8} + \dfrac{15}{16} + ...$ upto n terms can be written as

$$\left(1 - \frac{1}{2}\right) + \left(1 - \frac{1}{4}\right) + \left(1 - \frac{1}{8}\right) + \left(1 - \frac{1}{16}\right) .... \text{ upto n terms}$$

$$= n - \left(\frac{1}{2} + \frac{1}{4} + \frac{1}{8} + ... + n\,\text{terms}\right)$$

$$= n - \frac{\dfrac{1}{2}\left(1 - \dfrac{1}{2^n}\right)}{1 - \dfrac{1}{2}} = n + 2^{-n} - 1$$

Sol 3: (B) Let the common ratio of the G.P. be r. Then,

$Y = xr$ and $z = xr^2$

$\Rightarrow \ln y = \ln x + \ln r$ and $\ln z = \ln x + 2 \ln r$

Let $A = 1 + \ln x$, $D = \ln r$

Then, $\dfrac{1}{1 + \ln x} = \dfrac{1}{A}$,

$$\frac{1}{1 + \ln y} = \frac{1}{1 + \ln x + \ln r} = \frac{1}{A + D}$$

and $\dfrac{1}{1 + \ln z} = \dfrac{1}{1 + \ln x + 2\ln r} = \dfrac{1}{A + 2D}$

Therefore, $\dfrac{1}{1 + \ln x}, \dfrac{1}{1 + \ln y}, \dfrac{1}{1 + \ln z}$ are in H.P.

Sol 4: (A) Since A.M. $\ge$ G.M., then

$$\frac{(a+b) + (c+d)}{2} \ge \sqrt{(a+b)(c+d)}$$

$\Rightarrow M \le 1$

Also, $(a + b) + (c + d) > 0 (\because a, b, c, d > 0)$

$\therefore 0 < M \leq 1$

Sol 5: (D) Since a, b, c, d are in A.P.

$\Rightarrow \dfrac{a}{abcd}, \dfrac{b}{abcd}, \dfrac{c}{abcd}, \dfrac{d}{abcd}$ are in AP

$\Rightarrow \dfrac{1}{bcd}, \dfrac{1}{cda}, \dfrac{1}{abd}, \dfrac{1}{abc}$ are in A.P.

$\Rightarrow$ bcd, cda, abd, abc are in HP.

$\Rightarrow$ abc, abd, cda, bcd are in HP.

Sol 6: (D) Since a, b, c are in AP.

Let $a = A - D, b = A, c = A + D$

Given, $a + b + c = \dfrac{3}{2}$

$\Rightarrow (A - D) + A + (A + D) = \dfrac{3}{2} \Rightarrow 3A = \dfrac{3}{2} \Rightarrow A = \dfrac{1}{2}$

$\therefore$ The numbers are $\dfrac{1}{2} - D, \dfrac{1}{2}, \dfrac{1}{2} + D$

Also, $\left(\dfrac{1}{2} - D\right)^2, \dfrac{1}{4}, \left(\dfrac{1}{2} + D\right)^2$ are in GP.

$\Rightarrow \left(\dfrac{1}{4}\right)^2 = \left(\dfrac{1}{2} - D\right)^2 \left(\dfrac{1}{2} + D\right)^2 \Rightarrow \dfrac{1}{16} = \left(\dfrac{1}{4} - D^2\right)^2$

$\Rightarrow \dfrac{1}{4} - D^2 = \pm \dfrac{1}{4} \Rightarrow D^2 = \dfrac{1}{2} \Rightarrow D = \pm \dfrac{1}{\sqrt{2}}$

$\Rightarrow a = \dfrac{1}{2} + \dfrac{1}{\sqrt{2}}$ or $\dfrac{1}{2} - \dfrac{1}{\sqrt{2}}$

So, out of the given values, $a = \dfrac{1}{2} - \dfrac{1}{\sqrt{2}}$ is the right choice

Sol 7: (C) We know that, the sum of infinite term of G.P. is

$S_\infty = \begin{cases} \dfrac{a}{1-r}, & |r| < 1 \\ \infty, & |r| \leq 1 \end{cases}$

$\therefore S_\infty = \dfrac{x}{1-r} = 5$ (thus $|r| < 1$)

or $1 - r = \dfrac{x}{5} \Rightarrow r = \dfrac{5-x}{5}$ exists only when $|r| < 1$

i.e., $-1 < \dfrac{5-x}{5} < 1$

or $-10 < -x < 0 \Rightarrow 0 < x < 10$

Sol 8: (C) Let $S_n = cn^2$

$S_{n-1} = c(n-1)^2 = cn^2 + c - 2cn$

$\therefore T_n = 2cn - c \ (\because T_n = S_n - S_{n-1})$

$T_n^2 = (2cn - c)^2 = 4c^2n^2 + c^2 - 4c^2n$

$\therefore$ Sum $= ST_n^2$

$= \dfrac{4c^2 n(n+1)(2n+1)}{6} + nc^2 - 2c^2 n(n+1)$

$= \dfrac{2c^2 n(n+1)(2n+1) + 3nc^2 - 6c^2 n(n+1)}{3}$

$= \dfrac{nc^2(4n^2 + 6n + 2 + 3 - 6n - 6)}{3} = \dfrac{nc^2(4n^2 - 1)}{3}$

Sol 9: (B) Let $a, ar, ar^2, \ldots$

$a + ar = 12$ ... (i)

$ar^2 + ar^3 = 48$ ... (ii)

Dividing (ii) by (i), we have

$\dfrac{ar^2(1+r)}{a(r+1)} = 4$

$\Rightarrow r^2 = 4$ if $r \neq -1$

$\therefore r = -2$

Also, $a = -12$ (using (i)).

Sol 10: (B) Let $S = 1 + \dfrac{2}{3} + \dfrac{6}{3^2} + \dfrac{10}{3^3} + \dfrac{14}{3^4} + \ldots$ ... (i)

$\dfrac{1}{3}S = \dfrac{1}{3} + \dfrac{2}{3^2} + \dfrac{6}{3^3} + \dfrac{10}{3^4} + \ldots$ ... (ii)

Dividing (i) and (ii)

$S\left(1 - \dfrac{1}{3}\right) = 1 + \dfrac{1}{3} + \dfrac{4}{3^2} + \dfrac{4}{3^3} + \dfrac{4}{3^4} + \ldots$

$\dfrac{2}{3}S = \dfrac{4}{3} + \dfrac{4}{3^2}\left(1 + \dfrac{1}{3} + \dfrac{1}{3^2} + \ldots\right)$

$\Rightarrow \dfrac{2}{3}S = \dfrac{4}{3} + \dfrac{4}{3^2}\left(\dfrac{1}{1 - \dfrac{1}{3}}\right) = \dfrac{4}{3} + \dfrac{4}{3^2}\dfrac{3}{2} = \dfrac{4}{3} + \dfrac{2}{3} = \dfrac{6}{2}$

$\Rightarrow \dfrac{2}{3}S = \dfrac{6}{3} \Rightarrow S = 3$

Sol 11: (B) $M = \dfrac{\ell + n}{2}$

$\ell, G1, G2, G3, n$ are in G.P.

$r = \left(\dfrac{n}{\ell}\right)^{\frac{1}{4}}$

$$G_1 = \ell\left(\frac{n}{\ell}\right)^{\frac{1}{4}} \qquad G_2 = \ell\left(\frac{n}{\ell}\right)^{\frac{1}{2}} \qquad G_3 = \ell\left(\frac{n}{\ell}\right)^{\frac{3}{4}}$$

$$G_1^4 + 2G_2^4 + G_3^4$$

$$= \ell^4 \times \frac{n}{\ell} + 2\ell^4 \times \frac{n^2}{\ell^2} + \ell^4 \times \frac{n^3}{\ell^3}$$

$$= \ell^3 n + 2\ell^2 n^2 + \ell n^3$$

$$= n\ell(\ell^2 + 2n\ell + n^2)$$

$$= n\ell(\ell + n)^2$$

$$= 4m^2 n\ell$$

Sol 12: (B) $T_n = \dfrac{\dfrac{n^2(n+1)^2}{4}}{n^2}$

$$T_n = \frac{1}{4}(n+1)^2$$

$$T_n = \frac{1}{4}[n^2 + 2n + 1]$$

$$S_n = \sum_{n=1}^{n} T_n$$

$$S_n = \frac{1}{4}\left[\frac{n(n+1)(2n+1)}{6} + n(n+1) + n\right]$$

$$n = 9$$

$$S_9 = \frac{1}{4}\left[\frac{9 \times 10 \times 19}{6} + 9 \times 10 + 9\right] = \frac{1}{4}[285 + 90 + 9] = \frac{384}{4} = 96$$

Sol 13: (A) $a + d, a + 4d, a + 8d \to$ G.P

$$\therefore (a + 4d)^2 = a^2 + 9ad + 8d^2$$

$$\Rightarrow 8d^2 = ad \qquad \Rightarrow a = 8d$$

$$\therefore 9d, 12d, 16d \to \text{G.P.}$$

Common ratio $r = \dfrac{12}{9} = \dfrac{4}{3}$

Sol 14: (A)

$$\left(\frac{8}{5}\right)^2 + \left(\frac{12}{5}\right)^2 + \left(\frac{16}{5}\right)^2 + \left(\frac{20}{5}\right)^2 + \left(\frac{24}{5}\right)^2 + \ldots\ldots$$

$$\frac{8^2}{5^2} + \frac{12^2}{5^2} + \frac{16^2}{5^2} + \frac{20^2}{5^2} + \frac{24^2}{5^2} + \ldots\ldots$$

$$T_n = \frac{(4n+4)^2}{5^2}; \ S_n = \frac{1}{5^2}\sum_{n=1}^{10} 16(n+1)^2 = \frac{16}{25}\sum_{n=1}^{10}(n^2 + 2n + 1)$$

$$= \frac{16}{25}\left[\frac{10 \times 11 \times 21}{6} + \frac{2 \times 10 \times 11}{2} + 10\right] = \frac{16}{25} \times 505 = \frac{16}{5}m$$

$$\Rightarrow m = 101$$

Sol 15: (B) $a, ar, ar^2 \to$ G.P.

$a, 2ar, ar^2 \to$ A.P.

$$2 \times 2ar = a + ar^2$$

$$4r = 1 + r^2$$

$$\Rightarrow r^2 - 4r + 1 = 0$$

$$r = \frac{4 \pm \sqrt{16 - 4}}{2} = 2 \pm \sqrt{3}$$

$$\boxed{r = 2 + \sqrt{3}}$$

$r = 2 - \sqrt{3}$ is rejected

$$\therefore (r > 1)$$

G.P. is increasing.

Sol 16: (B) Statement-I has 20 terms whose sum is 8000 and statement-II is true and supporting statement-I.

$\therefore$ k^{th} bracket is $(k-1)^2 + k(k-1) + k^2 = 3k^2 - 3k + 1$.

Sol 17: (D) $100(T_{100}) = 50(T_{50}) \Rightarrow 2[a + 99d] = a + 49d$
$\Rightarrow a + 149d = 0 \Rightarrow T_{150} = 0$

Sol 18: (D) $f(x) = 2x^3 + 3x + k$

$$f'(x) = 6x^2 + 3 > 0 \qquad\qquad \forall x \in R$$

$\Rightarrow f(x)$ is strictly increasing function

$\Rightarrow f(x) = 0$ has only one real root, so two roots are not possible

Sol 19: (A) $x^2 + 2x + 3 = 0 \hfill \ldots \text{(i)}$

$ax^2 + bx + c = 0 \hfill \ldots \text{(ii)}$

Since equation (i) has imaginary roots

So equation (ii) will also have both roots same as (i). Thus

$$\frac{a}{1} = \frac{b}{2} = \frac{c}{3} \Rightarrow a = \lambda, b = 2\lambda, c = 3\lambda$$

Hence $1 : 2 : 3$

Sol 20: (C) $\dfrac{7}{10} + \dfrac{77}{100} + \dfrac{777}{10^3} + \ldots\ldots + $ up to 20 terms

$$= 7\left[\frac{1}{10} + \frac{11}{100} + \frac{111}{10^3} + \ldots\ldots + \text{up to 20 terms}\right]$$

$$= \frac{7}{9}\left[\frac{9}{10} + \frac{99}{100} + \frac{999}{1000} + \ldots\ldots + \text{up to 20 terms}\right]$$

$$= \frac{7}{9}\left[\left(1-\frac{1}{10}\right)+\left(1-\frac{1}{10^2}\right)+\left(1-\frac{1}{10^3}\right)+\ldots\ldots+ \text{up to 20 terms}\right]$$

$$= \frac{7}{9}\left[20 - \frac{\frac{1}{10}\left(1-\left(\frac{1}{10}\right)^{20}\right)}{1-\frac{1}{10}}\right] = \frac{7}{9}\left[20 - \frac{1}{9}\left(1-\left(\frac{1}{10}\right)^{20}\right)\right]$$

$$= \frac{7}{9}\left[\frac{179}{9}+\frac{1}{9}\left(\frac{1}{10}\right)^{20}\right] = \frac{7}{81}\left[179 + (10)^{-20}\right]$$

Sol 21: (A) $2y = x + z$

$2 \tan^{-1} y = \tan^{-1} x + \tan^{-1}(z)$

$$\tan^{-1}\left(\frac{2y}{1-y^2}\right) = \tan^{-1}\left(\frac{x+z}{1-xz}\right)$$

$$\frac{x+z}{1-y^2} = \frac{x+z}{1-xz}$$

$\Rightarrow y2 = xz$ or $x + z = 0 \Rightarrow x = y = z$

JEE Advanced/Boards

Exercise 1

Sol 1: (i) Let 2 numbers be a, b

Given H.M. $= \dfrac{2ab}{a+b} = 4 \Rightarrow a + b = \dfrac{ab}{2}$

We have A.M. $= \dfrac{a+b}{2}$ and G.M. $= \sqrt{ab}$

$(\text{G.M.})^2 = ab$ ... (i)

$2\text{A.M.} = a + b$ ... (ii)

$2A + G^2 = 27$

$a + b + ab = 27$ using (i) and (ii)

$\dfrac{3ab}{2} = 27 \Rightarrow ab = 18$

$a + b = 9$

$\Rightarrow a, b = 3, 6$

(ii) A.M. = G.M. + 15 = H.M. + 27

$\dfrac{a+b}{2} = \dfrac{2ab}{a+b} + 27 = \sqrt{ab} + 15$

$\text{HM} = \dfrac{G^2}{\text{AM}}$ (we know this)

$\Rightarrow (A - 27)(A) = (A - 15)^2$

$\Rightarrow 27A = 225 - 30A$

$A = 75$ ...(i)

$G.M. = 60$ ...(ii)

$H.M. = 48$

$a + b = 150$ using (i) and (ii)

$ab = 3600$

$a = 120$

$b = 30$

Sol 2: $H_{10} = 21$, $H_{21} = 10$

$$\frac{1}{H_{10}} = \frac{1}{H_1} + 9d = \frac{1}{21}$$

$$\frac{1}{H_{10}} = \frac{1}{H_1} + 20d = \frac{1}{10}$$

$$\Rightarrow 11d = \frac{11}{210} \Rightarrow d = \frac{1}{210}$$

$$\frac{1}{H_1} + \frac{9}{210} = \frac{10}{210}$$

$$\frac{1}{H_1} = \frac{1}{210}$$

$$\frac{1}{H_{210}} = \frac{1}{H_1} + 209d = \frac{1}{210} + \frac{209}{210} = \frac{210}{210} = 1$$

Sol 3: $\sin x$, $\sin^2 2x$, $\cos x \sin 4x$ are in GP.

$\sin^4 2x = \sin x \cos x \sin 4x$

$(2\sin x \cos x)^4 = \sin x \cos x \cdot 2 \sin 2x \cos 2x$

$16 \sin^4 x \cos^4 x = 4 \sin^2 x \cos^2 x \cos 2x$

$4 \sin^2 x \cos^2 x = \cos 2x$

$\sin^2 2x = \cos 2x$

$1 - \cos^2 2x = \cos 2x$

$\cos^2 2x + \cos 2x - 1 = 0$

$$\cos 2x = \frac{-1 \pm \sqrt{5}}{2}$$

$\therefore \cos\theta$ can never be equal to $\dfrac{-1-\sqrt{5}}{2}$ i.e

$$\therefore \cos 2x = \frac{-1+\sqrt{5}}{2}$$

$$\text{Common ratio} = \frac{\sin^2 2x}{\sin x} = \frac{4\sin^2 x \cos^2 x}{\sin x}$$

$= 4 \cos^2 x \sin x = 2 \cos x \sin 2x$

$= 2\sqrt{\dfrac{1+\cos 2x}{2}}\sqrt{1-\cos^2 2x}$

$= \sqrt{2}\sqrt{1-\dfrac{(6-2\sqrt{5})}{4}}\sqrt{1+\left(\dfrac{\sqrt{5}-1}{2}\right)}$

$= \sqrt{2}\sqrt{\dfrac{\sqrt{5}-1}{2}}\cdot\sqrt{\dfrac{\sqrt{5}+1}{2}} = \sqrt{2}\cdot\sqrt{\dfrac{4}{4}} = \sqrt{2}$

Sol 4: a, b, c, d, e be 5 numbers

a b c in AP, b c d in GP, c d e in HP

$2b = a + c$, $c^2 = bd$, $d = \dfrac{2ce}{c+e}$, Let b be b

c be br, d be br^2

$br^2 = \dfrac{2bre}{br+e} \Rightarrow br^2 + er = 2e \Rightarrow e = \dfrac{br^2}{2-r}$

$a = 2b - br = b(2-r)$

$ae = b^2r^2 = c^2$ hence a,c,e are in GP ...(i)

(ii) $\Rightarrow \dfrac{(2b-a)^2}{a} = \dfrac{c^2}{a} = \dfrac{b^2r^2}{b(2-r)} = \dfrac{br^2}{2-r} = e$

Hence proved.

(iii) $a = 2$ $e = 18 \Rightarrow c = \pm 6$

bc, d, e = (4, 6, 9); (−2, −6, −18)

$\Rightarrow b = 4, -2 \Rightarrow d = 9, -18$

Sol 5: $2\alpha^2$, α^4, 24 form A.P.

$\alpha^4 = \alpha^2 + 12$ ...(i)

$\alpha^4 - \alpha^2 = 12$

$\Rightarrow \alpha = 2, -2$ $(\alpha_1, \alpha_2 = 2, -2$

$(\beta^2)^2 = 1(6 - \beta^2)$

$\beta^4 + \beta^2 = 6$

$\beta^4 + \beta^2 - 6 = 0$

$\beta^4 + 3\beta^2 - 2\beta^2 - 6 = 0$

$\beta^2 = 2$

$\beta_1, \beta_2 = \sqrt{2}, -\sqrt{2}$

$\alpha_1^2 + \alpha_2^2 + \beta_1^2 + \beta_2^2 = 4 + 4 + 2 + 2 = 12$

Sol 6: 2G.P.s

$\dfrac{a_1}{1-r_1} = 1 \Rightarrow a_1 = 1 - r_1$

$\dfrac{a_2}{1-r_2} = 1 \Rightarrow a_2 = 1 - r_2$

$a_1 r_1 = a_2 r_2 \Rightarrow (1-r_1)r_1 = (1-r_2)r_2$

$\Rightarrow r_1 - r_2 = (r_1 - r_2)(r_1 + r_2)$

$\Rightarrow r_1 + r_2 = 1$

$a_1 r_1^2 = 1/8 \Rightarrow (1-r_1)r_1^2 = 1/8$

$\Rightarrow (2r_1 - 1)(4r_1^2 - 2r_1 - 1) = 0$

If $r_1 = \frac{1}{2}$ then $r_1 = r_2$

$\Rightarrow 4r_1^2 - 2r_1 - 1 = 0$

$\Rightarrow r_1 = \dfrac{1 \pm \sqrt{5}}{4}$

If $r_1 = \dfrac{1-\sqrt{5}}{4}$ then, $r_2 > 1$.

$\Rightarrow r_1 = \dfrac{1+\sqrt{5}}{4}$

$\therefore a_1 r_1 = (1-r_1)r_1 = \left(1 - \left(\dfrac{1+\sqrt{5}}{4}\right)\right)\left(\dfrac{1+\sqrt{5}}{4}\right)$

$= \left(\dfrac{3-\sqrt{5}}{4}\right)\left(\dfrac{1+\sqrt{5}}{4}\right) = \dfrac{\sqrt{5}-1}{8} = \dfrac{\sqrt{m}-n}{p}$

$\therefore 100m + 10n + p = 500 + 10 + 8 = 518$

Sol 7: $S = \displaystyle\sum_{n=1}^{99}\dfrac{5^{100}}{25^n + 5^{100}}$

$T_1 = \dfrac{5^{100}}{5^2 + 5^{100}}$

$T_{99} = \dfrac{5^{100}}{5^{2\times 99} + 5^{100}} = \dfrac{5^{100}}{\dfrac{5^{200}}{5^2} + 5^{100}} = \dfrac{5^2}{5^2 + 5^{100}}$

$T_1 + T_n = 1$

$S = T_1 + T_2 + \ldots + T_{99} = 1 + 1 \ldots T_{50}$

$= 49 + T_{50} = 49 + \dfrac{5^{100}}{5^{100} + 5^{100}} = 49 + 1/2$

$[S] = 49$

Sol 8: $ax^3 - ax^2 + 9bx - b = 0$

HM roots $= \dfrac{3}{\dfrac{1}{\alpha}+\dfrac{1}{\beta}+\dfrac{1}{\gamma}} = \dfrac{3\alpha\beta\gamma}{\alpha\beta + \beta\gamma + \gamma\alpha}$

$\alpha + \beta + \gamma = 1$

$\alpha\beta + \beta\gamma + \gamma\alpha = \dfrac{9b_1}{a_1}$

$$\alpha\beta\gamma = \frac{b_1}{a_1}$$

$$\text{H.M} = \frac{1}{3}$$

$$\frac{\alpha + \beta + \gamma}{3} \geq \text{H.M.}_{(abc)}$$

$$\frac{\alpha + \beta + \gamma}{3} \geq \frac{1}{3}$$

$$\alpha + \beta + \gamma \geq 1 \qquad \text{...(i)}$$

Its given $\alpha + \beta + \gamma = 1$

Equality of equation (i) holds only if $\alpha = \beta = \gamma$

i.e all the roots are $\dfrac{1}{3}$

$$\frac{b}{a} = \alpha^3 = \left(\frac{1}{3}\right)^3$$

$$b = 27a$$

$$b + a = 28a$$

$\therefore$ a is an integer, min $(a + b) = 28$

Sol 9: Let time taken to solve 1st problem be S time to solve second problem will be $\dfrac{S}{r}$

$$\frac{S}{r} + ... + \frac{S}{r^{n-1}} = 63.5 \qquad \text{...(i)}$$

$$S_n = 127 = S + \frac{S}{r} ... \frac{S}{r^{n-2}} \qquad \text{...(ii)}$$

$$31.5 = \frac{S}{r^2} + ... \frac{S}{r^{n-1}} \qquad \text{...(iii)}$$

$$\frac{S}{r} = 32 \qquad \text{...(iv)}$$

$$127 + \frac{S}{r^{n-1}} = 63.5 + S$$

$$63.5 + \frac{S}{r^{n-1}} = S$$

$$\frac{S}{r} - \frac{S}{r^n} = 63.5\left(1 - \frac{1}{r}\right)$$

$$32 - \frac{S}{r^n} = 63.5 - \frac{63.5}{r}$$

$$\frac{63.5}{r} \quad \frac{S}{r^n} = 31.5 \qquad \text{...(v)}$$

$$127 = \frac{S\left(1 - \dfrac{1}{r^{n-1}}\right)}{\left(1 - \dfrac{1}{r}\right)}$$

$$127 - \frac{127}{r} = S - \frac{S}{r^{n-1}}$$

$$127\left(1 - \frac{1}{r}\right) = 32r - r\left(\frac{63.5}{r} - 31.5\right)$$

from (v)

$$127 - \frac{127}{r} = 63.5\,r - 63.5$$

$$\Rightarrow r = 2$$

$$\therefore s = 64$$

$$\frac{s}{r^{n-1}} = \frac{1}{2}$$

$$2^{n-2} = 64$$

$$\Rightarrow n = 8$$

Sol 10: $a_n + 1 = a_n + 1$ for $n = 1 ... 97$

$$\Rightarrow a_2 = a_1 + 1$$

$$\Rightarrow a_3 = a_2 + 1 = a_1 + 2$$

$$\Rightarrow a_4 = a_1 + 3$$

$$a_n = a_1 + (n - 1)$$

$$\Rightarrow a_1 + a_2 ... a_{98} = 4949 = \frac{98}{2}[2a_1 + 97.1]$$

$$101 = 2a_1 + 97 \Rightarrow a_1 = 2$$

Now, we can write here Σa_{2k}

$$= a_2 + a_4 + a_6 ... a_{98} = a_1 + 1 + a_1 + 3 ... a_1 + 97$$

$$= \frac{49}{2}[2a_1 + 2 + 48 \times 2] = 49[a_1 + 49] = 49 \times 51$$

$$= 2499$$

Sol 11: $xyz = 55$ or $\dfrac{343}{55}$ acc to a, x, y, z, b in AP/HP

For a, x, y, z, b in AP

$$x = a + d;\ y = a + 2d\ z = a + 3d$$

$$b = a + 4d \Rightarrow d = \frac{b - a}{4}$$

$$(a + d)(a + 2d)(a + 3d) = 55 \qquad \text{...(i)}$$

For a, x, y, z, b in HP

$$\frac{1}{x} = \frac{1}{a} + d_H;\ \frac{1}{b} = \frac{1}{a} + 4d_H \Rightarrow d_H = \frac{1}{4}\left[\frac{1}{b} - \frac{1}{a}\right]$$

$$\frac{1}{y} = \frac{1}{a} + 2d_H$$

$$\frac{1}{z} = \frac{1}{a} + 3d_H$$

$$\frac{1}{xyz} = \left(\frac{1}{a} + d_H\right)\left(\frac{1}{a} + 2d_H\right)\left(\frac{1}{a} + 3d_H\right) = \frac{55}{343} \qquad \text{...(ii)}$$

Equation (i) can be written as

$$\left(a + \frac{b-a}{4}\right)\left(a + \frac{(b-a)2}{4}\right)\left(a + \frac{(b-a)3}{4}\right) = 55$$

$$\frac{(3a+b)(2a+2b)(a+3b)}{64} = 55$$

Equation (ii) can be written as

$$\left(\frac{3}{4a} + \frac{1}{4b}\right)\left(\frac{1}{2a} + \frac{1}{2b}\right)\left(\frac{1}{4a} + \frac{3}{4b}\right) = \frac{55}{343}$$

$$\frac{(3b+a)(2b+2a)(b+3a)}{64a^2b^2ab} = \frac{55}{343}$$

$$\Rightarrow a^3b^3 = 343$$

$$\Rightarrow ab = 7$$

a & b are integers

i.e $a = 1$, $b = 7$ or $a = 7$, $b = 1$

i.e $a^2 + b^2 = 50$

Sol 12: $10x^3 - cx^2 - 54x - 27 = 0$

Let α, β, γ be the roots

$$\alpha + \beta + \gamma = \frac{c}{10} \qquad \text{...(i)}$$

$$a\beta + by + \gamma\alpha = -\frac{54}{10} \qquad \text{...(ii)}$$

$$ab\gamma = \frac{27}{10} \qquad \text{...(iii)}$$

α β & γ are in harmonic progression

i.e $\beta = \dfrac{2\alpha\gamma}{\alpha+\gamma}$

$$b\alpha + b\gamma = 2\alpha\gamma \qquad \text{....(iv)}$$

Putting this in equation (iii)

$\beta = -3/2$ this in equation (iv)

$$(\alpha + \gamma)\left(\frac{-3}{2}\right) = \frac{-3.6}{10}$$

$$\alpha + \gamma = \frac{12}{5}$$

$$\Rightarrow \alpha = 3;\ \gamma = \frac{-3}{5}$$

The 3 roots are 3, $\dfrac{-3}{2}$, $\dfrac{-3}{5}$

$$\frac{c}{10} = \alpha + \beta + y = \frac{12}{5} - \frac{3}{2} = \frac{9}{10} \Rightarrow C = 9$$

Sol 13: We have $b^2 = ac$

Also, $\log_c a$, $\log_b c$ and $\log_a b$ are in AP

We can write $\log_a b = \log_c a + (3-1)d$

$$\Rightarrow d = \frac{\log_a b - \log_c a}{2}$$

$$= \frac{\log_a \sqrt{ac} - \log_c a}{2}$$

$$= \frac{1 + 3\log_a c}{4}$$

Given that $\log_a c = 4$

$$\therefore d = \frac{1 + 3 \times 4}{4} = \frac{13}{4}$$

Sol 14: $a = b - 2$

$ar^2 = b + 6$

$$\frac{a + ar^2}{2ar} = \frac{5}{3}$$

$$\frac{2b + 4}{2ar} = \frac{5}{3}$$

$$\frac{3}{5}(b + 2) = ar$$

$$\frac{9}{25}(b + 2)^2 = (b + 6)(b - 2)$$

$$\Rightarrow 9(b^2 + 4 + 4b) = 25(b^2 + 4b - 12)$$

$$16b^2 + 64b - 336 = 0$$

$$b^2 + 4b - 21 = 0$$

$$b^2 + 7b - 3b - 21 = 0$$

$b = 7, 3 \Rightarrow$ +ve integral value of b is 3.

Sol 15:
$$\frac{S_{1-11}}{S_{n-10-n}} = \frac{1}{8} \qquad \text{... (i)}$$

$$\frac{S_{10-n}}{S_{(n-8)-n}} = 2 \qquad \text{... (ii)}$$

$$S_{1-11} = a\frac{\left(r^{11} - 1\right)}{r-1} \qquad S_{10-n} = ar^9\frac{\left(r^{n-9} - 1\right)}{r-1}$$

$$S_{(n-10)-n} = ar^{n-11}\frac{\left(r^{11} - 1\right)}{r-1} \qquad S_{(n-8)-n} = a\frac{\left(r^{n-9} - 1\right)}{r-1}$$

Putting these values in equation (i) and equation (ii)

$$\frac{1}{r^{n-11}} = \frac{1}{8}$$

$$r^9 = 2 \Rightarrow r = 2^{\frac{1}{9}}$$

$$\Rightarrow r^{n-11} = 2^3$$

$$\Rightarrow 2^{\frac{n-11}{9}} = 2^3$$

$$\Rightarrow n = 11 + 27 = 38$$

Sol 16: $S_n = (1 + 2T_n)(1 - T_n)$

$$S_n = \frac{n}{2}[2a + (n-1)d]$$

$$= [1 + 2a + (n-1)\,2d]\,[1 - a - (n-1)d]$$

$$T_n = a + (n-1)d$$

$$S_n = 1 + T_n - 2T_n^2$$

$$S_1 = 1 + T_1 - 2T_1^2 = T_1$$

$$T_1 = \frac{1}{\sqrt{2}}$$

$$S_2 = 1 + T_2 - 2T_2^2 = T_1 + T_2$$

$$1 - \frac{1}{\sqrt{2}} = 2T_2^2$$

$$\left(\frac{\sqrt{2}-1}{2\sqrt{2}}\right) = T_2^2 = \frac{2-\sqrt{2}}{4} = \frac{\sqrt{4}-\sqrt{2}}{4}$$

$$\Rightarrow a = 4 \text{ and } b = 2$$

$$a + b = 6$$

Sol 17: Let number be abc $(a > b > c)$

$$a = a,\ b = \frac{a}{r},\ c = \frac{a}{r^2}$$

$$\Rightarrow 100a + \frac{10a}{r} + \frac{a}{r^2} - 100\frac{a}{r^2} - \frac{10a-a}{r} = 792$$

$$\Rightarrow 99a - \frac{99a}{r^2} = 792 \Rightarrow a - \frac{a}{r^2} = 8$$

$$\therefore \text{ New number } = 100(a-4) + 10b + c$$

$$\Rightarrow 2b = a - 4 + c \Rightarrow \frac{2a}{r} = a - 4 + \frac{a}{r^2}$$

$$\Rightarrow \frac{2a}{r} = a - 4 + a - 8$$

$$\Rightarrow 2a - 12 = \frac{2a}{r} \Rightarrow a = \frac{a}{r} + 6 = \frac{a}{r^2} + 8$$

$$\Rightarrow r = \frac{a}{a-6}$$

$$\Rightarrow \left(\frac{a}{a-6}\right)^2 = \frac{a}{a-8}$$

$$\Rightarrow a(a-8) = (a-6)^2 \Rightarrow a = 9,\ r = 3$$

So the number is 931

Sol 18: $S(\theta) = 1 + (1 + \sin\theta)\cos\theta$

$$+ (1 + \sin\theta + \sin^2\theta)\cos^2\theta \ldots \infty$$

$$= 1 + \cos\theta + \cos^2\theta \ldots$$

$$+ \sin\theta(\cos\theta + \cos^2\theta \ldots) + \sin^2\theta$$

$$= \frac{1}{1-\cos\theta} + \frac{\sin\theta}{1-\cos\theta}\cos\theta + \frac{\sin^2\theta\cos^2\theta}{1-\cos\theta}$$

$$= \frac{1}{1-\cos\theta}[1 + \sin\theta\cos\theta + \sin^2\theta\cos^2\theta \ldots]$$

$$S(\theta) = \frac{1}{(1-\sin\theta\cos\theta)(1-\cos\theta)}$$

$$S_{\left(\frac{\pi}{4}\right)} = \frac{1}{\left(1-\frac{1}{2}\right)}\left(1-\frac{1}{\sqrt{2}}\right) = \frac{2\sqrt{2}}{(\sqrt{2}-1)}$$

Sol 19: $\tan\left(\frac{\pi}{12}-x\right),\ \tan\frac{\pi}{12},\ \tan\left(\frac{\pi}{12}+x\right)$ are in GP

$$\tan^2\frac{\pi}{12} = \tan\left(\frac{\pi}{12}-x\right)\tan\left(\frac{\pi}{12}+x\right)$$

$$= \frac{\sin\left(\frac{\pi}{12}+x\right)\sin\left(\frac{\pi}{12}-x\right)}{\cos\left(\frac{\pi}{12}+x\right)\cos\left(\frac{\pi}{12}-x\right)} \Rightarrow \frac{\cos 2x - \cos\frac{\pi}{6}}{\cos 2x + \cos\frac{\pi}{6}} = \tan^2\frac{\pi}{12}$$

$$\cos 2x = \frac{\cos\frac{\pi}{6}\tan^2\frac{\pi}{12} + \cos\frac{\pi}{6}}{1-\tan^2\frac{\pi}{12}} = \frac{\cos\frac{\pi}{6}\left[\tan^2\frac{\pi}{12}+1\right]}{1-\tan^2\frac{\pi}{12}}$$

$$= \cos\frac{\pi}{6}\left[\frac{\sin^2\left(\frac{\pi}{12}\right) + \cos^2\left(\frac{\pi}{12}\right)}{\cos^2\left(\frac{\pi}{12}\right) - \sin^2\left(\frac{\pi}{12}\right)}\right] = \frac{\sqrt{3}}{2}\left[\frac{1}{\cos\left(\frac{\pi}{6}\right)}\right] = 1$$

$$= \frac{\sqrt{3}}{2} \times \frac{2}{\sqrt{3}} = 1$$

$$\therefore \cos 2x = 1$$

$\Rightarrow$ 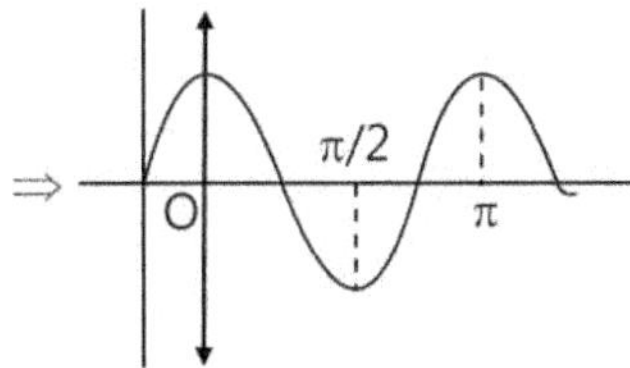

Solutions are O, π, 2π, 3π ... 99π

$$\frac{99}{2}[2\pi + 98\pi] = 50\,\pi \cdot 99 = 4950\,\pi$$

K = 4950

Exercise 2

Single Correct Choice Type

Sol 1: (A) A.M. = 9 + 99 ... 999999999/9

$\Rightarrow$ 9[1 + 11 + 111 + ... 111111111]/9

= 123456789

This does not contain 0

Sol 2: (D) Given $\displaystyle\sum_{k=1}^{360} \frac{1}{k\sqrt{k+1}+(k+1)\sqrt{k}} = \frac{m}{n}$

$$\sum_{k=1}^{360} = \frac{1}{\sqrt{k+1}\sqrt{k}}\left[\frac{1}{\sqrt{k}+\sqrt{k+1}}\frac{\sqrt{k+1}-\sqrt{k}}{\sqrt{k+1}-\sqrt{k}}\right]$$

$$\sum_{k=1}^{360} = \frac{\sqrt{k+1}-\sqrt{k}}{\sqrt{k+1}\sqrt{k}}$$

$$\sum_{k=1}^{360} = \sum\left[\frac{1}{\sqrt{k}} - \frac{1}{\sqrt{k+1}}\right]$$

$$\therefore \sum_{k=1}^{360} T_k = \frac{1}{\sqrt{1}} - \frac{1}{\sqrt{2}} + \frac{1}{\sqrt{2}} - \frac{1}{\sqrt{3}} \cdots - \frac{1}{\sqrt{361}}$$

$$= 1 - \frac{1}{19} = \frac{18}{19} = \frac{m}{n} \quad \text{(given)}$$

$\Rightarrow$ m + n = 18 + 19 = 37

Sol 3: (B) $\displaystyle\sum_{k=1}^{100}\frac{k}{k^4+k^2+1} = \sum\frac{k}{(k^2+1)^2-k^2}$

$$= \sum\frac{k}{(k^2+1+k)(k^2+1-k)}$$

$$= \sum\frac{k}{2k}\left[\frac{1}{k^2+k+1} + \frac{1}{k^2-k+1}\right]$$

$$= \sum\frac{1}{2}\left[\frac{1}{k^2-k+1} - \frac{1}{k^2+k+1}\right]$$

$$= \sum\frac{1}{2}\left[\frac{1}{1} - \frac{1}{3} + \frac{1}{3} - \frac{1}{7} \cdots \frac{1}{100^2+100+1}\right]$$

$$= \frac{1}{2}\left[\frac{100^2+100}{10101}\right] = \frac{5050}{10101}$$

Sol 4: (A)

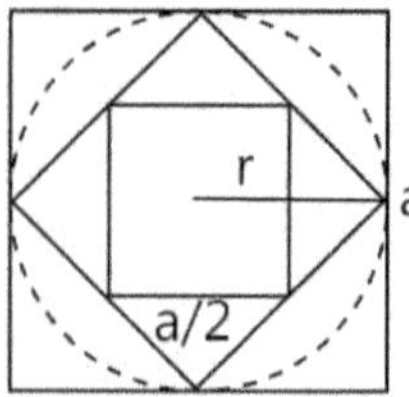

Circle inscribed in 1st circle = r = a/2

In 2nd circle = $r_1 = \dfrac{a}{2\sqrt{2}}$

In 3rd circle = $\dfrac{a}{2\times2}$ this is G.P. with common ratio $\dfrac{1}{\sqrt{2}}$

$$a_r = \frac{a}{2}\left(\frac{1}{\sqrt{2}}\right)^{n-1} = r\left(\frac{1}{2}\right)^{\frac{n-1}{2}} = r2^{\left(\frac{1-n}{2}\right)} \text{ [A]}$$

Assertion Reasoning Type

Sol 5: (D) Statement-I: If $(a + b + c)^3 \le 27\,abc$

$3a + 4b + 5c = 12$

Statement-II: $\Rightarrow$ A.M. $\ge$ G.M. (True)

We Know A.M. $\ge$ G.M.

For three numbers a + b + c

$$\Rightarrow \frac{a+b+c}{3} \ge (abc)^{1/3}$$

$\Rightarrow (a + b + c)^3 \ge 27\,abc$

Given $(a + b + c)^3 \le 27\,abc$

$\Rightarrow a = b = c$ & $(a + b + c)^3 = 27\,abc$

$3a + 4a + 5a = 12 \Rightarrow a = b = c = 1$

$$\frac{1}{a^2} + \frac{1}{b^3} + \frac{1}{b^5} = \frac{1}{1} + \frac{1}{1} + \frac{1}{1} = 3 \ne 10$$

So statement-II is false.

Multiple Correct Choice Type

Sol 6: (A, B, C, D) $a_1 = 25$, $b_1 = 75$, $a_{100} + b_{100} = 100$

$a_1 + 99d_a + b_1 + 99d_b = 100$

$\Rightarrow d_a + d_b = 0$

$\Rightarrow d_a = -d_B$

(a) Hence a is correct

(b) $a_n + b_n = a_1 + (n-1)d_a + b_1 + (n-1)d_B$

$= a_1 + b_1 + (n-1)(d_a + d_B) = a_1 + b_1 = 100$

Hence, correct

(c) $(a_1 + b_1)(a_2 + b_2)(a_3 + b_3)\ldots$

$= 100, 100, 100$ is in AP

(d) $\displaystyle\sum_{r=1}^{100}(a_r + b_r) = \sum_{r=1}^{100}(100) = 10000$

Sol 7: (B, C) $\sin(x-y), \sin x, \sin(x+y)$ are in HP

$\sin x = \dfrac{2\sin(x-y)\sin(x+y)}{\sin(x-y)+\sin(x+y)}$

$\Rightarrow 2\sin^2 x \cos y = 2\cos^2 y - 2\cos^2 x$

$\Rightarrow \sin^2 x(\cos y - 1) = \cos^2 y - 1$

$\Rightarrow \sin^2 x = 1 + \cos y = 2\cos^2 y/2$

$\Rightarrow \sin x \sec\left(\dfrac{y}{2}\right) = \sqrt{2}$

Sol 8: (A, B) Given series is $a\ ar\ ar^2\ ar^3\ldots$

Given that $ar - a = 6$ and $ar^3 - ar^2 = 54$

$\Rightarrow ar^2(r-1) = 54 \Rightarrow a(r-1) = 6$

Dividing these (2)

$r^2 = 9$

$r = \pm 3$

$a = 3$, for $r = 3$

$ar = 9$

$ar^2 = 27$

$a = 3$

sum = 39 [A]

for $r = -3$; $a = \dfrac{-3}{2}$

$ar = \dfrac{9}{2}$; $ar^2 = \dfrac{-27}{2}$

$\therefore$ sum $= \dfrac{-3}{2} + \dfrac{9}{2} - \dfrac{27}{2} = 3 - \dfrac{27}{2} = \dfrac{-21}{2}$

Sol 9: (A, C, D) $x^3 + px^2 + qx - 1$

Roots form increasing GP

Roots be $\dfrac{a}{r}, a, ar$

$\dfrac{a}{r} + a + ar = -p$

$\dfrac{a^2}{r} + a^2 r + a^2 = q$

$a^3 = 1 \Rightarrow a = 1$

$\left(r + 1 + \dfrac{1}{r}\right) = -p$ ... (i)

$\left(\dfrac{1}{r} + r + 1\right) = q \Rightarrow q = -p \Rightarrow p + q = 0$

$\dfrac{r + 1 + \dfrac{1}{r}}{3} \geq 1$ [AM $\geq$ GM]

$r + 1 + \dfrac{1}{r} \geq 3$

$p \in (-\infty, 3)$ [B is incorrect]

one root (a) is unity

one root is $\dfrac{1}{r}$ & other is r, so if 1 root is greater than 1

and other less than [ACD]

Sol 10: (B, D) $\log a, \log b, \log c, \log \dfrac{a}{2b}, \log \dfrac{2b}{3c},$

$\log \dfrac{3c}{a}$ are in AP

$a_1, a_2, a_3, a_4, a_5, a_6$

$2a_5 = a_4 + a_6$

$\dfrac{4b^2}{9c^2} - \dfrac{3c}{2b}$

$\dfrac{b}{c} = \dfrac{3}{2} \Rightarrow b = \dfrac{3c}{2}$

$a = \dfrac{3b}{2} = \dfrac{9c}{4}$

$a + b = \dfrac{15c}{4} > c$

$a + c = \dfrac{13c}{4} > \dfrac{3c}{2}$ (b)

$b + c = \dfrac{5c}{2} > \dfrac{9c}{4}$ (a)

Hence a, b, c can form Δ

$\log b - \log a = \log c - \log b$

$2\log b = \log a + \log c$

$b^2 = ac$ ie a, b, c are in GP[B]

$a = a, b = ar, c = ar^2$

$18(a + b + c)^2 - 18a^2 - 18b^2 - 18c^2$

$= 18(2ab + 2bc + 2ac)$

$= 36(ab + bc + ac) > ab$ so A is incorrect

Sol 11: (A, B, C, D) $x^2 - 3x + A = 0$

$x_1 + x_2 = 3$

$x_1 x_2 = A$

$x^2 - 12x + B = 0$

$x_3 x_4 = B$

$x_3 + x_4 = 12$

$x_1 = a$; $x_2 = ar$; $x_3 = ar^2$; $x_4 = ar^3$

$a^2 r = A$

$a(1 + r) = 3$

$a^2 r^5 = B$

$ar^2(1 + r) = 12$

$r = \pm 2$

$a = 1, -3$

$A = a^2 r = 1 \times 2 = 2$

$a^2 r^5 = 2^5 = 32 = B$

$x_1 + x_3 = a(1 + r^2) = 5$

$x_2 + x_4 = ar(1 + r^2) = 2.5 = 10$

Previous Years' Questions

Sol 1: (A, B, D) Since, first and $(2n - 1)$the terms are equal. Let first term be x and $(2n - 1)^{th}$ term by y. whose middle term is t_n.

Thus in arithmetic progression ; $t_n = \dfrac{x + y}{2} = a$

In geometric progression : $\sqrt{xy} = b$

In harmonic progression ; $t_n = \dfrac{2xy}{x + y} = c$

$\Rightarrow b^2 = ac$ and $a \geq b \geq c$ (using A.M. $\geq$ G.M. $\geq$ HM)

Here, equality holds (ie, a = b = c) only if all terms are same.

Sol 2: (B, C, D) Let a_n denotes the length of side of the square S_n.

We are given a_n = length of diagonal of S_{n+1}.

$\Rightarrow a_n = \sqrt{2} a_{n+1} \Rightarrow a_{n+1} = \dfrac{a_n}{\sqrt{2}}$

This show that a_1, a_2, a_3.... Form a G.P. with common ratio $1 / \sqrt{2}$

Therefore, $a_n = a_1 \left(\dfrac{1}{\sqrt{2}} \right)^{n-1}$

$\Rightarrow a_n = 10 \left(\dfrac{1}{\sqrt{2}} \right)^{n-1}$ (Qa_1 = 10 given)

$\Rightarrow a_n^2 = 100 \left(\dfrac{1}{\sqrt{2}} \right)^{2(n-1)}$

$\Rightarrow \dfrac{100}{2^{n-1}} \leq 1$ ($\because$ $a_n^2 \leq 1$given)

$\Rightarrow 100 \leq 2^{n-1}$

This is possible for n $\geq$ 8. So (b), (c), (d) are the answer.

Sol 3: (4) We have $S_k = \dfrac{\dfrac{k-1}{k!}}{1 - \dfrac{1}{k}} = \dfrac{1}{(k-1)!}$

Now, $(k^2 - 3k + 1) S_k = \{(k - 2)(k - 1) - 1\} \times S_k$

$= \dfrac{1}{(k-3)!} - \dfrac{1}{(k-1)!} \Rightarrow \sum\limits_{k=1}^{100} |(k^2 - 3k + 1)S_k|$

$= 1 + 1 + 2 - \left(\dfrac{1}{99!} + \dfrac{1}{98!} \right) = 4 - \dfrac{100^2}{100!}$

$\Rightarrow \dfrac{100^2}{100!} + \sum\limits_{k=1}^{100} |(k^2 - 3k + 1)S_k| = 4$

Sol 4: (0) $a_k = 2a_{k-1} - a_{k-2}$

$\Rightarrow a_1, a_2,...., a_{11}$ are in AP

$\therefore \dfrac{a_1^2 + a_2^2 + ... + a_{11}^2}{11} = \dfrac{11a^2 + 35 \times 11d^2 + 10ad}{11} = 90$

$\Rightarrow 225 + 35d^2 + 150d = 90$

$\Rightarrow 35d^2 + 150d + 135 = 0$

$\Rightarrow d = -3, -\dfrac{9}{7}$

Given $a_2 < \dfrac{27}{2}$ $\therefore$ d = -3 and d $\neq -\dfrac{9}{7}$

$\Rightarrow \dfrac{a_1 + a_2 + ... + a_{11}}{11} = \dfrac{11}{2}[30 - 10 \times 3] = 0$

Sol 5: $a_1, a_2, a_3, ...$ A_{100} is an A.P.

$a_1 = 3$, $Sp = \sum\limits_{i=1}^{p} a_1$, $1 \leq p \leq 100$

$\dfrac{S_m}{S_n} = \dfrac{S_{5n}}{S_n} = \dfrac{\dfrac{5n}{2}\left(6 + (5n - 1)d \right)}{\dfrac{n}{2}\left(6 - d + nd \right)}$

Sm is independent of n of $6 - d = 0 \Rightarrow d = 6$

$a_1 = 3$

$a_2 = 3 + 6 = 9$

$a_2 = 9$

Sol 6: (C) Let a and b are two numbers. Then,

$$A_1 = \frac{a+b}{3}; G_1 = \sqrt{ab}; H_1 = \frac{2ab}{a+b}$$

$$A_n = \frac{A_{n-1} + H_{n-1}}{2}, G_n = \sqrt{A_{n-1}H_{n-1}},$$

$$H_n = \frac{2A_{n-1}H_{n-1}}{A_{n-1} + H_{n-1}}$$

Clearly, $G_1 = G_2 = G_3 = \ldots = \sqrt{ab}$

Sol 7: (A) A_2 is A.M. of A_1 and H_1 and $A_1 > H_1$

$\Rightarrow A_1 > A_2 > H_1$

A_3 is A.M. of A_2 and H_2 and $A_2 > H_2$

$\Rightarrow A_2 > A_3 > A_4$

$: : :$

$\therefore A_1 > A_2 > A_3 > \ldots\ldots$

Sol 8: (B) As above $A_1 > H_2 > H_1$, $A_2 > H_3 > H_2$

$\therefore H_1 < H_2 < H_2 < \ldots\ldots$

Sol 9: (B) Here, $V_r = \frac{r}{2}[2r + (r-1)(2r-1)] = \frac{1}{2}(2r^3 - r^2 + r)$

$\therefore \Sigma V_r = \frac{1}{2}[2\Sigma r^3 - \Sigma r^2 + \Sigma r]$

$$= \frac{1}{2}\left[2\left(\frac{n(n+1)}{2}\right)^2 - \frac{n(n+1)(2n+1)}{6} + \frac{n(n+1)}{2}\right]$$

$$= \frac{n(n+1)}{12}[3n(n+1) - (2n+1) + 3]$$

$$= \frac{1}{12}n(n+1)(3n^2 + n + 2)$$

Sol 10: (D) $V_{r+1} - V_r = (r+1)^3 - r^3 - \frac{1}{2}[(r+1)^2 - r^2] + \frac{1}{2}(1)$

$= 3r^2 + 2r - 1$

$\therefore T_r = 3r^2 + 2r - 1 = (r+1)(3r-1)$

Which is a composite number.

Sol 11: (B) Since,

$T_r = 3r^2 + 2r - 1$

$\therefore T_{r+1} = 3(r+1)^2 + 2(r+1) - 1$

$\therefore Q_r = T_{r+1} - T_r = 3[2r+1] + 2[1]$

$Q_r = 6r + 5$

Sol 12: (A, D) $S_n < \lim\limits_{n\to\infty} S_n = \lim\limits_{n\to\infty} \sum\limits_{k=1}^{n} \frac{1}{n} \cdot \frac{1}{1 + k/n + (k/n)^2}$

$$= \int_0^1 \frac{dx}{1 + x + x^2} = \frac{\pi}{3\sqrt{3}}$$

Now, $T_n > \frac{\pi}{3\sqrt{3}}$ as $h\sum\limits_{k=0}^{n-1} f(kh) > \int_0^1 f(x)dx > h\sum\limits_{k=1}^{n} f(kh)$

Sol 13: (C) $b_1 = a_1$, $b_2 = a_1 + a_2$, $b_3 = a_1 + a_2 + a_3$, $b_4 = a_1 + a_2 + a_3 + a_4$

Hence b_1, b_2, b_3, b_4 are neither in A.P. nor in G.P. nor in H.P.

Sol 14: (C) $t_n = c\{n^2 - (n-1)^2\} = c(2n-1)$

$\Rightarrow t_n^2 = c^2(4n^2 - 4n + 1)$

$$\Rightarrow \sum\limits_{n=1}^{n} t_n^2 = c^2\left\{\frac{4n(n+1)(2n+1)}{6} - \frac{4n(n+1)}{2} + n\right\}$$

$$= \frac{c^2 n}{6}\{4(n+1)(2n+1) - 12(n+1) + 6\}$$

$$= \frac{c^2 n}{3}\{4n^2 + 6n + 2 - 6n - 6 + 3\} = \frac{c^2}{3}n(4n^2 - 1)$$

Sol 15: (3) $\sum\limits_{k=2}^{100} |(k^2 - 3k + 1)S_k|$

for $k = 2$ $|k^2 - 3k + 1) S_k| = 1$

$$\sum\limits_{k=3}^{100} \left|\frac{k-1}{(k-2)!} - \frac{k-1+1}{(k-1)!}\right|$$

$$\sum\limits_{k=3}^{100} \frac{1}{(k-3)!} + \frac{1}{(k-2)!} - \frac{1}{(k-2)!} - \frac{1}{(k-1)!}$$

$$\sum\limits_{k=3}^{100} \left(\frac{1}{(k-3)!} - \frac{1}{(k-1)!}\right)$$

$$S = 1 + \left(1 - \frac{1}{2!}\right) + \left(\frac{1}{1!} - \frac{1}{3!}\right) + \left(\frac{1}{2!} - \frac{1}{4!}\right) + \left(\frac{1}{3!} - \frac{1}{5!}\right) + \left(\frac{1}{4!} - \frac{1}{6!}\right) +$$

$$\cdots \left(\frac{1}{94!} - \frac{1}{96!}\right) + \left(\frac{1}{95!} - \frac{1}{97!}\right) + \left(\frac{1}{96!} - \frac{1}{98!}\right) + \left(\frac{1}{97!} - \frac{1}{99!}\right)$$

$$= 2 - \frac{1}{98!} - \frac{1}{99!}$$

$$\therefore E = \frac{100^2}{100!} + 3 - \frac{1}{98!} - \frac{1}{99.98!}$$

$$= \frac{100^2}{100!} + 3 - \frac{100}{99!} = \frac{100^2}{100.99!} + 3 - \frac{100}{99!} = 3$$

Sol 16: (0) $a_k = 2a_{k-1} - a_{k-2} \Rightarrow a_1, a_2, ..., a_{11}$ are in A.P.

$$\therefore \frac{a_1^2 + a_2^2 + ... + a_{11}^2}{11} = \frac{11a^2 + 35 \times 11d^2 + 10ad}{11} = 90$$

$\Rightarrow 225 + 35d2 + 150d = 90$

$35d2 + 150d + 135 = 0 \Rightarrow d = -3, -9/7$

Given $a^2 < \dfrac{27}{2} \therefore d = -3$ and $d \neq -9/7 \Rightarrow$

$$\frac{a_1 + a_2 + ... + a_{11}}{11} = \frac{11}{2}\left[30 - 10 \times 3\right] = 0$$

Sol 17: (B) $ax^2 + bx + c = 0 \Rightarrow x^2 + 6x - 7 = 0$

$\Rightarrow \alpha = 1, \beta = -7$

$$\sum_{n=0}^{\infty} \left(\frac{1}{\alpha} + \frac{1}{\beta}\right)^n = \sum_{n=0}^{\infty} \left(\frac{1}{1} - \frac{1}{7}\right)^n = 7$$

Sol 18: (D) Corresponding A.P.

$$\frac{1}{5}, ... \frac{1}{25} (20^{th} \text{ term})$$

$$\frac{1}{25} = \frac{1}{5} + 19d \quad \Rightarrow d = \frac{1}{19}\left(\frac{-4}{25}\right) = -\frac{4}{19 \times 25}$$

$a_n < 0$

$$\frac{1}{5} - \frac{4}{19 \times 25} \times (n-1) < 0$$

$$\frac{19 \times 5}{4} < n - 1$$

$n > 24.75$

Sol 19: (A, D) $S_n = \displaystyle\sum_{k=1}^{4n} (-1)^{\frac{k(k+1)}{2}} k^2$

$$= \sum_{r=0}^{(n-1)} \left((4r + 4)^2 + (4r + 3)^2 - (4r + 2)^2 - (4r + 1)^2\right)$$

$$= \sum_{r=0}^{(n-1)} (2(8r + 6) + 2(8r + 4)) = \sum_{r=0}^{(n-1)} (32r + 20)$$

$$= 16(n - 1)n + 20n = 4n(4n + 1) = \begin{cases} 1056 \text{ for } n = 8 \\ 1332 \text{ for } n = 9 \end{cases}$$

Sol 20: (6) $\dfrac{b}{a} = \dfrac{c}{b} = (\text{integer})$

$$b^2 = ac \Rightarrow c = \frac{b^2}{a}$$

$$\frac{a + b + c}{3} = b + 2$$

$$a + b + c = 3b + 6 \Rightarrow a - 2b + c = 6$$

$$a - 2b + \frac{b^2}{a} = 6 \Rightarrow 1 - \frac{2b}{a} + \frac{b^2}{a^2} = \frac{6}{a}$$

$$\left(\frac{b}{a} - 1\right)^2 = \frac{6}{a} \Rightarrow a = 6 \text{ only}$$

Sol 21: (9) Let seventh term be 'a' and common difference be 'd'

Given, $\dfrac{S_7}{S_{11}} = \dfrac{6}{11} \Rightarrow a = 15d$

Hence, $130 < 15d < 140$

$\Rightarrow d = 9$

Sol 22: (B) $\log(b_2) - \log(b_1) = \log(2)$

$\Rightarrow \dfrac{b_2}{b_1} = 2 \Rightarrow b_1, b_2, ...$ are in G.P. with common ratio 2

$\therefore t = b_1 + 2b_1 + ... + 250 b_1 = b_1 (251 - 1)$

$S = a_1 + a_2 + ... + a_{51}$

$$= \frac{51}{2}(a_1 + a_{51}) = \frac{51}{2}(b_1 + b_2) = \frac{51}{2}b_1(1 + 2^{50})$$

$$S - t = b_1\left(\frac{51}{2} + 51 \times 2^{49} - 2^{51} + 1\right) = b_1\left(\frac{53}{2} + 2^{49} \times 47\right)$$

$\Rightarrow S > t$

$b_{101} = 2_{100} b_1$

$a_{101} = a_1 + 100 d = 2(a_1 + 50d) - a_1$

$= 2a_{51} - a_1 = 2b_{51} - b_1 = (2 \times 2_{51} - 1) b_1 = (2_{51} - 1) b_1$

$\therefore b_{101} > a_{101}$

2. COMPLEX NUMBER

1. INTRODUCTION

The number system can be briefly summarized as $N \subset W \subset I \subset Q \subset R \subset C$, where N, W, I, Q, R and C are the standard notations for the various subsets of the numbers belong to it.

N - Natural numbers = {1, 2, 3 .... n}

W - Whole numbers = {0, 1, 2, 3 .... n}

I - Integers = {....2, -1, 0, 1, 2 .....}

Q – Rational numbers = $\left\{ \dfrac{1}{2}, \dfrac{3}{5} ..... \right\}$

IR – Irrational numbers = $\left\{ \sqrt{2}, \sqrt{3}, \pi \right\}$

C – Complex numbers

A complex number is generally represented by the letter "z". Every complex number z, can be written as, $z = x + iy$ where $x, y \in R$ and $i = \sqrt{-1}$.

x is called the real part of complex number, and

y is the imaginary part of complex number.

Note that the sign + does not indicate addition as normally understood, nor does the symbol "i" denote a number. These are parts of the scheme used to express numbers of a new class and they signify the pair of real numbers (x, y) to form a single complex number.

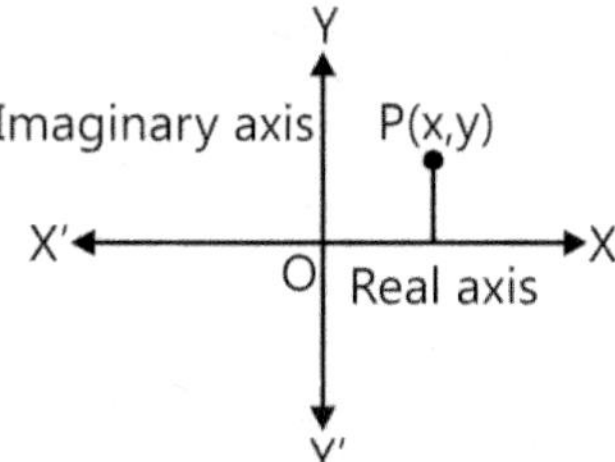

Figure 6.1: Representation of a complex number on a plane

Swiss-born mathematician Jean Robert Argand, after a systematic study on complex numbers, represented every complex number as a set of ordered pair (x, y) on a plane called complex plane.

All complex numbers lying on the real axis were called purely real and those lying on imaginary axis as purely imaginary.

Hence, the complex number $0 + 0i$ is purely real as well as purely imaginary but it is not imaginary.

Note

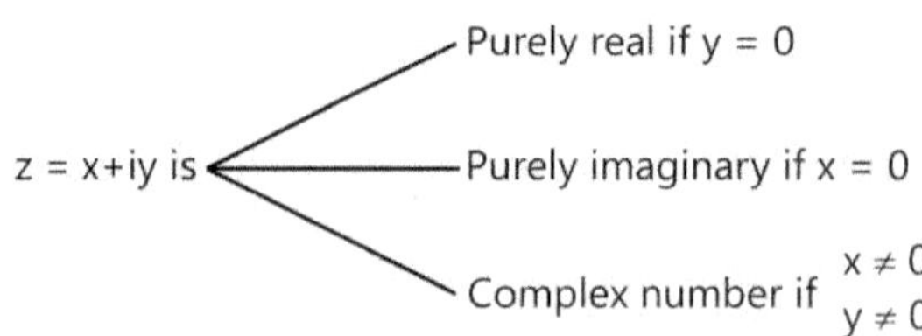

Figure 6.2: Classification of a complex number

(a) The symbol i combines itself with real number as per the rule of algebra together with

$i^2 = -1$; $i^3 = -i$; $i^4 = 1$; $i^{2014} = -1$; $i^{2015} = -i$ and so on.

In general, $i^{4n} = 1$, $i^{4n+1} = i$, $i^{4n+2} = -1$, $i^{4n+3} = -i$, $n \in I$ and $i^{4n} + i^{4n+1} + i^{4n+2} + i^{4n+3} = 0$

Hence, $1 + i^1 + i^2 + \ldots + i^{2014} + i^{2015} = 0$

(b) The imaginary part of every real number can be treated as zero. Hence, there is one-one mapping between the set of complex numbers and the set of points on the complex plane.

NOMORECLASS CONCEPTS

Complex number as an ordered pair: A complex number may also be defined as an ordered pair of real numbers and may be denoted by the symbol (a, b). For a complex number to be uniquely specified, we need two real numbers in a particular order.

2. ALGEBRA OF COMPLEX NUMBERS

(a) **Addition:** $(a + ib) + (c + id) = (a + c) + i(b + d)$

(b) **Subtraction:** $(a + ib) - (c + id) = (a - c) + i(b - d)$

(c) **Multiplication:** $(a + ib)(c + id) = (ac - bd) + i(ad + bc)$

(d) **Reciprocal:** If at least one of a, b is non-zero, then the reciprocal of $a + ib$ is given by

$$\frac{1}{a + ib} = \frac{a - ib}{(a + ib)(a - ib)} = \frac{a}{a^2 + b^2} - i\frac{b}{a^2 + b^2}$$

(e) **Quotient:** If at least one of c, d is non-zero, then quotient of $a + ib$ and $c + id$ is given by

$$\frac{a + ib}{c + id} = \frac{(a + ib)(c - id)}{(c + id)(c - id)} = \frac{(ac + bd) + i(bc - ad)}{c^2 + d^2} = \frac{ac + bd}{c^2 + d^2} + i\frac{bc - ad}{c^2 + d^2}$$

(f) Inequality in complex numbers is not discussed/defined. If a + ib > c + id is meaningful only if b = d = 0. However, equalities in complex numbers are meaningful. Two complex numbers z_1 and z_2 are said to be equal if $\text{Re}(z_1) = \text{Re}(z_2)$ and $\text{Im}(z_1) = \text{Im}(z_2)$. (Geometrically, the position of complex number z_1 on complex plane)

(g) In real number system if $p^2 + q^2 = 0$ implies, $p = 0 = q$. But if z_1 and z_2 are complex numbers then $z_1^2 + z_2^2 = 0$ does not imply $z_1 = z_2 = 0$. For e.g. $z_1 = i$ and $z_2 = 1$.

However if the product of two complex numbers is zero then at least one of them must be zero, same as in case of real numbers.

(h) In case x is real, then $|x| = \begin{cases} x, & \text{if } x \geq 0 \\ -x, & \text{if } x < 0 \end{cases}$ but in case of complex number z, $|z|$ means the distance of the point z from the origin.

NOMORECLASS CONCEPTS

- The additive inverse of a complex number $z = a + ib$ is $-z$ (i.e. $-a - ib$).
- For every non-zero complex number z, the multiplicative inverse of z is $\frac{1}{z}$.
- $|z| \geq |Re(z)| \geq Re(z)$ and $|z| \geq |Im(z)| \geq Im(z)$.
- $\frac{z}{|z|}$ is always a uni-modular complex number if $z \neq 0$.

Illustration 1: Find the square root of 5 + 12i. **(JEE MAIN)**

Sol: $z = 5 + 12i$

Let the square root of the given complex number be a + ib. Use algebra to simplify and get the value of a and b.

Let its square root $= a + ib \quad \Rightarrow 5 + 12i = a^2 - b^2 + 2abi$

$\Rightarrow a^2 - b^2 = 5$ ... (i)

$\Rightarrow 2ab = 12$ ... (ii)

$\Rightarrow (a^2 + b^2)^2 = (a^2 - b^2)^2 + 4a^2b^2 \quad \Rightarrow (a^2 + b^2)^2 = 25 + 144 = 169 \quad \Rightarrow a^2 + b^2 = 13$ ... (iii)

(i) + (iii) $\Rightarrow 2a^2 = 18 \Rightarrow a^2 = 9 \Rightarrow a = \pm 3$

If $a = 3 \Rightarrow b = 2$ If $a = -3 \Rightarrow b = -2$

$\therefore$ Square root $= 3 + 2i, -3 - 2i \quad \therefore$ Combined form $\pm(3 + 2i)$

Illustration 2: If $z = (x, y) \in C$. Find z satisfying $z^2 \times (1 + i) = (-7 + 17i)$. **(JEE MAIN)**

Sol: Algebra of Complex Numbers.

$(x + iy)^2 (1 + i) = -7 + 17i$

$\Rightarrow (x^2 - y^2 + 2xyi)(1 + i) = -7 + 17i; \quad x^2 - y^2 + i(x^2 - y^2) + 2xyi - 2xy = -7 + 17i$

$\Rightarrow (x^2 - y^2 - 2xy) + i(x^2 - y^2 + 2xy) = -7 + 17i \Rightarrow x = 3, y = 2 \quad \Rightarrow x = -3, y = -2$

$\Rightarrow z = -3 + i(-2) = -3 - 2i$

Illustration 3: If $x^2 + 2(1 + 2i)x - (11 + 2i) = 0$. Solve the equation. **(JEE ADVANCED)**

Sol: Use the quadratic formula to find the value of x.

$$\therefore x = \frac{-2(1 + 2i) \pm \sqrt{4 - 16 + 16i + 44 + 8i}}{2}$$

$\Rightarrow 2x = (-2)(1 + 2i) \pm \sqrt{32 + 24i}$

$\Rightarrow x = (-1)(1 + 2i) \pm \sqrt{8 + 6i} = -1 - 2i \pm (3 + i); \quad x = 2 - i, -4 - 3i$

Illustration 4: If $f(x) = x^4 - 4x^3 + 4x^2 + 8x + 44$. Find $f(3 + 2i)$. **(JEE ADVANCED)**

Sol: Let $x = 3+2i$, and square it to form a quadratic equation. Then try to represent f(x) in terms of this quadratic.

$x = 3 + 2i$

$\Rightarrow (x - 3)^2 = -4 \qquad\qquad \Rightarrow x^2 - 6x + 13 = 0$

$x^4 - 4x^3 + 4x^2 + 8x + 44 = x^2(x^2 - 6x + 13) + 2x^3 - 9x^2 + 8x + 44$

$\Rightarrow f(x) = x^2(x^2 - 6x + 13) + 2(x^3 - 6x^2 + 13x) + 3(x^2 - 6x + 13) + 5 \qquad \Rightarrow f(x) = 5$

3. IMPORTANT TERMS ASSOCIATED WITH COMPLEX NUMBER

Three important terms associated with complex number are conjugate, modulus and argument.

(a) Conjugate: If $z = x + iy$ then its complex conjugate is obtained by changing the sign of its imaginary part and denoted by $\bar{z}$ i.e. $\bar{z} = x - iy$ (see Fig 6.3).

The conjugate satisfies following basic properties

(i) $z + \bar{z} = 2\text{Re}(z)$

(ii) $z - \bar{z} = 2i\,\text{Im}(z)$

(iii) $z\bar{z} = x^2 + y^2$

(iv) If z lies in 1st quadrant then $\bar{z}$ lies in 4th quadrant and $-\bar{z}$ in the 2nd quadrant.

(v) If $x + iy = f(a + ib)$ then $x - iy = f(a - ib)$

For e.g. If $(2 + 3i)^3 = x + iy$ then $(2 - 3i)^3 = x - iy$

and, $\sin(\alpha + i\beta) = x + iy \Rightarrow \sin(\alpha - i\beta) = x - iy$

(vi) $z + \bar{z} = 0 \qquad \Rightarrow z$ is purely imaginary

(vii) $z - \bar{z} = 0 \qquad \Rightarrow z$ is purely real

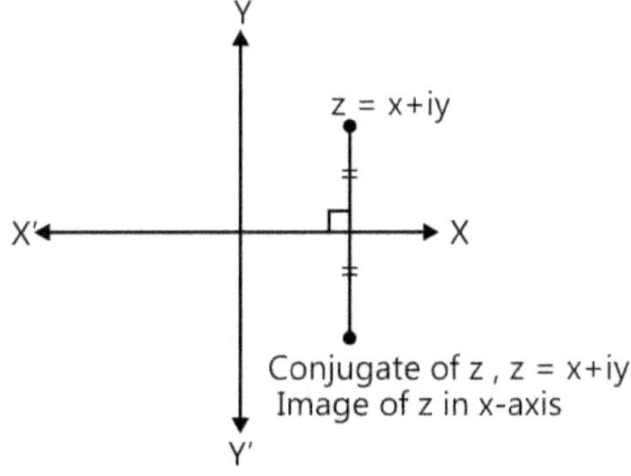

Figure 6.3: Conjugate of a complex number

(b) Modulus: If P denotes a complex number $z = x + iy$ then, $OP = |z| = \sqrt{x^2 + y^2}$. Geometrically, it is the distance of a complex number from the origin.

Hence, note that $|z| \geq 0, |i| = 1$ i.e. $\left|\sqrt{-1}\right| = 1$.

All complex number satisfying $|z| = r$ lie on the circle having centre at origin and radius equal to 'r'.

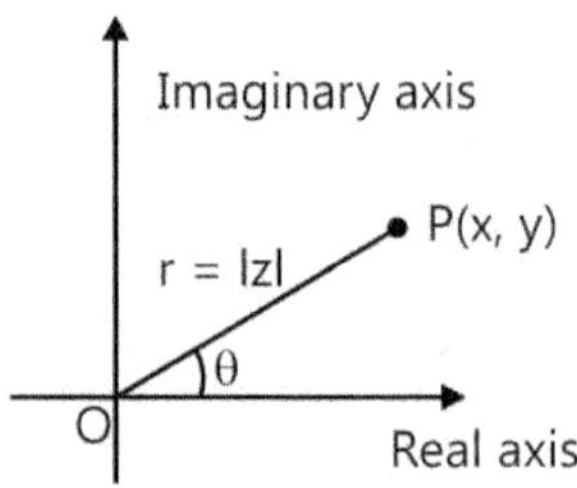

Figure 6.4: Modulus of a complex number

(c) Argument: If OP makes an angle θ (see Fig 6.4) with real axis in anticlockwise sense, then θ is called the argument of z. General values of argument of z are given by $2n\pi + \theta, n \in I$. Hence any two successive arguments differ by 2π.

Note: A complex number is completely defined by specifying both modulus and argument. However for the complex number $0 + 0i$ the argument is not defined and this is the only complex number which is completely defined by its modulus only.

(i) Amplitude (Principal value of argument): The unique value of θ such that $-\pi < \theta \leq \pi$ is called principal value of argument. Unless otherwise stated, amp z refers to the principal value of argument.

(ii) Least positive argument: The value of θ such that $0 < \theta \leq 2\pi$ is called the least positive argument.

If $\phi = \tan^{-1}\left|\dfrac{y}{x}\right|$.

- If $x > 0$, $y > 0$ (i.e. z is in first quadrant), then arg $z = \theta = \tan^{-1}\left|\dfrac{y}{x}\right|$.

- If $x < 0$, $y > 0$ (i.e. z is in 2nd quadrant, then arg $z = \theta = \pi - \tan^{-1}\left|\dfrac{y}{x}\right|$

- If $x < 0$, $y < 0$ (i.e. z is in 3rd quadrant), then arg $z = \theta = -\pi + \tan^{-1}\left|\dfrac{y}{x}\right|$

- If $x > 0$, $y < 0$ (i.e. z is in 4th quadrant), then arg $z = \theta = -\tan^{-1}\left|\dfrac{y}{x}\right|$

- If $y = 0$ (i.e. z is on the X-axis), then arg $(x + i0) = \begin{cases} 0, & \text{if } x > 0 \\ \pi, & \text{if } x < 0 \end{cases}$

- If $x = 0$ (i.e. z is on the Y-axis), then arg $(0 + iy) = \begin{cases} \dfrac{\pi}{2}, & \text{if } y > 0 \\ \dfrac{3\pi}{2}, & \text{if } y < 0 \end{cases}$

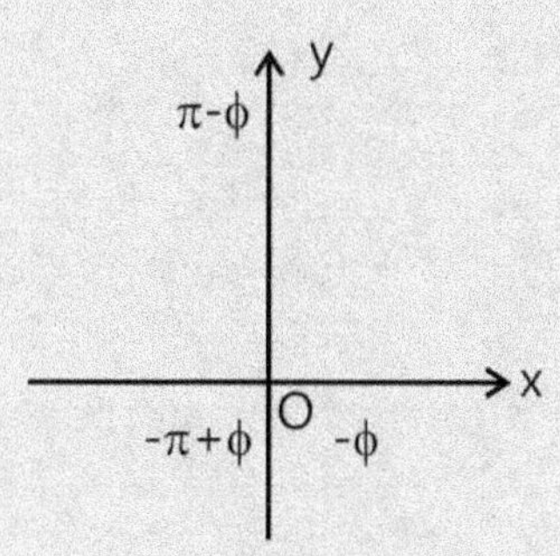

Figure 6.5

Illustration 5: For what real values of x and y, are $-3 + ix^2 y$ and $x^2 + y + 4i$ complex conjugate to each other? **(JEE MAIN)**

Sol: As $-3 + ix^2 y$ and $x^2 + y + 4i$ are complex conjugate of each other. Therefore $-3 + ix^2 y = \overline{x^2 + y + 4i}$.

$-3 + ix^2 y = x^2 + y - 4i$

Equating real and imaginary parts of the above question, we get

$-3 = x^2 + y \Rightarrow y = -3 - x^2$ ... (i)

and $x^2 y = -4$ ... (ii)

Putting the value of $y = -3 - x^2$ from (i) in (ii), we get

$x^2(-3 - x^2) = -4 \Rightarrow x^4 + 3x^2 - 4 = 0 \Rightarrow x^2 = \dfrac{-3 \pm \sqrt{9 + 16}}{2} = \dfrac{-3 \pm 5}{2} = \dfrac{2}{2}, \dfrac{-8}{2} = 1, -4$

$\therefore x^2 = 1 \Rightarrow x = \pm 1$

Putting value of $x = \pm 1$ in (i), we get $y = -3 - (1)^2 = -3 - 1 = -4$

Hence, $x = \pm 1$ and $y = -4$.

Illustration 6: Find the modulus of $\dfrac{1+i}{1-i} - \dfrac{1-i}{1+i}$.   **(JEE MAIN)**

Sol: As $|z| = \sqrt{x^2 + y^2}$, using algebra of complex number we will get the result.

Here, we have $\dfrac{1+i}{1-i} - \dfrac{1-i}{1+i} = \dfrac{(1+i)(1+i)}{(1-i)(1+i)} - \dfrac{(1-i)(1-i)}{(1+i)(1-i)}$

$= \dfrac{1 + i^2 + 2i}{1+1} - \dfrac{1 + i^2 - 2i}{1+1} = \dfrac{1 - 1 + 2i}{2} - \dfrac{1 - 1 - 2i}{2} = \dfrac{2i}{2} - \dfrac{(-2i)}{2} = i + i = 2i, \ \therefore \Rightarrow \left|\dfrac{1+i}{1-i} - \dfrac{1-i}{1+i}\right| = |2i| = 2.$

Illustration 7: Find the locus of z if $|z - 3| = 3|z + 3|$. (JEE MAIN)

Sol: Simply substituting $z = x + iy$ and by using formula $|z| = \sqrt{x^2 + y^2}$ we will get the result.

Let $z = x + iy$

$|x + iy - 3| = 3|x + iy + 3|$ $\quad\quad |(x - 3) + iy| = 3|(x + 3) + iy|$

$\sqrt{(x-3)^2 + y^2} = 3\sqrt{(x+3)^2 + y^2}$; $(x-3)^2 + y^2 = 9(x+3)^2 + 9y^2$.

Illustration 8: If α and β are different complex numbers with $|\beta| = 1$, then find $\left|\dfrac{\beta - \alpha}{1 - \bar{\alpha}\beta}\right|$. (JEE ADVANCED)

Sol: By using modulus and conjugate property, we can find out the value of $\left|\dfrac{\beta - \alpha}{1 - \bar{\alpha}\beta}\right|$.

We have, $|\beta| = 1 \Rightarrow |\beta|^2 = 1 \Rightarrow \beta\bar{\beta} = 1$

Now, $\left|\dfrac{\beta - \alpha}{1 - \bar{\alpha}\beta}\right| = \left|\dfrac{\beta - \alpha}{\beta\bar{\beta} - \bar{\alpha}\beta}\right| = \left|\dfrac{\beta - \alpha}{\beta\,(\bar{\beta} - \bar{\alpha})}\right| = \dfrac{|\beta - \alpha|}{|\beta||\overline{\beta - \alpha}|} = \dfrac{1}{|\beta|} = 1.$ $\left\{\text{as } |x + iy| = |\overline{x + iy}|\right\}$

Illustration 9: Find the number of non-zero integral solution of the equation $|1 - i|^x = 2^x$. (JEE ADVANCED)

Sol: As $|z| = \sqrt{x^2 + y^2}$, therefore by using this formula we can solve it.

We have, $|1 - i|^x = 2^x$

$\Rightarrow \left[\sqrt{1^2 + 1^2}\right]^x = 2^x$ $\quad\quad \Rightarrow \left(\sqrt{2}\right)^x = 2^x$ $\quad \Rightarrow 2^{\frac{x}{2}} = 2^x$ $\quad \Rightarrow \dfrac{x}{2} = 0 \Rightarrow x = 0$.

$\therefore$ The number of non zero integral solution is zero.

Illustration 10: If $\dfrac{a + ib}{c + id} = p + iq$. Prove that $\dfrac{a^2 + b^2}{c^2 + d^2} = p^2 + q^2$. (JEE MAIN)

Sol: Simply by obtaining modulus of both side of $\dfrac{a + ib}{c + id} = p + iq$.

We have, $\dfrac{a + ib}{c + id} = p + iq$

$\left|\dfrac{a + ib}{c + id}\right| = \sqrt{\dfrac{a^2 + b^2}{c^2 + d^2}}$ $\Rightarrow |p + iq| = \sqrt{p^2 + q^2}$; $\left|\dfrac{a + ib}{c + id}\right| = |p + iq|$ $\Rightarrow \dfrac{a^2 + b^2}{c^2 + d^2} = p^2 + q^2$.

Illustration 11: If $(x + iy)^{1/3} = a + ib$. Prove that $\dfrac{x}{a} + \dfrac{y}{b} = 4(a^2 - b^2)$. (JEE ADVANCED)

Sol: By using algebra of complex number. We have, $(x + iy)^{1/3} = a + ib$

$x + iy = (a + ib)^3$ $= a^3 + i^3 b^3 + 3a^2 ib + 3a(ib)^2 = a^3 - b^3 i + 3a^2 bi - 3ab^2$

$x + iy = (a^3 - 3ab^2) + (3a^2 b - b^3)i$; $x = a^3 - 3ab^2 = a\,(a^2 - 3b^2)$; $y = 3a^2 b - b^3$

$\dfrac{x}{a} + \dfrac{y}{b} = 4(a^2 - b^2)$.

4. REPRESENTATION OF COMPLEX NUMBER

4.1 Graphical Representation

Every complex number x + iy can be represented in a plane as a point P (x, y). X-coordinate of point P represents the real part of the complex number and y-coordinate represents the imaginary part of the complex number. Complex number x + 0i (real number) is represented by a point (x, 0) lying on the x-axis. Therefore, x-axis is called the real axis. Similarly, a complex number 0 + iy (imaginary number) is represented by a point on y-axis. Therefore, y-axis is called the imaginary axis.

The plane on which a complex number is represented is called complex number plane or simply complex plane or Argand plane (see Fig 6.6). The figure represented by the complex numbers as points in a plane is known as Argand Diagram.

4.2 Algebraic Form

If $z = x + iy$; then $|z| = \sqrt{x^2 + y^2}$; $\bar{z} = x - iy$, and $\theta = \tan^{-1}\left(\dfrac{y}{x}\right)$

Generally this form is useful in solving equations and in problems involving locus.

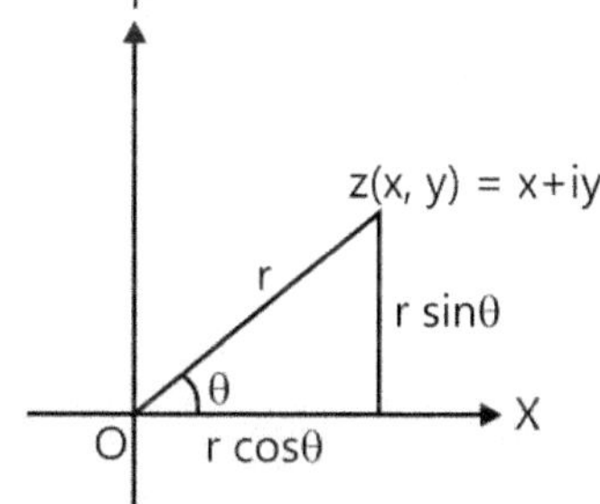

Figure 6.6: Graphical representation

4.3 Polar Form

Figure 6.7 shows the components of a complex number along the x and y-axes respectively. Then

$z = x + iy = r(\cos\theta + i\sin\theta) = r\,\text{cis}\,\theta$ where $|z| = r$; amp z = θ.

Aliter: $z = x + iy$

$$\Rightarrow z = \sqrt{x^2 + y^2}\left(\frac{x}{\sqrt{x^2 + y^2}} + i\frac{y}{\sqrt{x^2 + y^2}}\right)$$

$$\Rightarrow z = |z|\left(\cos\theta + i\sin\theta\right) = r\,\text{cis}\,\theta$$

Note: (a) $(\text{cis}\,\alpha)(\text{cis}\,\beta) = \text{cis}(\alpha + \beta)$

        (b) $(\text{cis}\,\alpha)(\text{cis}(-\beta)) = \text{cis}(\alpha - \beta)$

        (c) $\dfrac{1}{(\text{cis}\,\alpha)} = (\text{cis}\,\alpha)^{-1} = \text{cis}(-\alpha)$

Figure 6.7: Polar form

NOMORECLASS CONCEPTS

The unique value of θ such that $-\pi < \theta \leq \pi$ for which $x = r\cos\theta$ & $y = r\sin\theta$ is known as the principal value of the argument.

The general value of argument is $(2n\pi + \theta)$, where n is an integer and θ is the principal value of arg (z). While reducing a complex number to polar form, we always take the principal value.

The complex number $z = r(\cos\theta + i\sin\theta)$ can also be written as $r\,\text{cis}\,\theta$.

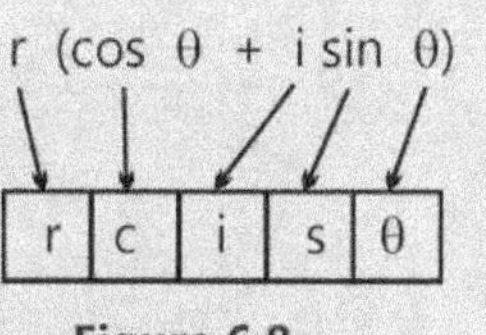

Figure 6.8

4.4 Exponential Form

Euler's formula, named after the famous mathematician Leonhard Euler, states that for any real number x, $e^{ix} = \cos x + i\sin x$.

Hence, for any complex number $z = r(\cos\theta + i\sin\theta)$, $z = re^{i\theta}$ is the exponential representation.

Note: (a) $\cos x = \dfrac{e^{ix} + e^{-ix}}{2}$ and $\sin x = \dfrac{e^{ix} - e^{-ix}}{2i}$ are known as Eulers identities.

(b) $\cos ix = \dfrac{e^{x} + e^{-x}}{2} = \cos hx$ is always positive real $\forall\ x \in R$ and is > 1.

and, $\sin ix = i\ \dfrac{e^{x} - e^{-x}}{2} = i\sin hx$ is always purely imaginary.

4.5 Vector Representation

The knowledge of vectors can also be used to represent a complex number

$z = x + iy$. The vector $\overrightarrow{OP}$, joining the origin O of the complex plane to the

point P (x, y), is the vector representation of the complex number z=x+iy,

(see Fig 6.9). The length of the vector $\overrightarrow{OP}$, that is, $|\overrightarrow{OP}|$ is the modulus of z.

The angle between the positive real axis and the vector $\overrightarrow{OP}$, more exactly, the

angle through which the positive real axis must be rotated to cause it to

have the same direction as $\overrightarrow{OP}$ (considered positive if the rotation is

counter-clockwise and negative otherwise) is the argument of the complex
number z.

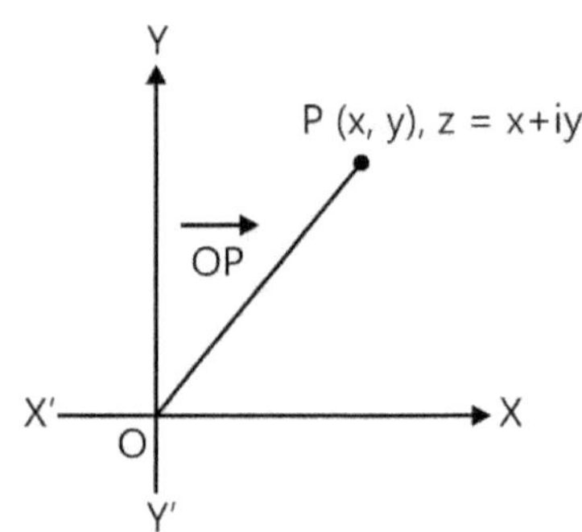

Figure 6.9 Vector representation

Illustration 12: Find locus represented by $\mathrm{Re}\left(\dfrac{1}{x+iy}\right) < \dfrac{1}{2}$. (JEE MAIN)

Sol: Multiplying numerator and denominator by $x - iy$.

We have, $\mathrm{Re}\left(\dfrac{1}{x+iy}\right) < \dfrac{1}{2}$ $\qquad \mathrm{Re}\left(\dfrac{x-iy}{x^2+y^2}\right) < \dfrac{1}{2}$

$\Rightarrow \dfrac{x}{x^2+y^2} < \dfrac{1}{2}$ $\qquad \Rightarrow x^2 + y^2 - 2x > 0$

Locus is the exterior of the circle with centre (1, 0) and radius = 1.

Illustration 13: If $z = 1 + \cos\dfrac{6\pi}{5} + i\sin\dfrac{6\pi}{5}$. Find r and amp z. (JEE MAIN)

Sol: By using trigonometric formula we can reduce given equation in the form of $z = r(\cos\theta + i\sin\theta)$.

$z = 2\cos^2\dfrac{3\pi}{5} + 2i\sin\dfrac{3\pi}{5}\cos\dfrac{3\pi}{5} = 2\cos\dfrac{3\pi}{5}\left[\cos\dfrac{3\pi}{5} + i\sin\dfrac{3\pi}{5}\right]$

$= -2\cos\dfrac{2\pi}{5}\left[-\cos\dfrac{2\pi}{5} + i\sin\dfrac{2\pi}{5}\right] = 2\cos\dfrac{2\pi}{5}\left[\cos\dfrac{2\pi}{5} - i\sin\dfrac{2\pi}{5}\right]$ Hence, $|z| = 2\cos\dfrac{2\pi}{5}$; amp z $= -\dfrac{2\pi}{5}$

Illustration 14: Show that the locus of the point $P(\omega)$ denoting the complex number $z + \dfrac{1}{z}$ on the complex plane is a standard ellipse where $|z| = a$, where $a \neq 0, 1$. **(JEE ADVANCED)**

Sol: Here consider $w = x + iy$ and $z = \alpha + i\beta$ and then solve this by using algebra of complex number.

Let $w = z + \dfrac{1}{z}$ where $z = \alpha + i\beta$, $\alpha^2 + \beta^2 = a^2$ (as $|z| = a$)

$$x + iy = \alpha + i\beta + \frac{1}{\alpha + i\beta} = \alpha + i\beta + \frac{\alpha - i\beta}{\alpha^2 + \beta^2} = \left(\alpha + \frac{\alpha}{a^2}\right) + i\left(\beta - \frac{\beta}{a^2}\right) \quad \therefore x = \alpha\left(1 + \frac{1}{a^2}\right); \; y = \beta\left(1 - \frac{1}{a^2}\right)$$

$$\therefore \frac{x^2}{\left(1 + \dfrac{1}{a^2}\right)^2} + \frac{y^2}{\left(1 - \dfrac{1}{a^2}\right)^2} = \alpha^2 + \beta^2 = a^2; \qquad\qquad \therefore \frac{x^2}{\left(a + \dfrac{1}{a}\right)^2} + \frac{y^2}{\left(a - \dfrac{1}{a}\right)^2} = 1.$$

5. IMPORTANT PROPERTIES OF CONJUGATE, MODULUS AND ARGUMENT

For z, z_1 and $z_2 \in C$,

(a) Properties of Conjugate:

    (i) $\quad z + \bar{z} = 2\text{Re}\,(z)$

    (ii) $\quad z - \bar{z} = 2i\,\text{Im}\,(z)$

    (iii) $\overline{(\bar{z})} = z$

    (iv) $\overline{z_1 + z_2} = \bar{z}_1 + \bar{z}_2$

    (v) $\overline{z_1 - z_2} = \bar{z}_1 - \bar{z}_2$

    (vi) $\overline{z_1 z_2} = \bar{z}_1 \cdot \bar{z}_2$

    (vii) $\overline{\left(\dfrac{z_1}{z_2}\right)} = \dfrac{\bar{z}_1}{\bar{z}_2}; \; z_2 \neq 0$

(b) Properties of Modulus:

    (i) $\quad |z| \geq 0; |z| \geq \text{Re}(z); |z| \geq \text{Im}(z); |z| = |\bar{z}| = |-z|$

    (ii) $\quad z\bar{z} = |z|^2; \text{ if } |z| = 1, \text{ then } z = \dfrac{1}{\bar{z}}$

    (iii) $|z_1 z_2| = |z_1| \cdot |z_2|$

    (iv) $\left|\dfrac{z_1}{z_1}\right| = \dfrac{|z_1|}{|z_2|}, \; z_2 \neq 0$

    (v) $|z^n| = |z|^n$

    (vi) $|z_1 + z_2|^2 + |z_1 - z_2|^2 = 2\left[|z_1|^2 + |z_2|^2\right]$

    (vii) $\big||z_1| - |z_2|\big| \leq |z_1 + z_2| \leq |z_1| + |z_2|$ [Triangle Inequality]

(c) Properties of Amplitude:

(i) $\text{amp}(z_1 \cdot z_2) = \text{amp } z_1 + \text{amp } z_2 + 2k\pi, \ k \in I$

(ii) $\text{amp}\left(\dfrac{z_1}{z_2}\right) = \text{amp } z_1 - \text{amp } z_2 + 2k\pi, \ k \in I$

(iii) $\text{amp}(z^n) = n \, \text{amp}(z) + 2k\pi$, where the value of k should be such that RHS lies in $(-\pi, \pi]$

Based on the above information, we have the following

- $|\text{Re}(z)| + |\text{Im}(z)| \le \sqrt{2}\,|z|$

- $\big||z_1| - |z_2|\big| \le |z_1 - z_2| \le |z_1| + |z_2|$. Thus $|z_1| + |z_2|$ is the greatest possible value of $|z_1 + z_2|$ and $\big||z_1| - |z_2|\big|$ is the least possible value of $|z_1 + z_2|$.

- If $\left|z + \dfrac{1}{z}\right| = a$, the greatest and least values of |z| are respectively $\dfrac{a + \sqrt{a^2 + 4}}{2}$ and $\dfrac{-a + \sqrt{a^2 + 4}}{2}$.

- $|z_1 + \sqrt{z_1^2 - z_2^2}\,| + |z_2 - \sqrt{z_1^2 - z_2^2}\,| = |z_1 + z_2| + |z_1 - z_2|$

- If $z_1 = z_2 \Leftrightarrow |z_1| = |z_2|$ and $\arg z_1 = \arg z_2$

- $|z_1 + z_2| = |z_1| + |z_2| \Leftrightarrow \arg(z_1) = \arg(z_2)$ i.e. z_1 and z_2 are parallel.

- $|z_1 + z_2| = |z_1| + |z_2| \Leftrightarrow \arg(z_1) - \arg(z_2) = 2n\pi$, where n is some integer.

- $|z_1 - z_2| = \big||z_1| - |z_2|\big| \Leftrightarrow \arg(z_1) - \arg(z_2) = 2n\pi$, where n is some integer.

- $|z_1 + z_2| = |z_1 - z_2| \Leftrightarrow \arg(z_1) - \arg(z_2) = (2n+1)\dfrac{\pi}{2}$, where n is some integer.

- If $|z_1| \le 1, |z_2| \le 1$, then $|z_1 + z_2|^2 \le \big(|z_1| - |z_2|\big)^2 + \big(\arg(z_1) - \arg(z_2)\big)^2$, and $|z_1 + z_2|^2 \ge \big(|z_1| + |z_2|\big)^2 - \big(\arg(z_1) - \arg(z_2)\big)^2$.

Illustration 15: If $z_1 = 3 + 5i$ and $z_2 = 2 - 3i$, then verify that $\overline{\left(\dfrac{z_1}{z_2}\right)} = \dfrac{\overline{z_1}}{\overline{z_2}}$ **(JEE MAIN)**

Sol: Simply by using properties of conjugate.

$$\dfrac{z_1}{z_2} = \dfrac{3+5i}{2-3i} = \dfrac{(3+5i)}{(2-3i)} \times \dfrac{(2+3i)}{(2+3i)} = \dfrac{6+9i+10i+15i^2}{4-9i^2} = \dfrac{6+19i+15(-1)}{4+9} = \dfrac{6+19i-15}{13} = \dfrac{-9+19i}{13} = \dfrac{-9}{13} + \dfrac{19}{13}i$$

$$\text{L.H.S.} = \overline{\left(\dfrac{z_1}{z_2}\right)} = \overline{\left(-\dfrac{9}{13} + \dfrac{19}{13}i\right)} = -\dfrac{9}{13} - \dfrac{19}{13}i$$

$$\text{R.H.S.} = \dfrac{\overline{z_1}}{\overline{z_2}} = \dfrac{\overline{3+5i}}{\overline{2-3i}} = \dfrac{3-5i}{2+3i} = \dfrac{(3-5i)}{(2+3i)} \times \dfrac{(2-3i)}{(2-3i)}$$

$$= \dfrac{6-9i-10i+15i^2}{4-9i^2} = \dfrac{6-19i+15(-1)}{4+9} = \dfrac{6-19i-15}{13} = -\dfrac{9}{13} - \dfrac{19}{13}i \qquad \therefore \ \overline{\left(\dfrac{z_1}{z_2}\right)} = \dfrac{\overline{z_1}}{\overline{z_2}}$$

Illustration 16: If z be a non-zero complex number, then show that $\overline{(z^{-1})} = (\overline{z})^{-1}$. **(JEE MAIN)**

Sol: By considering z = a + ib and using properties of conjugate we can prove given equation.

Let z = a + ib   Since, $z \neq 0$, we have $x^2 + y^2 > 0$

$$z^{-1} = \frac{1}{z} = \frac{1}{a+ib} = \frac{1}{a+ib} \times \frac{a-ib}{a-ib} = \frac{a}{a^2+b^2} - \frac{ib}{a^2+b^2} \Rightarrow \overline{\left(z^{-1}\right)} = \frac{a}{a^2+b^2} + \frac{ib}{a^2+b^2} \qquad \text{... (i)}$$

and $(\overline{z})^{-1} = \dfrac{1}{\overline{z}} = \dfrac{1}{\overline{a+ib}} = \dfrac{1}{a-ib} = \dfrac{1}{a-ib} \times \dfrac{a+ib}{a+ib} = \dfrac{a}{a^2+b^2} + i\dfrac{b}{a^2+b^2}$ $\qquad \text{... (ii)}$

From (i) and (ii), we get $\overline{(z^{-1})} = (\overline{z})^{-1}$.

Illustration 17: If $\dfrac{(a+i)^2}{2a-i} = p + iq$, then show that $p^2 + q^2 = \dfrac{(a^2+1)^2}{4a^2+1}$. **(JEE MAIN)**

Sol: Multiply given equation to its conjugate.

We have, p + iq = $\dfrac{(a+i)^2}{2a-i}$ $\qquad \text{... (i)}$

Taking conjugate of both sides, we get $\overline{p+iq} = \overline{\left(\dfrac{(a+i)^2}{(2a-i)}\right)}$

$$\Rightarrow p - iq = \frac{\overline{(a+i)^2}}{\overline{(2a-i)}} \quad \left[\because \overline{\left(\frac{z_1}{z_2}\right)} = \frac{\overline{z}_1}{\overline{z}_2}\right] \qquad \Rightarrow p - iq = \frac{(a-i)^2}{(2a+i)} \quad \text{... (ii)} \left[\text{using } \overline{(z^2)} = \overline{z \cdot z} = \overline{z} \cdot \overline{z} = (\overline{z})^2\right]$$

Multiplying (i) and (ii), we get (p + iq) (p – iq) = $\left(\dfrac{(a+i)^2}{2a-i}\right)\left(\dfrac{(a-i)^2}{2a+i}\right)$

$$\Rightarrow p^2 - i^2 q^2 = \frac{(a^2-i^2)^2}{4a^2-i^2} \qquad \Rightarrow p^2 + q^2 = \frac{(a^2+1)^2}{4a^2+1}.$$

Illustration 18: Let $z_1, z_2, z_3, \ldots \ldots z_n$ are the complex numbers such that $|z_1| = |z_2| = \ldots \ldots = |z_n| = 1$. If z =

$\left(\displaystyle\sum_{k=1}^{n} z_k\right)\left(\displaystyle\sum_{k=1}^{n} \dfrac{1}{z_k}\right)$ then prove that

(i) z is a real number $\qquad\qquad$ (ii) $0 < z \leq n^2$ **(JEE ADVANCED)**

Sol: Here $|z_1| = |z_2| = \ldots \ldots = |z_n| = 1$, therefore $z\overline{z} = 1 \Rightarrow \overline{z} = \dfrac{1}{z}$. Hence by substituting this to z = $\left(\displaystyle\sum_{k=1}^{n} z_k\right)\left(\displaystyle\sum_{k=1}^{n} \dfrac{1}{z_k}\right)$, we can solve above problem.

Now, z = $(z_1 + z_2 + z_3 + \ldots\ldots + z_n)\left(\dfrac{1}{z_1} + \dfrac{1}{z_2} + \ldots\ldots + \dfrac{1}{z_n}\right)$

$= (z_1 + z_2 + z_3 + \ldots\ldots + z_n)(\overline{z}_1 + \overline{z}_2 + \ldots\ldots + \overline{z}_n)$ $\qquad = (z_1 + z_2 + z_3 + \ldots\ldots + z_n)\left(\overline{z_1 + z_2 + \ldots\ldots + z_n}\right)$

$= |z_1 + z_2 + z_3 + \ldots\ldots + z_n|^2$ which is real

$\leq \left(|z_1| + |z_2| + |z_3| + \ldots\ldots + |z_n|\right)^2 = n^2$ $\qquad\qquad \therefore \qquad 0 < z \leq n^2$.

Illustration 19: Let x_1, x_2 are the roots of the quadratic equation $x^2 + ax + b = 0$ where a, b are complex numbers and y_1, y_2 are the roots of the quadratic equation $y^2 + |a| y + |b| = 0$. If $|x_1| = |x_2| = 1$, then prove that $|y_1| = |y_2| = 1$.

(JEE ADVANCED)

Sol: Solve by using modulus properties of complex number.

Let $x^2 + ax + b = 0$ where x_1 and x_2 are complex numbers

$$x_1 + x_2 = -a \qquad\qquad \text{... (i)}$$

$$\text{and } x_1 x_2 = b \qquad\qquad \text{... (ii)}$$

From (ii) $|x_1| |x_2| = |b| \Rightarrow |b| = 1$ $\qquad\qquad$ Also $|-a| = |x_1 + x_2|$

$\therefore \qquad |a| \le |x_1| + |x_2| \qquad$ or $\qquad |a| \le 2$

Now consider $y^2 + |a| y + |b| = 0$, $\begin{cases} y_1 \\ y_2 \end{cases}$ where y_1 and y_2 are complex numbers

$$y_{1,2} = \frac{-|a| \pm \sqrt{|a|^2 - 4|b|}}{2} = \frac{-|a| \pm \left(\sqrt{4 - |a|^2}\right) i}{2} \qquad \therefore \qquad |y_{1,2}| = \frac{\sqrt{|a|^2 + 4 - |a|^2}}{2} = 1$$

Hence, $|y_1| = |y_2| = 1$.

6. TRIANGLE ON COMPLEX PLANE

In a $\triangle ABC$, the vertices A, B and C are represented by the complex numbers z_1, z_2 and z_3 respectively, then

(a) Centroid: The centroid 'G' is given by $\dfrac{z_1 + z_2 + z_3}{3}$. Refer to Fig 6.10.

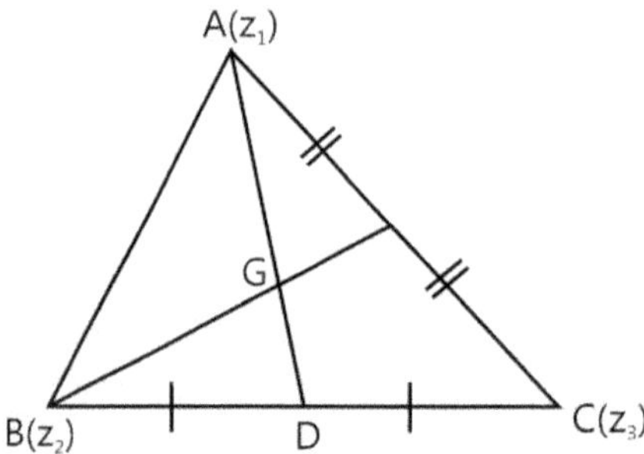

Figure 6.10: Centroid

(b) Incentre: The incentre 'I' is given by $\dfrac{a z_1 + b z_2 + c z_3}{a + b + c}$. Refer to Fig 6.11.

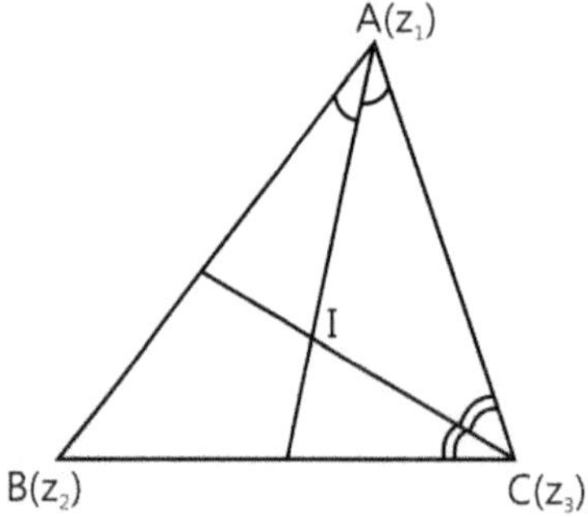

Figure 6.11: Incentre

(c) **Orthocentre:** The orhtocentre 'H' is given by $\dfrac{z_1 \tan A + z_2 \tan B + z_3 \tan C}{\sum \tan A}$.

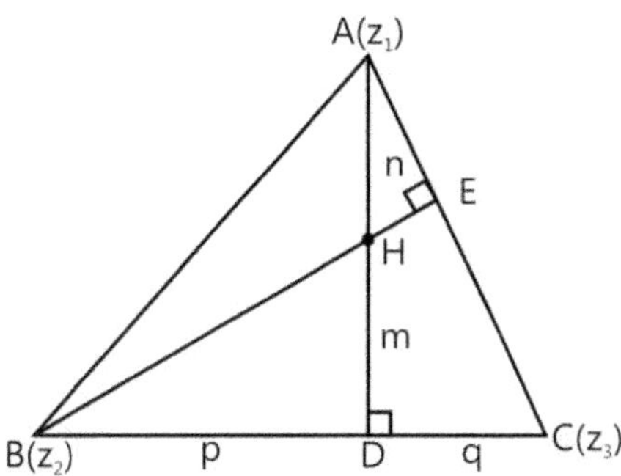

Figure 6.12: Orthocentre

Proof: From section formula, we have $z_D = \dfrac{p\, z_3 + q\, z_2}{a}$

In ΔABD and ΔACD, $p = c \cos B$ and $q = b \cos C$. Refer to Fig 6.12.

Therefore, $z_D = \dfrac{b \cos C\, z_2 + c \cos B\, z_3}{a}$

Now, $AE = c \cos A$; $n = AH = AE \operatorname{cosec} C = c \cos A \operatorname{cosec} C$

$\Rightarrow n = 2R \cos A$ \qquad\qquad [Using Sine Rule]

and $m = c \cos B \cot C$ \qquad or, $m = 2R \cos B \cos C$ \qquad\qquad [Using Sine Rule]

Hence, $z_H = \dfrac{mz_1 + n Z_D}{m + n}$.

$$= \frac{2R \cos B \cos C z_1 + 2R \cos A \left(\dfrac{b \cos C\, z_2 + c \cos B\, z_3}{a} \right)}{2R\,(\cos A + \cos B \cos C)}$$

$$= \frac{a \cos B \cos C z_1 + b \cos A \cos C z_2 + c \cos A \cos B z_3}{a(-\cos(B + C) + \cos B \cos C)}$$

$$= \frac{z_1 (\sin A \cos B \cos C) + z_2 (\sin B \cos C \cos A) + z_3 (\sin C \cos A \cos B)}{\sin A\,(\sin B \sin C)}$$

$$\therefore\ Z_H = \frac{z_1 \tan A + z_2 \tan B + z_3 \tan C}{\sum \tan A} \quad \text{or} \quad \frac{z_1 \tan A + z_2 \tan B + z_3 \tan C}{\prod \tan A}$$

[If $A + B + C = \pi$, then $\tan A + \tan B + \tan C = \tan A \tan B \tan C$]

(d) **Circumcentre:**

Let R be the circumradius and the complex number z_0 represent the circumcentre of the triangle as shown in Fig 6.11.

$\therefore\ |z_1 - z_0| = |z_2 - z_0| = |z_3 - z_0|$

Consider, $|z_1 - z_0|^2 = |z_2 - z_0|^2$

$(z_1 - z_0)(\overline{z}_1 - \overline{z}_0) = (z_2 - z_0)(\overline{z}_2 - \overline{z}_0)$

$$\overline{z}_1(z_1 - z_0) - \overline{z}_2(z_2 - z_0) = \overline{z}_0\left[(z_1 - z_0) - (z_2 - z_0)\right]$$

$$\overline{z}_1(z_1 - z_0) - \overline{z}_2(z_2 - z_0) = \overline{z}_0(z_1 - z_2) \qquad \text{... (i)}$$

Similarly 1st and 3rd gives

$$\overline{z}_1(z_1 - z_0) - \overline{z}_3(z_3 - z_0) = \overline{z}_0(z_1 - z_3) \qquad \text{... (ii)}$$

On dividing (i) by (ii), $\overline{z}_0$ gets eliminated and we obtain z_0.

Alternatively: From Fig 6.13, we have

$$\frac{BD}{DC} = \frac{m}{n} = \frac{Ar.\ \Delta ABD}{Ar.\ \Delta ADC} = \frac{Ar.\ \Delta PBD}{Ar.\ \Delta PDC}$$

$$\therefore \frac{m}{n} = \frac{Ar.\ \Delta ABD - Ar.\ \Delta PBD}{Ar.\ \Delta ADC - Ar.\ \Delta PDC} = \frac{\Delta_3}{\Delta_2}$$

$$\therefore \frac{m}{n} = \frac{\dfrac{R^2}{2}\sin 2C}{\dfrac{R^2}{2}\sin 2B} = \frac{\sin 2C}{\sin 2B}$$

Hence, $Z_D = \dfrac{\sin 2B\,(z_2) + \sin 2C\,(z_3)}{\sin 2B + \sin 2C}$

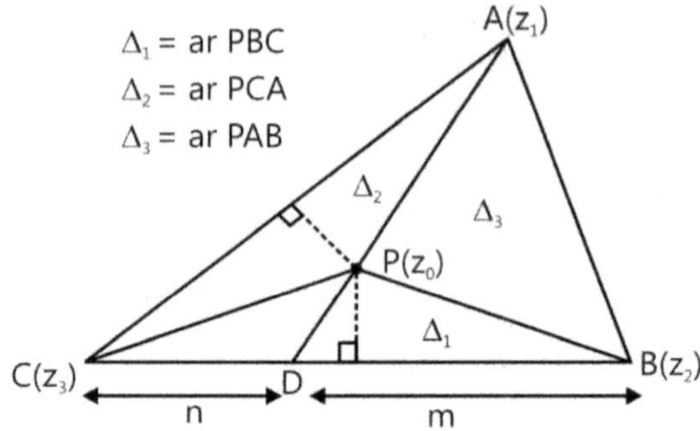

Figure 6.13: Circumcentre

Now, $\dfrac{PA}{PD} = \dfrac{l}{k} = \dfrac{\Delta ABP}{\Delta PBD} = \dfrac{\Delta APC}{\Delta CPD} = \dfrac{\Delta ABP + \Delta APC}{\Delta PBD + \Delta CPD} \quad \therefore \dfrac{l}{k} = \dfrac{\Delta_3 + \Delta_2}{\Delta_1} = \dfrac{\sin 2C + \sin 2B}{\sin 2A}$

Hence, $z_0 = \dfrac{kz_1 + l z_D}{k + l} = \dfrac{z_1 \sin 2A + z_2 \sin 2B + z_3 \sin 2C}{\sum \sin 2A}$

NOMORECLASS CONCEPTS

- The area of the triangle whose vertices are z, iz and z + iz is $\dfrac{1}{2}|z|^2$.

- The area of the triangle with vertices z, ωz and z + ωz is $\dfrac{\sqrt{3}}{4}|z|^2$.

- If z_1, z_2, z_3 be the vertices of an equilateral triangle and z_0 be the circumcentre, then $z_1^2 + z_2^2 + z_3^2 = 3z_0^2$.

- If $z_1, z_2, z_3, \ldots\ldots z_n$ be the vertices of a regular polygon of n sides and z_0 be its centroid, then $z_1^2 + z_2^2 + \ldots\ldots + z_n^2 = nz_0^2$.

- If z_1, z_2, z_3 be the vertices of a triangle, then the triangle is equilateral if $(z_1 - z_2)^2 + (z_2 - z_3)^2 + (z_3 - z_1)^2 = 0$

  or $z_1^2 + z_2^2 + z_3^2 = z_1 z_2 + z_2 z_3 + z_3 z_1$ or $\dfrac{1}{z_1 - z_2} + \dfrac{1}{z_2 - z_3} + \dfrac{1}{z_3 - z_1} = 0$.

- If z_1, z_2, z_3 are the vertices of an isosceles triangle, right angled at z_2 then $z_1^2 + 2z_2^2 + z_3^2 = 2z_2(z_1 + z_3)$.

- If z_1, z_2, z_3 are the vertices of a right-angled isosceles triangle, then $(z_1 - z_2)^2 = 2(z_1 - z_3)(z_3 - z_2)$.

- If z_1, z_2, z_3 be the affixes of the vertices A, B, C respectively of a triangle ABC, then its orthocentre is $\dfrac{a\,(\sec A)z_1 + b\,(\sec B)z_2 + c\,(\sec C)z_3}{a \sec A + b \sec B + c \sec C}$.

Illustration 20: If z_1, z_2, z_3 are the vertices of an isosceles triangle right angled at z_2 then prove that $z_1^2 + 2z_2^2 + z_3^2 = 2z_2(z_1 + z_3)$ **(JEE MAIN)**

Sol: Here $(z_1 - z_2) = (z_3 - z_2)e^{\frac{i\pi}{2}}$. Hence by squaring both side we will get the result.

$\Rightarrow (z_1 - z_2)^2 = i^2(z_3 - z_2)^2$

$\Rightarrow z_3^2 + z_2^2 - 2z_3z_2 = -z_1^2 - z_2^2 + 2z_1z_2 \Rightarrow z_1^2 + 2z_2^2 + z_3^2 = 2z_2(z_1 + z_3)$.

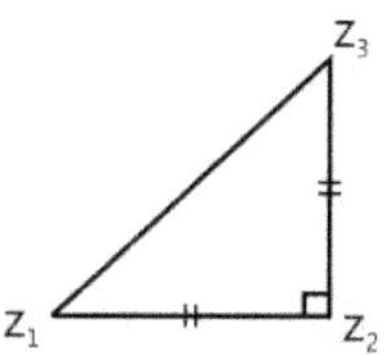

Figure 6.14

Illustration 21: A, B, C are the points representing the complex numbers z_1, z_2, z_3 respectively and the circumcentre of the triangle ABC lies at the origin. If the altitudes of the triangle through the opposite vertices meets the circumcircle at D, E, F respectively. Find the complex numbers corresponding to the points D, E, F in terms of z_1, z_2, z_3. **(JEE MAIN)**

Sol: Here the $\angle BOD = \pi - 2B$, hence $\overrightarrow{OD} = \overrightarrow{OB}\, e^{i(\pi - 2B)}$.

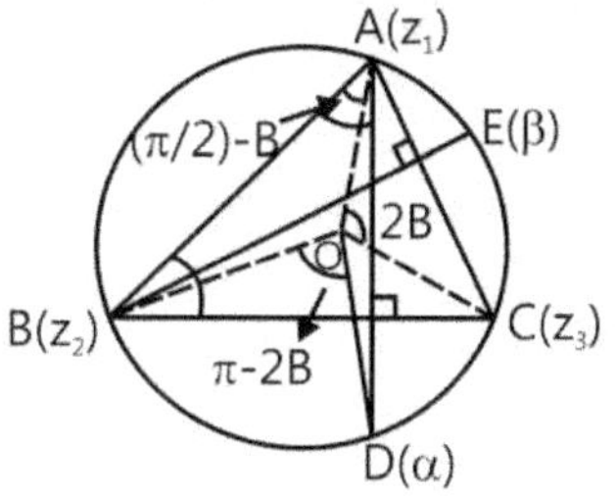

From Fig 6.13, we have $\overrightarrow{OD} = \overrightarrow{OB}\, e^{i(\pi - 2B)}$;

$\alpha = z_2\, e^{i(\pi - 2B)} = -z_2\, e^{-i2B}$ ... (i)

also, $z_1 = z_3\, e^{i\,2B}$ ... (ii)

$\therefore \alpha z_1 = -z_2 z_3 \qquad \Rightarrow \alpha = \dfrac{-z_2 z_3}{z_1}$

Similarly, $\beta = \dfrac{-z_3 z_1}{z_2}$ and $\gamma = \dfrac{-z_1 z_2}{z_3}$.

Figure 6.15

Illustration 22: If z_r $(r = 1, 2, ...,6)$ are the vertices of a regular hexagon then prove that $\sum_{r=1}^{6} z_r^2 = 6z_0^2$, where z_0 is the circumcentre of the regular hexagon.

(JEE MAIN)

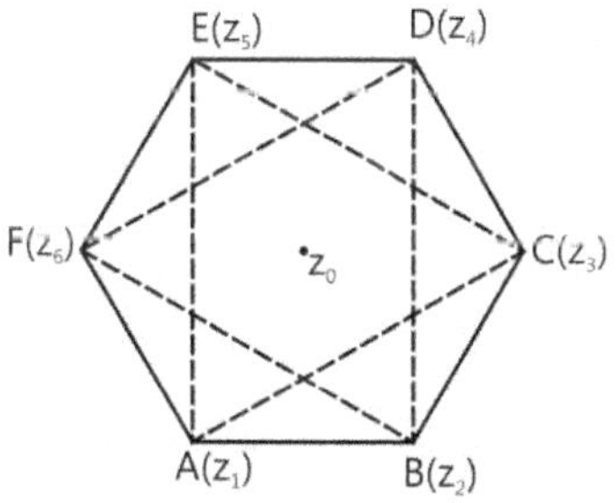

Sol: As we know If $z_1, z_2, z_3, ...z_n$ be the vertices of a regular polygon of n sides and z_0 be its centroid, then $z_1^2 + z_2^2 + ... + z_n^2 = nz_0^2$.

Here by the Fig 6.14,

$3z_0^2 = z_1^2 + z_3^2 + z_5^2$

and, $3z_0^2 = z_2^2 + z_4^2 + z_6^2 \qquad \Rightarrow 6z_0^2 = \sum_{r=1}^{6} z_r^2$.

Figure 6.16

Illustration 23: If z_1, z_2, z_3 are the vertices of an equilateral triangle then prove that $z_1^2 + z_2^2 + z_3^2 = z_1z_2 + z_2z_3 + z_3z_1$ and if z_0 is its circumcentre then $3z_0^2 = z_1^2 + z_2^2 + z_3^2$. **(JEE ADVANCED)**

Sol: By using triangle on complex plane we can prove

$z_1^2 + z_2^2 + z_3^2 = z_1z_2 + z_2z_3 + z_3z_1$ and by using $z_0 = \dfrac{z_1 + z_2 + z_3}{3}$ we can prove $3z_0^2 = z_1^2 + z_2^2 + z_3^2$.

To Prove, $z_1^2 + z_2^2 + z_3^2 = z_1z_2 + z_2z_3 + z_3z_1$

As seen in the Fig 6.17,

$\therefore \dfrac{z_1 - z_2}{z_2 - z_3} = \dfrac{(z_3 - z_2)\, e^{\frac{i\pi}{3}}}{(z_1 - z_3)\, e^{\frac{i\pi}{3}}} \Rightarrow (z_1 - z_2)(z_1 - z_3) = -(z_2 - z_3)^2$

$\Rightarrow z_1^2 - z_1z_3 - z_2z_1 + z_2z_3 + z_2^2 + z_3^2 - 2z_2z_3 = 0 \quad \therefore \sum z_1^2 = \sum z_1z_2$

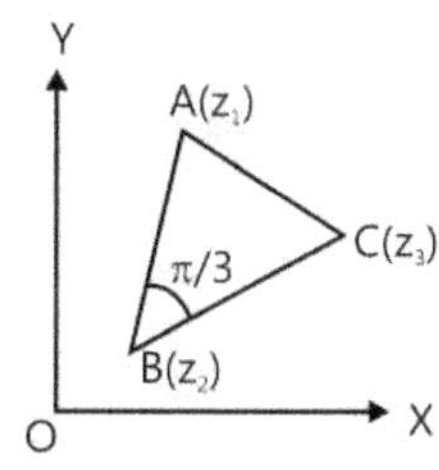

Figure 6.17

Now if z_0 is the circumcentre of the Δ, then we need to prove $3z_0^2 = z_1^2 + z_2^2 + z_3^2$.

Since in an equilateral triangle, the circumcentre coincides with the centroid, we have $z_0 = \dfrac{z_1 + z_2 + z_3}{3}$

$\Rightarrow (z_1 + z_2 + z_3)^2 = \left(3z_0\right)^2$

$\Rightarrow \sum z_1^2 + 2\sum z_1 z_2 = 9z_0^2 \quad \therefore 3\sum z_1^2 = 9z_0^2.$

Illustration 24: Prove that the triangle whose vertices are the points z_1, z_2, z_3 on the Argand plane is an equilateral triangle if and only if $\dfrac{1}{z_2 - z_3} + \dfrac{1}{z_3 - z_1} + \dfrac{1}{z_1 - z_2} = 0.$ **(JEE ADVANCED)**

Sol: Consider ABC is the equilateral triangle with vertices z_1, z_2 and z_3 respectively.

Therefore $|z_2 - z_3| = |z_3 - z_1| = |z_1 - z_2|$.

Let ABC be a triangle such that the vertices A, B and C are z_1, z_2 and z_3 respectively.

Further, let $\alpha = z_2 - z_3$, $\beta = z_3 - z_1$ and $\gamma = z_1 - z_2$. Then $\alpha + \beta + \gamma = 0$ ... (i)

As shown in Fig 6.16, let ΔABC be an equilateral triangle. Then, BC = CA = AB

$\Rightarrow |z_2 - z_3| = |z_3 - z_1| = |z_1 - z_2| \quad \Rightarrow |\alpha| = |\beta| = |\gamma|$

$\Rightarrow |\alpha|^2 = |\beta|^2 = |\gamma|^2 = \lambda \text{(say)}$

$\Rightarrow \alpha\bar{\alpha} = \beta\bar{\beta} = \gamma\bar{\gamma} = \lambda$

$\Rightarrow \bar{\alpha} = \dfrac{\lambda}{\alpha}, \bar{\beta} = \dfrac{\lambda}{\beta}, \bar{\gamma} = \dfrac{\lambda}{\gamma}$ ... (ii)

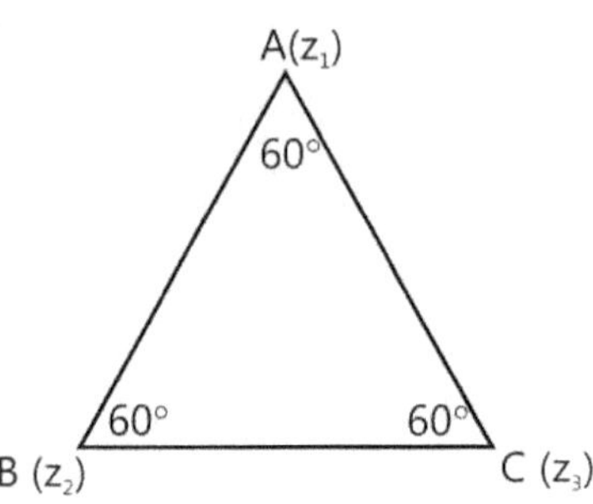

Figure 6.18

Now, $\alpha + \beta + \gamma = 0$ [from (i)]

$\Rightarrow \bar{\alpha} + \bar{\beta} + \bar{\gamma} = 0 \qquad \Rightarrow \dfrac{\lambda}{\alpha} + \dfrac{\lambda}{\beta} + \dfrac{\lambda}{\gamma} = 0 \qquad \text{[Using (ii)]}$

$\Rightarrow \dfrac{1}{\alpha} + \dfrac{1}{\beta} + \dfrac{1}{\gamma} = 0 \Rightarrow \dfrac{1}{z_2 - z_3} + \dfrac{1}{z_3 - z_1} + \dfrac{1}{z_1 - z_2} = 0$ which is the required condition.

Conversely, let ABC be a triangle such that

$\Rightarrow \dfrac{1}{z_2 - z_3} + \dfrac{1}{z_3 - z_1} + \dfrac{1}{z_1 - z_2} = 0 \qquad \text{i.e. } \Rightarrow \dfrac{1}{\alpha} + \dfrac{1}{\beta} + \dfrac{1}{\gamma} = 0$

Thus, we have to prove that the triangle is equilateral. We have, $\dfrac{1}{\alpha} + \dfrac{1}{\beta} + \dfrac{1}{\gamma} = 0$

$\Rightarrow \dfrac{1}{\alpha} = -\left(\dfrac{1}{\beta} + \dfrac{1}{\gamma}\right) \qquad \Rightarrow \dfrac{1}{\alpha} = -\left(\dfrac{\beta + \gamma}{\beta\gamma}\right) \qquad \Rightarrow \dfrac{1}{\alpha} = \dfrac{\alpha}{\beta\gamma} \qquad \Rightarrow \alpha^2 = \beta\gamma \qquad \Rightarrow |\alpha|^2 = |\beta\gamma|$

$\Rightarrow |\alpha|^2 = |\beta||\gamma| \qquad \Rightarrow |\alpha|^3 = |\alpha||\beta||\gamma|$

Similarly, $\Rightarrow |\beta|^3 = |\alpha||\beta||\gamma|$ and $|\gamma|^3 = |\alpha||\beta||\gamma|$

$\therefore \qquad |\alpha| = |\beta| = |\gamma|$

$\Rightarrow |z_2 - z_3| = |z_3 - z_1| = |z_1 - z_2| \quad \Rightarrow BC = CA = AB$

Hence, the given triangle is an equilateral triangle.

Illustration 25: Prove that the roots of the equation $\dfrac{1}{z-z_1}+\dfrac{1}{z-z_2}+\dfrac{1}{z-z_3}=0$ (where z_1,z_2,z_3 are pair wise distinct complex numbers) correspond to points on a complex plane, which lie inside a triangle with vertices z_1,z_2,z_3 excluding its boundaries. **(JEE ADVANCED)**

Sol: By using modulus and conjugate properties we can reduce given expression as $\dfrac{\overline{z}-\overline{z}_1}{|z-z_1|^2}+\dfrac{\overline{z}-\overline{z}_2}{|z-z_2|^2}+\dfrac{\overline{z}-\overline{z}_3}{|z-z_3|^2}$

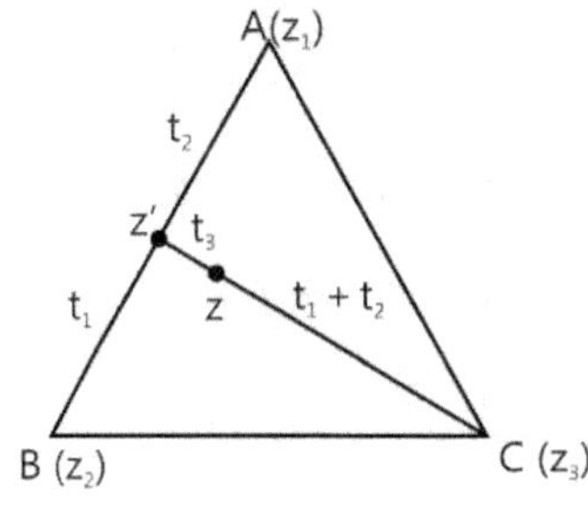

Figure 6.19

$=0$. Therefore by putting $|z-z_i|^2=\dfrac{1}{t_i}$, where i = 1, 2 and 3, we will get the result.

$t_1(\overline{z}-\overline{z}_1)+t_2(\overline{z}-\overline{z}_2)+t_3(\overline{z}-\overline{z}_3)=0$ where $|z-z_1|^2=\dfrac{1}{t_1}$ etc and $t_1,t_2,t_3\in R^+$

$t_1(z-z_1)+t_2(z-z_2)+t_3(z-z_3)=0$

$(t_1+t_2+t_3)z=t_1z_1+t_2z_2+t_3z_3$ $\Rightarrow z=\dfrac{t_1z_1+t_2z_2+t_3z_3}{t_1+t_2+t_3}$

$\Rightarrow z=\dfrac{t_1z_1+t_2z_2}{t_1+t_2}\cdot\dfrac{t_1+t_2}{t_1+t_2+t_3}+\dfrac{t_3z_3}{t_1+t_2+t_3}=\dfrac{t_1+t_2}{t_1+t_2+t_3}z'+\dfrac{t_3z_3}{t_1+t_2+t_3}$

$\Rightarrow z=\dfrac{(t_1+t_2)z'+t_3z_3}{t_1+t_2+t_3}$ $\Rightarrow z$ lies inside the $\Delta z_1z_2z_3$

If $t_1=t_2=t_3$ $\Rightarrow z$ is the centroid of the triangle.

Also, it implies $|z-z_1|=|z-z_2|=|z-z_3|$ $\Rightarrow z$ is the circumcentre.

Illustration 26: Let z_1 and z_2 be roots of the equation $z^2+pz+q=0$, where the coefficients p and q may be complex numbers. Let A and B represent z_1 and z_2 in the complex plane. If $\angle AOB=\alpha\neq0$ and OA = OB, where O is the origin, prove that $p^2=4q\cos^2\dfrac{\alpha}{2}$. **(JEE ADVANCED)**

Sol: Here $\overrightarrow{OB}=\overrightarrow{OA}e^{i\alpha}$. Therefore by using formula of sum and product of roots of quadratic equation we can prove this problem.

Since z_1 and z_2 are roots of the equation $z^2+pz+q=0$

$z_1+z_2=-p$ and $z_1z_2=q$ (1)

Since OA = OB. So $\overrightarrow{OB}$ is obtained by rotating $\overrightarrow{OA}$ in anticlockwise direction through angle α.

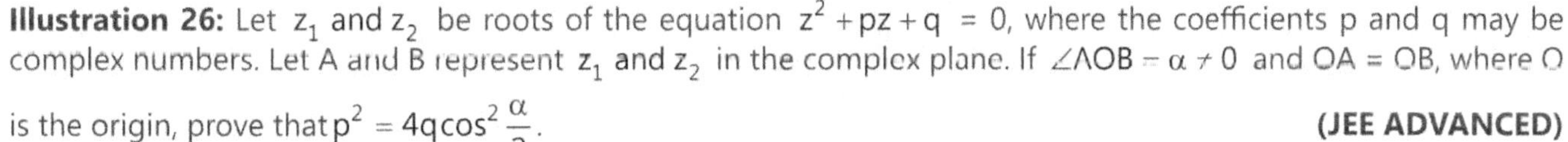

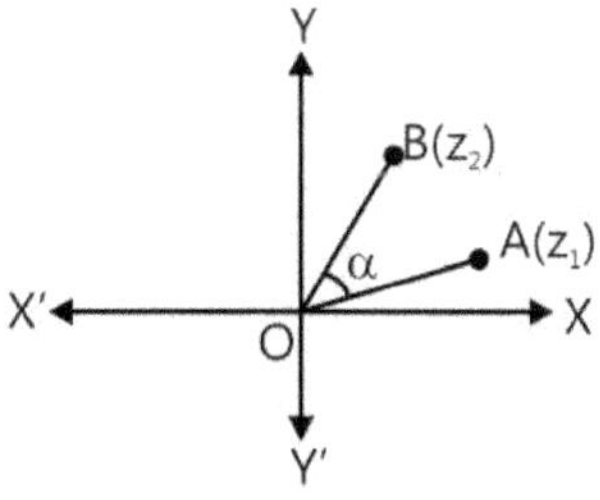

Figure 6.20

$\therefore \overrightarrow{OB}=\overrightarrow{OA}e^{i\alpha}$ $\Rightarrow z_2=z_1e^{i\alpha}$ $\Rightarrow\dfrac{z_2}{z_1}=e^{i\alpha}\Rightarrow\dfrac{z_2}{z_1}=\cos\alpha+i\sin\alpha$

$\Rightarrow\dfrac{z_2}{z_1}+1=1+\cos\alpha+i\sin\alpha\Rightarrow\dfrac{z_2+z_1}{z_1}=2\cos\dfrac{\alpha}{2}\left(\cos\dfrac{\alpha}{2}+i\sin\dfrac{\alpha}{2}\right)=2\cos\dfrac{\alpha}{2}e^{\frac{i\alpha}{2}}$

$\Rightarrow\dfrac{z_2+z_1}{z_1}=2\cos\dfrac{\alpha}{2}e^{\frac{i\alpha}{2}}\Rightarrow\left(\dfrac{z_2+z_1}{z_1}\right)^2=4\cos^2\dfrac{\alpha}{2}e^{i\alpha}$

$\Rightarrow\left(\dfrac{z_2+z_1}{z_1}\right)^2=4\cos^2\dfrac{\alpha}{2}\dfrac{z_2}{z_1}\Rightarrow\left(z_2+z_1\right)^2=4z_1z_2\cos^2\dfrac{\alpha}{2}$

$\Rightarrow(-p)^2=4q\cos^2\dfrac{\alpha}{2}$ $\Rightarrow p^2=4q\cos^2\dfrac{\alpha}{2}$.

Illustration 27: On the Argand plane z_1, z_2 and z_3 are respectively the vertices of an isosceles triangle ABC with AC = BC and equal angles are θ. If z_4 is the incentre of the triangle then prove that $(z_2 - z_1)(z_3 - z_1) = (1 + \sec\theta)(z_4 - z_1)^2$

(JEE ADVANCED)

Sol: Here by using angle rotation formula we can solve this problem. From Fig 6.21, we have

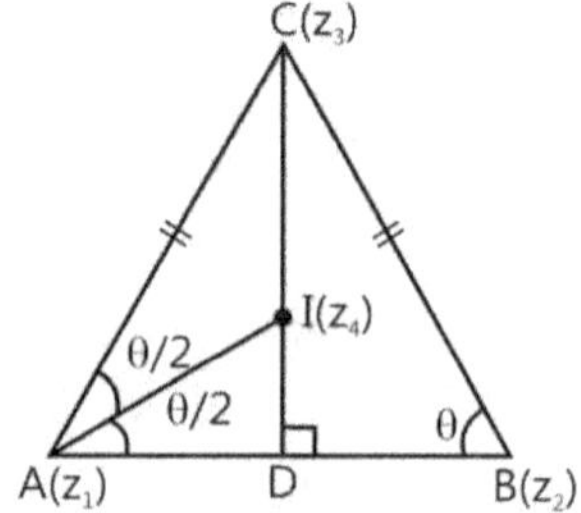

$$\frac{z_2 - z_1}{|z_2 - z_1|} = \frac{z_4 - z_1}{|z_4 - z_1|} e^{i\theta/2} \qquad \text{... (i) (clockwise)}$$

and

$$\frac{z_3 - z_1}{|z_3 - z_1|} = \frac{z_4 - z_1}{|z_4 - z_1|} e^{i\theta/2} \qquad \text{... (ii) (anticlockwise)}$$

Multiplying (i) and (ii)

$$\frac{(z_2 - z_1)(z_3 - z_1)}{(z_4 - z_1)^2} = \frac{|(z_2 - z_1)||(z_3 - z_1)|}{|z_4 - z_1|^2} = \frac{AB|AC|}{(AI)^2} = \frac{2(AD)(AC)}{(AI)^2} = \frac{2(AD)^2}{(AI)^2} \cdot \frac{AC}{AD}$$

$$= 2\cos^2\frac{\theta}{2}\sec\theta = (1 + \cos\theta)\sec\theta.$$

Figure 6.21

7. REPRESENTATION OF DIFFERENT LOCI ON COMPLEX PLANE

(a) $|z - (1 + 2i)| = 3$ denotes a circle with centre (1, 2) and radius 3 (see Fig 6.22).

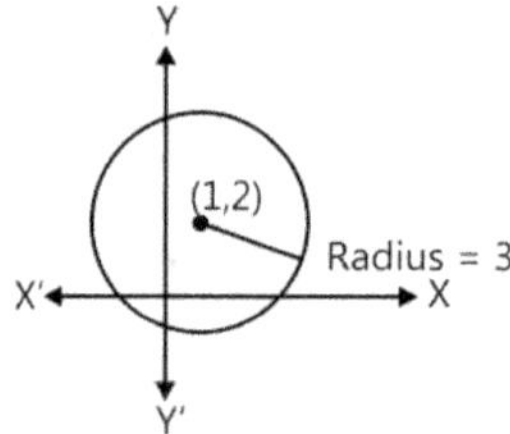

Figure 6.22: Circle on a complex plane

(b) $|z - 1| = |z - i|$ denotes the equation of the perpendicular bisector of join of (1, 0) and (0, 1) on the Argand plane (see Fig 6.24).

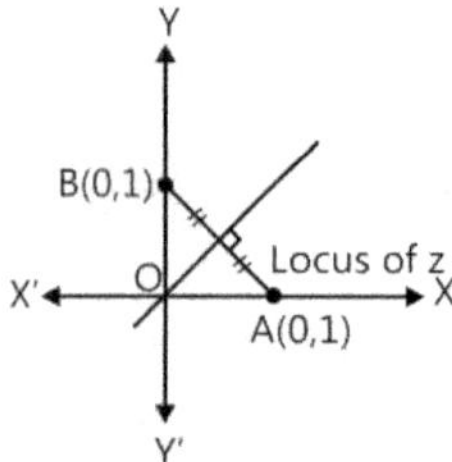

Figure 6.23: Perpendicular bisector complex plane

(c) $|z - 4i| + |z + 4i| = 10$ denotes an ellipse with foci at (0, 4) and (0, − 4); major axis 10; minor axis 6 with $e = \dfrac{4}{5}$ (see Fig 6.24).

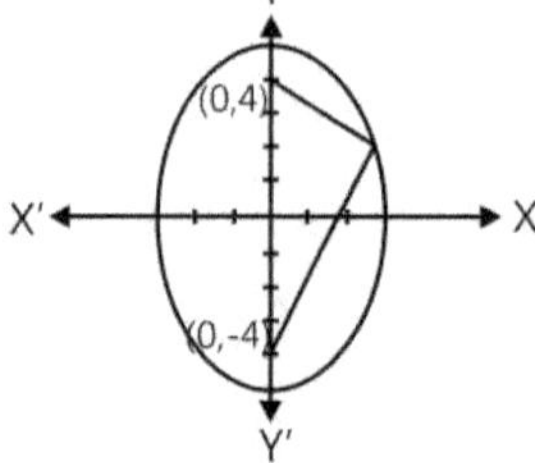

Figure 6.24: Ellipse on a complex plane

$$e^2 = 1 - \frac{36}{100} = \frac{64}{100} \quad \Rightarrow \quad e = \frac{4}{5} \quad \left[\frac{x^2}{9} + \frac{y^2}{25} = 1 \right]$$

(d) $|z - 1| + |z + 1| = 1$ denotes no locus. (Triangle inequality).

(e) $|z - 1| < 1$ denotes area inside a circle with centre (1, 0) and radius 1.

(f) $2 \leq |z - 1| < 5$ denotes the region between the concentric circles of radii 5 and 2. Centred at (1, 0) including the inner boundary (see Fig 6.25).

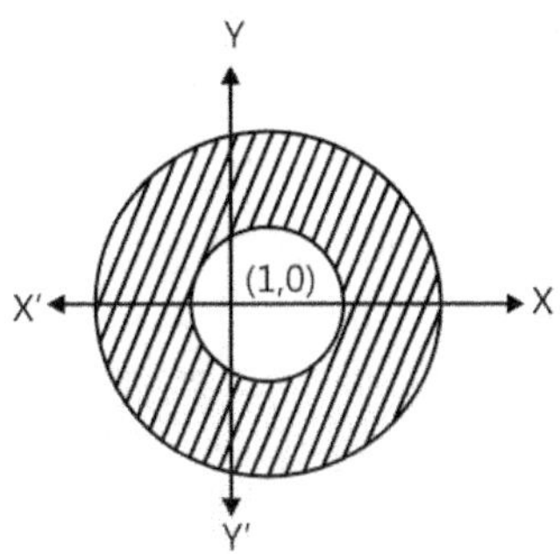

Figure 6.25: Circle disc on a complex plane

(g) $0 \leq \arg z \leq \dfrac{\pi}{4}$ $(z \neq 0)$ where z is defined by positive real axis and the part of the line x = y in the first quadrant. It includes the boundary but not the origin. Refer to Fig 6.26.

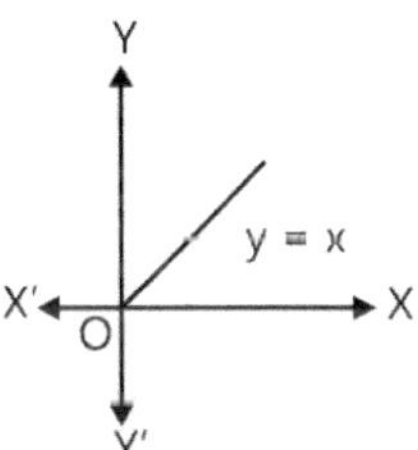

Figure 6.26

(h) $\mathrm{Re}\,(z^2) > 0$ denotes the area between the lines x = y and x = − y which includes the x-axis.

Hint: $(x^2 - y^2) + 2xyi = 0 \quad \Rightarrow x^2 - y^2 > 0 \qquad \Rightarrow (x - y)\,(x + y) > 0.$

Illustration 28: Solve for z, if $z^2 + |z| = 0$. (JEE MAIN)

Sol: Consider $z = x + iy$ and solve this using algebra of complex number.

Let $z = x + iy \quad \Rightarrow (x + iy)^2 + \sqrt{x^2 + y^2} = 0 \qquad \Rightarrow \left(x^2 - y^2 + \sqrt{x^2 + y^2} \right) + (2ixy) = 0$

$\Rightarrow$ Either x = 0 or y = 0; $\quad x = 0 \Rightarrow -y^2 + |y| = 0 \qquad \Rightarrow y = 0, 1, -1 \therefore z = 0, i, -i$

and, $y = 0 \Rightarrow x^2 + |x| = 0 \qquad \Rightarrow x = 0 \qquad \therefore z = 0$

Therefore, z = 0, z = i, z = − i.

Illustration 29: If the complex number z is to satisfy $|z| = 3, \left|z - \{a(1 + i) - i\}\right| \leq 3$ and $|z + 2a - (a + 1)\,i| > 3$ simultaneously for at least one z then find all $a \in R$. (JEE ADVANCED)

Sol: Consider z = x + iy and solve these inequalities to get the result.

All z at a time lie on a circle $|z| = 3$ but inside and outside the circles $|z - \{a(1 + i) - i\}| = 3$ and $|z + 2a - (a + 1)i| = 3$, respectively.

Let $z = x + iy$ then equation of circles are $x^2 + y^2 = 9$ ... (i)

$(x - a)^2 + (y - a + 1)^2 = 9$ ... (ii)

and $(x + 2a)^2 + (y - a - 1)^2 = 9$ ... (iii)

Circles (i) and (ii) should cut or touch then distance between their centres $\leq$ sum of their radii.

$$\Rightarrow \sqrt{(a - 0)^2 + (a - 1 - 0)^2} \leq 3 + 3 \Rightarrow a^2 + (a - 1)^2 \leq 36$$

$$\Rightarrow 2a^2 - 2a - 35 \leq 0 \Rightarrow a^2 - a - \frac{35}{2} \leq 0$$

$$\Rightarrow \left(a - \frac{1}{2}\right)^2 \leq \frac{71}{4} \quad \therefore \frac{1 - \sqrt{71}}{2} \leq a \leq \frac{1 + \sqrt{71}}{2} \quad \text{... (iv)}$$

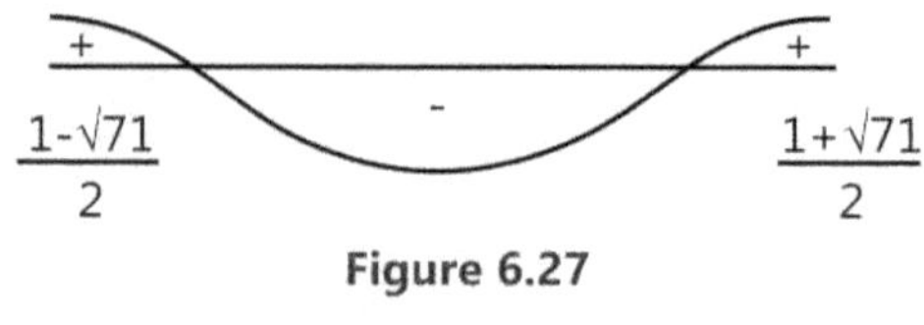

Figure 6.27

Again circles (i) and (iii) should not cut or touch then distance between their centres > sum of the radii

$$\Rightarrow \sqrt{(-2a - 0)^2 + (a + 1 - 0)^2} > 3 + 3 \Rightarrow \sqrt{5a^2 + 2a + 1} > 6 \qquad \Rightarrow 5a^2 + 2a + 1 > 36$$

$$\Rightarrow 5a^2 + 2a - 35 > 0 \qquad \Rightarrow a^2 + \frac{2a}{5} - 7 > 0$$

$$\text{Then } \left(a - \frac{-1 - 4\sqrt{11}}{5}\right)\left(a - \frac{-1 + 4\sqrt{11}}{5}\right) > 0$$

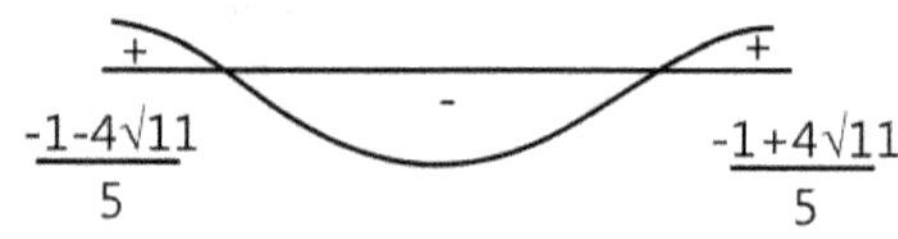

Figure 6.28

$$\therefore \qquad a \in \left(-\infty, \frac{-1 - 4\sqrt{11}}{5}\right) \cup \left(\frac{-1 + 4\sqrt{11}}{5}, \infty\right) \quad \text{... (v)}$$

The common values of a satisfying (iv) and v are

$$a \in \left(\frac{1 - \sqrt{71}}{2}, \frac{-1 - 4\sqrt{11}}{5}\right) \cup \left(\frac{-1 + 4\sqrt{11}}{5}, \frac{1 + \sqrt{71}}{2}\right)$$

8. DEMOIVRE'S THEOREM

Statement: $(\cos n\theta + i\sin n\theta)$ is the value or one of the values of $(\cos\theta + i\sin\theta)^n$, $\forall n \in Q$. Value if n is an integer. One of the values if n is rational which is not integer, the theorem is very useful in determining the roots of any complex quantity.

Note: We use the theory of equations to find the continued product of the roots of a complex number.

NOMORECLASS CONCEPTS

The theorem is not directly applicable to $(\sin\theta + i\cos\theta)^n$, rather

$$(\sin\theta + i\cos\theta)^n = \left[\cos\left(\frac{\pi}{2} - \theta\right) + i\sin\left(\frac{\pi}{2} - \theta\right)\right]^n = \cos n\left(\frac{\pi}{2} - \theta\right) + i\sin n\left(\frac{\pi}{2} - \theta\right)$$

8.1 Application

Cube root of unity

(a) The cube roots of unity are $1,\ \dfrac{-1+i\sqrt{3}}{2},\ \dfrac{-1-i\sqrt{3}}{2}$

[Note that $1 - i\sqrt{3} = -2$ and $1 + i\sqrt{3} = -2\omega^2$]

(b) If ω is one of the imaginary cube roots of unity then $1 + \omega + \omega^2 = 0$.

In general $1 + \omega^r + \omega^{2r} = 0$; where r = 1, and not a multiple of 3.

(c) In polar form the cube roots of unity are: $\cos 0 + i\sin 0$; $\cos\dfrac{2\pi}{3} + i\sin\dfrac{2\pi}{3}$; $\cos\dfrac{4\pi}{3} + i\sin\dfrac{4\pi}{3}$

(d) The three cube roots of unity when plotted on the argand plane constitute the vertices of an equilateral triangle.

[Note that the 3 cube roots of i lies on the vertices of an isosceles triangle]

(e) The following factorization should be remembered.

For a, b, c $\in R$ and ω being the cube root of unity,

    (i) $a^3 - b^3 = (a - b)(a - \omega b)(a - \omega^2 b)$

    (ii) $x^2 + x + 1 = (x - \omega)(x - \omega^2)$

    (iii) $a^3 + b^3 = (a + b)(a + \omega b)(a + \omega^2 b)$

    (iv) $a^3 + b^3 + c^3 - 3abc = (a + b + c)(a + \omega b + \omega^2 c)(a + \omega^2 b + \omega c)$

n^{th} roots of unity: If $1, \alpha_1, \alpha_2, \alpha_3, \ldots\ldots, \alpha_{n-1}$ are the n, n^{th} roots of unity then

    (i)   They are in G.P. with common ratio $e^{i\left(\frac{2\pi}{n}\right)} = \cos\dfrac{2\pi}{n} + i\sin\dfrac{2\pi}{n}$

    (ii) $1^p + \alpha_1^p + \alpha_2^p + \ldots\ldots + \alpha_{n-1}^p = 0$ if p is not an integral multiple of n

        $1^p + (\alpha_1)^p + (\alpha_2)^p + \ldots\ldots + (\alpha_{n-1})^p = n$ if p is an integral multiple of n.

    (iii) $(1 - \alpha_1)(1 - \alpha_2) \ldots\ldots (1 - \alpha_{n-1}) = n$.

Steps to determine nth roots of a complex number

    (i)   Represent the complex number whose roots are to be determined in polar form.

    (ii)   Add $2m\pi$ to the argument.

    (iii)   Apply De Moivre's TheoremT

    (iv)   Put m = 0, 1, 2, 3, ……. (n – 1) to get all the nth roots.

Explanation: Let $z = 1^{\frac{1}{n}} = (\cos 0 + i\sin 0)^{\frac{1}{n}} = (\cos 2m\pi + i\sin 2m\pi)^{\frac{1}{n}} = \left(\cos\dfrac{2m\pi}{n} + i\sin\dfrac{2m\pi}{n}\right)$

Put m = 0, 1, 2, 3, ……. (n – 1), we get

$\underbrace{1, \cos\dfrac{2\pi}{n} + i\sin\dfrac{2\pi}{n}}_{\alpha}, \cos\dfrac{4\pi}{n} + i\sin\dfrac{4\pi}{n}, \ldots\ldots, \cos\dfrac{2(n-1)\pi}{n} + i\sin\dfrac{2(n-1)\pi}{n}$ (n, n^{th} roots in G.P.)

Now, $S = 1^p + \alpha^p + \alpha^{2p} + \alpha^{3p} + \ldots\ldots + \alpha^{(n-1)p} = \dfrac{1-(\alpha^p)^n}{1-\alpha^p} = \dfrac{1-(\alpha^n)^p}{1-\alpha^p}$

$$= \dfrac{1-(\alpha^n)^p}{1-\alpha^p} = \begin{cases} \dfrac{0}{\text{non zero}} = 0, & \text{if } p \text{ is not an integral multiple of } n \\[2mm] \dfrac{0}{0} = \text{indeterminant}, & \text{if } p \text{ is an integral multiple of } n \end{cases}$$

Again, if x is one of the nth root of unity then $x^n - 1 = (x-1)(x-\alpha_1)(x-\alpha_2)\ldots\ldots(x-\alpha_{n-1})$

$1 + x + x^2 + \ldots\ldots + x^{n-1} = \dfrac{x^n - 1}{x-1} \equiv (x-\alpha_1)(x-\alpha_2)\ldots\ldots(x-\alpha_{n-1})$

Put $x = 1$, to get $(1-\alpha_1)(1-\alpha_2)\ldots\ldots(1-\alpha_{n-1}) = n$

Similarly put $x = -1$, is to get other result.

NOMORECLASS CONCEPTS

Square roots of $z = a + ib$ are $\pm\left[\sqrt{\dfrac{|z|+a}{2}} + i\sqrt{\dfrac{|z|-a}{2}}\right]$ for $b > 0$.

If $1, \alpha_1, \alpha_2, \alpha_3, \ldots\ldots, \alpha_{n-1}$ are the n, n^{th} roots of unity then

$(1+\alpha_1)(1+\alpha_2)\ldots\ldots(1+\alpha_{n-1}) = 0$ if n is even and 1 if n is odd.

$1 \cdot \alpha_1 \cdot \alpha_2 \cdot \alpha_3 \cdot \ldots\ldots \cdot \alpha_{n-1} = 1$ or -1 according as n is odd or even.

$$(\omega-\alpha_1)(\omega-\alpha_2)\ldots\ldots(\omega-\alpha_{n-1}) = \begin{cases} 0, & \text{if } n = 3k \\ 1, & \text{if } n = 3k+1 \\ 1+\omega, & \text{if } n = 3k+2 \end{cases}$$

Illustration 30: If $x = a+b$, $y = a\omega + b\omega^2$ and $z = a\omega^2 + b\omega$, then prove that $x^3 + y^3 + z^3 = 3(a^3 + b^3)$ **(JEE MAIN)**

Sol: Here $x + y + z = 0$. Take cube on both side.

$x + y + z = 0 \qquad \Rightarrow x^3 + y^3 + z^3 = 3xyz \therefore \text{LHS} = 3xyz$

$= 3(a+b)(a\omega + b\omega^2)(a\omega^2 + b\omega) = 3(a+b)(a\omega + b\omega^2)(a\omega^2 + b\omega\cdot\omega^3) = 3\omega^3(a+b)(a+b\omega)(a+b\omega^2) = 3(a^3 + b^3)$

Illustration 31: The value of expression $1(2-\omega)(2-\omega^2) + 2(3-\omega)(3-\omega^2) + \ldots + (n-1)(n-\omega)(n-\omega^2)$.

(JEE ADVANCED)

Sol: The given expression represent as $x^3 - 1 = (x-1)(x-\omega)(x-\omega^2)$. Therefore by putting $x = 2, 3, 4 \ldots n$, we will get the result.

$x^3 - 1 = (x-1)(x-\omega)(x-\omega^2)$

Put $x = 2$ $2^3 - 1 = 1 \cdot (2-\omega)(2-\omega)^2$ $\qquad$ Put $x = 3$ $3^3 - 1 = 2 \cdot (3-\omega)(3-\omega^2)$:

Put $x = n$ $n^3 - 1 = (n-1)(n-\omega)(n-\omega^2)$

$\therefore \text{LHS} = (2^3 + 3^3 + \ldots\ldots + n^3) - (n-1) \qquad = (1^3 + 2^3 + 3^3 + \ldots\ldots + n^3) - n = \left(\dfrac{n(n+1)}{2}\right)^2 - n$

9. SUMMATION OF SERIES USING COMPLEX NUMBER

(a) $\cos\theta + \cos 2\theta + \cos 3\theta + \ldots\ldots + \cos n\theta = \dfrac{\sin\left(\dfrac{n\theta}{2}\right)}{\sin\left(\dfrac{\theta}{2}\right)}\cos\left(\dfrac{n+1}{2}\right)\theta$

(b) $\sin\theta + \sin 2\theta + \sin 3\theta + \ldots\ldots + \sin n\theta = \dfrac{\sin\left(\dfrac{n\theta}{2}\right)}{\sin\left(\dfrac{\theta}{2}\right)}\sin\left(\dfrac{n+1}{2}\right)\theta$

Note: If $\theta = \dfrac{2\pi}{n}$, then the sum of the above series vanishes.

9.1 Complex Number and Binomial Coefficients

Try the following questions using the binomial expansion of $(1+x)^n$ and substituting the value of x according to the binomial coefficients in the respective question.

Find the value of the following

(i) $C_0 + C_4 + C_8 + \ldots\ldots$

(ii) $C_1 + C_5 + C_9 + \ldots\ldots$

(iii) $C_2 + C_6 + C_{10} + \ldots\ldots$

(iv) $C_3 + C_7 + C_{11} + \ldots\ldots$

(v) $C_0 + C_3 + C_6 + C_9 + \ldots\ldots$

Hint (v): In the expansion of $(1+x)^n$, put $x = 1, \omega,$ and ω^2 and add the three equations.

Illustration 32: If $1, \omega, \omega^2, \ldots\ldots, \omega^{n-1}$ are n^{th} roots of unity, then the value of $(5-\omega)(5-\omega^2)\ldots\ldots(5-\omega^{n-1})$ is equal to **(JEE MAIN)**

Sol: Here consider $x = (1)^{\frac{1}{n}}$, therefore $x^n - 1 = 0$ (has n roots i.e. $1, \omega, \omega^2, \ldots\ldots, \omega^{n-1}$).

$\Rightarrow x^n - 1 = (x-1)(x-\omega)(x-\omega^2)\ldots\ldots(x-\omega^{n-1})$

$\Rightarrow \dfrac{x^n - 1}{x-1} = (x-\omega)(x-\omega^2)\ldots\ldots(x-\omega^{n-1})$

$\Rightarrow$ Putting x = 5 in both sides, we get

$\therefore (5-\omega)(5-\omega^2)\ldots\ldots(5-\omega^{n-1}) = \dfrac{5^n - 1}{4}.$

10. APPLICATION IN GEOMETRY

10.1 Distance Formula

Distance between $A(z_1)$ and $B(z_2)$ is given by $AB = |z_2 - z_1|$. Refer Fig 6.29.

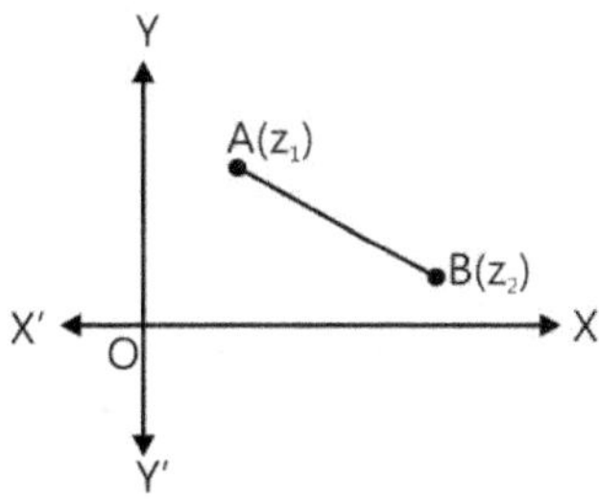

Figure 6.29

10.2 Section Formula

The point P(z) which divides the join of $A(z_1)$ and $B(z_2)$ in the ratio m: n is

given by $z = \dfrac{mz_2 + nz_1}{m+n}$. Refer Fig 6.30.

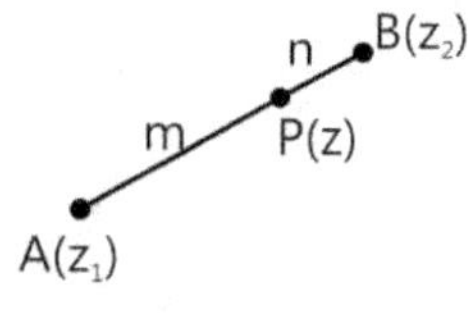

Figure 6.30

10.3 Midpoint Formula

Mid-point M(z) of the segment AB is given by $z = \dfrac{1}{2}(z_1 + z_2)$.

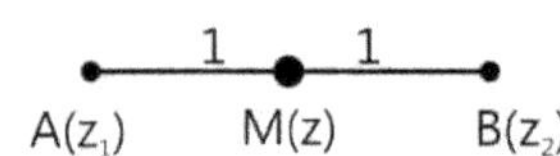

Figure 6.31 Mid point formula

10.4 Condition For Four Non-Collinear Points

Condition(s) for four non-collinear $A(z_1), B(z_2), C(z_3)$ and $D(z_4)$ to represent vertices of a

(a) Parallelogram: The diagonals AC and BD must bisect each other

$\Leftrightarrow \quad \dfrac{1}{2}(z_1 + z_3) = \dfrac{1}{2}(z_2 + z_4)$

$\Leftrightarrow \quad z_1 + z_3 = z_2 + z_4$

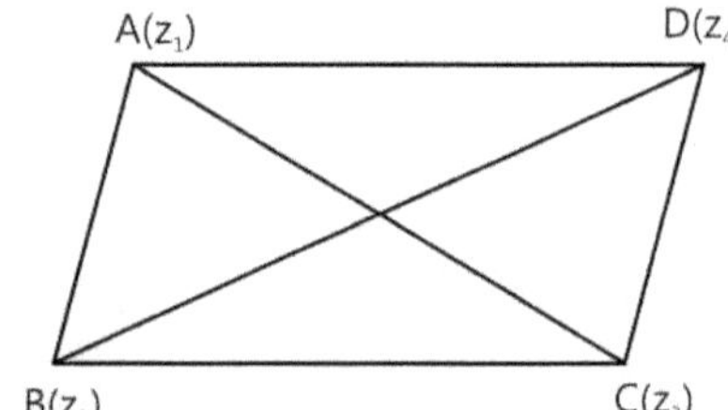

(b) Rhombus:

    (i) The diagonals AC and BD bisect each other

$\Leftrightarrow \quad z_1 + z_3 = z_2 + z_4$, and

    (ii) A pair of two adjacent sides are equal, for instance AD = AB

$\Leftrightarrow \quad |z_4 - z_1| = |z_2 - z_1|$

Figure 6.32

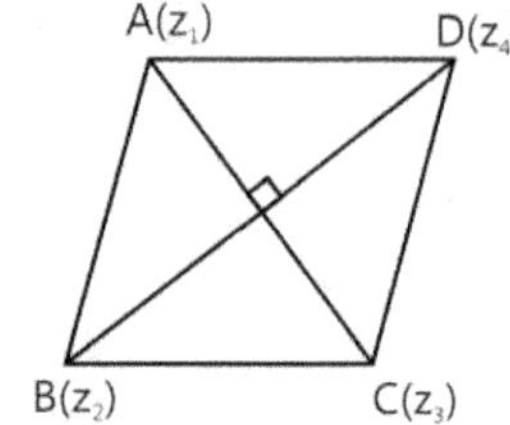

(c) Square:

    (i) The diagonals AC and BD bisect each other

$\Leftrightarrow \quad z_1 + z_3 = z_2 + z_4$

    (ii) A pair of adjacent sides are equal; for instance, AD = AB

$\Leftrightarrow \quad |z_4 - z_1| = |z_2 - z_1|$

    (iii) The two diagonals are equal, that is AC = BD

$\Leftrightarrow \quad |z_3 - z_1| = |z_4 - z_2|$

Figure 6.33

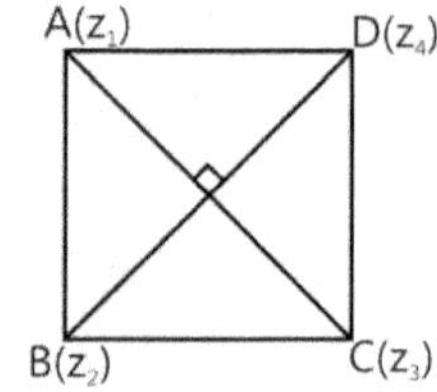

(d) Rectangle:

    (i) The diagonals AC and BD bisect each other

$\Leftrightarrow \quad z_1 + z_3 = z_2 + z_4$

    (ii) The diagonals AC and BD are equal

$\Leftrightarrow \quad |z_3 - z_1| = |z_4 - z_2|$

Figure 6.34

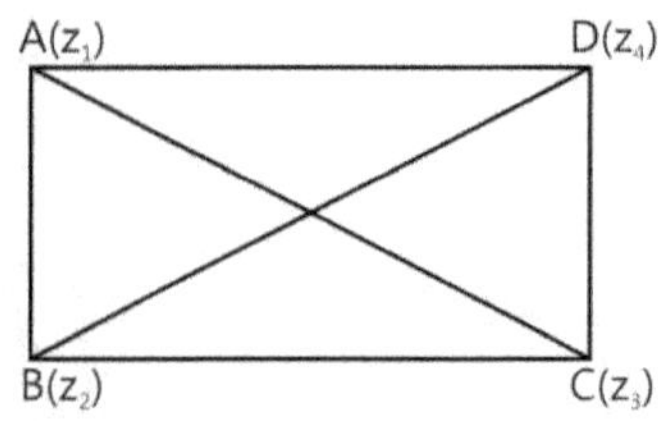

Figure 6.35

10.5 Triangle

In a triangle ABC, let the vertices A, B and C be represented by the complex numbers z_1, z_2, and z_3 respectively. Then

(a) **Centroid:** The centroid (G), is the point of intersection of medians of $\triangle ABC$. It is given by the formula

$$z = \frac{1}{3}(z_1 + z_2 + z_3)$$

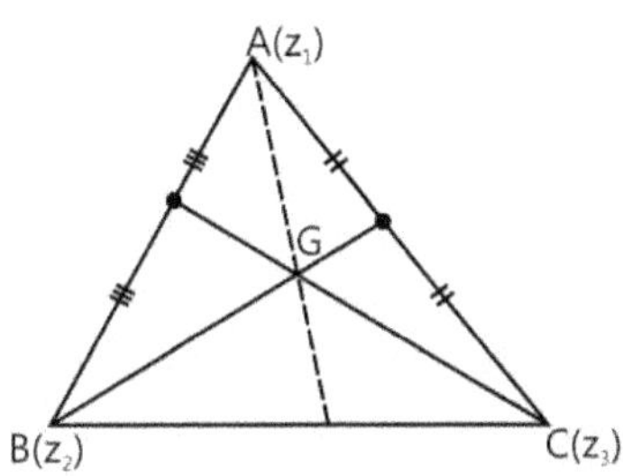

Figure 6.36 (a)

(b) **Incentre:** The incentre (I) of $\triangle ABC$ is the point of intersection of internal angular bisectors of angles of $\triangle ABC$. It is given by the formula

$$z = \frac{az_1 + bz_2 + cz_3}{a+b+c},$$

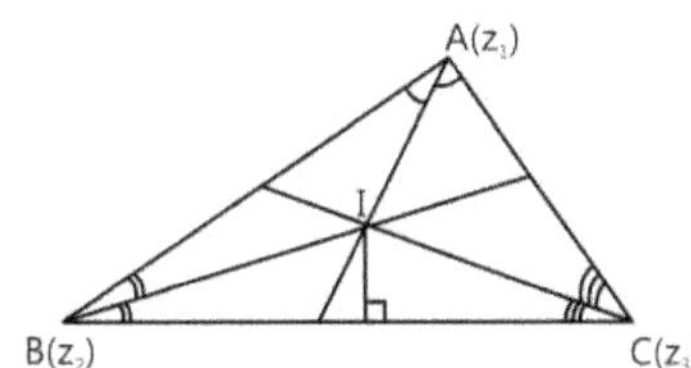

Figure 6.36 (b)

(c) **Circumcentre:** The circumcentre (S) of $\triangle ABC$ is the point of intersection of perpendicular bisectors of sides of $\triangle ABC$. It is given by the formula

$$z = \frac{|z_1|^2(z_2 - z_3) + |z_2|^2(z_3 - z_1) + |z_3|^2(z_1 - z_2)}{\overline{z}_1(z_2 - z_3) + \overline{z}_2(z_3 - z_1) + \overline{z}_3(z_1 - z_2)} = \frac{\begin{vmatrix} |z_1|^2 & z_1 & 1 \\ |z_2|^2 & z_2 & 1 \\ |z_3|^2 & z_3 & 1 \end{vmatrix}}{\begin{vmatrix} \overline{z}_1 & z_1 & 1 \\ \overline{z}_2 & z_2 & 1 \\ \overline{z}_3 & z_3 & 1 \end{vmatrix}}$$

Also, $z = \dfrac{z_1(\sin 2A) + z_2(\sin 2B) + z_3(\sin 2C)}{\sin 2A + \sin 2B + \sin 2C}$

Figure 6.36 (c)

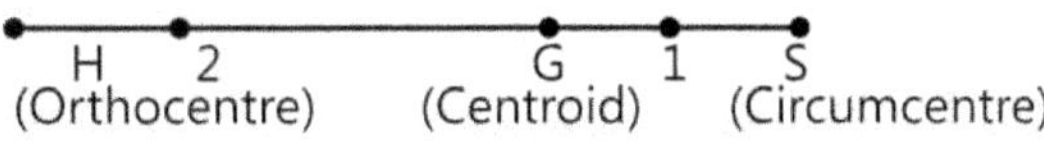

Figure 6.37

(d) **Euler's Line:** The orthocenter H, the centroid G and the circumcentre S of a triangle which is not equilateral lies on a straight line. In case of an equilateral triangle these points coincide.

G divides the join of H and S in the ratio 2 : 1 (see Fig 6.37).

Thus, $z_G = \dfrac{1}{3}(z_H + 2z_S)$

10.6 Area of a Triangle

Area of $\triangle ABC$ with vertices $A(z_1)$, $B(z_2)$ and $C(z_3)$ is given by

$$\Delta = \left| \frac{1}{4i} \begin{vmatrix} z_1 & \bar{z}_1 & 1 \\ z_2 & \bar{z}_2 & 1 \\ z_3 & \bar{z}_3 & 1 \end{vmatrix} \right| = \left| \frac{1}{2} \operatorname{Im}(\bar{z}_1 z_2 + \bar{z}_2 z_3 + \bar{z}_3 z_1) \right|$$

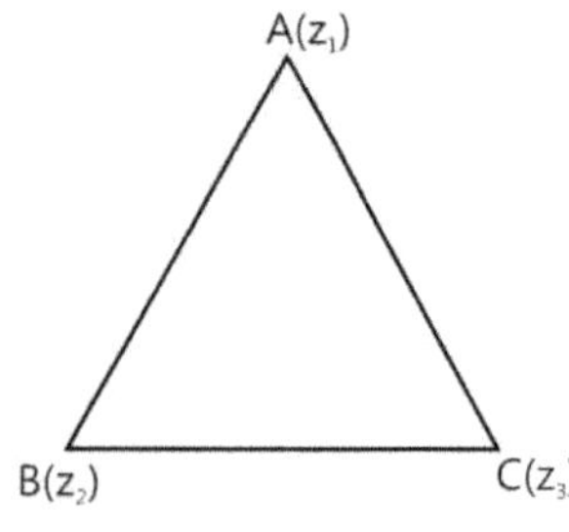

Figure 6.38

10.7 Conditions for Triangle to be Equilateral

The triangle ABC with vertices $A(z_1)$, $B(z_2)$ and $C(z_3)$ is equilateral

$$\text{iff } \frac{1}{z_2 - z_3} + \frac{1}{z_3 - z_1} + \frac{1}{z_1 - z_2} = 0$$

$$\Leftrightarrow z_1^2 + z_2^2 + z_3^2 = z_2 z_3 + z_3 z_1 + z_1 z_2 \quad \Leftrightarrow z_1 \bar{z}_2 = z_2 \bar{z}_3 = z_3 \bar{z}_1 \Leftrightarrow z_1^2 = z_2 z_3 \text{ and } z_2^2 = z_1 z_3$$

$$\Leftrightarrow \begin{vmatrix} 1 & z_2 & z_3 \\ 1 & z_3 & z_1 \\ 1 & z_1 & z_2 \end{vmatrix} = 0 \quad \Leftrightarrow \frac{z_2 - z_1}{z_3 - z_2} = \frac{z_3 - z_2}{z_1 - z_2}$$

$$\Leftrightarrow \frac{1}{z - z_1} + \frac{1}{z - z_2} + \frac{1}{z - z_3} = 0 \text{ where } z = \frac{1}{3}(z_1 + z_2 + z_3).$$

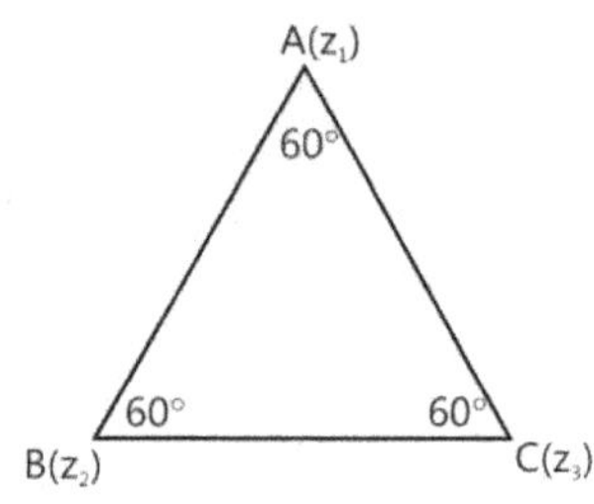

Figure 6.39

10.8 Equation of a Straight line

(a) **Non-parametric form:** An equation of a straight line joining the two points $A(z_1)$ and $B(z_2)$ is

$$\operatorname{Arg}\left(\frac{z - z_1}{z_2 - z_1} \right) = 0 \quad \begin{vmatrix} z & \bar{z} & 1 \\ z_1 & \bar{z}_1 & 1 \\ z_2 & \bar{z}_2 & 1 \end{vmatrix} = 0$$

$$\text{or} \quad \frac{z - z_1}{z_2 - z_1} = \frac{\bar{z} - \bar{z}_1}{\bar{z}_2 - \bar{z}_1}$$

$$\text{or } z(\bar{z}_1 - \bar{z}_2) - \bar{z}(z_1 - z_2) + z_1 \bar{z}_2 - z_2 \bar{z}_1 = 0$$

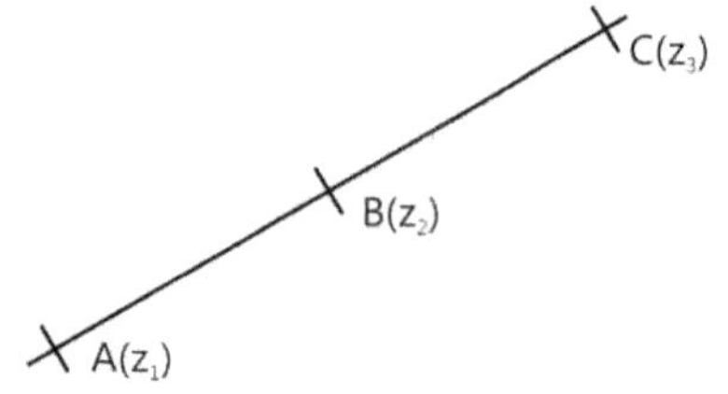

Figure 6.40

(b) **Parametric form:** An equation of the line segment between the points $A(z_1)$ and $B(z_2)$ is
$z = tz_1 + (1 - t)z_2,\ t(0,1)$ where t is a real parameter.

(c) **General equation of a straight line:** The general equation of a straight line is $\bar{a}z + a\bar{z} + b = 0$ where, a is non-zero complex number and b is a real number.

10.9 Complex Slope of a Line

If $A(z_1)$ and $B(z_2)$ are two points in the complex plane, then complex slope of AB is defined to be $\mu = \dfrac{z_1 - z_2}{\bar{z}_1 - \bar{z}_2}$

Two lines with complex slopes μ_1 and μ_2 are

(i) Parallel, if $\mu_1 = \mu_2$ $\qquad$ (ii) Perpendicular, if $\mu_1 + \mu_2 = 0$

The complex slope of the line $\bar{a}z + a\bar{z} + b = 0$ is given by $\left(\dfrac{-a}{\bar{a}} \right)$.

10.10 Length of Perpendicular from a Point to a Line

Length of perpendicular of point $A(\omega)$ from the line $\bar{a}z + a\bar{z} + b = 0$.

Where $a \in C - \{0\}$, and $b \in R$ is given by $p = \dfrac{|\bar{a}\omega + a\bar{\omega} + b|}{2|a|}$

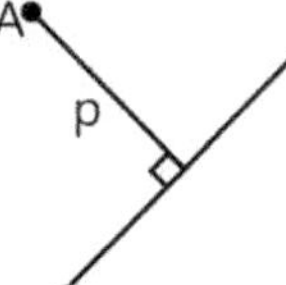

Figure 6.41

10.11 Equation of Circle

(a) An equation of the circle with centre z_0 and radius r is $|z - z_0| = r$ or

$z = z_0 + re^{i\theta}, 0 \le \theta < 2\pi$ (parametric form) or $z\bar{z} - z_0\bar{z} - \bar{z}_0 z + z_0\bar{z}_0 - r^2 = 0$

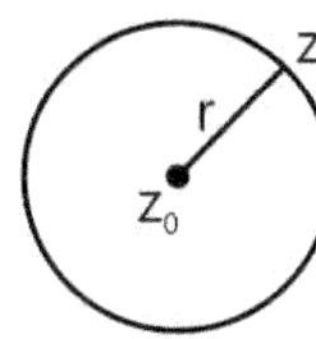

Figure 6.42

(b) General equation of a circle is $z\bar{z} + a\bar{z} + \bar{a}z + b = 0$ ... (i)

Where a is a complex number and b is a real number such that $a\bar{a} - b \ge 0$. Centre of (i) is $-a$ and its radius is $\sqrt{a\bar{a} - b}$

(c) Diameter form of a circle: An equation of the circle one of whose diameter is the segment joining $A(z_1)$ and $B(z_2)$ is $(z - z_1)(\bar{z} - \bar{z}_2) + (\bar{z} - \bar{z}_1)(z - z_2) = 0$

(d) An equation of the circle passing through two points $A(z_1)$ and $B(z_2)$

is $(z - z_1)(\bar{z} - \bar{z}_2) + (\bar{z} - \bar{z}_1)(z - z_2) + i k \begin{vmatrix} z & \bar{z} & 1 \\ z_1 & \bar{z}_1 & 1 \\ z_2 & \bar{z}_2 & 1 \end{vmatrix} = 0$ where k is a real parameter.

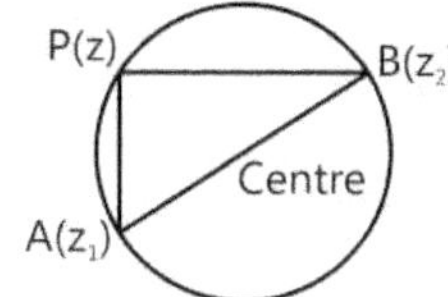

Figure 6.43

(e) Equation of a circle passing through three non-collinear points.

Let three non-collinear points be $A(z_1)$, $B(z_2)$ and $C(z_3)$ and $P(z)$ be any point on the circle through A, B and C.

Then either $\angle ACB = \angle APB$ [when angles are in the same segment]

or, $\angle ACB + \angle APB = \pi$ [when angles are in the opposite segment] (see Fig 6.44).

$\Rightarrow \arg\left(\dfrac{z_3 - z_2}{z_3 - z_1}\right) - \arg\left(\dfrac{z - z_2}{z - z_1}\right) = 0$ or, $\arg\left(\dfrac{z_3 - z_2}{z_3 - z_1}\right) + \arg\left(\dfrac{z - z_1}{z - z_2}\right) = \pi$

$\Rightarrow \arg\left[\left(\dfrac{z_3 - z_2}{z_3 - z_1}\right)\left(\dfrac{z - z_1}{z - z_2}\right)\right] = 0$

or, $\arg\left[\left(\dfrac{z_3 - z_2}{z_3 - z_1}\right)\left(\dfrac{z - z_1}{z - z_2}\right)\right] = \pi$

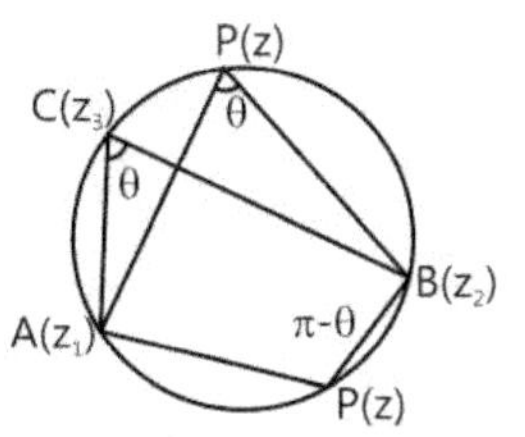

Figure 6.44

In any case, we get $\dfrac{(z - z_1)(z_3 - z_2)}{(z - z_2)(z_3 - z_1)}$ is purely real.

$\Leftrightarrow \dfrac{(z - z_1)(z_3 - z_2)}{(z - z_2)(z_3 - z_1)} = \dfrac{(\bar{z} - \bar{z}_1)(\bar{z}_3 - \bar{z}_2)}{(\bar{z} - \bar{z}_2)(\bar{z}_3 - \bar{z}_1)}$

(f) Condition for four points to be concyclic.

Four points z_1, z_2, z_3 and z_4 will lie on the same circle if and only if $\dfrac{(z_4 - z_1)(z_3 - z_2)}{(z_4 - z_2)(z_3 - z_1)}$ is purely real.

$\Leftrightarrow \dfrac{(z_4 - z_1)(z_3 - z_2)}{(z_4 - z_2)(z_3 - z_1)} = \dfrac{(\bar{z}_4 - \bar{z}_1)(\bar{z}_3 - \bar{z}_2)}{(\bar{z}_4 - \bar{z}_2)(\bar{z}_3 - \bar{z}_1)}$

Three points z_1, z_2 and z_3 are collinear if $\begin{vmatrix} z_1 & \overline{z}_1 & 1 \\ z_2 & \overline{z}_2 & 1 \\ z_3 & \overline{z}_3 & 1 \end{vmatrix} = 0$.

If three points $A(z_1)$, $B(z_2)$ and $C(z_3)$ are collinear then slope of AB = slope of BC = slope of AC

$$\Rightarrow \frac{z_1 - z_2}{\overline{z}_1 - \overline{z}_2} = \frac{z_2 - z_3}{\overline{z}_2 - \overline{z}_3} = \frac{z_1 - z_3}{\overline{z}_1 - \overline{z}_3}$$

Illustration 33: If the imaginary part of $\dfrac{2z+1}{iz+1}$ is -4, then the locus of the point representing z in the complex plane is

(a) A straight line   (b) A parabola   (c) A circle   (d) An ellipse   **(JEE MAIN)**

Sol: Put $z = x + iy$ and then equate its imaginary part to -4.

Let $z = x + iy$, then $\dfrac{2z+1}{iz+1} = \dfrac{2(x+iy)+1}{i(x+iy)+1} = \dfrac{(2x+1)+2iy}{(1-y)+ix} = \dfrac{[(2x+1)+2iy]\,[(1-y)-ix]}{(1-y)^2 + x^2}$

As $\text{Im}\left(\dfrac{2z+1}{iz+1}\right) = -4$, we get $\dfrac{2y(1-y) - x(2x+1)}{x^2 + (1-y)^2} = -4$

$\Rightarrow 2x^2 + 2y^2 + x - 2y = 4x^2 + 4(y^2 - 2y + 1)$ $\Rightarrow 2x^2 + 2y^2 - x - 6y + 4 = 0$. It represents a circle.

Illustration 34: The roots of $z^5 = (z-1)^5$ are represented in the argand plane by the points that are

(a) Collinear   (b) Concyclic

(c) Vertices of a parallelogram   (d) None of these   **(JEE MAIN)**

Sol: Apply modulus on both the side of given expression.

Let z be a complex number satisfying $z^5 = (z-1)^5$.

$\Rightarrow |z^5| = |(z-1)^5|$ $\Rightarrow |z|^5 = |z-1|^5$ $\Rightarrow |z| = |z-1|$

Thus, z lies on the perpendicular bisector of the segment joining the origin and $(1 + i0)$ i.e. z lies on $\text{Re}(z) = \dfrac{1}{2}$.

Illustration 35: Let z_1 and z_2 be two non-zero complex numbers such that $\dfrac{z_1}{z_2} + \dfrac{z_2}{z_1} = 1$, then the origin and points represented by z_1 and z_2

(a) Lie on straight line   (b) Form a right triangle

(c) Form an equilateral triangle   (d) None of these   **(JEE ADVANCED)**

Sol: Here consider $z = \dfrac{z_1}{z_2}$ and z_1 and z_2 are represented by A and B respectively and O be the origin.

Let $z = \dfrac{z_1}{z_2}$, then $z + \dfrac{1}{z} = 1$ $\Rightarrow z^2 - z + 1 = 0$

$\Rightarrow z = \dfrac{1 \pm \sqrt{3}i}{2}$ $\Rightarrow \dfrac{z_1}{z_2} = \dfrac{1 \pm \sqrt{3}i}{2}$

If z_1 and z_2 are represented by A and B respectively and O be the origin, then

$$\frac{OA}{OB} = \frac{|z_1|}{|z_2|} = \left|\frac{1 \pm \sqrt{3}i}{2}\right| = \sqrt{\frac{1}{4} + \frac{3}{4}} = 1 \;\Rightarrow\; OA = OB$$

Also, $\dfrac{AB}{OB} = \dfrac{|z_2 - z_1|}{|z_2|} = \left|1 - \dfrac{z_1}{z_2}\right| = \left|1 - \left(\dfrac{1}{2} \pm \dfrac{\sqrt{3}}{2}i\right)\right| = \left|\dfrac{1}{2} \mp \dfrac{\sqrt{3}}{2}i\right| = \sqrt{\dfrac{1}{4} + \dfrac{3}{4}} = 1$

$\Rightarrow AB = OB \qquad$ Thus, $OA = OB = AB \therefore \qquad \triangle OAB$ is an equilateral triangle.

Illustration 36: If z_1, z_2, z_3 are the vertices of an isosceles triangle, right angled at the vertex z_2, then the value of $(z_1 - z_2)^2 + (z_2 - z_3)^2$ is

(a) -1 $\qquad\qquad$ (b) 0 $\qquad\qquad$ (c) $(z_1 - z_3)^2$ $\qquad\qquad$ (d) None of these $\qquad\qquad$ **(JEE ADVANCED)**

Sol: Here use distance and argument formula of complex number to solve this problem.

As ABC is an isosceles right angled triangle with right angle at B,

$$BA = BC \text{ and } \angle ABC = 90° \Rightarrow |z_1 - z_2| = |z_3 - z_2| \text{ and } \arg\left(\frac{z_3 - z_2}{z_1 - z_2}\right) = \frac{\pi}{2}$$

$$\Rightarrow \frac{z_3 - z_2}{z_1 - z_2} = \frac{|z_3 - z_2|}{|z_1 - z_2|}\left[\cos\left(\frac{\pi}{2}\right) + i\sin\left(\frac{\pi}{2}\right)\right] = i$$

$$\Rightarrow (z_3 - z_2)^2 = -(z_1 - z_2)^2 \qquad\qquad \Rightarrow (z_1 - z_2)^2 + (z_2 - z_3)^2 = 0.$$

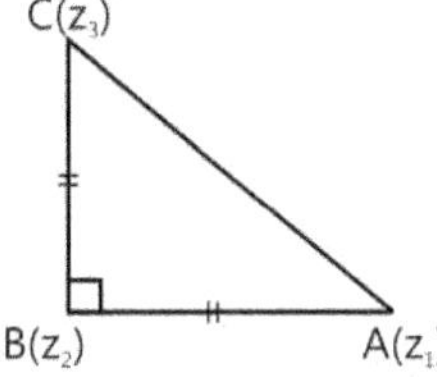

Figure 6.45

11. CONCEPTS OF ROTATION OF COMPLEX NUMBER

Let z be a non-zero complex number. We can write z in the polar form as follows:

$z = r(\cos\theta + i\sin\theta) = re^{i\theta}$ where $r = |z|$ and $\arg(z) = \theta$ (see Fig 6.46).

Consider a complex number $ze^{i\alpha}$.

$$ze^{i\alpha} \qquad = (re^{i\theta})e^{i\alpha} \qquad = re^{i(\theta + \alpha)}$$

Thus, $ze^{i\alpha}$ represents the complex number whose modulus is r and argument is $\theta + \alpha$.

Geometrically, $ze^{i\alpha}$ can be obtained by rotating the line segment joining

O and P(z) through an angle α in the anticlockwise direction.

Corollary: If $A(z_1)$ and $B(z_2)$ are two complex number such that

$\angle AOB = \theta$, then $z_2 = \dfrac{|z_2|}{|z_1|}z_1 e^{i\theta}$ (see Fig 6.47).

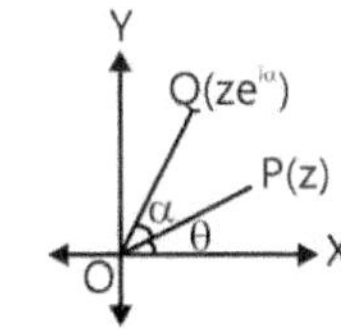

Figure 6.46

Let $z_1 = r_1 e^{i\alpha}$ and $z_2 = r_2 e^{i\beta}$ $\qquad$ where $\qquad |z_1| = r_1, |z_2| = r_2$.

Then $\dfrac{z_2}{z_1} = \dfrac{r_2 e^{i\beta}}{r_1 e^{i\alpha}} = \dfrac{r_2}{r_1}e^{i(\beta - \alpha)}$

Thus, $\dfrac{z_2}{z_1} = \dfrac{r_2}{r_1}e^{i\theta}$ $(\because \beta - \alpha = \theta)$ $\qquad \Rightarrow z_2 = \dfrac{|z_2|}{|z_1|}z_1 e^{i\theta}$

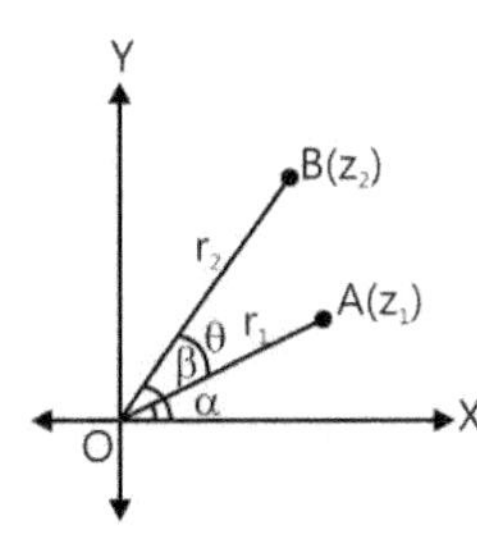

Figure 6.47

Multiplication of a complex number, z with i.

Let $z = r(\cos\theta + i\sin\theta)$ and $i = \left(\cos\dfrac{\pi}{2} + i\sin\dfrac{\pi}{2}\right)$, then $iz = r\left[\cos\left(\dfrac{\pi}{2}+\theta\right) + i\sin\left(\dfrac{\pi}{2}+\theta\right)\right]$.

Hence, iz can be obtained by rotating the vector z by right angle in the positive sense. And so on, to multiply a vector by -1 is to turn it through two right angles.

Thus, multiplying a vector by $(\cos\theta + i\sin\theta)$ is to turn it through the angle θ in the positive sense.

Illustration 37: Suppose $A(z_1)$, $B(z_2)$ and $C(z_3)$ are the vertices of an equilateral triangle inscribed in the circle $|z| = 2$. If $z_1 = 1 + \sqrt{3}i$, then z_2 and z_3 are respectively.

(a) $-2, 1-\sqrt{3}i$ (b) $-1+\sqrt{3}i, -2$

(c) $-2, -1+\sqrt{3}i$ (d) $-2, 2+\sqrt{3}i$ **(JEE ADVANCED)**

Sol: As we know $x + iy = re^{i\theta}$. Hence by using this formula we can obtain z_2 and z_3.

$$z_1 = 1 + \sqrt{3}i = 2e^{\frac{i\pi}{3}}$$

Since, $\angle AOC = \dfrac{2\pi}{3}$ and $\angle BOC = \dfrac{2\pi}{3}$, $z_2 = z_1 e^{\frac{2\pi i}{3}}$ and $z_3 = z_2 e^{\frac{2\pi i}{3}}$

$$\Rightarrow z_3 = 2e^{\pi i} = 2(\cos\pi + i\sin\pi) = -2 \text{ and } z_3 = 2e^{\frac{5\pi i}{3}}$$

$$= 2\left[\cos\left(2\pi - \dfrac{\pi}{3}\right) + i\sin\left(2\pi - \dfrac{\pi}{3}\right)\right]$$

$$= 2\left[\cos\dfrac{\pi}{3} - i\sin\dfrac{\pi}{3}\right] = 2\left[\dfrac{1}{2} - \dfrac{\sqrt{3}}{2}i\right] = 1 - \sqrt{3}i.$$

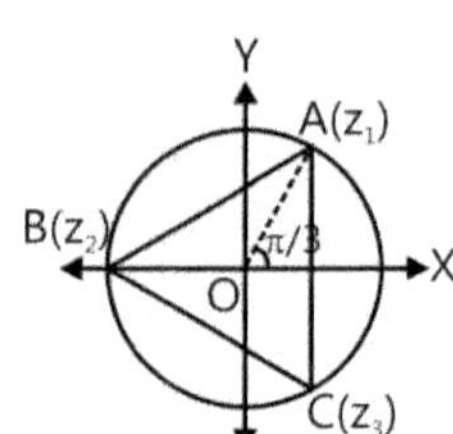

Figure 6.48

PROBLEM-SOLVING TACTICS

(a) On a complex plane, a complex number represents a point.

(b) In case of division and modulus of a complex number, the conjugates are very useful.

(c) For questions related to locus and for equations, use the algebraic form of the complex number.

(d) Polar form of a complex number is particularly useful in multiplication and division of complex numbers. It directly gives the modulus and the argument of the complex number.

(e) Translate unfamiliar statements by changing z into x+iy.

(f) Multiplying by $\cos\theta$ corresponds to rotation by angle θ about O in the positive sense.

(g) To put the complex number $\dfrac{a+ib}{c+id}$ in the form A + iB we should multiply the numerator and the denominator by the conjugate of the denominator.

(h) Care should be taken while calculating the argument of a complex number. If z = a + ib, then arg(z) is not always equal to $\tan^{-1}\left(\dfrac{b}{a}\right)$. To find the argument of a complex number, first determine the quadrant in which it lies, and then proceed to find the angle it makes with the positive x-axis.

For example, if z = − 1 − i, the formula $\tan^{-1}\left(\dfrac{b}{a}\right)$ gives the argument as $\dfrac{\pi}{4}$, while the actual argument is $\dfrac{-3\pi}{4}$.

FORMULAE SHEET

(a) Complex number $z = x + iy$, where $x, y \in R$ and $i = \sqrt{-1}$.

(b) If z = x + iy then its conjugate $\bar{z}$ = x − iy.

(c) Modulus of z, i.e. $|z| = \sqrt{x^2 + y^2}$

(d) Argument of z, i.e. $\theta = \begin{cases} \tan^{-1}\left|\dfrac{y}{x}\right| & x > 0, \quad y > 0 \\[2mm] \pi - \tan^{-1}\left|\dfrac{y}{x}\right| & x < 0, \quad y > 0 \\[2mm] -\pi + \tan^{-1}\left|\dfrac{y}{x}\right| & x < 0, \quad y < 0 \\[2mm] -\tan^{-1}\left|\dfrac{y}{x}\right| & x > 0, \quad y < 0 \end{cases}$

where, $\phi = \tan^{-1}\left|\dfrac{y}{x}\right|$

$\theta = \pi - \phi \qquad \theta = \phi$

$\theta = -\pi + \phi \qquad \theta = -\phi$

(e) If y=0, then argument of z, i.e. $\theta = \begin{cases} 0, & \text{if } x > 0 \\ \pi, & \text{if } x < 0 \end{cases}$

(f) If x=0, then argument of z, i.e. $\theta = \begin{cases} \dfrac{\pi}{2}, & \text{if } y > 0 \\[2mm] \dfrac{3\pi}{2}, & \text{if } y < 0 \end{cases}$

(g) In polar form $x = r\cos\theta$ and $y = r\sin\theta$, therefore $z = r(\cos\theta + i\sin\theta)$

(h) In exponential form complex number $z = re^{i\theta}$, where $e^{i\theta} = \cos\theta + i\sin\theta$.

(i) $\cos x = \dfrac{e^{ix} + e^{-ix}}{2}$ and $\sin x = \dfrac{e^{ix} - e^{-ix}}{2i}$

(j) Important properties of conjugate

    (i) $z + \bar{z} = 2\,\mathrm{Re}(z)$ and $z - \bar{z} = 2\,\mathrm{Im}(z)$

    (ii) $z = \bar{z} \Leftrightarrow z$ is purely real

    (iii) $z + \bar{z} = 0 \Leftrightarrow z$ is purely imaginary

    (iv) $z\bar{z} = [\mathrm{Re}(z)]^2 + [\mathrm{Im}(z)]^2$

    (v) $\overline{z_1 + z_2} = \bar{z}_1 + \bar{z}_2$

    (vi) $\overline{z_1 - z_2} = \bar{z}_1 - \bar{z}_2$

    (vii) $\overline{z_1 z_2} = \bar{z}_1 \bar{z}_2$

    (viii) $\overline{\left(\dfrac{z_1}{z_2}\right)} = \dfrac{\bar{z}_1}{\bar{z}_2}$ if $z_2 \neq 0$

(k) Important properties of modulus

    If z is a complex number, then

    (i) $|z| = 0 \Leftrightarrow z = 0$

    (ii) $|z| = |\bar{z}| = |-z| = |-\bar{z}|$

    (iii) $-|z| \leq \mathrm{Re}(z) \leq |z|$

    (iv) $-|z| \leq \mathrm{Im}(z) \leq |z|$

    (v) $z\bar{z} = |z|^2$

    If z_1, z_2 are two complex numbers, then

    (i) $|z_1 z_2| = |z_1||z_2|$

    (ii) $\left|\dfrac{z_1}{z_2}\right| = \dfrac{|z_1|}{|z_2|}$, if $z_2 \neq 0$

    (iii) $|z_1 + z_2|^2 = |z_1|^2 + |z_2|^2 + \bar{z}_1 z_2 + z_1 \bar{z}_2 = |z_1|^2 + |z_2|^2 + 2\,\mathrm{Re}(z_1 \bar{z}_2)$

    (iv) $|z_1 - z_2|^2 = |z_1|^2 + |z_2|^2 - \bar{z}_1 z_2 - z_1 \bar{z}_2 = |z_1|^2 + |z_2|^2 - 2\,\mathrm{Re}(z_1 \bar{z}_2)$

(l) Important properties of argument

    (i) $\arg(\bar{z}) = -\arg(z)$

    (ii) $\arg(z_1 z_2) = \arg(z_1) + \arg(z_2)$

    In fact $\arg(z_1 z_2) = \arg(z_1) + \arg(z_2) + 2k\pi$

$$\text{where, } k = \begin{cases} 0, & \text{if } -\pi < \arg(z_1) + \arg(z_2) \leq \pi \\ 1, & \text{if } -2\pi < \arg(z_1) + \arg(z_2) \leq -\pi \\ -1, & \text{if } \pi < \arg(z_1) + \arg(z_2) \leq 2\pi \end{cases}$$

    (iii) $\arg(z_1 \bar{z}_2) = \arg(z_1) - \arg(z_2)$

    (iv) $\arg\left(\dfrac{z_1}{z_2}\right) = \arg(z_1) - \arg(z_2)$

(v) $|z_1 + z_2| = |z_1 - z_2| \qquad \Leftrightarrow \arg(z_1) - \arg(z_2) = \dfrac{\pi}{2}$

(vi) $|z_1 + z_2| = |z_1| + |z_2| \qquad \Leftrightarrow \arg(z_1) = \arg(z_2)$

If $z_1 = r_1(\cos\theta_1 + i\sin\theta_1)$ and $z_2 = r_2(\cos\theta_2 + i\sin\theta_2)$, then

(vii) $|z_1 + z_2|^2 = |z_1|^2 + |z_2|^2 + 2|z_1||z_2|\cos(\theta_1 - \theta_2) = r_1^2 + r_2^2 + 2r_1r_2\cos(\theta_1 - \theta_2)$

(viii) $|z_1 - z_2|^2 = |z_1|^2 + |z_2|^2 - 2|z_1||z_2|\cos(\theta_1 - \theta_2) = r_1^2 + r_2^2 - 2r_1r_2\cos(\theta_1 - \theta_2)$

(m) Triangle on complex plane

(i) Centroid (G), $z_G = \dfrac{z_1 + z_2 + z_3}{3}$

(ii) Incentre (I), $z_I = \dfrac{az_1 + bz_2 + cz_3}{a + b + c}$

(iii) Orthocentre (H), $z_H = \dfrac{z_1\tan A + z_2\tan B + z_3\tan C}{\sum \tan A}$

(iv) Circumcentre (S), $z_S = \dfrac{z_1(\sin 2A) + z_2(\sin 2B) + z_3(\sin 2C)}{\sin 2A + \sin 2B + \sin 2C}$

(n) $(\cos\theta + i\sin\theta)^n = \cos n\theta + i\sin n\theta$

(o) $\sqrt{z} = \sqrt{x + iy} = \pm\left[\sqrt{\dfrac{|z| + x}{2}} + i\sqrt{\dfrac{|z| - x}{2}}\right]$ for $y > 0$

(p) Distance between $A(z_1)$ and $B(z_2)$ is given by $|z_2 - z_1|$

(q) Section formula: The point P (z) which divides the join of the segment AB in the ratio m : n

is given by $z = \dfrac{mz_2 + nz_1}{m + n}$.

(r) Midpoint formula: $z = \dfrac{1}{2}(z_1 + z_2)$.

(s) Equation of a straight line

(i) Non-parametric form: $z(\bar{z}_1 - \bar{z}_2) - \bar{z}(z_1 - z_2) + z_1\bar{z}_2 - z_2\bar{z}_1 = 0$

(ii) Parametric form: $z = tz_1 + (1 - t)z_2$

(iii) General equation of straight line: $\bar{a}z + a\bar{z} + b = 0$

(t) Complex slope of a line, $\mu = \dfrac{z_1 - z_2}{\bar{z}_1 - \bar{z}_2}$. Two lines with complex slopes μ_1 and μ_2 are

(i) Parallel, if $\mu_1 = \mu_2$

(ii) Perpendicular, if $\mu_1 + \mu_2 = 0$

(u) Equation of a circle: $|z - z_0| = r$

JEE Main/Boards

Example 1: f z_1 and z_2 are $1 - i, -2 + 4i$ respectively. Find $\text{Im}\left(\dfrac{z_1 z_2}{\overline{z}_1}\right)$.

Sol: $\dfrac{z_1 z_2}{\overline{z}_1} = \dfrac{(1-i)(-2+4i)}{1+i} = \dfrac{-2+2i+4i+4}{1+i}$

$= \dfrac{2+6i}{1+i} \times \dfrac{1-i}{1-i} = \dfrac{2+6i-2i+6}{2} = 4+2i$

$\therefore \text{Im}\left(\dfrac{z_1 z_2}{\overline{z}_1}\right) = 2$.

Example 2: Find the square root of $z = -7 - 24i$.

Sol: Consider $z_0 = x + iy$ be a square root then $z_0^2 = -7 - 24i$.

$-7 - 24i = x^2 - y^2 + 2ixy$

Equating real and imaginary parts we get

$x^2 - y^2 = -7$ ... (i)

and $2xy = -24$ ... (ii)

$(x^2 + y^2)^2 = (x^2 - y^2)^2 + 4x^2 y^2$

$= (-7)^2 + (-24)^2 = 625$

$\therefore \qquad x^2 + y^2 = 25$ ... (iii)

Solving (i) and (iii), we get,

$(x, y) = (3, -4); (-3, 4)$ by (ii)

$\therefore \qquad z_0 = \pm(3 - 4i)$.

Example 3: If n is a positive integer and ω be an imaginary cube root of unity, prove that

$1 + \omega^n + \omega^{2n} \begin{cases} 3, \text{ when n is a mulitple of 3} \\ 0, \text{ when n is not a mulitple of 3} \end{cases}$

Sol: Case I: $n = 3m; \ m \in I$

$\therefore 1 + \omega^n + \omega^{2n} = 1 + \omega^{3m} + \omega^{6m}$

$= 1 + 1 + 1 \ [\because \omega^3 = 1] \qquad = 3$

Case II: $n = 3m + 1$ or $3m + 2; \ m \in I$

(a) Let $n = 3m + 1$

$\therefore \ \text{L.H.S} = 1 + \omega^{3m+1} + \omega^{6m+2} = 1 + \omega + \omega^2 = 0$

(b) Let $n = 3m + 2$

$1 + \omega^{3m+2} + \omega^{6m+4} = 1 + \omega^2 + \omega^4 = 1 + \omega^2 + \omega = 0.$

Example 4: Show that $\left|\dfrac{z-3}{z+3}\right| = 2$ represents a circle.

Sol: Consider $z = x + iy$ and then by taking modulus we will get the result.

Let $z = x + iy$

$\therefore \ \left|\dfrac{z-3}{z+3}\right| = 2 \ \Rightarrow \ \left|\dfrac{x-3+iy}{x+3+iy}\right| = 2$

$\therefore \ |x-3+iy|^2 = 2^2 \, |x+3+iy|^2$

or $\quad (x-3)^2 + y^2 = 4\left((x+3)^2 + y^2\right)$

$\Rightarrow 3x^2 + 3y^2 + 30x + 27 = 0$

which represents a circle.

Example 5: If $|z_1| = |z_2| = \ldots\ldots = |z_n| = 1$

prove that $|z_1 + z_2 + \ldots\ldots + z_n| = \left|\dfrac{1}{z_1} + \dfrac{1}{z_2} + \ldots\ldots + \dfrac{1}{z_n}\right|$

Sol: $|z_j| = 1 \ \Rightarrow z_j \overline{z}_j = 1 \ \forall \ j = 1, \ldots\ldots, n$

$(\because \qquad z\overline{z} = |z^2|)$

L.H.S.

$|z_1 + z_2 + \ldots\ldots + z_n| = \left|\dfrac{1}{\overline{z}_1} + \dfrac{1}{\overline{z}_2} + \ldots\ldots + \dfrac{1}{\overline{z}_n}\right| =$

$\left|\overline{\dfrac{1}{z_1} + \dfrac{1}{z_2} + \dfrac{1}{z_3} + \ldots\ldots + \dfrac{1}{z_n}}\right|$

$= \left|\dfrac{1}{z_1} + \dfrac{1}{z_2} + \dfrac{1}{z_3} + \ldots\ldots + \dfrac{1}{z_n}\right| = \text{R.H.S.}$

Example 6: If $|z_1 + z_2| = |z_1 - z_2|$, prove that

$\arg z_1 - \arg z_2 = \text{odd multiple of } \dfrac{\pi}{2}$.

Sol: As we know $|z| = z.\overline{z}$. Apply this formula and consider $z = r(\cos\theta + i\sin\theta)$.

$|z_1 + z_2|^2 = |z_1 - z_2|^2$

$\Rightarrow (z_1 + z_2)(\overline{z}_1 + \overline{z}_2) = (z_1 - z_2)(\overline{z}_1 - \overline{z}_2)$ or

$z_1\overline{z}_1 + z_2\overline{z}_2 + z_2\overline{z}_1 + z_1\overline{z}_2 = z_1\overline{z}_1 + z_2\overline{z}_2 - z_2\overline{z}_1 - z_1\overline{z}_2$

or $\quad 2(z_2\overline{z}_1 + z_1\overline{z}_2) = 0$; $\text{Re}(z_1\overline{z}_2) = 0$

Let $z_1 = r_1(\cos\theta_1 + i\sin\theta_1)$ and $z_2 = r_2(\cos\theta_2 + i\sin\theta_2)$;
then $z_1\overline{z}_2 = r_1 r_2\left(\cos(\theta_1 - \theta_2) + i\sin(\theta_1 - \theta_2)\right)$

$\therefore \cos(\theta_1 - \theta_2) = 0\left(\text{as Re}(z_1\overline{z}_2) = 0\right)$

$\theta_1 - \theta_2 = $ odd multiple of $\dfrac{\pi}{2}$.

Example 7: If $|z - 1| < 3$, prove that $|iz + 3 - 5i| < 8$.

Sol: Here we have to reduce $iz + 3 - 5i$ as the sum of two complex numbers containing $z - 1$. because we have to use

$|z - 1| < 3$.

$|iz + 3 - 5i| = |iz - i + 3 - 4i|$

$= |3 - 4i + i(z - 1)| \le |3 - 4i| + |i(z-1)|$

(by triangle inequality) $< 5 + 1 \cdot 3 = 8$

Example 8: If $(1 + x)^n = a_0 + a_1 x + a_2 x^2 + \ldots\ldots\ldots + a_n x^n$, then show that

(a) $a_0 - a_2 + a_4 + \ldots = 2^{\frac{n}{2}} \cos\dfrac{n\pi}{4}$

(b) $a_1 - a_3 + a_5 + \ldots = 2^{\frac{n}{2}} \sin\dfrac{n\pi}{4}$

Sol: Simply put $x = i$ in the given expansion and then by using formula

$z = r(\cos\theta + i\sin\theta)$ and $(\cos\theta + i\sin\theta)^n$

$= \cos n\theta + i\sin n\theta$, we can solve this problem.

Put $x = i$ in the given expansion

$(1 + i)^n = a_0 + a_1 i + a_2 i^2 + \ldots + a_n i^n$.

$\left[\sqrt{2}\left(\cos\dfrac{\pi}{4} + i\sin\dfrac{\pi}{4}\right)\right]^n$

$= (a_0 - a_2 + a_4 - \ldots) + i(a_1 - a_3 + a_5 - \ldots)$

$2^{n/2}\left(\cos\dfrac{n\pi}{4} + i\sin\dfrac{n\pi}{4}\right)$

$= (a_0 - a_2 + a_4 + \ldots\ldots) + i(a_1 - a_3 + a_5 + \ldots\ldots)$

Equating real and imaginary parts.

$2^{\frac{n}{2}} \cos\dfrac{n\pi}{4} = a_0 - a_2 + a_4 + \ldots\ldots$

$2^{\frac{n}{2}} \sin\dfrac{n\pi}{4} = a_1 - a_3 + a_5 + \ldots\ldots$

Example 9: Solve the equation $z^{n-1} = \overline{z} : n\in N$

Sol: Apply modulus on both side.

$Z^{n-1} = \overline{z}; \qquad |z|^{n-1} = |\overline{z}| = |z|$

$\therefore |z| = 0$ or $|z| = 1 \quad$ If $|z| = 0$ then $z = 0$,

Let $|z| = 1$; then, $z^n = z\overline{z} = 1$

$\therefore z = \cos\dfrac{2m\pi}{n} + i\sin\dfrac{2m\pi}{n} : m = 0, 1, \ldots, n - 1$

Example 10: If $z = x + iy$ and $\omega = \dfrac{1 - iz}{z - i}$
with $|\omega| = 1$, show that, z lies on the real axis.

Sol: Substitute value of ω in $|\omega| = 1$.

$|\omega| = \left|\dfrac{1 - iz}{z - i}\right| = 1 \Rightarrow |1 - iz| = |z - i|$

or, $|1 - ix + y| = |x + i(y - 1)|$

or, $(1 + y)^2 + x^2 = x^2 + (y - 1)^2$ or, $4y = 0$

Hence z lies on the real axis.

Example 11: If a complex number z lies in the interior or on the boundary of a circle of radius as 3 and centre at $(0, -4)$ then greatest and least value of $|z + 1|$ are-

(A) $3 + \sqrt{17}, \sqrt{17} - 3$ (B) 6, 1

(C) $\sqrt{17}, 1$ (D) 3, 1

Sol: Greatest and least value of $|z + 1|$ means maximum and minimum distance of circle from the point $(-1, 0)$. In circle greatest and least distance of it from any point is along the normal.

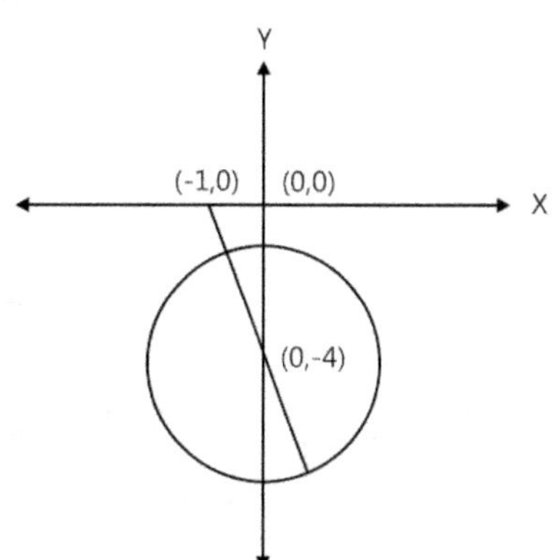

$\therefore$ Greatest distance $= 3 + \sqrt{1^2 + 4^2} = 3 + \sqrt{17}$

Least distance $= \sqrt{1^2 + 4^2} - 3 = \sqrt{17} - 3$

Example 12: Find the equation of the circle for which $\arg\left(\dfrac{z - 6 - 2i}{z - 2 - 2i}\right) = \pi / 4$.

Sol: $\arg\left(\dfrac{z - 6 - 2i}{z - 2 - 2i}\right) = \pi / 4$ represent a major arc of circle of which Line joining (6, 2)) and (2, 2) is a chord that subtends an angle $\dfrac{\pi}{4}$ at circumference.

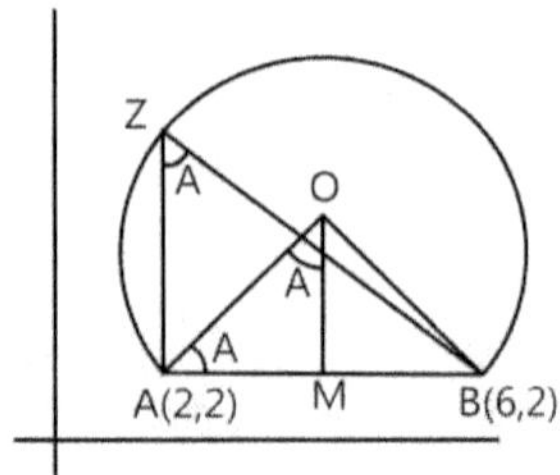

Clearly AB is parallel to real (x) axis, M is mid-point,

$M \equiv (4, 2)$, OM = AM = 2

$\therefore$ O = (4, 4) and $OA^2 = OM^2 + AM^2 = 2\sqrt{2}$ Equation of required circle is

$|z - 4 - 4i| = 2\sqrt{2}$

Example 13: If $|z| \geq 3$, prove that the least value of $\left|z + \dfrac{1}{z}\right|$ is $\dfrac{8}{3}$.

Sol: Here $\left|z + \dfrac{1}{z}\right| \geq |z| - \dfrac{1}{|z|}$.

Now $|z| \geq 3$

$\therefore \quad \dfrac{1}{|z|} \leq \dfrac{1}{3}$ or $-\dfrac{1}{|z|} \geq -\dfrac{1}{3}$ ... (i)

Adding the two like inequalities

$|z| - \dfrac{1}{|z|} \geq 3 - \dfrac{1}{3} = \dfrac{8}{3}$ ... (ii)

Hence from (i) and (ii), we get $\left|z + \dfrac{1}{z}\right| \geq \dfrac{8}{3}$

$\therefore$ Least value is $\dfrac{8}{3}$

Example 14: If z_1, z_2, z_3 are non-zero complex numbers such that $z_1 + z_2 + z_3 = 0$ and $z_1^{-1} + z_2^{-1} + z_3^{-1} = 0$ then prove that the given points are the vertices of an equilateral triangle. Also show that $|z_1| = |z_2| = |z_3|$.

Sol: Use algebra to solve this problem.

Given $z_1 + z_2 + z_3 = 0$, and from 2nd relation $z_2 z_3 + z_3 z_1 + z_1 z_2 = 0$

$\therefore z_2 z_3 = -z_1 (z_2 + z_3) = -z_1 (-z_1) = z_1^2$

$\therefore z_1^3 = z_1 z_2 z_3 = z_2^3 = z_3^3$

$\therefore |z_1|^3 = |z_2|^3 = |z_3|^3$

Above shows that distance of origin from A, B, C is same.

Origin is circumcentre, but $z_1 + z_2 + z_3 = 0$

implies that centroid is also at the origin so that the triangle must be equilateral.

JEE Advanced/Boards

Example 1: For constant $c \geq 1$, find all complex numbers z satisfying the equation $z + c\,|z + 1| + i = 0$

Sol: Solve this by putting $z = x + iy$.

Let $z = x + iy$.

The equation $z + c\,|z + 1| + i = 0$ becomes

$x + iy + c\,\sqrt{(x + 1)^2 + y^2} + i = 0$

or $x + c\,\sqrt{(x + 1)^2 + y^2} + i(y + 1) = 0$

Equating real and imaginary parts, we get

$y + 1 = 0 \Rightarrow y = -1$ ... (i)

and $x + c\,\sqrt{(x + 1)^2 + y^2} = 0 : x < 0$ ...(ii)

Solving (i) and (ii), we get

$x + c\,\sqrt{(x + 1)^2 + 1} = 0$ or $x^2 = c^2[(x + 1)^2 + 1]$

or $(c^2 - 1)\,x^2 + 2c^2 x + 2c^2 = 0$

If c = 1, then x = − 1. Let c > 1 ; then,

$x = \dfrac{-2c^2 \pm \sqrt{4c^4 - 8c^2(c^2 - 1)}}{2(c^2 - 1)} = \dfrac{-c^2 \pm c\sqrt{2 - c^2}}{c^2 - 1}$

As x is real and c > 1, we have: $1 < c \leq \sqrt{2}$

(Thus, for $c > \sqrt{2}$, there is no solution). Since both values of x satisfy (ii), both values are admissible.

Example 2: Find the sixth roots of $z = 64i$.

Sol: Here $i = \cos\dfrac{\pi}{2} + i\sin\dfrac{\pi}{2}$ and sixth root of z

i.e. $z_r = z^{1/6}$.

$$z = 64\left(\cos\dfrac{\pi}{2} + i\sin\dfrac{\pi}{2}\right) \qquad \therefore \ z_r = z^{1/6}$$

$$= 2\left[\cos\dfrac{2r\pi + \dfrac{\pi}{2}}{6} + i\sin\dfrac{2r\pi + \dfrac{\pi}{2}}{6}\right]$$

Where $r = 0, 1, 2, 3, 4, 5$

The roots $z_0, z_1, z_2, z_3, z_4, z_5$ are given by

$$z_0 = 2\left(\cos\dfrac{\pi}{12} + i\sin\dfrac{\pi}{12}\right)$$

$$z_1 = 2\left(\cos\dfrac{5\pi}{12} + i\sin\dfrac{5\pi}{12}\right)$$

$$z_2 = 2\left(\cos\dfrac{9\pi}{12} + i\sin\dfrac{9\pi}{12}\right)$$

$$z_3 = 2\left(\cos\dfrac{13\pi}{12} + i\sin\dfrac{13\pi}{12}\right) = -2\left(\cos\dfrac{\pi}{12} + i\sin\dfrac{\pi}{12}\right)$$

$$z_4 = 2\left(\cos\dfrac{17\pi}{12} + i\sin\dfrac{17\pi}{12}\right) = -2\left(\cos\dfrac{5\pi}{12} + i\sin\dfrac{5\pi}{12}\right)$$

$$z_5 = 2\left(\cos\dfrac{21\pi}{12} + i\sin\dfrac{21\pi}{12}\right) = -2\left(\cos\dfrac{9\pi}{12} + i\sin\dfrac{9\pi}{12}\right)$$

Example 3: Locate the region in the Argand plane for the complex number z satisfying

(a) $|z - 4| < |z - 2|$ (b) $\dfrac{\pi}{6} \le \arg z \le \dfrac{\pi}{4}$

Sol: Consider $z = x + iy$ and solve by using properties of modulus and argument.

(a) Let $z = x + iy$

$|x + iy - 4| < |x + iy - 2|$

$(x - 4)^2 + y^2 < (x - 2)^2 + y^2$ or $-4x + 12 < 0$

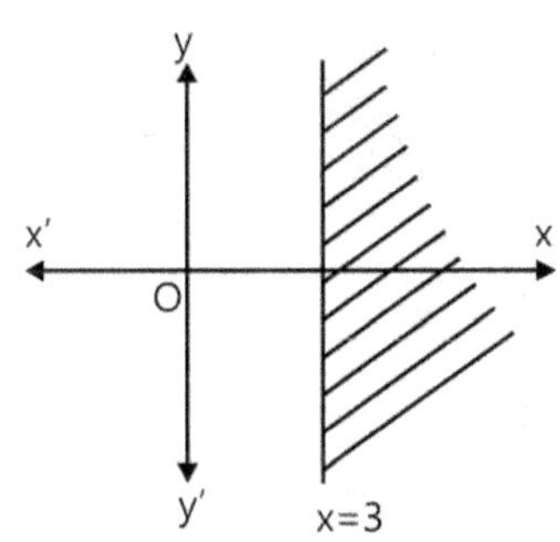

$\mathrm{Re}(z) > 3$ (see the Figure above)

(b) Let $z = x + iy$, then, $x > 0$ and $y > 0$

$$\arg z = \tan^{-1}\dfrac{y}{x} \quad \tan\dfrac{\pi}{6} \le \dfrac{y}{x} \le \tan\dfrac{\pi}{4}$$

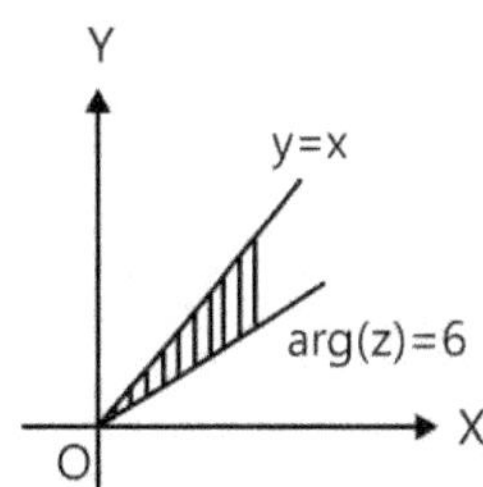

$$\dfrac{1}{\sqrt{3}} \le \dfrac{y}{x} \le 1; \quad x \le \sqrt{3}y \text{ and } y \le x$$

Hence the given inequality represents the region bounded by the rays $y = x$ and $y = \dfrac{1}{\sqrt{3}}x$ except the origin.

Example 4: If $z_1^2 + z_2^2 - 2z_1z_2\cos\theta = 0$, show that the points z_1, z_2 and the origin, in the argand plane, are the vertices of an isosceles triangle.

Sol: By using formula of roots of quadratic equation we can solve it.

$z_1^2 + z_2^2 - 2z_1z_2\cos\theta = 0$

$$\Rightarrow \left(\dfrac{z_1}{z_2}\right)^2 - 2\left(\dfrac{z_1}{z_2}\right)\cos\theta + 1 = 0$$

$$\Rightarrow \left(\dfrac{z_1}{z_2}\right) = \dfrac{2\cos\theta \pm \sqrt{4\cos^2\theta - 4}}{2}$$

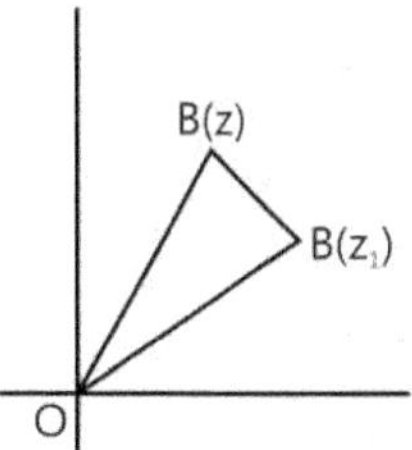

$= \cos\theta \pm i\sin\theta$

$$\Rightarrow \left|\dfrac{z_1}{z_2}\right| = |\cos\theta \pm i\sin q| - 1$$

$$\Rightarrow \left|\dfrac{z_1}{z_2}\right| = 1 \Rightarrow |z_1| = |z_2| \text{ or } OA = OB$$

Hence points $A(z)$, $B(z)$ and the origin are the vertices of an isosceles triangle.

Example 5: Let three vertices A, B, C (taken in clock wise order) of an isosceles right angled triangle with right angle at C, be affixes of complex numbers z_1, z_2, z_3 respectively. Show that $(z_1 - z_2)^2 = 2(z_1 - z_3)(z_3 - z_2)$.

Sol: Here $\dfrac{z_2 - z_3}{z_1 - z_3} = e^{-i\pi/2}$. Therefore solve it using algebra method.

Given CB = CA and angle $\angle C = \dfrac{\pi}{2}$

$$\dfrac{z_2 - z_3}{z_1 - z_3} = e^{-i\pi/2} \qquad \text{or} \qquad (z_3 - z_2)^2 = i^2(z_1 - z_3)^2$$

$$(z_3 - z_2)^2 = -(z_1 - z_3)^2$$

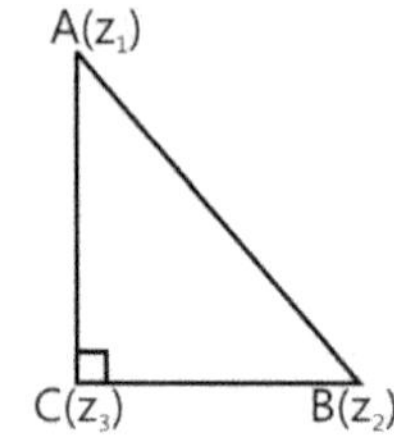

or

$$z_3^2 + z_2^2 - 2z_2z_3 + z_1^2 + z_3^2 - 2z_1z_3 = 0$$

Add and subtract $2z_1z_2$, we get

$$z_1^2 + z_2^2 - 2z_1z_2 + 2z_3^2 - 2z_2z_3 - 2z_1z_3 + 2z_1z_2 = 0, \text{ or}$$

$$(z_1 - z_2)^2 + 2[z_3(z_3 - z_2) - z_1(z_3 - z_2)] = 0 \text{ or}$$

$$(z_1 - z_2)^2 + 2(z_3 - z_1)(z_3 - z_2) = 0, \text{ or}$$

$$(z_1 - z_2)^2 = 2(z_1 - z_3)(z_3 - z_2).$$

Example 6: If A, B, C be the angles of triangle then

prove that $\begin{vmatrix} e^{2iA} & e^{-iC} & e^{-iB} \\ e^{-iC} & e^{2iB} & e^{-iA} \\ e^{-iB} & e^{-iA} & e^{2iC} \end{vmatrix}$ is purely real.

Sol: Here $A+B+C = \pi$, therefore $e^{pi} = \cos \pi + i \sin \pi = -1$. And by using properties of matrices we can solve this problem.

$$e^{-pi} = -1 \qquad \qquad \ldots \text{(i)}$$

$$e^{i(B+C)} = e^{i(\pi - A)} = e^{pi} e^{-iA} = -e^{-iA}$$

$$e^{-i(B+C)} = -e^{iA} \qquad \qquad \ldots \text{(ii)}$$

Take e^{iA}, e^{iB} and e^{iC} common from R_1, R_2 and R_3 respectively. $\Delta = e^{i(A+B+C)}$

$$\begin{vmatrix} e^{iA} & e^{-i(A+C)} & e^{-i(A+B)} \\ e^{-i(B+C)} & e^{iB} & e^{-i(B+A)} \\ e^{-i(B+C)} & e^{-i(C+A)} & e^{iC} \end{vmatrix}$$

$$= -1 \begin{vmatrix} e^{iA} & -e^{iB} & -e^{iC} \\ -e^{iA} & e^{iB} & -e^{iC} \\ -e^{iA} & -e^{iB} & e^{iC} \end{vmatrix}, \text{ by (2)}$$

Take e^{iA}, e^{iB} and e^{iC} common from C_1, C_2 and C_3 and again put $e^{i(A+B+C)} = e^{i\pi} = -1$.

$$\therefore \Delta = (-1)(-1) \begin{vmatrix} 1 & -1 & -1 \\ -1 & 1 & -1 \\ -1 & -1 & 1 \end{vmatrix}$$

Now make two zeros and expand

$\Delta = -4$ which is purely real.

Example 7: Prove that $|a + b|^2 + |a - b|^2$

$= 2(|a|^2 + |b|^2)$. Interpret the result geometrically and deduce that $\left| c + \sqrt{c^2 - d^2} \right| + \left| c - \sqrt{c^2 - d^2} \right| = |c + d| + |c - d|$; all numbers involved being complex

Sol: By using algebra of complex number and modulus property we can prove given expresson. And then by using Appolonius theorem we can interpret the result geometrically.

$$S = |a + b|^2 + |a - b|^2$$

$$= |a|^2 + |b|^2 + 2\text{Re}(a\bar{b}) + |a|^2 + |b|^2 - 2\text{Re}(a\bar{b}) = 2(|a|^2 + |b|^2) \text{ (proved)}$$

Now $|a + b|^2 + |a - b|^2 = 2(|a|^2 + |b|^2)$

This is nothing but Appolonius theorem. In $\triangle OAB$, M is midpoint of AB on applying Appolonius theorem we get

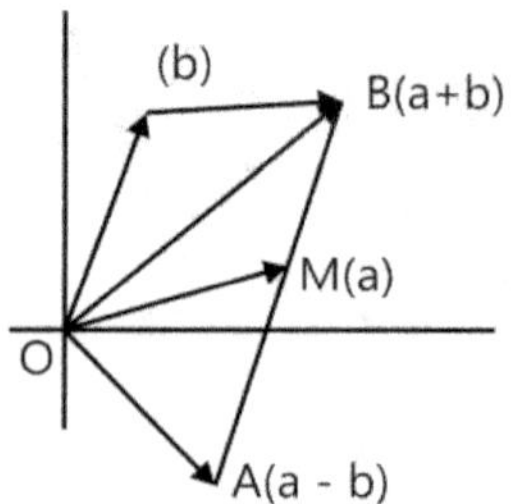

$$OA^2 + OB^2 = 2(AM^2 + OM^2)$$

Now take $a = \sqrt{\dfrac{c + d}{2}}$ and $b = \sqrt{\dfrac{c - d}{2}}$

Then using result

$$|a + b|^2 + |a - b|^2 = 2(|a|^2 + |b|^2)$$

$$\text{RHS} = 2\left(\left| \sqrt{\dfrac{c + d}{2}} \right|^2 + \left| \sqrt{\dfrac{c - d}{2}} \right|^2 \right) = |c + d| + |c - d|$$

L.H.S. $= \left(\left|\sqrt{\dfrac{c+d}{2}} + \sqrt{\dfrac{c-d}{2}}\right|\right)^2 + \left(\left|\sqrt{\dfrac{c+d}{2}} - \sqrt{\dfrac{c-d}{2}}\right|\right)^2$

On simplifying we get

L.H.S. $= \left|c + \sqrt{c^2 - d^2}\right| + \left|c - \sqrt{c^2 - d^2}\right|$

Example 8: Show that the triangles whose vertices are z_1, z_2, z_3 and a, b, c are

similar if $\begin{vmatrix} z_1 & a & 1 \\ z_2 & b & 1 \\ z_3 & c & 1 \end{vmatrix} = 0.$

Sol: Consider triangle ABC and DEF are similar, therefore $\dfrac{AB}{DE} = \dfrac{BC}{EF}$ and $\angle ABC = \angle DEF$.

Suppose z_1, z_2, z_3 are given by A, B, C respectively and a, b, c are given by D, E, F respectively. Since the triangle

ABC and DEF are similar $\dfrac{AB}{DE} = \dfrac{BC}{EF}$ and

$\angle ABC = \angle DEF = \alpha \text{(say)}$

We have $\angle B = \arg\left(\dfrac{z_1 - z_2}{z_3 - z_2}\right) = \arg\left(\dfrac{a-b}{c-b}\right)$

$\Rightarrow \dfrac{z_1 - z_2}{z_3 - z_2} = \dfrac{AB}{BC}(\cos\alpha + i\sin\alpha)$ ... (i)

and $\dfrac{a-b}{c-b} = \dfrac{DE}{EF}(\cos\alpha + i\sin\alpha)$ ... (ii)

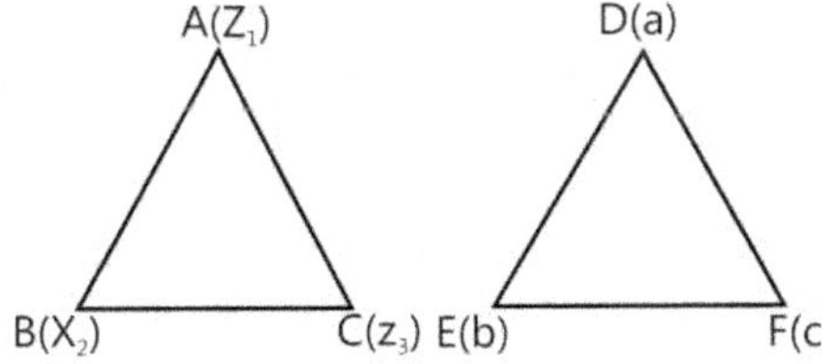

Since $\dfrac{AB}{DE} = \dfrac{BC}{EF}$ we have $\dfrac{AB}{BC} = \dfrac{DE}{EF}$

Thus, from (i) and (ii) we get

$\dfrac{z_1 - z_2}{z_3 - z_2} = \dfrac{a-b}{c-b} \Rightarrow \begin{vmatrix} z_1 - z_2 & a-b \\ z_3 - z_2 & c-b \end{vmatrix} = 0$

$\Rightarrow \begin{vmatrix} z_1 - z_2 & a-b & 0 \\ z_2 & b & 1 \\ z_3 - z_2 & c-b & 0 \end{vmatrix} = 0$

Applying $R_1 \rightarrow R_1 + R_2$ and $R_3 \rightarrow R_3 + R_2$

we get $\begin{vmatrix} z_1 & a & 1 \\ z_2 & b & 1 \\ z_3 & c & 1 \end{vmatrix} = 0$

Example 9: If $b_1 + b_2 + b_3 + b_4 = 0$ where b_1 etc. are non-zero real numbers, sum of no two being zero, and $b_1 z_1 + b_2 z_2 + b_3 z_3 + b_4 z_4 = 0$ where no three of the points z_1, z_2, z_3, z_4 are collinear then prove that the four point concyclic if $b_1 b_2 |z_1 - z_2|^2 = b_3 b_4 |z_3 - z_4|^2$.

Sol: Here the four points A, B, C, D will be concyclic if PA.PB = PC.CD. therefore obtain PA, PB, PC and CD and simplify.

Here $b_1 + b_2 = -(b_3 + b_4)$

Also $b_1 z_1 + b_2 z_2 = -(b_3 z_3 + b_4 z_4)$

Dividing these, $\dfrac{b_1 z_1 + b_2 z_2}{b_1 + b_2} = \dfrac{b_3 z_3 + b_4 z_4}{b_3 + b_4}$

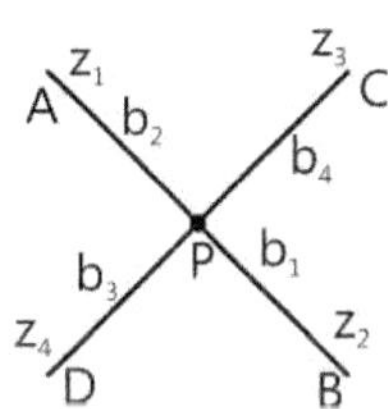

The left side gives the point that divides the line segment joining $A(z_1)$, $B(z_2)$ in the ratio $b_2 : b_1$, and the right side gives the point that divides the line segment joining the points $C(z_3)$, $D(z_4)$ in the ratio $b_4 : b_3$. So the line segments intersect at P which is

Represented by $\dfrac{b_1 z_1 + b_2 z_2}{b_1 + b_2}$ as well as $\dfrac{b_3 z_3 + b_4 z_4}{b_3 + b_4}$

Now, $AB = |z_1 - z_2|$

$\therefore PA = \dfrac{b_2}{b_1 + b_2}(z_1 - z_2)$ & $PB = \dfrac{b_1}{b_1 + b_2}(z_1 - z_2)$

Also, $CD = |z_3 - z_4|$

$\therefore PC = \dfrac{b_4}{b_3 + b_4}(z_3 - z_4)$ & $PD = \dfrac{b_3}{b_3 + b_4}(z_3 - z_4)$

The four points A, B, C, D will be concyclic if PA.PB = PC.CD

i.e. $\dfrac{b_1 b_2}{(b_1 + b_2)^2}|z_1 - z_2|^2 = \dfrac{b_3 b_4}{(b_3 + b_4)^2}|z_3 - z_4|^2$

i.e. $b_1 b_2 |z_1 - z_2|^2 = b_3 b_4 |z_3 - z_4|^2$

$(\because b_1 + b_2 = -(b_3 + b_4))$

Example 10: Show that all the roots of the equation $z^n \cos q_0 + z^{n-1} \cos q_1 + z^{n-2} \cos q_2 + \ldots\ldots + z \cos q_{n-1} + \cos q_n = 2$ lie outside the circle $|z| = \dfrac{1}{2}$ where q_0, q_1 etc. are real.

Sol: By using triangle inequality.

Here $|z^n \cos q_0 + z^{n-1} \cos q_1 + z^{n-2} \cos q_2 + \ldots + z \cos q_{n-1} + \cos q_n| = 2$ .... (i)

By triangle inequality.

$2 = |z^n \cos \theta_n + 2^{n-1} \cos q_1 + 2^{n-1} \cos q_2 + \ldots + z \cos q_{n-1} + \cos q_n| \le |z^n \cos q_n| + |z^{n-1} \cos q_1| +$

$|z^{n-2} \cos q_2| + \ldots + + |z \cos q_{n-1}| + |\cos q_n|$

$= |z_n| |\cos q_n| + |z^{n-1}| |\cos q_n| + \ldots\ldots + |z| |\cos q_{n-1}| + |\cos q_n| \le |z|^n + |z|^{n-1} + \ldots + |z| + 1$

($\because |\cos q_1| \le 1$ and $|z^{n+1}| = |z|^{n+1}$)

$= \dfrac{1 - |z|^{n+1}}{1 - |z|} < \dfrac{1}{1 - |z|} \quad \therefore 2 < \dfrac{1}{1 - |z|}$

So $1 - |z|$ is positive and $1 - |z| < \dfrac{1}{2}$

$\therefore |z| > 1 - \dfrac{1}{2} = \dfrac{1}{2}$

$\therefore$ All z satisfying (i) lie outside the circle $|z| = \dfrac{1}{2}$

Example 11: If $z + \dfrac{1}{z} = 2\cos\theta$, prove that $\left| \dfrac{z^{2n} - 1}{z^{2n} + 1} \right| = |\tan nq|$.

Sol: By using formula of roots of quadratic equation, we can solve this problem.

Here $z + \dfrac{1}{z} = 2 \cos \theta$;

$\therefore z^2 - 2 \cos \theta . z + 1 = 0$

$\therefore z = \dfrac{2\cos\theta \pm \sqrt{4\cos^2\theta - 4}}{2} = \cos \theta \pm i \sin \theta$

Taking positive sign, $z = \cos \theta + i \sin \theta$

$\therefore \dfrac{1}{z} = (\cos \theta + i \sin \theta)^{-1} = \cos \theta - i \sin \theta$

$\therefore \dfrac{z^{2n} - 1}{z^{2n} + 1} = \dfrac{z^n - \dfrac{1}{z^n}}{z^n + \dfrac{1}{z^n}} = \dfrac{(\cos\theta + i\sin\theta)^n - (\cos\theta - i\sin\theta)^n}{(\cos\theta + i\sin\theta)^n + (\cos\theta - i\sin\theta)^n}$

$= \dfrac{\cos n\theta + i\sin n\theta - (\cos n\theta - i\sin n\theta)}{\cos n\theta + i\sin n\theta + (\cos n\theta - i\sin n\theta}$

$= \dfrac{2i \sin n\theta}{2\cos n\theta} = i\tan n\theta$. Taking negative sign,

similarly we get $\dfrac{z^{2n} - 1}{z^{2n} + 1} = \dfrac{-2i\sin n\theta}{2\cos n\theta} = -i\tan n\theta$

$\therefore \left| \dfrac{z^{2n} - 1}{z^{2n} + 1} \right| = |\pm i \tan nq| = |\tan nq|,$

For $|\pm i| = 1$.

Example 12: Find the complex number z which satisfies the condition $|z - 2 + 2i| = 1$ and has the least absolute value.

Sol: Here $z - 2 + 2i = \cos \theta + i \sin \theta$, therefore by obtaing modulus of z we can solve above problem.

$|z - 2 + 2i| = 1$

$\Rightarrow z - 2 + 2i = \cos \theta + i \sin \theta$

Where θ is some real number.

$\Rightarrow z = (2 + \cos \theta) + (\sin \theta - 2)i$

$\Rightarrow |z| = [(2 + \cos \theta)^2 + (\sin \theta - 2)^2]^{1/2}$

$= [8 + \cos^2 \theta + \sin^2 \theta + 4(\cos \theta - \sin \theta)]^{1/2}$

$= \left[9 + 4\sqrt{2} \cos\left(\theta + \dfrac{\pi}{4} \right) \right]^{1/2}$

$|z|$ will be least if $\cos (\theta + \pi/4)$ is least, that is, if $\cos (\theta + \pi/4) = -1$ or $\theta = \dfrac{3\pi}{4}$. Thus, least value of $|z|$ is

$\left(9 - 4\sqrt{2} \right)^{1/2}$ for $z = \left(2 - \dfrac{1}{\sqrt{2}} \right) + i\left(\dfrac{1}{\sqrt{2}} - 2 \right)$

Example 13: For every real number $c \ge 0$, find all the complex numbers z which satisfy the equation. $2|z| - 4 cz + 1 + ic = 0$.

Sol: Substitute $z = x + iy$ and equate real and imaginary part to zero.

$2\sqrt{x^2 + y^2} - 4c (x + iy) + 1 + ic = 0$

$\therefore -4 cy + c = 0 \Rightarrow y = \dfrac{1}{4}$ ... (i)

$2\sqrt{x^2 + \dfrac{1}{16}} - 4cx + 1 = 0$ or

$4\left(x^2 + \dfrac{1}{16} \right) = (4 cx - 1)^2$

$4x^2 (4c^2 - 1) - 8 cx + \dfrac{3}{4} = 0$

$$\therefore \ x = \frac{8c \pm \sqrt{64c^2 - 12(4c^2 - 1)}}{8(4c^2 - 1)}$$

$$\text{or } x = \frac{4c \pm \sqrt{4c^2 + 3}}{4(4c^2 - 1)} \qquad \qquad \dots \text{(ii)}$$

x is real as $c \geq 0$, z = (x, y) as given by (i) and (ii), $c \geq 0$.

Example 14: Consider a square ABCD such that $z_1, z_2, z_3,$ and z_4 represent its vertices A, B, C and D respectively. Express 'z_3' and 'z_4' in terms of z_1 & z_2.

Sol: Consider the rotation of AB about A through an angle $\dfrac{\pi}{4}$.

Therefore $\dfrac{z_3 - z_1}{z_2 - z_1} = \left|\dfrac{z_3 - z_1}{z_2 - z_1}\right| e^{i\frac{\pi}{4}}$.

$$\frac{z_3 - z_1}{z_2 - z_1} = \left|\frac{z_3 - z_1}{z_2 - z_1}\right| e^{i\frac{\pi}{4}} = \sqrt{2}\left(\cos\frac{\pi}{4} + i\sin\frac{\pi}{4}\right)$$

$$\Rightarrow z_3 = z_1 + (z_2 - z_1)(1 + i)$$

Similarly $z_4 = z_1 + i(z_3 - z_1)$

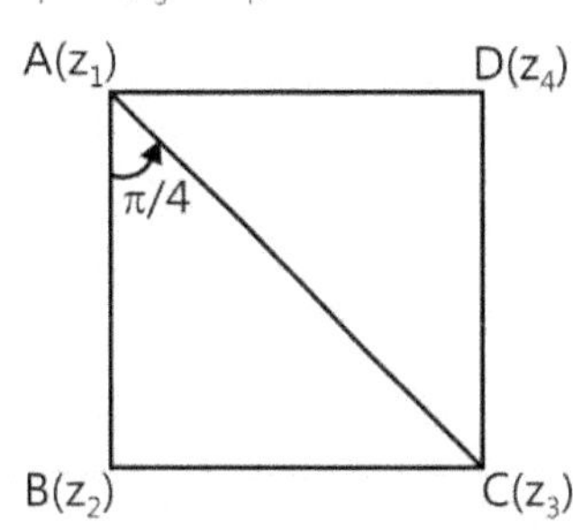

JEE Main/Boards

Exercise 1

Q.1 Find all non-zero complex numbers z satisfying $\bar{z} = iz^2$.

Q.2 Express $\dfrac{1 + 2i + 3i^2}{1 - 2i + 3i^2}$ in the form A + iB.

Q.3 Find x and y if (x + iy)(2 − 3i) = (4 + i)

Q.4 Find x and y if $\dfrac{(1+i)x - 2i}{3+i} + \dfrac{(2-3i)y + i}{3-i} = i$

Q.5 If $x = a + b$, $y = a\alpha + b\beta$ and $z = a\beta + b\alpha$, where α and β are complex cube roots of unity, show that $xyz = a^3 + b^3$.

Q.6 $\dfrac{1 + 7i}{(2 - i)^2}$ in the polar form.

Q.7 Find the square root of $- 8 - 6i$.

Q.8 Find the value of smallest positive integer n, for which $\left(\dfrac{1+i}{1-i}\right)^n = 1$.

Q.9 Show that the complex number z = x + iy which satisfies the equation $\left|\dfrac{z - 5i}{z + 5i}\right| = 1$ lies on the x-axis.

Q.10 If z = 1 + i tan α, where $\pi < \alpha < \dfrac{3\pi}{2}$. find the value of |z| cos α.

Q.11 If 1, ω, ω^2 be the cube roots of unity, find the roots of the equation $(x - 1)^3 + 8 = 0$.

Q.12 If |z| < 4, prove that $|i\,z + 3 - 4i| < 9$.

Q.13 2 + i 3 is a vertex of square inscribed in circle |z − 1| = 2. Find other vertices.

Q.14 Find the centre and radius of the circle formed by the points represented by z = x +iy satisfying the relation $\dfrac{|z - \alpha|}{|z - \beta|} = k(k \neq 1)$ where α & β are constant complex number's given by $\alpha = \alpha_1 + i\alpha_2$ & $\beta = \beta_1 + i\beta_2$

Q.15 Prove that there exists no complex number z such that $|z| < \dfrac{1}{3}$ and $\sum\limits_{r=1}^{a} a_r z^r = 1$ where $|a_r| < 2$.

Q.16 Let a complex number α, $\alpha \neq 11$, be a root of the equation $zp + q - zp - zq + 1 = 0$, where p, q are distinct primes. Show that either $1 + \alpha + \alpha^2 + \dots + \alpha^{p-1} = 0$ or $1 + \alpha + \alpha^2 + \dots + \alpha^{q-1} = 0$, but not both together.

Q.17 Show that the area of the triangle on the Argand diagram formed by the complex numbers: z, iz and $z + iz$ is: $\frac{1}{2}|z|^2$.

Q.18 If $iz^3 + z^2 - z + i = 0$ then show that $|z| = 1$.

Q.19 Find the value of the expression

$1(2 - \omega)(2 - \omega^2) + 2(3 - \omega)(3 - \omega^2) + \ldots$

$+ (n - 1)(n - \omega)(n - \omega^2)$ where ω is an imaginary cube root of unity.

Q.20 If $x = \frac{1}{2}\left(5 - \sqrt{3}i\right)$, then find the value of

$x^4 - x^3 - 12x^2 + 23x + 12$.

Q.21 Let the complex numbers z_1, z_2 and z_3 be the vertices of an equilateral triangle. Let z_0 be the circumcentre of the triangle. Then prove that: $z_1^2 + z_2^2 + z_3^2 = 3z_0^2$.

Q.22 If z_1, z_2, z_3 are the vertices of an isosceles triangle, right angled at z_2, prove that $z_1^2 + 2z_2^2 + z_3^2 = 2z_2(z_1 + z_3)$.

Q.23 Show that the equation

$$\frac{A^2}{x-a} + \frac{B^2}{x-b} + \frac{C^2}{x-c} + \ldots \ldots \frac{H^2}{x-h} = x + \ell,$$

Where A, B, C, ..., a, b, c, ... and ℓ are real, cannot have imaginary roots.

Q.24 Find the common roots of the equation

$z^3 + 2z^2 + 2z + 1 = 0$ and $z^{1985} + z^{100} + 1 = 0$.

Q.25 If n is an odd integer greater than 3 but not a multiple of 3, prove that $[(x + y)^n - x^n - y^n]$ is divisible by $xy(x + y)(x^2 + xy + y^2)$.

Q.26 If α and β are any two complex numbers,

show that $\left|\alpha + \sqrt{\alpha^2 - \beta^2}\right| + \left|\alpha - \sqrt{\alpha^2 - \beta^2}\right|$

$= |\alpha + \beta| + |\alpha - \beta|$

Q.27 Let $z_1 = 10 + 6i$ and $z_2 = 4 + 6i$. If z is any complex number such that the argument of $\frac{z - z_1}{z - z_2}$ is $\frac{\pi}{4}$, then prove that $|z - 7 - 9i| = 3\sqrt{2}$.

Q.28 If $|z| \le 1$, $|w| \le 1$, show that

$|z - w|^2 \le (|z| - |w|)^2 + (\arg z - \arg w)^2$.

Q.29 Let A and B be two complex numbers such that $\frac{A}{B} + \frac{B}{A} = 1$, prove that the origin and the points represented by A and B form the vertices of an equilateral triangle.

Q.30 Let z_1, z_2, z_3 be three complex numbers and a, b, c be real number not all zero, such that $a + b + c = 0$ and $az_1 + bz_2 + cz_3 = 0$.

Show that z_1, z_2, z_3 are collinear.

Q.31 If $|z - 4 + 3i| \le 2$, find the least and the greatest values of $|z|$ and hence find the limits between which $|z|$ lies.

Q.32 If $|z_1| < 1$ and $\left|\frac{z_1 - z_2}{1 - \bar{z}_1 z_2}\right| < 1$, then show that $|z_2| < 1$.

Q.33 Find the locus of points z if $\log_{\sqrt{3}} \frac{|z|^2 - |z| + 1}{2 + |z|} < 2$.

Q.34 For complex numbers z and ω, prove that $|z|^2 \omega - |\omega|^2 z = z - \omega$ if and only if $z = w$ or $z\bar{\omega} = 1$.

Exercise 2

Single Correct Choice Type

Q.1 $|z + 4| \le 3$, $Z \in C$: then the greatest and least value of $|z + 1|$ are:

(A) (7, 1)   (B) (6, 1)   (C) (6, 0)   (D) None

Q.2 The maximum & minimum values of $|z + 1|$ when $|z + 3| \le 3$ are

(A) (5, 0)   (B) (6, 0)   (C) (7, 1)   (D) (5, 1)

Q.3 The points $z_1 = 3 + \sqrt{3}i$ and $z_2 = 2\sqrt{3} + 6i$ are given on a complex plane. The complex number lying on the bisector of the angle formed by the vectors z_1 and z_2 is:

(A) $z = \frac{(3 + 2\sqrt{3})}{2} + \frac{\sqrt{3} + 2}{2}i$

(B) $z = 5 + 5i$

(C) $z = -1 - I$

(D) None of these

Q.4 If z_1, z_2, z_3, z_4 are the vertices of a square in that order, then which of the following do(es) not hold good?

(A) $\dfrac{z_1 - z_2}{z_3 - z_2}$ is purely imaginary

(B) $\dfrac{z_1 - z_3}{z_2 - z_4}$ is purely imaginary

(C) $\dfrac{z_1 - z_2}{z_3 - z_4}$ is purely imaginary

(D) None of these

Q.5 Let z_1, z_2 and z_3 be the complex numbers representing the vertices of a triangle ABC respectively and a, b, c are lengths of BC, CA, AB. If P is a point representing the complex number z_0 satisfying: $a(z_1 - z_0) + b(z_2 - z_0) + c(z_3 - z_0) = 0$, then w.r.t. the triangle ABC, the point P is its:

(A) Centroid　　　　　(B) Orthocentre

(C) Circumcentre　　　(D) Incentre

Q.6 Three complex numbers α, β & γ are represented in the Argand diagram by the three points A, B, C respectively. The complex number represented by D where A, B, C, D form a parallelogram with BD on a diagonal is:

(A) $\alpha - \beta + \gamma$　　　　(B) $-\alpha + \beta + \gamma$

(C) $\alpha + \beta - \gamma$　　　　(D) $\alpha - \beta - \gamma$

Q.7 If the complex number z satisfies the condition $|z| \geq 3$, then the least value of $\left| z + \dfrac{1}{z} \right|$ is

(A) $\dfrac{5}{3}$　　(B) $\dfrac{8}{3}$　　(C) $\dfrac{11}{3}$　　(D) None of these

Q.8 Point z_1 & z_2 are adjacent vertices of a regular octagon. The vertex z_3 adjacent to $z_2(z_3 \neq z_1)$ can be represented by:

(A) $z_2 + \dfrac{1}{\sqrt{2}}(1 \pm i)(z_1 + z_2)$

(B) $z_2 + \dfrac{1}{\sqrt{2}}(1 \pm i)(z_1 - z_2)$

(C) $z_2 + \dfrac{1}{\sqrt{2}}(1 \pm i)(z_2 - z_1)$

(D) None of these

Q.9 If q_1, q_2, q_3 are the roots of the equation, $x^3 + 64 = 0$, then the value of the determinant $\begin{vmatrix} q_1 & q_2 & q_3 \\ q_2 & q_3 & q_1 \\ q_3 & q_1 & q_2 \end{vmatrix}$ is:

(A) 1　　　　　　　　　(B) 4

(C) 10　　　　　　　　(D) none of these

Q.10 $z = (3 + 7i)(p + iq)$ where $p, q \in I - \{0\}$ purely imaginary then minimum value of $|z|^2$ is

(A) 0　　(B) 58　　(C) $\dfrac{3364}{3}$　　(D) 3364

Q.11 On the complex plane triangles OAP & OQR are similar and $\ell\,(OA) = 1$. If the points P and Q denotes the complex numbers z_1 & z_2 then the complex number 'z' denoted by the point R is given by:

(A) $z_1 z_2$　　(B) $\dfrac{z_1}{z_2}$　　(C) $\dfrac{z_2}{z_1}$　　(D) $\dfrac{z_1 + z_2}{z_2}$

Q.12 If A and B be two complex numbers satisfying $\dfrac{A}{B} + \dfrac{B}{A} = 1$. Then the two points represented by A and B and the origin form the vertices of

(A) An equilateral triangle

(B) An isosceles triangle which is not equilateral

(C) An isosceles triangle which is not right angled

(D) A right angled triangle

Q.13 The solutions of the equation in z, $|z|^2 - (z + \bar{z}) + i(z - \bar{z}) + 2 = 0$ are:

(A) $2 + i, 1 - i$　　　　(B) $1 + i, 1 - i$

(C) $1 + 2i, -1 - I$　　　(D) $1 + i, 1 + i$

Q.14 If $z_1 = -3 + 5i$: $z_2 = -5 - 3i$ and z is a complex number lying on the line segment joining z_1 & z_2 then arg z can be:

(A) $-\dfrac{3\pi}{4}$　　(B) $-\dfrac{\pi}{4}$　　(C) $\dfrac{\pi}{6}$　　(D) $\dfrac{5\pi}{6}$

Q.15 The points of intersection of the two curves $|z - 3| = 2$ and $|z| = 2$ in an argand plane are:

(A) $\dfrac{1}{2}(7 \pm i\sqrt{3})$　　　　(B) $\dfrac{1}{2}(3 \pm i\sqrt{7})$

(C) $\dfrac{3}{2} \pm i\sqrt{\dfrac{7}{2}}$　　　　(D) $\dfrac{7}{2} \pm i\sqrt{\dfrac{3}{2}}$

Q.16 Let z to be complex number having the argument θ, $0 < \theta < \dfrac{\pi}{2}$ and satisfying the equality $|z - 3i| = 3$. Then $\cot \theta - \dfrac{6}{z}$ is equal to:

(A) 1   (B) – 1   (C) i   (D) – i

Q.17 The locus represented by the equation, $|z - 1| + |z + 1| = 2$ is:

(A) An ellipse with foci (1, 0): (– 1, 0)

(B) One of the family of circles passing through the points of intersection of the circles $|z + 1| = 1$

(C) The radical axis of the circles $|z - 1| = 1$ and $|z + 1| = 1$

(D) The portion of the real axis between the points (1, 0) and (– 1, 0)

Q.18 Let P denotes a complex number z on the Argand's plane, and Q denotes a complex number $\sqrt{2|z|^2} \cos\left(\dfrac{\pi}{4} + \theta\right)$ where θ = amp z if 'O' is the origin, then the $\triangle OPQ$ is:

(A) Isosceles but not right angled

(B) Right angled but not isosceles

(C) Right isoscles

(D) Equilateral

Q.19 Let z_1, z_2, z_3 be three distinct complex numbers satisfying $|z_1 - 1| = |z_2 - 1| = |z_3 - 1|$.

If $z_1 + z_2 + z_3 = 3$ then z_1, z_2, z_3 must represent the vertices of:

(A) An equilateral triangle

(B) An isoseles triangles which is not equilateral

(C) A right triangle

(D) Nothing definite can be said

Q.20 If $p = a + b\omega + c\omega^2$; $q = b + c\omega + a\omega^2$; and $r = c + a\omega + b\omega^2$ where a, b, c $\neq$ 0 and ω is the complex cube root of unity, then:

(A) $p + q + r = a + b + c$

(B) $p^2 + q^2 + r^2 = a^2 + b^2 + c^2$

(C) $p^2 + q^2 + r^2 = 2(pq + qr + rp)$

(D) None of these

Previous Years' Questions

Q.1 The smallest positive integer n for which $\left(\dfrac{1+i}{1-i}\right)^n = 1$, is   *(1980)*

(A) 8   (B) 16

(C) 12   (D) None of these

Q.2 The complex numbers z = x + iy which satisfy the equation $\left|\dfrac{z - 5i}{z + 5i}\right| = 1$ lie on   *(1981)*

(A) The x-axis

(B) The straight line y = 5

(C) A circle passing through the origin

(D) None of these

Q.3 If z = x + iy and w = (1 – iz) / (z – i), then $|w| = 1$ implies that, in the complex plane   *(1983)*

(A) z lies on the imaginary axis

(B) z lies on the real axis

(C) z lies on the unit circle

(D) None of these

Q.4 The points z_1, z_2, z_3, z_4 in the complex plane are the vertices of a parallelogram taken in order, if and only if   *(1983)*

(A) $z_1 + z_4 = z_2 + z_3$ (B) $z_1 + z_3 = z_2 + z_4$

(C) $z_1 + z_2 = z_3 + z_4$ (D) None of these

Q.5 If z_1 and z_2 are two non-zero complex numbers such that $|z_1 + z_2| = |z_1| + |z_2|$, then arg (z_1) – arg (z_2) is equal to   *(1987)*

(A) $-\pi$ (B) $-\dfrac{\pi}{2}$ (C) 0   (D) $\dfrac{\pi}{2}$

Q.6 The complex numbers sin x + i cos 2x and cos x – i sin 2x are conjugate to each other, for   *(1988)*

(A) $x = n\pi$ (B) x = 0

(C) $x = \left(n + \dfrac{1}{2}\right)\pi$ (D) No value of x

Q.7 If ω ($\neq$ 1) is a cube root of unity and $(1 + \omega)^7 = A + B\omega$, then A and B are respectively   *(1995)*

(A) 0, 1   (B) 1, 1   (C) 1, 0   (D) – 1, 1

Q.8 Let z and ω be two non-zero complex numbers such that $|z| = |\omega|$ and arg (z) + arg(ω) = π, then z equals **(1995)**

(A) ω (B) $-\omega$ (C) $\overline{\omega}$ (D) $-\overline{\omega}$

Q.9 If ω is an imaginary cube root of unity, then $(1 + \omega - \omega^2)^7$ is equal to **(1998)**

(A) $128\,\omega$ (B) $-128\,\omega$ (C) $128\,\omega^2$ (D) $-128\,\omega^2$

Q.10 The value of sum $\sum_{n=1}^{13}\left(i^n + i^{-n+1}\right)$ where $i = \sqrt{-1}$ equals **(1998)**

(A) i (B) $i - 1$ (C) $-i$ (D) 0

Q.11 If $\begin{vmatrix} 6i & -3i & 1 \\ 4 & 3i & -1 \\ 20 & 3 & i \end{vmatrix} = x + iy$, then **(1998)**

(A) $x = 3, y = 1$ (B) $x = 1, y = 1$

(C) $x = 0, y = 3$ (D) $x = 0, y = 0$

Q.12 If z_1, z_2 and z_3 are complex numbers such that

$$|z_1| = |z_2| = |z_3| = \left|\frac{1}{z_1} + \frac{1}{z_2} + \frac{1}{z_3}\right| = 1,$$

then $|z_1 + z_2 + z_3|$ is **(2000)**

(A) Equal to 1 (B) Less than 1

(C) Greater than 3 (D) Equal to 3

Q.13 Let $\omega = -\dfrac{1}{2} + i\dfrac{\sqrt{3}}{2}$, then value of the determinant

$$\begin{vmatrix} 1 & 1 & 1 \\ 1 & -1-\omega^2 & \omega^2 \\ 1 & \omega^2 & \omega \end{vmatrix} \text{ is} \qquad \textbf{(2002)}$$

(A) 3ω (B) $3\,\omega\,(\omega - 1)$ (C) $3\omega^2$ (D) $3\,\omega\,(1 - \omega)$

Q.14 If ω ($\neq 1$) be a cube root of unity and $(1 + \omega^2)^n = (1 + \omega^4)^n$, then the least positive value of n is **(2004)**

(A) 2 (B) 3 (C) 5 (D) 6

Q.15 A man walks a distance of 3 units from the origin towards the North-West (N 45° E) direction. From there, he walks a distance of 4 units towards the North-West (N 45° W) direction to reach a point P. Then, the position of P in the Argand plane is **(2007)**

(A) $3e^{i\pi/4} + 4i$ (B) $(3 - 4i)e^{i\pi/4}$

(C) $(4 + 3i)e^{i\pi/4}$ (D) $(3 + 4i)e^{i\pi/4}$

Q.16 If $z \neq 1$ and $\dfrac{z^2}{z-1}$ is real, then the point represented by the complex number z lies **(2012)**

(A) Either on the real axis or on a circle passing through the origin.

(B) On a circle with centre at the origin.

(C) Either on the real axis or on a circle not passing through the origin.

(D) On the imaginary axis.

Q.17 If z is a complex number of unit modulus and argument θ, then arg $\left(\dfrac{1+z}{1+\overline{z}}\right)$ equals **(2013)**

(A) $\dfrac{\pi}{2} - \theta$ (B) θ (C) $\pi - \theta$ (D) $-\theta$

Q.18 If z is a complex number such that $|z| \geq 2$, then the minimum value of $\left|z + \dfrac{1}{2}\right|$ **(2014)**

(A) Is equal to $\dfrac{5}{2}$

(B) Lies in the interval (1, 2)

(C) Is strictly greater than $\dfrac{5}{2}$

(D) Is strictly greater than $\dfrac{3}{2}$ but less than $\dfrac{5}{2}$

Q.19 A complex number z is said to be unimodular if $|z| = 1$. Suppose z_1 and z_2 are complex numbers such that $\dfrac{z_1 - 2z_2}{2 - z_1\overline{z_2}}$ is unimodular and z_2 is not unimodular. Then the point z_1 lies on a: **(2015)**

(A) Straight line parallel to y-axis

(B) Circle of radius 2

(C) Circle of radius $\sqrt{2}$

(D) Straight line parallel to x-axis

Q.20 A value of θ for which $\dfrac{2 + 3i \sin \theta}{1 - 2i \sin \theta}$ is purely imaginary, is: **(2016)**

(A) $\dfrac{\pi}{6}$ (B) $\sin^{-1}\left(\dfrac{\sqrt{3}}{4}\right)$

(C) $\sin^{-1}\left(\dfrac{1}{\sqrt{3}}\right)$ (D) $\dfrac{\pi}{3}$

Exercise 1

Q.1 Prove that with regard to the quadratic equation $z^2 + (p + ip')z + q + iq' = 0$ where p, p', q, q' are all real.

(i) If the equation has one real root then $q'^2 - pp'q' + qp'^2 = 0$

(ii) If the equation has two equal roots then $p^2 - q'^2 = 4q$ and $pp' = 2q'$.

state whether these equal roots are real or complex.

Q.2 Let $z = 18 + 26i$ where $z_0 = x_0 + iy_0$ (x_0, $y_0 \in R$) is the cube roots of z having least positive argument. Find the value of $x_0 y_0 (x_0 + y_0)$.

Q.3 Show that the locus formed by z in the equation $z^3 + iz = 1$ never crosses the coordinate axes in the Argand's plane.

Further show that $|z| = \sqrt{\dfrac{-\mathrm{Im}(z)}{2\mathrm{Re}(z)\mathrm{Im}(z)+1}}$

Q.4 Consider the diagonal matrix $A_n = \mathrm{dia}\,(d_1, d_2, d_3, \dots d_n)$ of order where

$D_i = a^{i-1}$, $1 \le i \le n$ and $\alpha = e^{\frac{i2\pi}{n}}$; $i = \sqrt{-1}$, is the nth root of unity.

Let L: represent the value of Tr. $(A_7)^7$.

M: denotes the value of det (A_{2n+1}) + det (A_{2n}).

Find the value of $(L + M)$.

[Note: $T_r(A)$ denotes trace of square matrix A]

Q.5 Let $z_1, z_2 \in C$ such that $z_1^2 + z_2^2 \in R$. If $z_1(z_1^2 - 3z_2^2) = 10$ and $z_2(3z_1^2 - z_2^2) = 30$. Find the value of $(z_1^2 + z_2^2)$.

Q.6 If the equation $(z + 1)^7 + z^7 = 0$ has roots $z_1, z_2, \dots z_7$, find the value of

(a) $\displaystyle\sum_{r=1}^{7} \mathrm{Re}(Z_r)$ and $\displaystyle\sum_{r=1}^{7} \mathrm{Im}(Z_r)$

Q.7 If z is one of the imaginary 7th roots of unity, then find the equation whose roots are $(z + z^4 + z^2)$ and $(z^6 + z^3 + z^5)$.

Q.8 If the expression $z^5 - 32$ can be factorised into linear and quadratic factors over real coefficients as $(z^5 - 32) = (z - 2)(z^2 - pz + 4)(z^2 - zq + 4)$ then find the value of $(p^2 + 2p)$.

Q.9 Let z_1 & z_2 be any two arbitrary complex numbers then prove that:
$$|z_1| + |z_2| \ge \frac{1}{2}(|z_1| + |z_2|)\left|\frac{z_1}{|z_1|} + \frac{z_2}{|z_2|}\right|.$$

Q.10 Let z_i ($i = 1, 2, 3, 4$) represent the vertices of a square all of which lie on the sides of the triangle with vertices $(0, 0)$, $(2, 1)$ and $(3, 0)$. If z_1 and z_2 are purely real, then area of triangle formed by z_3, z_4 and origin is m (where m and n are in their lowest form). Find the value of $(m + n)$.

Q.11 (i) Let C_r's denotes the combinatorial coefficients in the expansion of $(1 + x)^n$, $n \in N$. If the integers

$a_n = C_0 + C_3 + C_6 + C_9 + \dots$

$b_n = C_1 + C_4 + C_7 + C_{10} + \dots$

and $c_n = C_2 + C_3 + C_8 + C_{11} + \dots$

then prove that

(a) $a_n^3 + b_n^3 + c_n^3 - 3a_n b_n c_n = 2^n$.

(b) $(a_n - b_n)^2 + (b_n - c_n)^2 + (c_n - a_n)^2 = 2$

(ii) Prove the identity:

$(C_0 - C_2 + C_4 - C_6 + \dots)^2$
$+ (C_1 - C_3 + C_5 - C_7 + \dots)^2 = 2^n$.

Q.12 Let z_1, z_2, z_3, z_4 be the vertices A, B, C, D respectively of a square on the Argand diagram taken in anticlockwise direction then prove that:

(i) $2z_2 = (1 + i)z_1 + (1 - i)z_3$ & (ii) $2z_4 = (1 - i)z_1 + (1 + i)z_3$

Q.13 A function f is defined on the complex number by $f(z) = (a + bi)z$, where 'a' and 'b' are positive numbers. This function has the property that the image of each point in the complex plane is equidistant from that point and the origin. Given that $|a + bi| = 8$ and that $b^2 = \dfrac{u}{v}$ where u and v are co-primes. Find the value of $(u + v)$.

Q.14 Prove that

(a) $\cos x + {}^nC_1 \cos 2x + {}^nC_2 \cos 3x + ... + {}^nC_n \cos(n+1)$

$x = 2^n . \cos^n \dfrac{x}{2} . \cos \left(\dfrac{n+2}{2} \right) x$

(b) $\sin x + {}^nC_1 \sin 2x + {}^nC_2 \sin 3x + ... + {}^nC_n \sin(n+1)$

$x = 2^n . \cos^n \dfrac{x}{2} . \sin \left(\dfrac{n+2}{2} \right) x.$

Q.15 Let $f(x) = ax^3 + bx^2 + cx + d$ be a cubic polynomial with real coefficients satisfying $f(i) = 0$ and $f(1 + i) = 5$. Find the value of $a^2 + b^2 + c^2 + d^2$.

Q.16 Let $\omega_1, \omega_2, \omega_3, ... \omega_n$ be the complex numbers. A line L on the complex plane is called a mean line for the points $\omega_1, \omega_2, \omega_3, ... \omega_n$ if L contains the points (complex numbers) $z_1, z_2, z_3, ... z_n$ such that $\sum\limits_{k=1}^{n}(z_k - \omega_k) = 0$.

Now for the complex number $\omega_1 = 32 + 170i$, $\omega_2 = -7 + 64i$, $\omega_3 = -9 + 200i$, $\omega_4 = 1 + 27i$ and $\omega_5 = -14 + 43i$, there is a unique mean line with y-intercept 3. Find the slope of the line.

Q.17 A particle start to travel from a point P on the curve C_1: $|z - 3 - 4i| = 5$, where $|z|$ is maximum. From P, the particle moves through an angle $\tan^{-1}\dfrac{3}{4}$ in anticlockwise direction on $|z - 3 - 4i| = 5$ and reaches at point Q. From Q, it comes down parallel to imaginary axis by 2 units and reaches at point R. Find the complex number corresponding to point R in the Argand plane.

Q.18 Evaluate: $\sum\limits_{p=1}^{32}(3p-2)\left(\sum\limits_{q=1}^{10}\left(\sin\dfrac{2q\pi}{11} - i\cos\dfrac{2q\pi}{11} \right) \right)^p$

Q.19 Let a, b, c be distinct complex numbers

such that $\dfrac{a}{1-b} = \dfrac{b}{1-c} = \dfrac{c}{1-a} = k.$

Find the value of k.

Q.20 Let α, β be fixed complex numbers and z is a variable complex number such that

$|z - \alpha|^2 + |z - \beta|^2 = k$

Find out the limits for 'k' such that the locus of z is a circle. Find also the centre and radius of the circle.

Q.21 C is the complex number. $f: C \to R$ is defined by $f(z) = |z^3 - z + 2|$. Find the maximum value of $f(z)$ if $|z| = 1$.

Q.22 Let a, b, c are distinct integers and ω, ω^2 are the imaginary cube roots of unity. If minimum value of $|a + b\omega + c\omega^2| + |a + b\omega^2 + c\omega|$ is $n^{\frac{1}{4}}$ where $n \in N$, then find the value of n.

Q.23 If the area of the polygon whose vertices are the solutions (in the complex plane) of the equation

$x^7 + x^6 + x^5 + x^4 + x^3 + x^2 + x + 1 = 0$

can be expressed in the simplest form as $\dfrac{a\sqrt{b} + c}{d}$, find the value of $(a + b + c + d)$.

Q.24 If a and b are positive integer such that $N = (a + ib)^3 - 107i$ is a positive integer.

Find N.

Q.25 If the biquadratic $x^4 + ax^3 + bx^2 + cx + d = 0$ (a, b, c, d $\in$ R) has 4 non real roots, two with sum $3 + 4i$ and the other two with product $13 + i$. Find the value of 'b'.

Q.26 Resolve $z^5 + 1$ into linear and quadratic

factors with real coefficients. Deduce that:

$4\sin\dfrac{\pi}{10}\cos\dfrac{\pi}{5} = 1$.

Q.27 If $x = 1 + i\sqrt{3} : y = 1 - i\sqrt{3}$ & $z = 2$,

prove that $x^p + y^p = z^p$ for every prime p > 3.

Q.28 Dividing f(z) by z – i, we get the remainder i and dividing it by z + i we get the remainder 1 + i. Find the remainder upon the division of f(z) by $z^2 + 1$.

Q.29 (a) Let z = a + b be a complex number, where x and y are real numbers. Let A and B be the sets defined by

$A = \{z| \ |z| \le 2\}$ and

$B = \{z| \ (1 - i)z + (1 + i) \bar{z} \ge 4\}$.

Find the area of the region $A \cap B$.

(b) For all real numbers x, let the mapping $f(x) = \dfrac{1}{x-i}$, where $i = \sqrt{-1}$. If there exist real number a, b, c and d for which f(a), f(b), f(c) and f(d) form a square on the complex plane. Find the area of the square.

Q.30

Column I	Column II
(A) Let ω be a non-real cube root of unity then the number of distinct elements in the set $\left\{(1+\omega+\omega^2+\ldots+\omega^n)^m \mid m,n\in N\right\}$	(p) 4
(B) Let 1, ω, ω^2 be the cube root of unity. The least possible degree of a polynomial with real coefficients having roots $2\omega,(2+3\omega),(2+3\omega^2),(2-\omega-\omega^2)$, is	(q) 5
(C) $\alpha=6+4i$ and $\beta=(2+4i)$ are two complex numbers on the complex plane. A complex number z. A complex number z satisfying $\text{amp}\left(\dfrac{z-\alpha}{z-\beta}\right) = \dfrac{\pi}{6}$ moves on the major segment of a circle whose radius is	(r) 6
	(s) 7

Exercise 2

Single Correct Choice Type

Q.1 The set of points on the complex plane such that $z^2 + z + 1$ is real and positive. (where $z = x + iy$, $x, y \in R$) is:

(A) Complete real axis only

(B) Complete real axis or all points on the line $2x + 1 = 0$

(C) Complete real axis or a line segment joining points $\left(-\dfrac{1}{2}, \dfrac{\sqrt{3}}{2}\right) \& \left(-\dfrac{1}{2}, -\dfrac{\sqrt{3}}{2}\right)$ excluding both.

(D) Complete real axis or set of points lying inside the rectangle formed by the lines.

$2x + 1 = 0$; $2x - 1 = 0$; $2y - \sqrt{3} = 0$ & $2y + \sqrt{3} = 0$

Q.2 The complex numbers whose real and imaginary parts are integers and satisfy the relation $z\bar{z}^3 + z^3\bar{z} = 350$ form a rectangle on the Argand plane, the length of whose diagonal is-

(A) 5   (B) 10   (C) 15   (D) 25

Q.3 Let z_1 and z_2 be non-zero complex numbers satisfying the equation, $z_1^2 - 2z_1z_2 + 2z_2^2 = 0$.

The geometrical nature of the triangle whose vertices are the origin and the points representing z_1 & z_2.

(A) An isosceles right angled triangle

(B) A right angled triangle which is not isosceles

(C) An equilateral triangle

(D) An isosceles triangle which is not right angled

Q.4 The set of points on the Argand diagram which satisfy both $|z| \le 4$ & $\text{Arg}\, z = \dfrac{\pi}{3}$ is:

(A) A circle and line   (B) A radius of a circle

(C) A sector of a circle   (D) An infinite part line

Q.5 If z_1 & z_2 are two complex numbers & if $\text{arg}\dfrac{z_1+z_2}{z_1-z_2} = \dfrac{\pi}{2}$ but $|z_1 + z_2| \ne |z_1 - z_2|$ then the figure formed by the points represented by $0, z_1, z_2$ & $z_1 + z_2$ is:

(A) A parallelogram but not a square rectangle or a rhombus

(B) A rectangle but not a square

(C) A rhombus but not a square

(D) A square

Q.6 If z_1, z_2, z_3 are the vertices of the $\triangle ABC$ on the complex plane & are also the roots of the equation $z^3 - 3\alpha z^2 + 3\beta z + x = 0$, then the condition for the $\triangle ABC$ to be equilateral triangle is:

(A) $\alpha^2 = \beta$ (B) $\alpha = \beta^2$

(C) $\alpha^2 = 2\beta$ (D) $\alpha = 3\beta^2$

Q.7 Let A, B, C represent the complex numbers z_1, z_2, z_3 respectively on the complex plane. If the circumcentre of the triangle. ABC lies at the origin then the orthocenter is represented by the complex number:

(A) $z_1 + z_2 - z_3$ (B) $z_2 + z_3 - z_1$

(C) $z_3 + z_1 - z_2$ (D) $z_1 + z_2 + z_3$

Q.8 Which of the following represents a point in an argand's plane, equidistant from the roots of equation $(z + 1)^4 = 16z^4$?

(A) $(0, 0)$ 　(B) $\left(-\dfrac{1}{3}, 0\right)$ 　(C) $\left(\dfrac{1}{3}, 0\right)$ 　(D) $\left(0, \dfrac{2}{\sqrt{5}}\right)$

Q.9 The equation of the radical axis of the two circles presented by the equations.

$|z - 2| = 3$ and $|z - 2 - 3i| = 4$ on the complex plane is:

(A) $3y + 1 = 0$ 　　　(B) $3y - 1 = 0$

(C) $2y - 1 = 0$ 　　　(D) None of these

Q.10 Number of real solution of the equation, $z^3 + iz - 1 = 0$ is

(A) Zero 　(B) One 　(C) Two 　(D) Three

Q.11 A point 'z' moves on the curve $|z - 4 - 3i| = 2$ in an argand plane. The maximum and minimum values of $|z|$ are:

(A) 2, 1 　(B) 6, 5 　(C) 4, 3 　(D) 7, 3

Q.12 Let $z = 1 - \sin\alpha + i\cos\alpha$ where $\alpha \in \left(0, \dfrac{\pi}{2}\right)$, then the modulus and the principle value of the argument of z are respectively:

(A) $\sqrt{2(1 - \sin\alpha)}, \left(\dfrac{\pi}{4} + \dfrac{\alpha}{2}\right)$ 　(B) $\sqrt{2(1 - \sin\alpha)}, \left(\dfrac{\pi}{4} - \dfrac{\alpha}{2}\right)$

(C) $\sqrt{2(1 + \sin\alpha)}, \left(\dfrac{\pi}{4} + \dfrac{\alpha}{2}\right)$ 　(D) $\sqrt{2(1 + \sin\alpha)}, \left(\dfrac{\pi}{4} - \dfrac{\alpha}{2}\right)$

Q.13 z_1 and z_2 are complex numbers. Then

Equ. $\left|\dfrac{z_1 + z_2}{2} + \sqrt{z_1 z_2}\right| + \left|\dfrac{z_1 + z_2}{2} - \sqrt{z_1 z_2}\right| =$

(A) $2\left|\sqrt{z_1} + \sqrt{z_2}\right|$ 　　(B) $2\left|\sqrt{z_1} - \sqrt{z_2}\right|$

(C) $2\left(\left|\sqrt{z_1}\right|^2 + \left|\sqrt{z_2}\right|^2\right)$ 　(D) $\left(\left|\sqrt{z_1}\right|^2 + \left|\sqrt{z_2}\right|^2\right)$

Q.14 If α, β be the roots of the equation $u^2 - 2u + 2 = 0$ and if $\cot\theta = x + 1$, then $\dfrac{(x + \alpha)^n - (x + \beta)^n}{\alpha - \beta}$ is equal to:

(A) $\dfrac{\sin n\theta}{\sin^n\theta}$ 　(B) $\dfrac{\cos n\theta}{\cos^n\theta}$ 　(C) $\dfrac{\sin n\theta}{\cos^n\theta}$ 　(D) $\dfrac{\cos n\theta}{\sin^n\theta}$

Q.15 If A_r ($r = 1, 2, 3, \ldots, n$) are the vertices of a regular polygon inscribed in a circle of radius R, then

$(A_1A_2)^2 + (A_1A_3)^2 + (A_1A_4)^2 + \ldots + (A_1A_n)^2 =$

(A) $\dfrac{nR^2}{2}$ 　　　　(B) $2nR^2$

(C) $4R^2 \cot\dfrac{\pi}{2n}$ 　　(D) $(2n - 1) R^2$

Q.16 If the equation $z^4 + a_1z^3 + a_2z^2 + a_3z + a_4 = 0$, where a_1, a_2, a_3, a_4 are real coefficients different from zero has a pure imaginary root then the expression $\dfrac{a_3}{a_1 a_2} + \dfrac{a_1 a_4}{a_2 a_3}$ has the value equal to

(A) 0 　(B) 1 　(C) -2 　(D) 2

Q.17 All roots of the equation $(1 + z)^6 + z^6 = 0$

(A) Lie on a unit circle with centre at the origin

(B) Lie on a unit circle with centre at $(-1, 0)$

(C) Lie on the vertices of a regular polygon with centre at the origin

(D) Are collinear

Q.18 Number of roots of the equation $z^{10} - z^5 - 992 = 0$ with real part negative is:

(A) 3 　(B) 4 　(C) 5 　(D) 6

Q.19 z_1 and z_2 are two distinct points in argand plane. If $a|z_1| = b|z_2|$, then the point $\dfrac{az_1}{bz_2} + \dfrac{bz_2}{az_1}$ is a point on the $(a, b \in R)$

(A) Line segment $[-2, 2]$ of the real axis

(B) Line segment $[-2, 2]$ of the imaginary axis

(C) Unit circle $|z| = 1$

(D) The line with arg $z = \tan^{-1} 2$

Q.20 If ω is an imaginary cube root of unity, then the value of $(p + q)^3 + (p\omega + q\omega^2)^3 + (p\omega^2 + q\omega)^3$ is

(A) $p^3 + q^3$

(B) $3(p^3 + q^3)$

(C) $3(p^3 + q^3) - pq(p + q)$

(D) $3(p^3 + q^3) + pq(p + q)$

Q.21 The solution set of the equation $\left(1+i\sqrt{3}\right)^x - 2^x = 0$

(A) Form an A.P.   (B) Form a G.P.

(C) Form an H.P.   (D) Is a empty set

Multiple Correct Choice Type

Q.22 In the quadratic equation $x^2 + (p + iq)x + 3i = 0$, p and q are real. If the sum of the squares of the roots is 8 then

(A) $p = 3, q = -1$ (B) $p = 3, q = 1$

(C) $p = -3, q = -1$ (D) $p = -3, q = 1$

Q.23 Let z_1 and z_2 be complex numbers such that $z_1 \neq z_2$ and $|z_1| - |z_2|$. If z_1 has positive real part and z_2 has negative imaginary part, then $\dfrac{z_1 + z_2}{z_1 - z_2}$ may be

(A) Zero   (B) Real & positive

(C) Real and negative   (D) Purely imaginary

Q.24 Given $a, b, x, y \in R$ then which of the following statement(s) hold good?

(A) $(a + ib)(x + iy)^{-1} = a - ib \Rightarrow x^2 + y^2 = 1$

(B) $(1 - ix)(1 + ix)^{-1} = a - ib \Rightarrow a^2 + b^2 = 1$

(C) $(a + ib)(a - ib)^{-1} = x - iy \Rightarrow |x + iy| = 1$

(D) $(y - ix)(a + ib)^{-1} = y + ix \Rightarrow |a - ib| = 1$

Q.25 If $z = x = iy = r(\cos\theta + i\sin\theta)$ then the values of $\sqrt{z}$ is equal to:

(A) $\pm\dfrac{1}{\sqrt{2}}\left(\sqrt{r+x} + i\sqrt{r-x}\right)$ for $y \geq 0$

(B) $\pm\dfrac{1}{\sqrt{2}}\left(\sqrt{r+x} - i\sqrt{r-x}\right)$ for $y \geq 0$

(C) $\pm\dfrac{1}{\sqrt{2}}\left(\sqrt{r+x} + i\sqrt{r-x}\right)$ for $y \leq 0$

(D) $\pm\dfrac{1}{\sqrt{2}}\left(\sqrt{r+x} - i\sqrt{r-x}\right)$ for $y \leq 0$

Q.26 For two complex numbers z_1 and z_2

$$\left(az_1 + b\bar{z}_1\right)\left(cz_2 + d\bar{z}_2\right) = \left(cz_1 + d\bar{z}_1\right)\left(az_2 + b\bar{z}_2\right)$$

If $(a, b, c, d \in R)$:

(A) $\dfrac{a}{b} = \dfrac{c}{d}$ (B) $\dfrac{a}{d} = \dfrac{b}{c}$

(C) $|z_1| = |z_2|$ (D) $\arg z_1 = \arg z_2$

Q.27 Let z_1, z_2 be two complex numbers represented by point on the circle $|z_1| = 1$ and $|z_2| = 2$ respectively, then:

(A) Max $|2z_1 + z_2| = 4$ (B) Min $|z_1 - z_2| = 1$

(C) $\left| z_2 + \dfrac{1}{z_1} \right| \leq 3$ (D) None of these

Q.28 If α, β any two complex numbers such that $\left| \dfrac{\alpha - \beta}{1 - \bar{\alpha}\beta} \right| = 1$, then

(A) $|\alpha| = 1$ (B) $|\beta| = 1$

(C) $\alpha = e^{i\theta}, \theta \in R$ (D) $\beta = e^{i\theta}, \theta \in R$

Q.29 On the argand plane, let $\alpha = -2 + 3z, \beta = -2 - 3z$ and $|z| = 1$. Then the correct statement is:

(A) α moves on the circle, centre at $(-2, 0)$ and radius 3

(B) α and β describe the same locus

(C) α and β move on different circles

(D) $\alpha - \beta$ moves on a circle concentric with $|z| = 1$

Q.30 The value of $i^n + i^{-n}$, for $i = \sqrt{-1}$ and $n \in I$ is:

(A) $\dfrac{2^n}{(1-i)^{2n}} + \dfrac{(1+i)^{2n}}{2^n}$ (B) $\dfrac{(1+i)^{2n}}{2^n} + \dfrac{(1-i)^{2n}}{2^n}$

(C) $\dfrac{(1+i)^{2n}}{2^n} + \dfrac{2^n}{(1-i)^{2n}}$ (D) $\dfrac{2^n}{(1+i)^{2n}} + \dfrac{2^n}{(1-i)^{2n}}$

Q.31 A complex number z satisfying the equation,

$$\log_{14}(13\,|z^2 - 4i|) + \log_{196}\dfrac{1}{(13 + |z^2 + 4i|)^2} = 0$$

(A) Can be purely real

(B) Can be purely imaginary

(C) Must be imaginary

(D) Must be real or purely imaginary

Q.32 Let S be the set of real values of x satisfying the inequality

$$1 - \log_2 \frac{|x+1+2i|-2}{\sqrt{2}-1} \geq 0, \text{ then S contains:}$$

(A) $[-3, -1)$ (B) $(-1, 1]$

(C) $[-2, 2)$ (D) $[1, 2]$

Q.33 If $x = \cos \alpha$; $y = \cos \beta$; $z = \cos \gamma$; Where $\alpha, \beta, \gamma \in$ R, then

(A) $\sum x = \Pi x \Rightarrow \cos(\alpha - \beta) = 1$

(B) $\Pi \dfrac{x-y}{z} = 8\Pi \cos \dfrac{\alpha - \beta}{2}$

(C) $\Pi \dfrac{x+y}{z}$ is real

(D) $\sum (Re)x = \cos\left(\sum \alpha\right), \sum Im(x) = \sin\left(\sum \alpha\right)$

Q.34 If $x_r = \cos\left(\dfrac{\pi}{2^r}\right)$ for $1 \leq r \leq n$; $r, n \in$ N then:

(A) $\displaystyle \lim_{n\to\infty} Re\left(\prod_{r=1}^{n} x_r\right) = -1$ (B) $\displaystyle \lim_{n\to\infty} Re\left(\prod_{r=1}^{n} x_r\right) = 0$

(C) $\displaystyle \lim_{n\to\infty} Im\left(\prod_{r=1}^{n} x_r\right) = 1$ (D) $\displaystyle \lim_{n\to\infty} Im\left(\prod_{r=1}^{n} x_r\right) = 0$

Q.35 If $1, z_1, z_2, z_3 \ldots\ldots z_{n-1}$ be the nth roots of unity and ω be a non real complex cube root of unity then the product $\displaystyle \prod_{r=1}^{n-1}(\omega - z_r)$ can be equal to:

(A) 0   (B) 1   (C) -1 (D) 2

Q.36 Identify the correct statements(s).

(A) No non zero complex number z satisfies the equation, $\bar{z} = -4z$

(B) $\bar{z} = z$ implies that z is purely real

(C) $\bar{z} = -z$ implies that z is purely imaginary

(D) If z_1, z_2 are the roots of the quadratic equation $az^2 + bz + c = 0$ such that Im $(z_1 z_2) \neq 0$ then a, b, c must be real numbers

Q.37 If the complex numbers z_1, z_2, z_3 & z_1', z_2' and z_3' are representing the vertices of two triangles such that $z_3 = (1 - z_0) z_1 + z_0 z_2$ and $z_3' = (1 - z_0) z_1' + z_0 z_2'$ where z_0 is also a complex number then:

(A) $\begin{vmatrix} z_1 & z_1' & 1 \\ z_2 & z_2' & 1 \\ z_3 & z_3' & 1 \end{vmatrix} = 0$

(B) The two triangles are congruent

(C) The two triangles are similar

(D) The two triangles have the same area.

Q.38 If 'z' be any complex number in a plane $(|z| \neq 0)$ then the complex number z for which the multiplication inverse is equal to the additive inverse is:

(A) $0 + i$ (B) $0 - i$ (C) $1 - i$ (D) $1 + i$

Q.39 Given $z = a + bi = \dfrac{1-ix}{1+ix}$; a, b, x $\in$ R, then which of the following holds good?

(A) $-\dfrac{\pi}{2} < \arg z \leq 0$ (B) $-\pi < \arg z \leq 0$

(C) $|z| = 1$ (D) $\arg z = \pi$; $|z| = 1$

Previous Years' Questions

Q.1 If z is any complex number satisfying $|z - 3 - 2i| \leq 2$, then the minimum value of $|2z - 6 + 5i|$ is ...... **(2011)**

Q.2 Let $\omega - e^{\frac{i2\pi}{3}}$ and a, b, c, x, y, z be non-zero complex number such that $a + b + c = x$, $a + b\omega + c\omega^2 = y$, $a + b\omega^2 + c\omega = z$. Then, the value of

$$\frac{|x|^2 + |y|^2 + |z|^2}{|a|^2 + |b|^2 + |c|^2} \text{ is ......} \qquad \textbf{(2011)}$$

Paragraph (for Q.3, 4, 5)

Read the following passage and answer the following questions.

Let A, B, C be three sets of complex number as defined below

$A = \{z: |I_m z \geq 1|\}$

$B = \{z: |z - 2 - i| = 3\}$

$C = \{z: Re\ ((1 - i)z) + \sqrt{2})$ **(2008)**

Q.3 The number of elements in the set $A \cap B \cap C$ is

(A) 0   (B) 1   (C) 2   (D) ∞

Q.4 Let z be any point in $A \cap B \cap C$. The $|z + 1 - i|^2 + |z - 5 - i|^2$ lies between

(A) 25 and 29   (B) 30 and 34

(C) 35 and 39   (D) 40 and 44

Q.5 Let z be any point in $A \cap B \cap C$ and let w any point satisfying $|w - 2 - i| < 3$. Then $|z| - |w| + 3$ lies between

(A) – 6 and 3   (B) – 3 and 6

(C) – 6 and 6   (D) – 3 and 9

Q.6 If $z = \left(\dfrac{\sqrt{3}}{2} + \dfrac{i}{2}\right)^5 + \left(\dfrac{\sqrt{3}}{2} - \dfrac{i}{2}\right)^5$, then   *(1982)*

(A) Re (z) = 0

(B) Im (z) = 0

(C) Re (z) > 0, Im (z) > 0

(D) Re (z) > 0, Im (z) < 0

Q.7 The inequality $|z - 4| < |z - 2|$ represents the region given by   *(1982)*

(A) Re (z) $\geq$ 0   (B) Re (z) < 0

(C) Re (z) > 0   (D) None of these

Q.8 If a, b, c and u, v, w are the complex numbers representing the vertices of two triangles such that c= (1–r) a+rb and w = (1 – r) u + rv, where r is a complex number, then the two triangles   *(1985)*

(A) Have the same area   (B) Are similar

(C) Are congruent   (D) None of these

Q.9 The value of $\displaystyle\sum_{k=1}^{6}\left(\sin\dfrac{2\pi k}{7} - i\cos\dfrac{2\pi k}{7}\right)$ is   *(1987)*

(A) – 1   (B) 0   (C) – i   (D) i

Q.10 Let z and w be two complex numbers such that $|z| \leq 1$, $|w| \leq 1$ and $|z + iw| = |z - i\overline{w}| = 2$, then z equals   *(1995)*

(A) 1 or i   (B) i or – i   (C) 1 or – 1   (D) i or –1

Q.11 For positive integers n_1, n_2 the value of expression

$(1+i)^{n_1} + (1+i^3)^{n_1} + (1+i^5)^{n_2} + (1+i^7)^{n_2}$,Here $i = \sqrt{-1}$

is a real number, if and only if   *(1996)*

(A) $n_1 = n_2 + 1$ (B) $n_1 = n_2 - 1$

(C) $n_1 = n_2$ (D) $n_1 > 0, n_2 > 0$

Q.12 If $i = \sqrt{-1}$, then

$$4 + 5\left(-\dfrac{1}{2} + \dfrac{i\sqrt{3}}{2}\right)^{334} + 3\left(-\dfrac{1}{2} + \dfrac{i\sqrt{3}}{2}\right)^{365}$$

is equal to   *(1999)*

(A) $1 - i\sqrt{3}$ (B) $-1 + i\sqrt{3}$ (C) $i\sqrt{3}$ (D) $-i\sqrt{3}$

Q.13 If arg (z) < 0, then arg (– z) – arg (z) equal *(2000)*

(A) π (B) $-\pi$ (C) $-\dfrac{\pi}{2}$ (D) $\dfrac{\pi}{2}$

Q.14 If z_1 = a + ib and z_2 = c + id are complex numbers such that $|z_1| = |z_2| = 1$ and Re $\left(z_1 \overline{z}_2\right) = 0$, then the pair of complex numbers w_1 = a + ic and w_2 = b + id satisfied by   *(1985)*

(A) $|w_1| = 1$ (B) $|w_2| = 1$

(C) Re $\left(w_1 \overline{w}_2\right) = 0$ (D) None of these

Q.15 Let z_1 and z_2 be two distinct complex numbers and let z = (1 – t) z_1 + tz_2 for some real number t with 0 < t < 1. If arg (w) denotes the principal argument of a non-zero complex number w, then   *(2010)*

(A) $|z - z_1| + |z - z_2| = |z_1 - z_2|$

(B) arg (z – z_1) = arg (z – z_2)

(C) $\begin{vmatrix} z - z_1 & \overline{z} - \overline{z}_1 \\ z_2 - z_1 & \overline{z}_2 - \overline{z}_1 \end{vmatrix} = 0$

(D) arg (z – z_1) = arg (z_2 – z_1)

Q.16 Match the statements in column I with those in column II.   *(2010)*

[Note: Here z takes values in the complex plane and Im z and Re z denote, respectively, the imaginary part and the real part of z.]

Column I	Column II
(A) The set of points z satisfying $|z - i|z|| = |z + i|z||$ Is contained in or equal to	(p) An ellipse with eccentricity $\dfrac{4}{5}$
(B) The set of points z satisfying $|z+4|+|z-4|=10$ is contained in or equal to	(q) The set of points z satisfying Im = z = 0
(C) If $|w|=2$, then the set of points $z = w - \dfrac{1}{w}$ is contained in or equal to	(r) The set of points z satisfying $|$ Im $z| \leq 1$

(D) If $|w| = 1$, then the set of points $z = w + \dfrac{1}{w}$ is contained in or equal to	(s) The set of points z satisfying $|\text{Re } z| \le 2$
	The set of points z satisfying $|z| \le 3$

Q.17 Let ω be the complex number $\cos\dfrac{2\pi}{3} + i\sin\dfrac{2\pi}{3}$. Then the number of distinct complex number z satisfying **(2010)**

$$\begin{vmatrix} z+1 & \omega & \omega^2 \\ \omega & z+\omega^2 & 1 \\ \omega^2 & 1 & z+\omega \end{vmatrix} = 0 \text{ is equal to}$$

Q.18 Let z_1 and z_2 be two distinct complex numbers and let $z = (1-t)z_1 + tz_2$ for some real number t with $0 < t < 1$. If Arg (w) denotes the principal argument of a nonzero complex number w, then **(2010)**

(A) $|z - z_1| + |z - z_2| = |z_1 - z_2|$

(B) $\text{Arg}(z - z_1) = \text{Arg}(z - z_2)$

(C) $\begin{vmatrix} z - z_1 & \bar{z} - \bar{z}_1 \\ z_2 - z_1 & \bar{z}_2 - \bar{z}_1 \end{vmatrix} = 0$

(D) $\text{Arg}(z - z_1) = \text{Arg}(z_2 - z_1)$

Q.19 if z is any complex number satisfying $|z - 3 - 2i| \le 2$, then the minimum value of $|2z - 6 - 5i|$ is **(2011)**

Q.20 Let $\omega = e^{i\pi/3}$, and a, b, c, x, y, z be non-zero complex numbers such that

$a + b + c = x$

$a + b\omega + c\omega^2 = y$

$a + b\omega^2 + c\omega = z$

Then the value of $\dfrac{|x|^2 + |y|^2 + |z|^2}{|a|^2 + |b|^2 + |c|^2}$ is **(2011)**

Q.21 Let $\omega \ne 1$ be a cube root of unity and S be the set of all non-singular matrices of the form

$$\begin{bmatrix} 1 & a & b \\ \omega & 1 & c \\ \omega^2 & \omega & 1 \end{bmatrix}$$ where each if a, b and c is either ω or ω^2.

Then the number of distinct matrices in the set S is **(2011)**

(A) 2   (B) 6   (C) 4   (D) 8

Q.22 Let z be a complex number such that the imaginary part of z is nonzero and $a = z^2 + z + 1$ is real. Then a cannot take the value **(2012)**

(A) -1   (B) $\dfrac{1}{3}$ (C) $\dfrac{1}{2}$ (D) $\dfrac{3}{4}$

Q.23 Let complex numbers α and $\dfrac{1}{\alpha}$ lies on circle $(x - x_0)^2 + (y - y_0)^2 = r^2$ and $(x - x_0)^2 + (y - y_0)^2 = 4r^2$, respectively. If $z_0 = x_0 + iy_0$ satisfies the equation $2|z_0|^2 = r^2 + 2$, then $|\alpha| =$ **(2013)**

(A) $\dfrac{1}{\sqrt{2}}$ (B) $\dfrac{1}{2}$ (C) $\dfrac{1}{\sqrt{7}}$ (D) $\dfrac{1}{3}$

Q.24 Let $\omega = \dfrac{\sqrt{3} + i}{2}$ and $P = \{\omega^n : n = 1, 2, 3, ...\}$. Further $H_1 = \left\{z \in C : \text{Re} z > \dfrac{1}{2}\right\}$ and $H_2 = \left\{z \in C : \text{Re} z < \dfrac{-1}{2}\right\}$, where C is the set of all complex numbers. If $z_1 \in P \cap H_1, z_2 \in P \cap H_2$ and O represents the origin, then $\angle z_1 O z_2 =$ **(2013)**

(A) $\dfrac{\pi}{2}$ (B) $\dfrac{\pi}{6}$ (C) $\dfrac{2\pi}{3}$ (D) $\dfrac{5\pi}{6}$

Q.25 Let ω be a complex cube root of unity with $\omega \ne 1$ and $P = [p_{ij}]$ be a $n \times n$ matrix with $p_{ij} = \omega^{i+j}$. Then $P^2 \ne 0$, when n = **(2013)**

(A) 57   (B) 55   (C) 58   (D) 56

Paragraph (for Q.26 and Q.27) **(2013)**

Let $S = S_1 \cap S_2 \cap S_3$, where

$S_1 = \{z \in C : |z| < 4\}, S_2 = \left\{z \in C : \text{Im}\left[\dfrac{z - 1 + \sqrt{3}i}{1 - \sqrt{3}i}\right] > 0\right\}$

and $S_3 = \{z \in C : \text{Re } Z > 0\}$

Q.26 Area of S =

(A) $\dfrac{10\pi}{3}$ (B) $\dfrac{20\pi}{3}$ (C) $\dfrac{16\pi}{3}$ (D) $\dfrac{32\pi}{3}$

Q.27 $\underset{z \in S}{\text{Min}} \left| 1 - 3i - z \right| =$

(A) $\dfrac{2 - \sqrt{3}}{2}$ (B) $\dfrac{2 + \sqrt{3}}{2}$ (C) $\dfrac{3 - \sqrt{3}}{2}$ (D) $\dfrac{3 + \sqrt{3}}{2}$

Q.28 Match the following: *(2014)*

Column I	Column II
(i) The number of polynomials f(x) with non-negative integer coefficients of degree ≤ 2, satisfying $f(0) = 0$ and $\displaystyle\int_0^1 f(x)dx = 1$, is	(p) 8
(ii) The number of points in the interval $\left[-\sqrt{13}, \sqrt{13} \right]$ at which $f(x) = \sin\left(x^2\right) + \cos\left(x^2\right)$ attains its maximum value, is	(q) 2
(iii) $\displaystyle\int_{-2}^{2} \dfrac{3x^2}{\left(1 + e^x\right)} dx$ equals	(r) 4
(iv) $\dfrac{\left(\displaystyle\int_{-1/2}^{1/2} \cos 2x.\log\left(\dfrac{1+x}{1-x}\right)dx\right)}{\left(\displaystyle\int_{0}^{1/2} \cos 2x.\log\left(\dfrac{1+x}{1-x}\right)dx\right)}$ equals	(s) 0

Codes:

	(i)	(ii)	(iii)	(iv)
(A)	r	q	s	p
(B)	q	r	s	p
(C)	r	q	p	s
(D)	q	r	p	s

Q.29 For any integer k, let $\alpha_k = \cos\left(\dfrac{k\pi}{7}\right) + i \sin\left(\dfrac{k\pi}{7}\right)$, where $i = \sqrt{-1}$. The value of the expression.

$$\dfrac{\displaystyle\sum_{k=1}^{12} \left| \alpha_{k+1} - \alpha_k \right|}{\displaystyle\sum_{k=1}^{3} \left| \alpha_{4k-1} - \alpha_{4k-2} \right|} \text{ is} \qquad (2015)$$

Q.30 Let $z = \dfrac{-1 + \sqrt{3}i}{2}$, where $i = \sqrt{-1}$, and $r, s \in \{1, 2, 3\}$.

Let $P = \begin{bmatrix} (-z)^r & z^{2s} \\ z^{2s} & z^r \end{bmatrix}$ and I be the identity matrix of order 2. Then the total number of ordered pairs (r, s) for which $P^2 = -I$ is   *(2016)*

Q.31 Let a, b $\in \mathbb{R}$ and $a^2 + b^2 \neq 0$.

Suppose $S = \left\{ z \in \mathbb{C} : z = \dfrac{1}{a + ibt}, \, t \in \mathbb{R}, \, t \neq 0 \right\}$, where $i = \sqrt{-1}$. If $z = x + iy$ and $z \in S$, then (x, y) lies on   *(2016)*

(A) The circle with radius $\dfrac{1}{2a}$ and centre $\left(\dfrac{1}{2a}, 0\right)$ for $a > 0, b \neq 0$

(B) The circle with radius $-\dfrac{1}{2a}$ and centre $\left(-\dfrac{1}{2a}, 0\right)$ for $a < 0, b \neq 0$

(C) The x-axis for $a \neq 0, b = 0$

(D) The y-axis for $a = 0, b \neq 0$

Important Questions

JEE Main/Boards

Exercise 1

Q.6	Q.9	Q.15
Q.18	Q.22	Q.24
Q.28	Q.31	Q.34

Exercise 2

Q.2	Q.8	Q.10
Q.13	Q.16	Q.18

Previous Years' Questions

Q.2	Q.4	Q.7
Q.10	Q.13	Q.15

JEE Advanced/Boards

Exercise 1

Q.7	Q.11	Q.13
Q.16	Q.18	Q.25
Q.29	Q.30	

Exercise 2

Q.2	Q.6	Q.9
Q.15	Q.19	Q.22
Q.25	Q.27	Q.31
Q.33	Q.36	Q.39

Previous Years' Questions

Q.2	Q.4	Q.8
Q.11	Q.14	Q.15

Answer Key

JEE Main/Boards

Exercise 1

Q.1 $z = 0, i, \pm \dfrac{\sqrt{3}}{2} - \dfrac{i}{2}$

Q.2 $-i$

Q.3 $x = \dfrac{5}{13}, y = \dfrac{14}{13}$

Q.4 $x = 3, y - 1$

Q.6 $\sqrt{2}\left[\cos\left(\dfrac{3\pi}{4}\right) + i\sin\left(\dfrac{3\pi}{4}\right)\right]$

Q.7 $\pm (1 - 3i)$

Q.8 $n = 4$

Q.10 -1

Q.11 $-1, 1 - 2\omega, 1 - 2\omega^2$

Q.13 $z_1 = (1 - \sqrt{3}) + i$; $z_2 = -i\sqrt{3}$; $z_3 = (1 + \sqrt{3}) - i$

Q.14 Centre $\left[\dfrac{\alpha - K^2\beta}{1 - K^2}\right]$, radius $\dfrac{|\alpha + \beta - K|}{1 - K^2}$

Q.19 $\dfrac{(n-1)n}{4}[n^2 + 3n + 4]$

Q.20 5

Q.31 $3 \leq |z| \leq 7$

Q.24 ω, ω^2

Q.33 Interior of circle $x^2 + y^2 = 25$

Exercise 2

Single Correct Choice Type

Q.1 C	**Q.2** D	**Q.3** B	**Q.4** C	**Q.5** D	**Q.6** A	**Q.7** B
Q.8 C	**Q.9** D	**Q.10** D	**Q.11** A	**Q.12** A	**Q.13** D	**Q.14** D
Q.15 B	**Q.16** C	**Q.17** D	**Q.18** C	**Q.19** A	**Q.20** C	

Previous Years' Questions

Q.1 D	**Q.2** A	**Q.3** B	**Q.4** B	**Q.5** C	**Q.6** D	**Q.7** B
Q.8 D	**Q.9** D	**Q.10** B	**Q.11** D	**Q.12** A	**Q.13** B	**Q.14** B
Q.15 D	**Q.16** A	**Q.17** B	**Q.18** B	**Q.19** B	**Q.20** C	

JEE Advanced/Boards

Exercise 1

Q.2 12 **Q.4** 7 **Q.5** 10 **Q.6** (a) $-\dfrac{7}{2}$, (b) zero **Q.7** $x^2 + x + 2 = 0$ **Q.8** 4

Q.10 41 **Q.13** 259 **Q.15** 26 **Q.16** 163 **Q.17** $(3 + 7i)$ **Q.18** $48(1 - i)$

Q.19 $-\omega$ or $-\omega^2$ **Q.20** $k > |\alpha - \beta|^2$

Q.21 If (z) is maximum when $z = \omega$, when ω is the cube root unity v and If (z) $= \sqrt{13}$

Q.22 144 **Q.23** 8 **Q.24** 198 **Q.25** 51

Q.26 $(z + 1)(z^2 - 2z \cos 36° + 1)(z^2 - 2z \cos 108° + 1)$ **Q.28** $\dfrac{iz}{2} + \dfrac{1}{2} + i$

Q.29 (a) $\pi - 2$; (b) $\dfrac{1}{2}$ **Q.30** $A \to s;\ B \to q;\ C \to p$

Exercise 2

Single Correct Choice Type

Q.1 C	**Q.2** B	**Q.3** A	**Q.4** B	**Q.5** C	**Q.6** A	**Q.7** D
Q.8 C	**Q.9** B	**Q.10** A	**Q.11** D	**Q.12** A	**Q.13** D	**Q.14** A
Q.15 B	**Q.16** B	**Q.17** D	**Q.18** A	**Q.19** A	**Q.20** B	**Q.21** A

Solutions

JEE Main/Boards

Exercise 1

Sol 1: $\bar{z} = i(z^2) \Rightarrow$ Let $x = a + ib$

$\Rightarrow a - ib = i(a^2 - b^2 + 2abi)$

$\Rightarrow a = -2ab$ and $-b = a^2 - b^2$

$a(1 + 2b) = 0$ and $a^2 = b^2 - b$

$a = 0$ or $b = -\dfrac{1}{2}$

if $a = 0 \Rightarrow b = 0, 1$

if $b = -\dfrac{1}{2}$, $a = \pm\dfrac{\sqrt{3}}{2}$

Complex numbers are $z = 0, i, \pm\dfrac{\sqrt{3}}{2} - \dfrac{i}{2}$

Sol 2: $\dfrac{1 + 3i^2 + 2i}{1 + 3i^2 - 2i} = \dfrac{-2 + 2i}{-2 - 2i} = \dfrac{1 - i}{1 + i}$

$= \left(\dfrac{1 - i}{1 + i}\right)\left(\dfrac{1 - i}{1 - i}\right) = \dfrac{1 + i^2 - 2i}{1 - i^2} = -i$

Sol 3: $(x + iy)(2 - 3i) = 4 + i$

$\Rightarrow (2x + 3y) + i(2y - 3x) = 4 + i$

$\Rightarrow 2x + 3y = 4$ and $2y - 3x = 1$

$\Rightarrow x = \dfrac{5}{13}$ and $y = \dfrac{14}{13}$

Sol 4: $\dfrac{(1 + i)x - 2i}{3 + i} + \dfrac{(2 - 3i)y + i}{3 - i} = i$

$= \dfrac{[x + i(x - 2)][3 - i] + [3 + i][2y + i(1 - 3y)]}{10}$

$= \dfrac{[3x + x - 2 + i(3x - 6 - x)] + [6y + 3y - 1 + i(2y + 3 - 9y)]}{10}$

$= \dfrac{4x - 2 + i(2x - 6) + (9y - 1) + i(-7y + 3)}{10}$

$= \dfrac{4x + 9y - 3 + i(2x - 7y - 3)}{10} = i$

$2x - 7y - 3 = 10$ and $4x + 9y - 3 = 0$

$\Rightarrow x = 3$ and $y = -1$

Sol 5: $x = a + b$

$y = \alpha a + b\beta$

$z = a\beta + b\alpha$

α and β are complex cube roots of unity

$\Rightarrow \alpha\beta = 1, \alpha^2 = \beta, \beta^2 = \alpha$

(as $\alpha = \omega, \beta = \omega^2, \alpha^2 = \beta$) ..... (i)

$xyz = (a + b)(\alpha a + b\beta)(a\beta + b\alpha)$

$= (\alpha a^2 + \alpha ab + \beta ab + b^2\beta)(a\beta + b\alpha)$

$= \alpha a^3\beta + \alpha a^2 b\beta + \alpha^2 b\beta^2 + ab^2\beta^2$

$\qquad + \alpha^2 a^2 b + \alpha^2 ab^2 + ab^2 a\beta + b^3\alpha\beta$

$= \alpha\beta(a^3 + b^3 + a^2 b + ab^2) + \alpha^2(a^2 b + b^2 a)$

$\qquad\qquad\qquad + \beta^2 (a^2 b + b^2 a)$

from eq^n. (i)

$= a^3 + b^3 + a^2 b + ab^2 + (a^2 b + b^2 a)(\alpha^2 + b^2)$

$= a^3 + b^3 + a^2 b + ab^2 + (a^2 b + b^2 a)(-1)$

$= a^3 + b^3 + a^2 b + ab^2 - a^2 b - ab^2$

$= a^3 + b^3$ hence proved.

Sol 6: $\dfrac{1+7i}{(2-i)^2} \times \dfrac{(2+i)^2}{(2+i)^2} = \dfrac{(1+7i)(2+i)^2}{25}$

$= \dfrac{(1+7i)(3+4i)}{25} = \dfrac{-25 + 25i}{25}$

$z = -1 + i$

$z(\theta) = |z|e^{i q}$

$|z| = \sqrt{1+1} = \sqrt{2}$

$\tan\theta = \dfrac{1}{-1} = -1$

$\Rightarrow \theta = \tan^{-1}(-1) = \dfrac{3\pi}{4}$

$\Rightarrow z(\theta) = \sqrt{2}e^{i\frac{3\pi}{4}} = \sqrt{2}\left[\cos\left(\dfrac{3\pi}{4}\right) + i\sin\left(\dfrac{3\pi}{4}\right)\right]$

Sol 7: $\sqrt{-8 - 6i} = a + ib$

$-8 - 6i = a^2 - b^2 + i(2ab)$

$\Rightarrow a^2 - b^2 = -8$ ...... (i)

$2ab = -6$

$\Rightarrow ab = -3$ ...... (ii)

Solving (i) and (ii), we get

$a = -1$ & $b = +3$

$a = 1$ & $b = -3$

So square root

$= (-1 + 3i)$ and $(1 - 3i)$

Sol 8: $\left(\dfrac{1+i}{1-i}\right)^n = 1$

$\dfrac{(1+i)^{2n}}{[(1+i)(1-i)]^n} = 1$

$\dfrac{(1+i)^{2n}}{2^n} = 1 \Rightarrow \dfrac{\left[(1+i)^2\right]^n}{2^n} = 1 \Rightarrow \dfrac{[2i]^n}{2^n} = 1$

$\Rightarrow i^n = 1 \Rightarrow n = 4, 8, 12$

Minimum value of n is 4

Sol 9: $\left|\dfrac{z - 5i}{z + 5i}\right| = 1$

$z = x + iy$

$\Rightarrow |x + i(y - 5)| = |x + i(y+5)|$

$\Rightarrow x^2 + (y - 5)^2 = x^2 + (y + 5)^2$

$\Rightarrow y = 0$

i.e. complex part of z is zero.

z is pure real i.e. it lies on x axis.

Sol 10: $z = 1 + i\tan\alpha$

$|z| = \sqrt{1 + \tan^2\alpha} = \sqrt{\sec^2\alpha} = |\sec\alpha|$

For $\alpha \in \left(\dfrac{\pi}{2}, \dfrac{3\pi}{2}\right)$

$\sec\alpha < 0 \Rightarrow |\sec\alpha| = -\sec\alpha$

$|z| = -\sec\alpha$

$|z|\cos\alpha = -1$

Sol 11: $(x - 1)^3 = -8$

$x - 1 = (-8)^{1/3}$

$x - 1 = (8^{1/3})(-1)^{1/3} \Rightarrow 1 - x = (8)^{1/3}(1)^{1/3}$

$(-x + 1) = 2, 2\omega, 2\omega^2$

$x = -1, 1 - 2\omega, 1 - 2\omega^2$

Sol 12: $|z| < 4$

$|3 + i(z - 4)| < 9$ (To prove)

We know that $|z_1 + z_2| \le |z_1| + |z_2|$

$|iz + (3 - 4i)| \le |z| + |3 - i4|$

$|z| < 4$

$\Rightarrow |iz + (3 - 4i)| < 4 + 5$

$\Rightarrow |iz + (3 - 4i)| < 9$ Hence proved

Sol 13: $2 + i\sqrt{3}$ is vertex of square inscribed in $|z - 1| = 2$

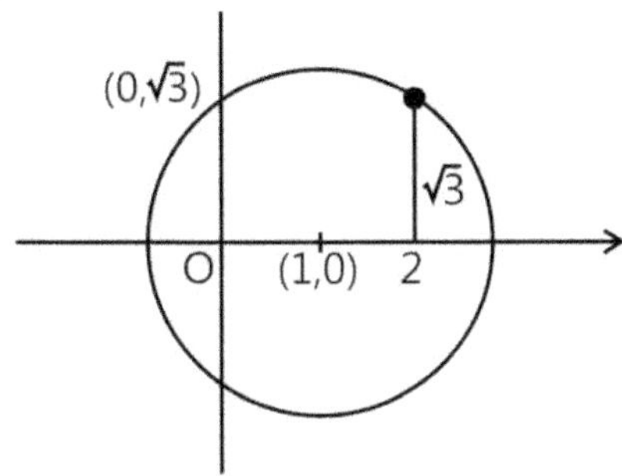

Here one vertex is $A(2, \sqrt{3})$ and equation of circle is $(x-1)^2 + y^2 = 4$

Radius of the circle is 2. Hence side of the square will be $2\sqrt{2}$.

Points that lie on the circle and are at a distance $2\sqrt{2}$ from A are $B(1 - \sqrt{3}, 1)$ and $D(1 + \sqrt{3}, -1)$.

The point C will be the other end of diameter of A. Hence $C(0, -\sqrt{3})$.

Hence the four vertices are

$2 + i\sqrt{3}$, $1 - \sqrt{3} + i$, $-\sqrt{3}i$ and $1 + \sqrt{3} - i$

Sol 14: $\left|\dfrac{z - \alpha}{z - \beta}\right| = k$

$\Rightarrow |z - \alpha|^2 = k^2|z - \beta|$

$\alpha = \alpha_1 + i\alpha_2$, $\beta = b_1 + i\beta_2$

$\Rightarrow (x - \alpha_1)^2 + (y - \alpha_2)^2 = k^2[(x - B_1)^2 + (y - \beta_2)^2]$

$\Rightarrow x^2 + \alpha_1{}^2 + y^2 + \alpha_2{}^2 - 2x\alpha_1 - 2y\alpha_2$

$= k^2x^2 + k^2y^2 + k^2\beta_1{}^2 + k^2\beta_2{}^2 - 2x\beta_1 k^2 - 2y\beta_2 k^2$

$\Rightarrow x^2(k^2-1) + y^2(k^2-1) + x(2\alpha_1 - 2\beta_1 k^2)$

$+ y(2\alpha_2 - 2\beta_2 k^2) + k^2\beta_1{}^2 + k^2\beta_2{}^2 - \alpha_1{}^2 - \alpha_2{}^2 = 0$

$\Rightarrow x^2 + y^2 + x\left[\dfrac{2\alpha_1 - 2\beta_1 k^2}{k^2 - 1}\right] + y\left[\dfrac{2\alpha_2 - 2\beta_2 k^2}{k^2 - 1}\right]$

$+ \left[\dfrac{k^2\beta_1{}^2 + k^2\beta_2{}^2 - \alpha_1{}^2 - \alpha_2{}^2}{k^2 - 1}\right] = 0$

Eq^n. of circle

with centre as $\left[\dfrac{\beta_1 k^2 - \alpha_1}{k^2 - 1}, \dfrac{\beta_2 k^2 - \alpha_2}{k^2 - 1}\right]$

or $\left[\dfrac{\alpha - k^2\beta}{1 - k^2}\right]$

and radius $= \dfrac{(\alpha_1 + \beta_1 - k)^2 + (\alpha_2 + \beta_2)^2}{1 - k^2} = \dfrac{|\alpha + \beta - k|}{1 - k^2}$

Sol 15: $|z| < \dfrac{1}{3}$ and $\displaystyle\sum_{r=1}^{n} a_r z^r = 1$ ... (i)

$\Rightarrow |a_1 z^1 + a_2 z^2 + a_3 z^3 \dots a_n z^n| = 1$

$|z_1 + z_2 + z_3 + \dots z_n| \le |z_1| + |z_2| + \dots |z_n|$ ... (ii)

$\Rightarrow 1 \le |a_1 z| + |a_2 z^2| + |a_3 z^3| \dots |a_n z^n|$

$\Rightarrow |a_1||z| + |a_2||z|^2 + |a_3||z|^3 \dots |a_n||z|^n \ge 1$

$\Rightarrow |z| + |z|^2 \dots |z|^n \ge \dfrac{1}{2}$

$\Rightarrow$ (Limiting case $n \to \infty$)

$\dfrac{|z|}{1 - |z|} \ge \dfrac{1}{2} \Rightarrow |z| \ge \dfrac{1}{3} - (2)$

From (i) and (ii), we can say that there is no 'z' satisfying both conditions.

Sol 16: $z^{p+q} - z^p - z^q + 1 = 0$

$z^p(z^q - 1).(z^q - 1) = 0$

$(z^p - 1).(z^q - 1) = 0$

$z^p = 1$ or $z^q = 1$

If α is roots, then $\alpha^\circ, \alpha, \alpha^2, \dots \alpha^{n-1}$ also roots of $z^p = 1$ or $z^q = 1$

Sum of roots $= 1 + \alpha + \alpha^2 + \dots + \alpha^{n-1} = 0$

Sol 17: $z, iz, z + iz$

$(x, y)\ (-y, x)\ (x - y, y + x)$

$\Delta = \dfrac{1}{2}\begin{vmatrix} x & y & 1 \\ -y & x & 1 \\ x-y & x+y & 1 \end{vmatrix} = \dfrac{1}{2} \ |-x^2 - y^2|$

$\Rightarrow |D| = \dfrac{1}{2}(x^2 + y^2) = \dfrac{1}{2}|z|^2$

Sol 18: $iz^3 + i + z^2 - z = 0$

$iz(z^2 + i) + i(z^2 + i) = 0$

$(iz + 1)(z^2 + i) = 0$

Either $z = i$ or $z^2 = i$

If $z = 1 \Rightarrow |z| = |i| = 1$

If $z^2 = i \Rightarrow |z|^2 = 1 \Rightarrow |z| = 1$

Hence, $|z| = 1$

Sol 19: $1(2 - \omega)(2 - \omega^2) + 2(3 - \omega)(3 - \omega^2) + \ldots\ldots (n - 1)(n - \omega)(n - \omega^2)$

$T_n = n(n + 1 - \omega)(n + 1 - \omega^2)$

$= (n^2 + n - n\omega)(n + 1 - \omega^2)$

$= n^3 + n^2 - n^2\omega^2 + n^3 + n - n\omega^2 - n^2\omega - n\omega + n\omega^3$

$= n^3 + n^2(2 - \omega^2 - \omega) + n(1 - \omega - \omega^2 + 1)$

$= n^3 + 3n^2 + 3n$

$S_n = \Sigma T_n = \Sigma(n^3 + 3n^2 + 3n)$

$= \Sigma n^3 + 3\Sigma n^2 + 3\Sigma n$

$= \left[\dfrac{n(n+1)}{2}\right]^2 + \dfrac{3n(n+1)(2n+1)}{6} + \dfrac{3n(n+1)}{2}$

$= \dfrac{n(n+1)}{2}\left[\dfrac{n(n+1)}{2} + (2n+1) + 3\right]$

$= \dfrac{n(n+1)}{4}[n^2 + n + 6 + 4n + 2]$

$Sn = \Sigma T_n = \dfrac{n(n+1)}{4}[n^2 + 5n + 8]$

$S_{n-1} = \dfrac{(n-1)n}{4}[n^2 + 1 - 2n + 5n - 5 + 8]$

$= \dfrac{(n-1)n}{4}[n^2 + 3n + 4]$

Sol 20: $x = \dfrac{1}{2}(5 - i\sqrt{3}) = \dfrac{1}{2}(6 - 1 - i\sqrt{3}) = 3 + w$

$x^4 - x^3 - 12x^2 + 23x + 12$

$= x^3(x - 3) + 2x^2(x - 3) - 6x(x - 3)$

$+ 5(x - 3) + 27$

$= (x - 3)[x^3 + 2x^2 - 6x + 5] + 27$

$= (x - 3)[(x - 3)(x^2 + 5x + 9) + 32] + 27$

$= (x - 3)^2[x^2 + 5x + 9] + 32(x - 3) + 27$

$= (x - 3)^2[x^2 - 3x + 8x - 24 + 33] + 32(x - 3) + 27$

$= (x - 3)^3(x + 8) + 33(x - 3)^2 + 32(x - 3) + 27$

$= \omega^3(x + 8) + 33(\omega^2) + 32\omega + 27$

$= \omega + 1 + 32\omega + 33\omega^2 + 27$

$= \omega + \omega^2 + 32\omega + 32\omega^2 + 38$

$= -1 - 32 + 38 = 5$

Sol 21: $z_2 - z_1 = (z_3 - z_1)e^{-i\pi/3}$

$z_3 - z_2 = (z_1 - z_2)e^{-i\pi/3}$

$\dfrac{z_2 - z_1}{z_3 - z_2} = \dfrac{z_3 - z_1}{z_1 - z_2}$

$\Rightarrow z_1^2 + z_2^2 + z_3^2 = z_1 z_2 + z_2 z_3 + z_3 z_1$

$\left(\dfrac{z_1 + z_2 + z_3}{3}\right)^2 = z_0^2$

$\Rightarrow \dfrac{z_1^2 + z_2^2 + z_3^2 + 2(z_1^2 + z_2^2 + z_3^2)}{9} = z_0^2$

$\Rightarrow z_1^2 + z_2^2 + z_3^2 = 3\,z_0^2$

Sol 22: $z_1\, z_2\, z_3$ of an isosceles ΔL at z_2

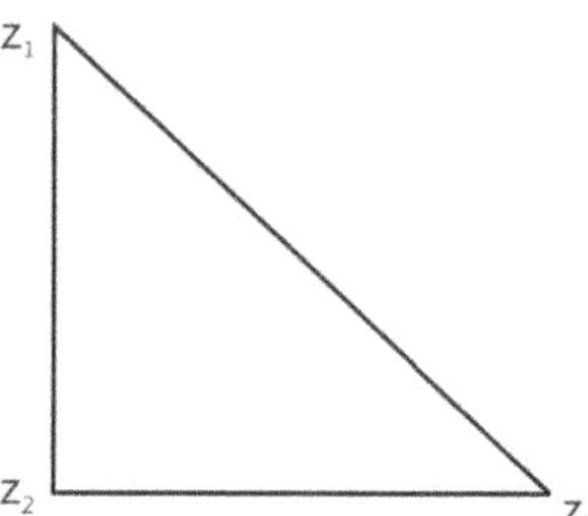

Considering z_2 at origin

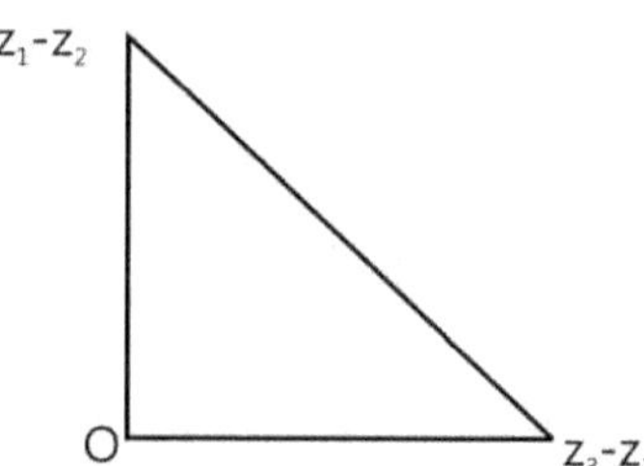

$(z_1 - z_2) = (z_3 - z_2)e^{i\pi/2}$

$z_1 - z_2 = (z_3 - z_2)i$

$(z_1 - z_2)^2 = -(z_3 - z_1)^2$

$\Rightarrow z_1^2 + z_3^2 + 2z_2^2 = 2z_1z_3 + 2z_1z_2$

$\Rightarrow z_1^2 + 2z_2^2 + z_3^2 = 2z_2(z_1 + z_3)$

Sol 23: $\dfrac{A^2}{x-a} + \dfrac{B^2}{x-b} .... \dfrac{H^2}{x-H} = x + \ell$ Let $x = p + iq$

$(p + \ell) + iq = \sum \dfrac{A^2}{(p-a)+iq} = \sum \dfrac{A^2[(p-a)-iq]}{(p-a)^2 + q^2}$

$= \sum \dfrac{A^2(p-a)}{(p-a)^2+q^2} - i\sum \dfrac{A^2q}{(p-a)^2+q^2}$

Equating the imaginary terms we get

$q + \sum \dfrac{A^2q}{(p-a)^2+q^2} = 0$

$\Rightarrow q\left(1 + \sum \dfrac{A^2}{(p-a)^2+q^2}\right) = 0$

$\Rightarrow q = 0$ or $\sum \dfrac{A^2}{(p-a)^2+q^2} = -1$

Here A, p, a and q are all real hence q = 0 is the only solution.

$\Rightarrow$ x cannot have imaginary roots.

Sol 24: $z^3 + 2z^2 + 2z + 1 = 0$

$\Rightarrow z = -1, \omega, \omega^2$

$z^{1985} + z^{100} + 1 = 0$

$\Rightarrow z = \omega, \omega^2$

Common roots are ω, ω^2

Sol 25: Let $f(x,y) = (x+y)^n - x^n - y^n$

$xy(x+y)(x^2+xy+y^2) = (x-0)(y-0)(x+y)(x-wy)(x-w^2y)$

$f(x=0)=0$ and $f(y=0)=0$

$f(y=-x)=(x-x)^n-(x)^n-(-x)^n=0$

$f(x=wy)=(wy+y)n-w^ny^n-y^n$; $= y^n[w2n + wn +1] = 0$

Similarly $f(x=w^2y)=0$

$\therefore f(x)$ is divisible by $xy(x+y)(x^2+xy+y^2)$

Sol 26: $T = \left|\alpha + \sqrt{\alpha^2-\beta^2}\right| + \left|\alpha - \sqrt{\alpha^2-\beta^2}\right|$

$T^2 = \left|\alpha + \sqrt{\alpha^2-\beta^2}\right|^2 + \left|\alpha - \sqrt{\alpha^2-\beta^2}\right|^2 +$

$2\left|\alpha + \sqrt{\alpha^2-\beta^2}\right|\left|\alpha - \sqrt{\alpha^2-\beta^2}\right|$

$= \left(\alpha + \sqrt{\alpha^2+\beta^2}\right)\overline{\left(\alpha + \sqrt{\alpha^2-\beta^2}\right)} + \left(\alpha + \sqrt{\alpha^2-\beta^2}\right)$

$\overline{\left(\alpha - \sqrt{\alpha^2-\beta^2}\right)} + 2|\alpha^2 - a^2 + \beta^2|$

$= 2\left[|\alpha^2| + \left|\sqrt{\alpha^2-\beta^2}\right|^2\right] + 2|\beta|^2$

$= 2|\alpha|^2 + 2|\beta|^2 + 2|\alpha^2 - \beta^2|$

$= 2|\alpha|^2 + 2|\beta|^2 + 2|\alpha+\beta||\alpha-\beta|$

$= \alpha\bar\alpha + \beta\bar\beta + \alpha\bar\beta + \bar\alpha\beta + \alpha\bar\alpha + \beta\bar\beta -$

$\qquad\qquad \alpha\bar\beta - \beta\bar\alpha + 2|\alpha+\beta||\alpha-\beta|$

$= |\alpha+\beta|^2 + |\alpha-\beta|^2 + 2|\alpha+\beta||\alpha-\beta|$

$= \big[|\alpha+\beta| + |\alpha-\beta|\big]^2$

$T = |\alpha+\beta| + |\alpha-\beta|$ Hence proved

Sol 27: $z_1 = 10 + 6i; z_2 = 4 + 6i$

$\dfrac{z-z_1}{z-z_2} = \dfrac{x-10+(y-6)i}{x-4+(y-6)i}\left(\dfrac{(x-4)-i(y-6)}{(x-4)-i(y-6)}\right)$

$= \dfrac{(x-10)(x-4)+(y-6)^2 + i\big[(y-6)(x-4)+(x-10)(6-y)\big]}{(x-4)^2+(y-6)^2}$

$\arg\dfrac{z-z_1}{z-z_2} = \dfrac{\pi}{4}$

$\Rightarrow (y-6)(x-4)+(x-10)(6-y)$

$= (x-10)(x-4)+(y-6)^2$

$\Rightarrow (y-6)(x-4-x+10)=x^2-14x+40+(y-6)^2$

$\Rightarrow 6(y-6) = x^2-14x+40+y^2+36-12y$

$\Rightarrow x^2 + y^2 - 18y - 14x + 112 = 0$

$\Rightarrow (x-7)^2 + (y-9)^2 = 18$

$\Rightarrow |z - 7 - 9i| = 3\sqrt{2}$

Sol 28: $|z-w|^2 = (z-w)(\bar z - \bar w)$

$= |z|^2 + |w|^2 - \bar z w - z\bar w + 2|z||w| - 2|z||w|$

$= (|z|-|w|)^2 - \bar z w - z\bar w + 2|z||w|$

Let $z = r_1\cos\theta_1; w = r_2\cos\theta_2$

$$= \left(|z| - |w|\right)^2 - 2r_1 r_1 \cos\left(\theta_1 - \theta_2\right) + 2r_1 r_2$$

$$= \left(|z| - |w|\right)^2 + 2r_1 r_1 \left(2\sin^2\left(\frac{\theta_1 - \theta_2}{2}\right)\right)$$

$$= \left(|z| - |w|\right)^2 + 4r_1 r_1 \left(\sin\frac{\theta_1 - \theta_2}{2}\right)^2$$

We know, $\sin\theta \le \theta$ and $r_1, r_2, \le 1$

$$\le \left(|z| - |w|\right)^2 + 4\times\left(\frac{\theta_1 - \theta_2}{2}\right)^2$$

$$\le \left(|z| - |w|\right)^2 + \left(\theta_1 - \theta_2\right)^2$$

$$\le \left(|z| - |w|\right)^2 + \left(\arg|z| - \arg|w|\right)^2$$

Sol 29: $\dfrac{A}{B} + \dfrac{B}{A} = 1$

$A^2 + B^2 = AB$

$$\frac{A}{B} = 1 - \frac{B}{A}$$

$$\frac{|A|}{|B|} = \frac{|A - B|}{|A|}$$

$$\frac{|A^2|}{|B|} = |A - B|$$

$$\frac{|B^2|}{|A|} = |B - A|$$

$$|A - B| = |B - A| \Rightarrow \left|\frac{A^2}{B}\right| = \left|\frac{B^2}{A}\right|$$

$\Rightarrow |A^3| = |B^3|$

$\Rightarrow |A| = |B| = |A - B|$

i.e. all sides are equal it forms an equilateral D

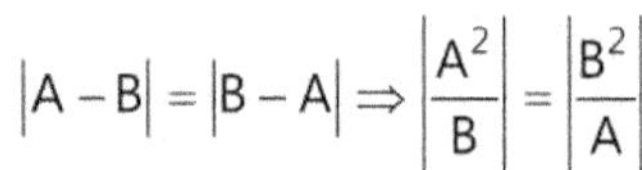

Sol 30: $a + b + c = 0 \Rightarrow b = -a - c$

$az_1 + bz_2 + cz_3 = 0$

$az_1 + cz_3 - (a + c)z_2 = 0$

$$z_2 = \frac{az_1 + cz_3}{a + c}$$

$\Rightarrow a(z_1 - z_2) + c(z_3 - z_2) = 0$

$\Rightarrow (z_1 - z_2) = \dfrac{-c}{a}(z_3 - z_2)$

$z_1 - z_2 = k_1(z_3 - z_2)$

$\Rightarrow z_1, z_2$ and z_3 are collinear (by vector)

Sol 31: $|z - 4 + 3i| \le 2$

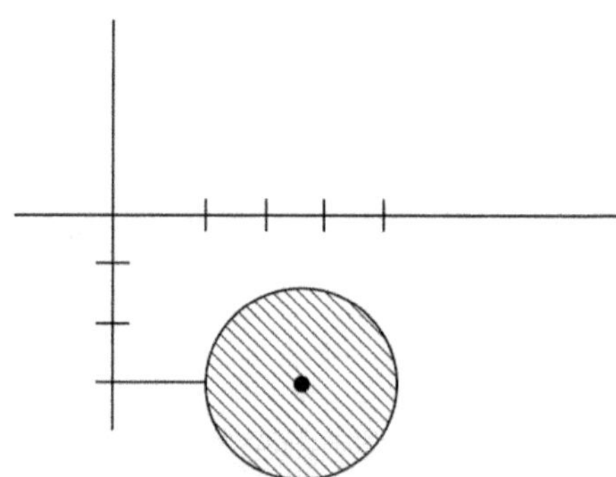

The shaded area show z

Min and max value = Distance of centre from origin $\pm$ radius

$= 5 \pm 2 = 3, 7$

$\Rightarrow 3 \le |z| \le 7$

Sol 32: $\left|\dfrac{z_1 - z_2}{1 - \overline{z}_1 z_2}\right| < 1$

$$\left|z_1 - z_2\right|^2 < \left|1 - \overline{z}_1 z_2\right|^2$$

$$(z_1 - z_2)(\overline{z}_1 - \overline{z}_2) < (1 - \overline{z}_1 z_2)(1 - z_1 \overline{z}_2)$$

$\Rightarrow |z_1|^2 + |z_2|^2 - z_1\overline{z}_2 - \overline{z}_1 z_2$

$ < -z_1\overline{z}_2 - z_2\overline{z}_1 + |z_1|^2|z_2|^2 + 1$

$\Rightarrow |z_1|^2 - 1 < (|z_1|^2 - 1)|z_2|^2$

$$|z_2|^2 < \frac{|z_1|^2 - 1}{|z_1|^2 - 1}$$

$|z_2|^2 < 1 \Rightarrow |z_2| < 1$

Sol 33: $\log_{\sqrt{3}}\left[\dfrac{|z|^2 - |z| + 1}{|z| + 2}\right] < 2$

$$\frac{|z|^2 - |z| + 1}{|z| + 2} < 3$$

$\Rightarrow |z|^2 - |z| + 1 < 3|z| + 6$

$\Rightarrow |z^2| - 4|z| - 5 < 0$

$\Rightarrow (|z| + 1)(|z| - 5) < 0$

$\Rightarrow |z| + 1 \geq 0$

So $(|z| - 5) < 0$

Interior of circle $x^2 + y^2 = 25$

Sol 34: $|z|^2\,\omega - |\omega^2|z = z - \omega$

$\Rightarrow z\bar{z}\omega - \omega\bar{\omega}z = z = \bar{\omega}$

$\Rightarrow z\omega(\bar{z} - \bar{\omega}) = z - \omega$

$\Rightarrow z(\bar{z}\omega - 1) = \omega(z\bar{\omega} - 1)$ ..... (i)

$\Rightarrow z(z\bar{\omega} - 1) = \bar{\omega}(\bar{z}\,\bar{\omega} - 1)$

$\Rightarrow$ Either $z\bar{\omega} = \bar{z}\omega = 1$ or $\bar{z} = \bar{\omega}$

$\dfrac{x}{\omega} = \dfrac{\bar{\omega}}{z} \Rightarrow z\bar{z} = \omega\bar{\omega} \Rightarrow |z| = |\omega|$

$|z|^2(\omega - z) = (z - \omega)$

$(\omega - z)(|z|^2 + 1) = 0 \Rightarrow \omega = z$

Exercise 2

Single Correct Choice Type

Sol 1: (C) $|z + 4| \leq 3$ Least and greatest value of $|z + 1|$

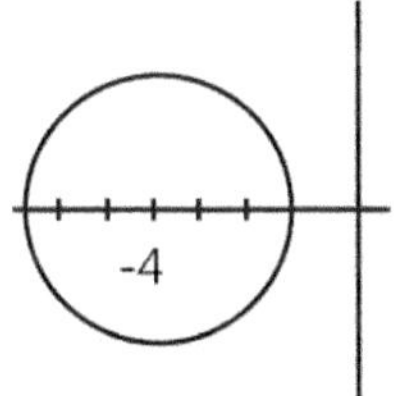

i.e. distance of z from $(-1, 0)$

Least is 0; maximum is $2r = 6$

Sol 2: (D) $|z + 3| \leq 3$ Least & greatest value of $|z + 1|$ ie its minimum and maximum distance from $(-1,0)$ is 1 and 5.

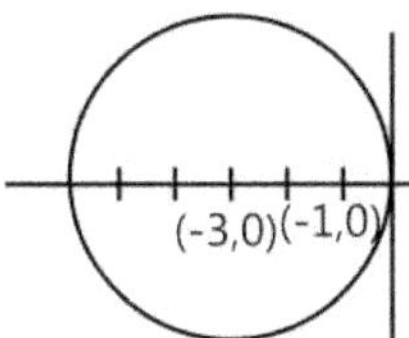

Sol 3: (B) $z_1 = 3 + i\sqrt{3}$

$z_2 = 2\sqrt{3} + 6i$

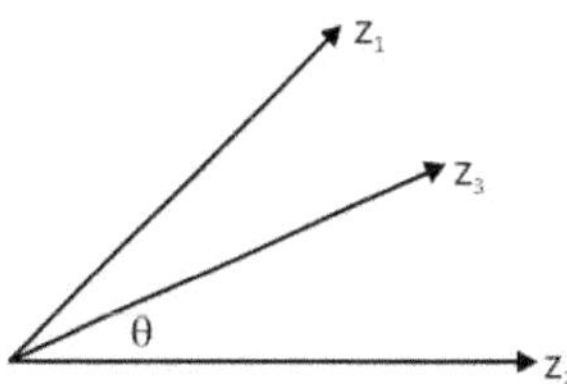

$\hat{z}_3 = \hat{z}_2 e^{i\theta}; \ \hat{z}_1 = \hat{z}_3 e^{i\theta}$

$\hat{z}_3^2 = \hat{z}_1\hat{z}_2 = \dfrac{24i}{24}$

$\hat{z}_3 = \sqrt{i} = \dfrac{1+i}{\sqrt{2}}$

Unit vector along bisector is $\dfrac{1+i}{\sqrt{2}}$, complex number lying along this vector is $5(1+i)$

Sol 4: (C) z_1, z_2, z_3, z_4 vertices of square

$z_1 - z_2 = i2\mathrm{Im}(z_1)$

$z_3 - z_2 = -2\mathrm{Re}(z_1)$

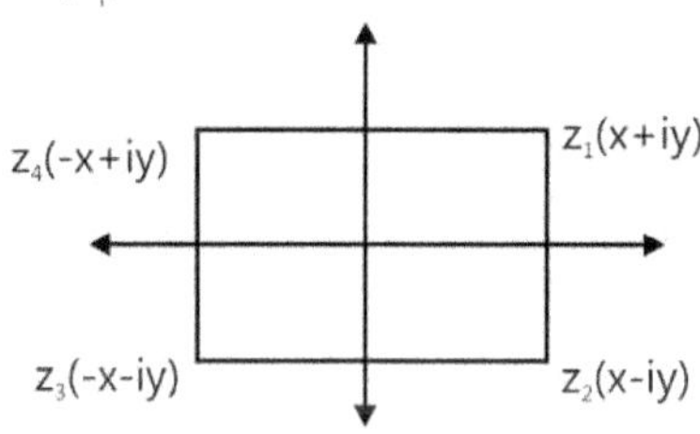

$\dfrac{z_1 - z_2}{z_3 - z_2}$ is imaginary

$z_2 - z_4 = [-x + iy - (x - iy)]$

$\qquad = 2\mathrm{Re}(z) - i2\mathrm{Im}(z)$

$z_1 - z_3 = 2x + 2iy = 2\mathrm{Re}(z) + i2\mathrm{Im}(z)$

$\dfrac{z_1 - z_3}{z_2 - z_4} = \dfrac{2x + 2iy}{2x - 2iy} = \dfrac{x + iy}{x - iy} = \dfrac{x^2 - y^2 + 2ixy}{x^2 + y^2}$

($x = y$ as it is aqueous)

$= \dfrac{2ixy}{x^2 + y^2}$ (Purely Imaginary)

Sol 5: (D) $\dfrac{az_1 + bz_2 + cz_3}{a + b + c} = z_0$

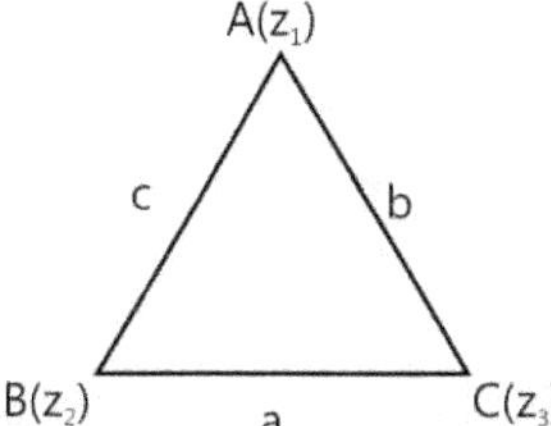

z_0 is incentre.

Sol 6: (A) $\delta - \gamma = \alpha - \beta$ [parallel vector]

$\delta = \alpha + \gamma - \beta$

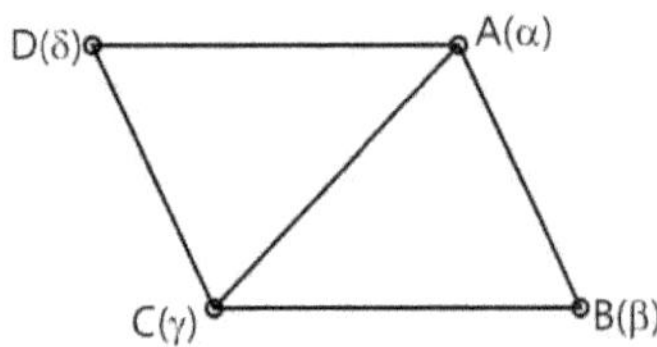

Sol 7: (B) $|z| \geq 3$

$$\left| z + \frac{1}{z} \right| \geq |z| - \frac{1}{|z|} = 3 - \frac{1}{3} = \frac{8}{3}$$

Sol 8: (C)

$$z_2 = z_1 e^{\pm \frac{i\pi}{4}} \qquad z_3 = z_2 e^{\pm \frac{i\pi}{4}}$$

$$z_3 - z_2 = (z_2 - z_1) \frac{(1 \pm i)}{\sqrt{2}}$$

$$z_3 = z_2 + (z_2 - z_1) \frac{(1 \pm i)}{\sqrt{2}}$$

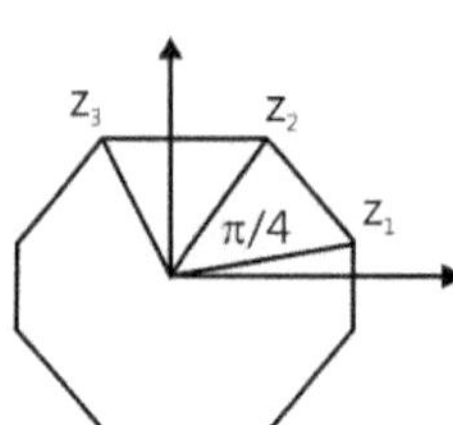

Sol 9: (D) $x^3 = (4)^3 (-1)^{1/3}$

$x = -4, -4\omega, -4\omega^2$

$$\begin{vmatrix} q_1 & q_2 & q_3 \\ q_2 & q_3 & q_1 \\ q_3 & q_1 & q_2 \end{vmatrix} = \begin{vmatrix} -4 & -4\omega & -4\omega^2 \\ -4\omega & -4\omega^2 & -4 \\ -4\omega^2 & -4 & -4\omega \end{vmatrix}$$

$$= -64 \begin{vmatrix} 1 & \omega & \omega^2 \\ \omega & \omega^2 & 1 \\ \omega^2 & 1 & \omega \end{vmatrix} = -64 \begin{vmatrix} 0 & 0 & 0 \\ \omega & \omega^2 & 1 \\ \omega^2 & 1 & \omega \end{vmatrix} = 0$$

Sol 10: (D) $z = (3 + 7i)(p + iq)$ is purely imaginary.

$\Rightarrow 3p - 7q = 0$

$\Rightarrow p = \dfrac{7q}{3}$

$|z|^2 = |3 + 7i|^2 |p + iq|^2 = 58 \left(\sqrt{p^2 + q^2} \right)^2$

for minimum $|z|$, $q = 3$, $p = 7$

$|z|^2 = 58(49+9) = 3364$

Sol 11: (A)

$$\frac{z_1}{z_3} = \frac{z}{z_2} \Rightarrow z = \frac{z_1 z_2}{z_3}$$

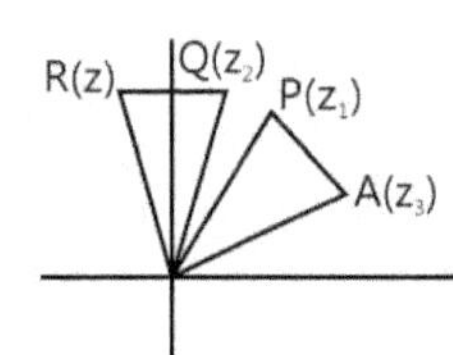

$|z_3| = 1 \Rightarrow |z| = |z_1 z_2|$

$\Rightarrow z = z_1 z_2$

Sol 12: (A) $\dfrac{A}{B} + \dfrac{B}{A} = 1$

Let $y = \dfrac{A}{B}$

$y + \dfrac{1}{y} = 1$

$\Rightarrow y^2 - y + 1 = 0$

$\Rightarrow y = \dfrac{1 + i\sqrt{3}}{2}$

$\Rightarrow \dfrac{A}{B} = \dfrac{1 + i\sqrt{3}}{2}$

$\Rightarrow \left| \dfrac{A}{B} \right| = 1$

From Rotation Theorem

$$\frac{A}{B} = \left| \frac{A}{B} \right| e^{i\theta}$$

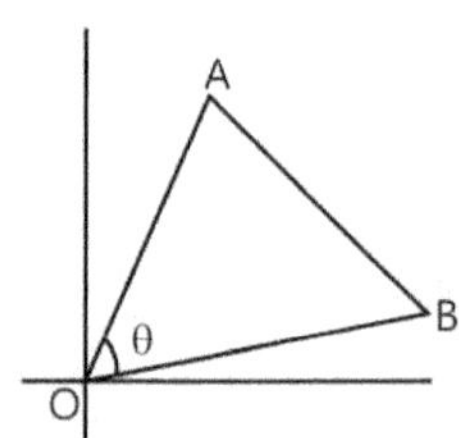

$\Rightarrow e^{i\theta} = \dfrac{1 + i\sqrt{3}}{2}$

$\Rightarrow \theta = 60^\circ$

and $|A| = |B| \Rightarrow \angle OAB = \angle OBA = 60^\circ$

$\Rightarrow$ AOB is equilateral triangle

Sol 13: (D) $|z|^2 - (z + \bar{z}) + i(z - \bar{z}) + 2 = 0$

$z\bar{z} - (z + \bar{z}) + i(z - \bar{z}) + 2 = 0$

Let $z = a + ib$

$a^2 + b^2 - 2a + i(i2b) + 2 = 0$

$a^2 - 2a + 1 + b^2 - 2b + 1 = 0$

$(a - 1)^2 + (b - 1)^2 = 0$

$\Rightarrow a = b = 1$

$\Rightarrow z = 1 + i$

Sol 14: (D) $z_1 = -3 + 5i$

$z_2 = -5 - 3i$

Eq^n of line $y = 4x + 17$

$\arg z = \tan^{-1} \left(\dfrac{4x + 17}{x} \right) = \tan^{-1} \left(4 + \dfrac{17}{x} \right)$

$x \in [-3, -5] \Rightarrow \arg z \in \left[\tan^{-1}\dfrac{-5}{3}, \tan^{-1}\dfrac{-3}{5}\right]$

Only option left is $\left(\dfrac{5\pi}{6}\right)$

Sol 15: (B) $|z-3| = 2 : (x-3)^2 + y^2 = 4$

$|z| = 2 : x^2 + y^2 = 4$

Points of intersection lie on the radical axis $S_1 - S_2 = 0$

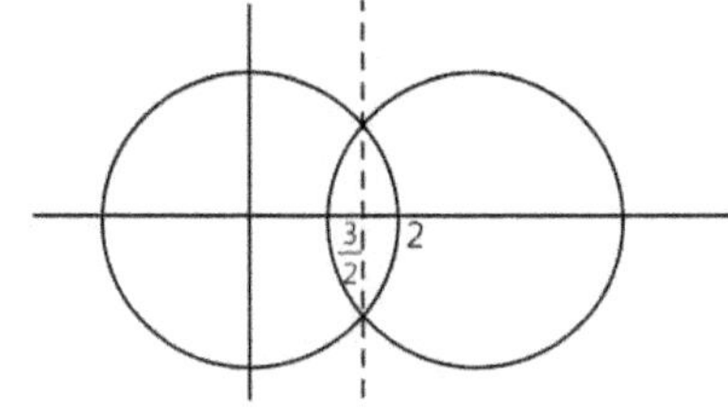

Radical axis is $x = \dfrac{3}{2}$

$x^2 + y^2 = 4$

$\Rightarrow \dfrac{9}{4} + y^2 = 4 \Rightarrow y^2 = 4 - \dfrac{9}{4}$

$\Rightarrow y^2 = \dfrac{7}{4} \Rightarrow y = \dfrac{\pm\sqrt{7}}{2}$

Complex no. is $\dfrac{3}{2} \pm \dfrac{i\sqrt{7}}{2}$

Sol 16: (C) $|z - 3i| = 3 \Rightarrow (x)^2 + (y-3)^2 = 9$

$\Rightarrow \tan\theta = \dfrac{y}{x}$

$\cot\theta \quad \dfrac{6}{z} = \dfrac{x}{y} - \dfrac{6(x-iy)}{(x+iy)(x-iy)} = \dfrac{x}{y} - \dfrac{6(x-iy)}{x^2+y^2}$

$x^2 + y^2 = 6y$

$\Rightarrow \dfrac{x}{y} - \dfrac{6(x-iy)}{6y} = i$

Sol 17: (D) $|z - 1| + |z + 1| = 2$

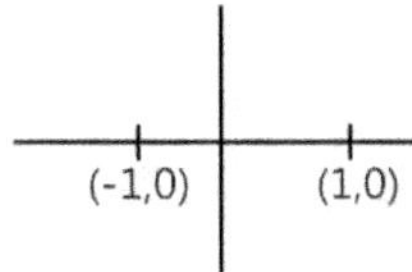

The portion of real axis between $(-1,0)$ & $(1,0)$ as the distance between both the point is 2

Sol 18: (C) $Q = \sqrt{2|z|^2}\ \text{cis}\left(\dfrac{\pi}{4} + \theta\right)$

$= |z|\ \sqrt{2}\left[\cos\left(\dfrac{\pi}{4} + \theta\right) + i\sin\left(\dfrac{\pi}{4} + \theta\right)\right]$

$= |z|[(\cos\theta - \sin\theta) + i(\sin\theta + \cos\theta)]$

$= |z|\left[(\cos\theta + i\sin\theta) + \dfrac{|z|i}{i}(-\sin\theta + i\cos\theta)\right]$

$= |z|\ [\cos\theta + i\sin\theta] + \dfrac{|z|}{i}[-\cos\theta - i\sin\theta]$

$= P + iP$

$\Rightarrow \arg\left(\dfrac{Q-P}{P}\right) = \dfrac{\pi}{2}$

This is right angle triangle with $|P| = |Q{-}P|$ and right angle at P.

Sol 19: (A) $|z_1 - 1| = |z_2 - 1| = |z_3 - 1|$

$z_1 + z_2 + z_3 = 3$

Centroid is at $z = 1$

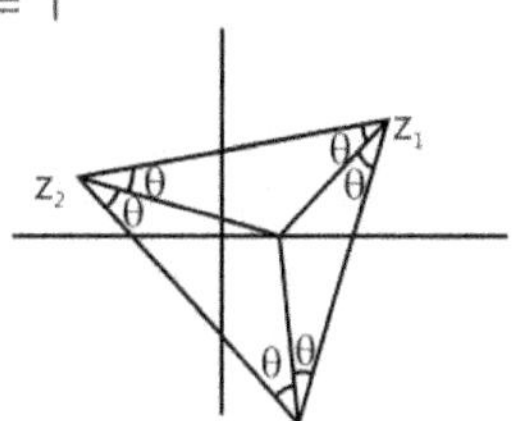

Distance of vertical z_1, z_2 and z_3 from centroid is same, which mean centroid coincides with circumcenter.

Therefore, Δ is equilateral.

Sol 20: (C) $p = a + b\omega + c\omega^2$

$q = b + c\omega + a\omega^2$

$r = c + a\omega + b\omega^2$

$p + q + r = a(1 + \omega + \omega^2) + b(1 + \omega + \omega^2)$

$+ c(1 + \omega + \omega^2)$

$= (a + b + c)(1 + \omega + \omega^2) = 0$

$p^2 + q^2 + r^2 = (p + q + r)^2 - 2pq - 2qr - 2rp$

$= -2(pq + qr + rp)$

$pq + qr + rp = ab + ac\omega + a^2\omega^2 + b^2\omega^2 + b^2\omega + bc\omega^2 + ab\omega^3 + cb\omega^2 + c^2\omega^3 + ca\omega + \dots\dots$

$= ab + 2ac\omega + ab + c^2 + (a^2 + 2bc)\omega^2 + b^2\omega$

$= 2ab + c^2 + 2ac\omega + (b^2 + 2bc)\omega^2 + a^2\omega + 2bc + a^2 + 2ba\omega + (b^2 + 2ac)\omega^2 + c^2\omega + 2ac + b^2 + 2bc\omega + (c^2 + a^2 + 2ab)\omega^2$

$= (2ab + 2bc + 2ac)(1 + \omega + \omega^2)$

$+ (c^2 + b^2 + a^2)(1 + \omega^2 + \omega)$

$p^2 + q^2 + r^2 = 0 = 2(pq + qr + rp)$

Previous Years' Questions

Sol 1: (D) Since, $\left(\dfrac{1+i}{1-i}\right)^n = 1$

$\Rightarrow \left(\dfrac{1+i}{1-i} \times \dfrac{1+i}{1+i}\right)^n = 1$

$\Rightarrow \left(\dfrac{2i}{2}\right)^n = 1 \Rightarrow i^n = 1$

The smallest positive integer n for which $i^n = 1$ is 4

$\therefore n = 4$

Sol 2: (A) Given, $\left|\dfrac{z-5i}{z+5i}\right| = 1$

$\Rightarrow |z - 5i| = |z + 5i|$

(if $|z - z_1| = |z - z_2|$,

Then it is a perpendicular bisector of z_1

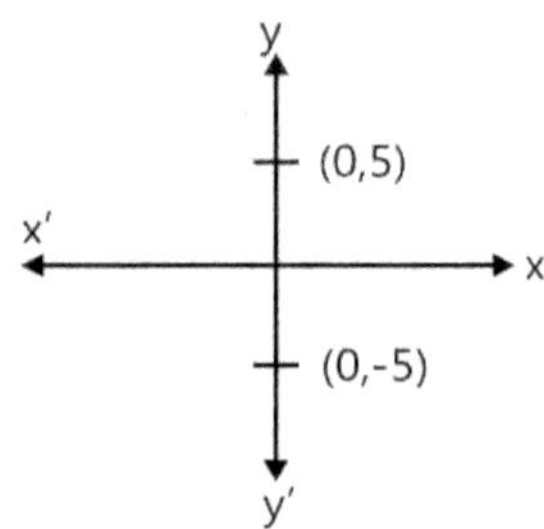

$\therefore$ Perpendicular bisector of (0, 5) and (0, – 5) is x-axis.

Sol 3: (B) Since, $|w| = 1$

$\Rightarrow \left|\dfrac{1-iz}{z-i}\right| = 1 \Rightarrow |z - i| = |1 - iz|$

$\Rightarrow |z - i| = |z + i|$

$(\because |1 - iz| = |-i| \, |z + i| = |z + i|)$

$\therefore$ It is a perpendicular bisector of (0, 1) and (0, –1)

i.e., x-axis

Thus, z lies on real axis.

Sol 4: (B) Since, z_1, z_2, z_3, z_4 are the vertices of parallelogram.

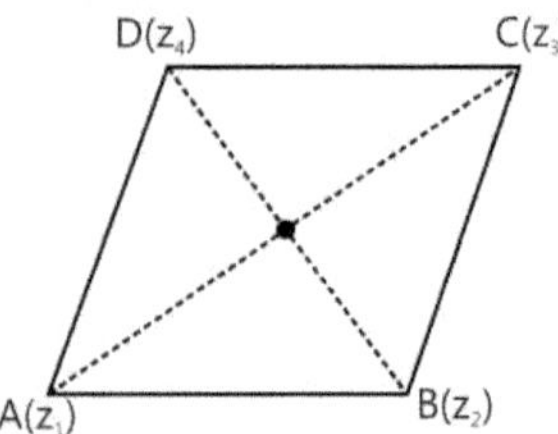

$\therefore$ Mid-point of AC = mid-point of BD

$\Rightarrow \dfrac{z_1 + z_3}{2} = \dfrac{z_2 + z_4}{2} \Rightarrow z_1 + z_3 = z_2 + z_4$

Sol 5: (C) Given, $|z_1 + z_2| = |z_1| + |z_2|$

On squaring both sides, we get

$|z_1|^2 + |z_2|^2 + 2|z_1| \, |z_2|\cos (\arg z_1 - z_2)$

$= |z_1|^2 + |z_2|^2 + 2|z_1| \, |z_2|$

$\Rightarrow 2|z_1| \, |z_2| \cos (\arg z_1 - \arg z_2) = 2 \, |z_1| \, z_2|$

$\Rightarrow \cos (\arg z_1 - \arg z_2) = 1$

$\Rightarrow \arg (z_1) - \arg (z_2) = 0$

Sol 6: (D) Since, $\overline{\sin x + i \cos 2x}$

$= \cos x - i \sin 2x$

$\Rightarrow \sin x - i \cos 2x = \cos x - i \sin 2x$

$\Rightarrow \sin x = \cos x$ and $\cos 2x = \sin 2x < x$

$\Rightarrow \tan x = 1$ and $\tan 2x = 1$

$\Rightarrow x = \dfrac{\pi}{4}$ and $x = \dfrac{\pi}{8}$ which is not possible at same time.

Hence, no such value exists.

Sol 7: (B) $(1 + \omega)^7 = (1 + \omega)(1 + \omega)^6$

$= (1 + \omega)(-\omega^2)^6$

$= 1 + \omega$

$\Rightarrow A + B\omega = 1 + 0 \Rightarrow A = 1, B = 1$

Sol 8: (D) Since, $|z| = |\omega|$ and $\arg (z) = \pi - \arg(\omega)$

Let $\omega = re^{i\theta}$, then $\overline{\omega} = re^{-i\theta}$

$\therefore z = re^{(\pi-\theta)} = re^{i\pi} \cdot e^{-i\theta}$

$= - re^{-i\theta} = - \overline{\omega}$

$\therefore$ z lies on the perpendicular bisector of the line joining $-i\omega$ and $- i\overline{\omega}$ is the mirror image of $- i\omega$ in the x-axis, the locus of z is the x-axis

Let $z = x + iy$ and $y = 0$

Now, $|z| \leq 1 \Rightarrow x^2 + 0^2 \leq 1$

$\Rightarrow - 1 \leq x \leq 1$ ..... (i)

$\therefore$ z may take values given in (i).

Sol 9: (D)

$(1 + \omega - \omega^2)^7 = (-\omega^2 - \omega^2)^7 \qquad (\because 1 + \omega + \omega^2 = 0)$

$= (-2\omega^2)^7 = (-2)^7 \omega^{14} = -128\omega^2$

Sol 10: (B)

$$\sum_{n=1}^{13}(i^n + i^{n+1}) = \sum_{n=1}^{13} i^n(1+i) = (1+i)\sum_{n=1}^{13} i^n$$

$$= (1+i)(i + i^2 + i^3 + \ldots + i^{13}) = (1+i)\left[\frac{i(1-i)}{1-i}\right]$$

$$= (1+i)i = -1+i$$

Alternate solution:

Since, sum of any four consecutive powers of iota is zero.

$$\therefore \sum_{n=1}^{13}(i^n + i^{n+1}) = (i + i^2 + \ldots + i^{13}) + (i^2 + i^3 + \ldots + i^{14})$$

$$= i + i^2 = i - 1$$

Sol 11: (D)

Given, $\begin{vmatrix} 6i & -3i & 1 \\ 4 & 3i & -1 \\ 20 & 3 & i \end{vmatrix} = x + iy$

$\Rightarrow -3i \begin{vmatrix} 6i & 1 & 1 \\ 4 & -1 & -1 \\ 20 & i & i \end{vmatrix} = x + iy$

$\Rightarrow \quad x + iy = 0 \qquad (\because C_2 \text{ and } C_3 \text{ are identical})$

$\Rightarrow \quad x = 0, y = 0$

Sol 12: (A)

Given, $|z_1| = |z_2| = |z_3| = 1$

Now, $|z_1| = 1$

$\Rightarrow |z_1|^2 = 1$

$\Rightarrow z_1 \overline{z_2} = 1, z_3 \overline{z_3} = 1$

Again now, $\left| \dfrac{1}{z_1} + \dfrac{1}{z_2} + \dfrac{1}{z_3} \right| = 1$

$\Rightarrow |\overline{z_1} + \overline{z_2} + \overline{z_3}| = 1$

$\Rightarrow |\overline{z_1 + z_2 + z_3}| = 1$

$\Rightarrow |z_1 + z_2 + z_3| = 1$

Sol 13: (B)

Let $\Delta = \begin{vmatrix} 1 & 1 & 1 \\ 1 & -1-\omega^2 & \omega^2 \\ 1 & \omega^2 & \omega \end{vmatrix}$

Applying $R_2 \to R_1 : R_3 \to R_3 - R_1$

$$= \begin{vmatrix} 1 & 1 & 1 \\ 0 & -2-\omega^2 & \omega^2-1 \\ 0 & \omega^2-1 & \omega-1 \end{vmatrix}$$

$$= (-2-\omega^2)(\omega-1) - (\omega^2-1)^2$$

$$= -2\omega + 2 - \omega^3 + \omega^2 - (\omega^4 - 2\omega^2 + 1)$$

$$= 3\omega^2 - 3\omega = 3\omega(\omega-1) \qquad (\because \omega^4 = \omega)$$

Sol 14: (B)

Given, $(1+\omega^2)^n = (1+\omega^4)^n$

$\Rightarrow (-\omega)^n = (-\omega^2)^n \qquad (\because \omega^3 = 1 \text{ and } 1 + \omega + \omega^2 = 0)$

$\Rightarrow \omega^n = 1$

$\Rightarrow n = 3$ is the least positive value of n.

Sol 15: (D) Let OA=3, so that the complex number associated with A is $3e^{i\pi/4}$. If z is the complex number associated with P, then

$$\frac{z - 3e^{i\pi/4}}{0 - 3e^{i\pi/4}} = \frac{4}{3}e^{-i\pi/2} = -\frac{4i}{3}$$

$$\Rightarrow 3z - 9e^{i\pi/4} = 12ie^{i\pi/4}$$

$$\Rightarrow z = (3 + 4i)e^{i\pi/4}$$

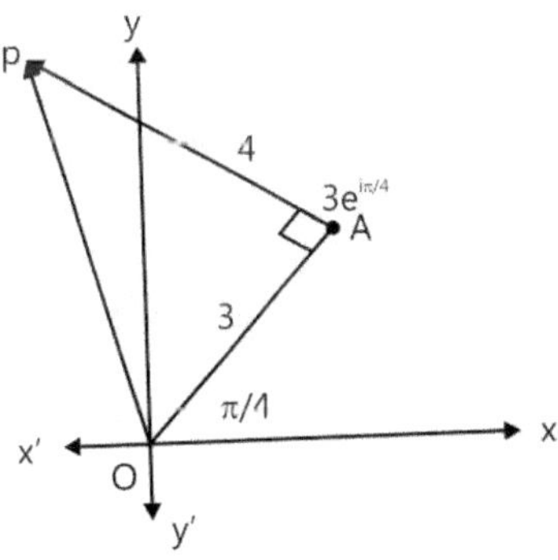

Sol 16: (A) Given: $\dfrac{z^2}{z-1}$ is real

Let $z = a + ib$, then

$$\frac{z^2}{z-1} = \frac{(a+ib)^2}{a+ib-1} = \frac{a^2 - b^2 + 2aib}{[(a-1)+ib]} \times \frac{[(a-1)-ib]}{[(a-1)-ib]}$$

$$\Rightarrow \frac{-b(a^2-b^2) + 2ab(a-1)}{(a-1)^2 + b^2} = 0$$

$$\Rightarrow -a^2 b + b^3 + 2a^2 b - 2ab = 0$$

$$\Rightarrow a^2 b + b^3 - 2ab = 0$$

$$\Rightarrow b(a^2 + b^2 - 2a) = 0$$

$\Rightarrow b = 0$ or $a^2 + b^2 - 2a = 0$

$\Rightarrow$ Either real axis or circle passing through origin.

Sol 17: (B) Let $\theta = \arg\left(\dfrac{1+z}{1+\bar{z}}\right)$

$\Rightarrow \theta = \arg\left(\dfrac{1+z}{1+\dfrac{1}{z}}\right) \quad \{|z| = 1 \Rightarrow z\,\bar{z} = 1\}$

$\Rightarrow \theta = \arg(z)$

Sol 18: (B) Given: The expression $\left|z + \dfrac{1}{2}\right|$ and $|z| \geq 2$

Using triangle in equality

$|z_1 + z_2| \geq \big|\,|z_1| - |z_2|\,\big|$

$\Rightarrow \left|z + \dfrac{1}{2}\right| \geq \left|\,|z| - \left|\dfrac{1}{2}\right|\,\right|$

$\Rightarrow \left|z + \dfrac{1}{2}\right| \geq \left|z - \dfrac{1}{2}\right|$

$\Rightarrow \left|z + \dfrac{1}{2}\right| \geq \dfrac{3}{2}$ lies in (1, 2)

Sol 19: (B) $\dfrac{z_1 - 2z_2}{2 - z_1\bar{z}_2} = 1$

$\Rightarrow |z_1 - 2z_2| = |2 - z_1\bar{z}_2|$

$\Rightarrow (z_1 - 2z_2)(\bar{z}_1 - 2\bar{z}_2) = (2 - z_1\bar{z}_2)(2 - \bar{z}_1 z_2)$

$\Rightarrow |z_1|^2 - 2z_1\bar{z}_2 - 2\bar{z}_1 z_2 + 4|z_2|^2$
$= 4 - 2\bar{z}_1 z_2 - 2z_1\bar{z}_2 + |z_1|^2|z_2|^2$

$\Rightarrow |z_1|^2 + 4|z_2|^2 = 4 + |z_2|^2|z_1|^2$

$\Rightarrow |z_1|^2 + \left(1 - |z_2|^2\right) - 4\left(1 - |z_2|^2\right) = 0$

$\Rightarrow \left(1 - |z_2|^2\right)\left(|z_1|^2 - 4\right) = 0$

$\Rightarrow |z_1|^2 = 4 \quad \{|z_2| \neq 1\}$

$\Rightarrow |z_1| = 2$

Sol 20: (C) $\dfrac{2 + 3i\sin\theta}{1 - 2i\sin\theta}$

$= \dfrac{(2 + 3i\sin\theta)(1 + 2i\sin\theta)}{1 + 4\sin^2\theta}$

$= \dfrac{2 + 4i\sin\theta + 3i\sin\theta - 6\sin^2\theta}{1 + 4\sin^2\theta}$

Given $\dfrac{2 - 6\sin^2\theta}{1 + 4\sin^2\theta} = 0$

$\Rightarrow \sin^2\theta = \dfrac{1}{3}$

$\Rightarrow \sin\theta = \pm\dfrac{1}{\sqrt{3}}$

$\Rightarrow \theta = \sin^{-1}\left(\dfrac{1}{\sqrt{3}}\right), -\sin^{-1}\left(\dfrac{1}{\sqrt{3}}\right)$

JEE Advanced/Boards

Exercise 1

Sol 1: $z^2 + (p + ip')z + (q + iq') = 0$

One real root

$(z^2 + pz + q) + i(p'z + q') = 0$

If z is real

$P'z = -q' \Rightarrow z = \dfrac{-q'}{p'}$

$z^2 + pz + q = 0$

$\dfrac{q'^2}{p'^2} - \dfrac{pq'}{p'} + q = 0$

$q'^2 - pp'q' + qp'^2 = 0$

If eq^n. has 2 equal roots

$(p + ip')^2 = 4(q + iq')$

$p^2 - p'^2 = 4q$ and $p'^2 = 4q^2$

The roots are $\dfrac{-(p + ip')}{2}$ i.e. roots (equal) are imaginary.

Sol 2: $z = 18 + 26i$

$z_0 = x_0 + iy_0$ is cube root of z

$z_0 = (r\{\cos\theta + i\sin\theta\})^{1/3} = r^{1/3}\, e^{i\theta/3}$

$= r^{1/3}\left(\cos\dfrac{\theta}{3} + i\sin\dfrac{\theta}{3}\right)$

$r = (1000)^{1/2}$

$r^{1/3} = (10^{3/2})^{1/3} = 10^{1/2}$

$\cos\theta = \dfrac{18}{\sqrt{1000}} \quad \sin\theta = \dfrac{26}{\sqrt{1000}}$

$\cos\theta = 0.57,\ \sin\theta = 0.82$

$\cos\theta = 3\cos\dfrac{\theta}{3} - 4\cos^3\left(\dfrac{\theta}{3}\right)$

$0.57 = 3t - 4t^3$

$\cos\left(\dfrac{\theta}{3}\right) = 0.2,\ 0.74,\ 0.95$

$\cos\dfrac{\theta}{3} = 0.95$

$\sin\dfrac{\theta}{3} = 0.3162$

$z = \sqrt{10}\,(0.95 + i\,0.3162) = 3 + i$

$x_0 = 3,\ y_0 = 1$

$x_0 y_0 (x_0 + y_0) = 3.4 = 12$

Sol 3: $z^3 + iz = 1$

$z(z^2 + i) = 1$

z can never be Purely real as $(z^2 + i)$ will be imaginary which multiplied with z will not be a real number, hence its locus will never cut x-coo axis Z can never be imaginary (pure) as $(z^2 + i)$ will be of form $(a + ib)$ which multiplied by z, cannot form 1 (real) therefore its locus can never cut y axis.

$z(z^2 + i) = 1$

$(x + iy)\,[x^2 - y^2 + i(2xy + 1)] = 1$

$x^3 - xy^2 + i(2x^2y + x + x^2y - y^3) - 2xy^2 - y = 1$

$x^3 - xy^2 - 2xy^2 - y + i(3x^2y + x - y^3) = 1$

$x^3 - 3xy^2 - y = 1\ \&\ 3x^2y + x - y^3 = 0$

$x(x^2 - 3y^2) = 1 + y\ \&\ y(y^2 - 3x^2) = x$

$|z|^2 = -\dfrac{1}{2}\left[\dfrac{1+y}{x} + \dfrac{x}{y}\right] \Rightarrow |z|^2 = -\dfrac{1}{2}\left[\dfrac{y + x^2 + y^2}{xy}\right]$

$|z|^2 = -\dfrac{1}{2xy}\left[y + |z|^2\right] \Rightarrow |z|^2\left[\dfrac{2xy + 1}{2xy}\right] = \dfrac{-y}{2xy}$

$\Rightarrow |z| = \sqrt{\dfrac{-\mathrm{Im}(z)}{2\,\mathrm{Re}(z)\,\mathrm{Im}(z) + 1}}$

Sol 4: $A_n = \mathrm{dia}(d_1,\ d_2,\ ......d_n)$

$d_i = a^{i-1};\ \alpha = e^{i2\pi/n}$

$\Rightarrow d_n = e^{i\frac{2\pi}{n}(n-1)}$

$L = \mathrm{Tr}(A_7)^7$

$M = \det A_{(2n+1)} + \det(A_{2n})$

$(A_7)^7 = e^{i\frac{2\pi 7(0)}{7}} + e^{i2\pi(1)} + ...e^{i2\pi(6)}$

$= \cos 0 + i\sin 0 + \cos 2\pi + i\sin 2\pi ... \cos 6\pi + i\sin 6p$

$= [1 + 1 + ...\ 7\ \text{times} + i\,(0 + 0..........)] = 7$

$\det A_{2n} = \prod e^{i\frac{2\pi}{2n}(K-1)} = e^{i\frac{\pi}{n}\left(2n\frac{(2n+1)}{2} - 2n\right)}$

$= e^{\frac{\pi i}{n}\left(\frac{4n^2 - 2n}{2}\right)} = e^{\frac{\pi}{n}i(2n^2 - n)} = e^{i\pi(2n-1)}$

$\det A_{2n+1} = e^{i2pn}$

$\det A_{2n} + \det A_{2n+1} = e^{i\pi(2n-1)} + e^{i2pn}$

$= \cos(2n-1)\pi + i\sin(2n-1)\pi + \cos 2n\pi + i\sin 2np$

$M = 0 \Rightarrow L + M = 7$

Sol 5: $z_1(z_1^2 - 3z_2^2) = 10$

$z_2(3z_1^2 - z_2^2) = 30$

$z_1 = a + ib,\ Z_2 = c + id$

$ab + cd = 0\ \text{as } z_1^2 + z_2^2 \text{ is real}$

$\Rightarrow z_1^3 - 3z_1z_2^2 + 3z_1^2z_2 - z_2^3 = 40$

$\Rightarrow z_1^2(z_1 + 3z_2) - z_2^2(z_2 + 3z_1) = 40$

$\Rightarrow (z_1 + iz_2)^3 = z_1^3 + iz_2^3 - 3z_1^2iz_2 - 3z_1z_2^2$

$= 10 - 30i$

$\Rightarrow (z_1 + iz_2)^3 = 10 + 30i$

$\Rightarrow ((z_1 + iz_2)^2)^3 = 1000 \Rightarrow z_1^2 + z_2^2 = 10$

Sol 6: $(z + 1)^7 + z^7 = 0$ has roots $z_1 .... z_7$

$\mathrm{Re}(z_1) + \mathrm{Re}(z_2) .... \mathrm{Re}(z_7) = ?$

$(z + 1)^7 = (-z)^7$

$\Rightarrow \left((a + 1) + ib\right)^7 = \left(-a - ib\right)^7$

One of the solution is

$\Rightarrow a + 1 = -a \Rightarrow a = \dfrac{-1}{2}$ and $b = 0$

$\Rightarrow z = \dfrac{-1}{2}$

$z_1 + z_2 + z_3 .... z_7 = \dfrac{-7}{2} + i(0)$

$\mathrm{Re}(z_1 + ...+ z_7) = \dfrac{-7}{2}$

$\mathrm{Im}\,(z_1 + z_2 ...+ z_7) = 0$

Sol 7: $z = (1)^{1/7}$

$\alpha + \beta = (z + z^2 + z^3 + z^4 + z^5 + z^6) = (-1) = -1$

$\alpha\beta = (z + z^4 + z^2)(z^3 + z^5 + z^6)$

$= z^4 + z^6 + z^7 + + z^7 + z^9 + z^{10} + z^5 + z^7 + z^8$

$= z^4 + z^6 + 3 + z^5 + z + z^3 + z^2$

$= 3 + z + z^4 + z^6 + z^5 + z^3 + z^2 = 3 - 1 = 2$

Eq$^n \to x^2 + x + 2 = 0$

Sol 8: $z^5 - 32 = z^5 - 2^5$

$= (z - 2)(z^2 - pz + 4)(z^2 - qz + 4)$

$= z^5 + z^4(-q - p - 2) + z^3(+pq + 8 + 2p + 2q) + z^2(-4p - 4q + (pq + 8) - 2)$

$+ z(16 + 8p + 8q) - 32 = 0$

$\Rightarrow p + q + 2 = 0$

$2p + 2q + 8 + pq = 0$

$\Rightarrow pq + 4 = 0 \Rightarrow pq = -4$

$p + q = -2 \Rightarrow p - \dfrac{4}{p} = -2$

$\Rightarrow p^2 + 2p - 4 = 0$

$\Rightarrow p^2 + 2p = 4$

Sol 9: $|z_1| + |z_2| \geq \dfrac{1}{2}\left(|z_1| + |z_2|\right)\left|\dfrac{z_1}{|z_1|} + \dfrac{z_2}{|z_2|}\right|$

RHS

$\dfrac{1}{2}\left(|z_1| + |z_2|\right)\left|\dfrac{z_1}{|z_1|} + \dfrac{z_2}{|z_2|}\right|$

$\left|\dfrac{z_1}{|z_1|} + \dfrac{z_2}{|z_2|}\right| \leq \left|\dfrac{z_1}{|z_1|}\right|\left|\dfrac{z_2}{|z_2|}\right| = 2$

$\dfrac{1}{2} \times 2\left(|z_1| + |z_2|\right) \geq \dfrac{1}{2}\left(|z_1| + |z_2|\right)\left|\dfrac{z_1}{|z_1|} + \dfrac{z_2}{|z_2|}\right|$

$|z_1| + |z_2| \geq \dfrac{1}{2}\left(|z_1| + |z_2|\right)\left|\dfrac{z_1}{|z_1|} + \dfrac{z_2}{|z_2|}\right|$

Sol 10:

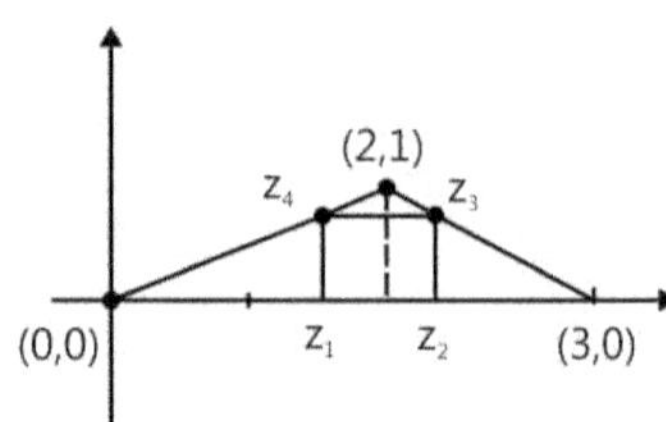

$z_1 = (a, 0),\ z_2 = (b, 0),\ z_3 = (b, c),\ z_4 = (a, c)$

$b - a = c$(square)

$a = 2c;\ b = 3c;\ b + c = 3$

$4c = 3 \Rightarrow c = \dfrac{3}{4},\ a = \dfrac{3}{2},\ b = \dfrac{9}{4}$

$\Delta(0, z_3, z_4) = \dfrac{1}{2}\begin{vmatrix} 0 & 0 & 1 \\ \dfrac{9}{4} & \dfrac{3}{4} & 1 \\ \dfrac{3}{2} & \dfrac{3}{4} & 1 \end{vmatrix}$

$= \dfrac{1}{2} \times \dfrac{3}{4} \times \dfrac{3}{4} = \dfrac{9}{32} = \dfrac{m}{n}$

$\Rightarrow m + n = 32 + 9 = 41$

Sol 11: $(1 + x)^n$

$a_n = C_0 + C_3 + C_6 + \ldots\ldots C_9$

$b_n = C_1 + C_4 + \ldots\ldots\ldots$

$c_n = C_2 + C_5 + C_8 + \ldots\ldots\ldots$

$a^3 + b^3 + c^3 - 3abc$

$(1+x)^n = C_0 + C_1 x + C_2 x^2 + C_3 x^3 \ldots\ldots\ldots C_n x^n$

$2^n = C_0 + C_1 + C_2 + \ldots\ldots C_n$

$(1+\omega)^n = C_0 + \omega(C_1 + C_4 + C_7 + \ldots\ldots)$

$+ \omega^2(C_2 + C_5 + C_8 + \ldots\ldots) + C_3 + C_6 \ldots\ldots$

$(1 + \omega^2)^n = C_0 + \omega^2(C_1 + C_4 + C_7 \ldots\ldots)$

$+ \omega(C_2 + C_5 + C_8 + \ldots\ldots) + C_3 + C_6 \ldots\ldots$

$2^n + (1 + \omega)^n + (1 + \omega^2)^n = 3(C_0 + C_3 + C_6 \ldots)$

$2^n + (-\omega^2)^n + (-\omega)^n = 3(C_0 + C_3 + C_6 \ldots)$

$2^n \omega^2 + \omega(-\omega^2)^n + (-\omega)^n = 3\omega^2(C_1 + C_2 + C_7 \ldots)$

$a_n = \dfrac{2^n + (-\omega^2)^n + (-\omega)^n}{3}$

$b_n = \dfrac{2^n \omega^2 + \omega(-\omega^2)^n + (-\omega)^n}{3\omega^2}$

$c_n = \dfrac{\omega^2 2^n + (-\omega^2)^n + \omega(-\omega)^n}{3\omega^2}$

Sol 12: $\dfrac{z_1 - z_2}{z_3 - z_2} = e^{i\pi/2} = i$

$\dfrac{z_1 - iz_3}{1 - i} = z_2 \Rightarrow Z_2 = \dfrac{z_1(1 + i) + z_3(1 - i)}{2}$

$\dfrac{z_1 - z_4}{z_3 - z_4} e^{-i\pi/2} = -i \Rightarrow z_1 - z_4 = -i(z_3 - z_4)$

$$\Rightarrow z_4 = \frac{z_1(1-i) + z_3(1+i)}{2}$$

Sol 13: $f(z) = (a + ib)(z) = c + id$

$z = (x + iy)$

Image $= c - id$

$(c - x)^2 + (y + d)^2 = c^2 + d^2$

$x^2 + y^2 - 2cx + 2dy = 0$

$a^2 + b^2 = 64$

$c = ax - by$

$d = bx - ay$

$\Rightarrow x^2 + y^2 - 2x(ax - by) + 2y(bx - ay) = 0$

$\Rightarrow x^2(1 - 2a) + y^2(1 - 2a) - 2bxy + 2bxy = 0$

$\Rightarrow (x^2 + y^2)(1 - 2a) = 0$

$\Rightarrow a = \dfrac{1}{2} \Rightarrow b^2 = 64 - \dfrac{1}{4} = \dfrac{255}{4} = \dfrac{u}{v}$

$\Rightarrow u + v = 259$

Sol 14:

$\cos x + {}^nC_1\cos 2x + {}^nC_2\cos 3x + \ldots\ldots + {}^nC_n\cos(n + 1)x$

$= \cos x + {}^nC_1\cos^2 x + {}^nC_2\cos^3 3x + \ldots\ldots + {}^nC_n\cos^{n+1}x$

$= \cos\bar{x}\,[1 + {}^nC_1\cos x + {}^nC_2\cos^2 x + \ldots\ldots + {}^nC_n\cos^n x]$

$= \cos x[1 + \cos x]^n$

$= \cos x\left[2\cos\dfrac{x}{2}\cos\dfrac{x}{2}\right]^n$

$= 2^n\cos^n\dfrac{x}{2}\cos\dfrac{n}{2}\cos$

$= 2^n\cos^n\dfrac{x}{2}\cos\left[\dfrac{n+2}{2}\right]x$

$= 2^n\cos^n\dfrac{x}{2}\left[\cos\dfrac{(n+2)x}{2} + i\sin\left[\dfrac{n+2}{2}\right]x\right]$

By comparing, we get the desired results. Hence, proved

Sol 15: $f(x) = ax^3 + bx^2 + cx + d \quad f(i) = 0$

$\Rightarrow -ai - b + ci + d = 0 + io \quad \Rightarrow -a + c = 0 \text{ and } d - b = 0$

$\Rightarrow a = c \text{ and } \quad b = d \qquad\qquad \ldots(i)$

And $f(1 + i) = 5$

$a(1+i)^3 + b(1+i)^2 + c(1+i) + d = 5$

$\Rightarrow a\left(1 + i^3 + 3i^2 + 3i\right) + b\left(1 + i^2 + 2i\right) + c(1+i) + d = 5$

$\Rightarrow a(1 - i - 3 + 3i) + b(1 - 1 + 2i) + c(1+i) + d = 5$

$\Rightarrow a(-2 + 2i) + b(2i) + c(1+i) + d = 5$

$\Rightarrow -2a + c + d = 5 \text{ and } 2a + 2b + c = 0$

From (i), we have

$d - a = 5 \text{ and } 3a + 2d = 0$

$\Rightarrow a = c = -2 \text{ and } b = d = 3$

$\therefore a^2 + b^2 + c^2 + d^2 = (-2)^2 + 3^2 + (-2)^2 + 3^2$

$= 4 + 9 + 4 + 9$

$= 26$

Sol 16: $\displaystyle\sum_{k=1}^{n}\left(z_k - w_k\right) = 0$

$\Rightarrow z_1 + z_2 + z_3 + z_4 + z_5$

$= 32 + 170i - 7 + 64i - 9 + 200i + 1 + 27i - 14 + 43i$

$= 3 + 504i$

If y-intercept is 3, then eq. of line

$y = mx + 3$

$yi = mxi + 3 \text{ for } i = 1,2\ldots\ldots 5$

Now, $z_1 + \ldots\ldots\ldots\ldots z_5 = 3 + 504i$

$(x_1 + x_2 + \ldots\ldots + x_5) + i(y_1 + y_2 + \ldots\ldots y_5) = 3 + 504i +$

$i\{m(x_1 + x_2 \ldots\ldots\ldots + x_5) + 15\} = 3 + 504i$

$x_1 + x_2 + \ldots\ldots\ldots + x_5 = 3$

$m(x_1 + x_2 + \ldots\ldots\ldots + x_5) + 15 = 504$

$\Rightarrow 3m + 15 = 504$

$\Rightarrow m = 163$

Sol 17:

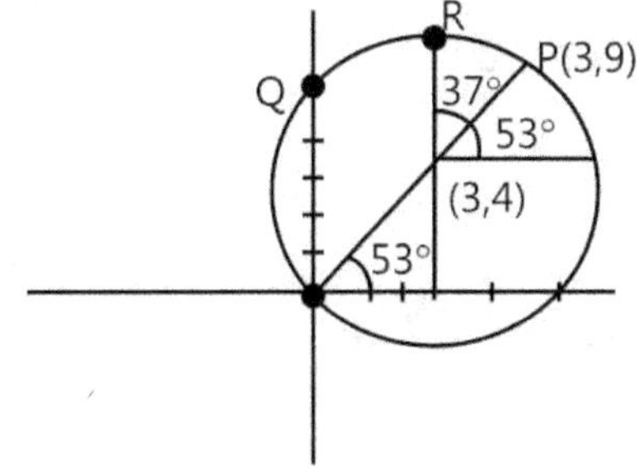

$|z - 3 - 4i| = 5 \tan^{-1}\dfrac{3}{4}$ (37°) in clockwise direction

It reaches Q (3, 9) above centre

Then it moves 2 unit downwards i.e. R (3, 7)

Sol 18: $\displaystyle\sum_{p=1}^{32}(3p+2)\left[\sum_{q=1}^{10}\left(\sin\frac{2q\pi}{11}-i\cos\frac{2q\pi}{11}\right)\right]^{p}$

$= \displaystyle\sum(3p+2)\left[-i\left(\sum_{q=0}^{10}e^{\frac{2q\pi i}{11}}-1\right)\right]^{p}$

$= \displaystyle\sum_{p=1}^{32}(3p+2)i^{p} = \sum_{p=1}^{32}(3p)i^{p}\left[2\sum_{1}^{32}i^{p}=0\right]$

$= 3i(1-3+\cdots-31)-3(2-4+\cdots-32)=48(1-i)$

Sol 19: $\dfrac{a}{1-b}=\dfrac{b}{1-c}=\dfrac{c}{1-a}=k$

$a = k - kb$

$b = k - kc$

$c = k - ka$

$a = k - k^2 + k^2(k - ka)$

$a = k - k^2 + k^3 - k^3 a)$

$a = \dfrac{k-k^2+k^3}{1+k^3}=b=c$

but $a \neq b \neq c$

i.e. $k^3 = -1 \Rightarrow k = -w, -w^2$

Sol 20: $|z-\alpha|^2 + |z-\beta|^2 = k$

Locus of z is a circle

$(x-\alpha_1)^2+(x-\beta_1)^2+(y-\alpha_2)^2+(y-\beta_2)^2 = k$

$\Rightarrow 2x^2 + 2y^2 - 2x(\alpha_1 + \beta_1) - 2y_2(\alpha_2 + \beta_2)$

$+ \alpha_1^2 + \alpha_2^2 + \beta_1^2 + \beta_2^2 - k = 0$

$\Rightarrow x^2 + y^2 - 2x\left(\dfrac{\alpha_1+\beta_1}{2}\right)-2y\left(\dfrac{\alpha_2+\beta_2}{2}\right)$

$+ \left(\dfrac{\alpha_1^2+\alpha_2^2+\beta_1^2+\beta_2^2-k}{2}\right)=0$

$r > 0$ i.e. $g^2 + f^2 - c > 0$

$\left(\dfrac{\alpha_1+\beta_1}{2}\right)^2+\left(\dfrac{\alpha_2+\beta_2}{2}\right)^2-2\left(\dfrac{\alpha_1^2+\alpha_2^2+\beta_1^2+\beta_2^2-k}{4}\right)>0$

$\Rightarrow -\alpha_1^2 + \alpha_2^2 - \beta_1^2 - \beta_2^2 + 2\alpha_1\beta_1 + 2\alpha_2\beta_2 + k > 0$

$\Rightarrow k > \alpha_1^2 + \alpha_2^2 + \beta_1^2 + \beta_2^2 - 2\alpha_1\beta_1 - 2\alpha_2\beta_2$

$\Rightarrow k > (\alpha_1 - \beta_1)^2 + (\alpha_2 - \beta_2)^2 \Rightarrow k > |\alpha - \beta|^2$

Centre $= \left(\dfrac{\alpha_1+\beta_1}{2},\dfrac{\alpha_2+\beta_2}{2}\right)$

Radius $= \sqrt{\dfrac{2k + 2\alpha_1\beta_1 + 2\alpha_2\beta_2 - \alpha_1^2 - \alpha_2^2 - \beta_1^2 - \beta_2^2}{4}}$

Sol 21: $f(z) = |z^3 - z + 2|$ $|z| = 1$

$(f(z))^2 = |z^3 - z + 2|^2 = (z^3 - z + 2)(\bar{z}^3 - \bar{z} + 2)$

$= 1 - z^2 + 2z^3 - \bar{z}^2 + z\bar{z} - 2z + 2\bar{z}^3 - 2\bar{z} + 4$

$= 6 - (z^2 - \bar{z}^2) - 2(z + \bar{z}) + 2(Z^3 + \bar{z}^3)$

$= 6 - 2(a^2 - b^2) - 4(a) + 2(z + \bar{z})(z^2 + \bar{z}^2 - z\bar{z})$

$= 6 - 2(a^2 - b^2) - 4(a) + 4a(2a^2 - 2b^2 - 1)$

$= 6 - 8a - 2(2a^2 - 1) + 8a(2a^2 - 1)$

$f(z) = 16a^3 - 4a^2 - 16a + 8$

$f'(z) = 48a^2 - 8a - 16 = 8(6a^2 - a - 2)$

$= 8(6a^2 - 4a + 3a - 2)$

$= 8(2a(3a - 2) + 1(3a - 1))$

$= 8(2a + 1)(3a - 2)$

$a = \dfrac{-1}{2}\quad a = \dfrac{2}{3}$

For $a = \dfrac{2}{3}$

$f(z) = 16 \times \dfrac{8}{27} - 4 \times \dfrac{4}{9} - 16 \times \dfrac{2}{3} + 8$

$= \dfrac{128 - 48 - 288 + 202}{27} < 0$

For $a = \dfrac{-1}{2}$; $f(z) = -2 - 1 + 8 + 8 = 13$

Maximum value of $f(z) = \sqrt{13}$

Sol 22: $|a + b\omega + c\omega^2| + |a + b\omega^2 + c\omega| \geq$

$|a + b\omega + c\omega^2 + a + b\omega^2 + c\omega|$

$= |2a - b - c| = |a - b + a - c|$

Minimum value will occur when

$a - b = k, a - c = -k$

i.e. $k = 1$

$a - b = 1$ and $a - c = -1$

with a & b being least possible value integer values

$\Rightarrow a = 2, b = 1, c = 3$

$|a + b\omega + c\omega^2| + |a + b\omega^2 + c\omega|$

$= 2|2 + \omega + 3\omega^2| = 2|1 + 2\omega^2|$

$= 2|\omega^2 - \omega| = 2|(-\sqrt{3})| = 2\sqrt{3}$

$= \sqrt{12} = n^{1/4} \Rightarrow n = 144$

Sol 23: $x^7 + x^6 + \ldots\ldots 1 = 0$

$$\left(\frac{x^8 - 1}{x - 1}\right) = 0$$

Total area = unshaded area + shaded area

$$\text{Unshaded area} = 6 \times 1 \begin{vmatrix} 0 & 0 & 1 \\ 0 & 1 & 1 \\ \dfrac{1}{\sqrt{2}} & \dfrac{1}{\sqrt{2}} & 1 \end{vmatrix}$$

$$= \frac{6}{2} \times \frac{1}{\sqrt{2}} = \frac{6}{2\sqrt{2}}$$

$$\text{Shaded area} = 2 \times \frac{1}{2} \times \frac{1}{\sqrt{2}} \times \frac{1}{\sqrt{2}} = \frac{1}{2}$$

$$\text{Total area} = \frac{6}{2\sqrt{2}} + \frac{1}{2} = \frac{6 + \sqrt{2}}{2\sqrt{2}} = \frac{6\sqrt{2} + 2}{4}$$

$$= \frac{3\sqrt{2} + 1}{2} = \frac{a\sqrt{b} + c}{d}$$

$$a + b + c + d = 3 + 2 + 1 + 2 = 8$$

Sol 24: $N = (a + ib)^3 - 107i$

$$= a^3 - ib^3 + 3a^2ib - 3ab^2 - 107i$$

$$N = (-b^3 + 3a^2b - 107)i + a^3 - 3ab^2$$

$$\Rightarrow 3a^2b - b^3 = 107 \Rightarrow b(3a^2 - b^2) = 107$$

$$\Rightarrow b = 1 \text{ and } 3a^2 - b^2 = 107 \Rightarrow a^2 = 36$$

$$\Rightarrow N = a^3 - 3ab^2 = 216 - 18 = 198$$

Sol 25: $x^4 + ax^3 + bx^2 + cx + d$ has 4 non-real roots.

$$\alpha + \beta = 3 + 4i$$

$$\gamma\delta = 13 + i\,(\gamma = \bar{\alpha},\ \delta = \bar{\beta})$$

$$\Rightarrow \bar{\alpha}\bar{\beta} = 13 + i,\ \alpha\beta = 13 - i$$

$$\alpha + \beta = 3 + 4i$$

$$\Rightarrow \alpha\beta + b\gamma + g\delta + d\alpha + a\gamma + b\delta = b$$

$$\Rightarrow \alpha\beta + \beta\bar{\alpha} + \bar{\alpha}\bar{\beta} + \alpha\bar{\alpha} + \beta\bar{\beta} + \alpha\bar{\beta} = b$$

$$\Rightarrow 13 - i + 13 + i + (\bar{\alpha} + \bar{\beta})(\alpha + \beta) = b$$

$$= 26 + (3 + 4i)(3 - 4i) = b$$

$$\Rightarrow b = 26 + 25 = 51$$

Sol 26: $z^5 + 1$

$$= (z^3 + 1)z^2 + (1 - z^2)$$

$$= (z + 1)(z^2 - z + 1)z^2 + (1 - z^2)$$

$$= (1 + z)[(z^2 - z + 1)z^2 + 1 - z]$$

$$= (1 + z)[z^4 - z^3 + z^2 - z + 1]$$

$$= (1 + z)[z^2 + az + 1][z^2 + bz + 1]$$

$$\Rightarrow b + a = -1$$

$$ba + 2 = 1$$

$$ab = -1$$

$$\Rightarrow a - \frac{1}{a} = -1 \Rightarrow a^2 + a - 1 = 0$$

$$\Rightarrow a = \frac{-1 \pm \sqrt{5}}{2} \Rightarrow a = \frac{-1 + \sqrt{5}}{2} = -2\cos 108°$$

$$\text{And } b = \frac{-1 - \sqrt{5}}{2} = -2\cos 36°$$

$\therefore$ Factors are $(1 + z)(z^2 - 2z\cos 36° + 1)(z^2 - 2\cos 108° + 1)$

Since $ab = -1$

$$\Rightarrow 4\cos 36°\cos 108° = -1 \Rightarrow 4\cos\frac{\pi}{5}\cos\frac{\pi}{10} = 1$$

Sol 27: $x = 1 + i\sqrt{3}$

$$y = 1 - i\sqrt{3}$$

$$z = 2$$

$$x = -2\omega y = -2\omega^2 z = 2$$

$$x^p + y^p p > 3 \text{ prime (P is odd)}$$

$$= -2^p\omega^p - 2^p\,\omega^{2p} = -2^p(\omega^p + \omega^{2p}) = 2^p = z^p$$

Sol 28: $f(z) = a(z - i) + i \Rightarrow f(i) = i$ $\qquad$ ... (i)

$f(z) = b(z + i) + (1 + i) \Rightarrow f(-i) = 1 + i$ $\qquad$ ... (ii)

$f(z) = c(z^2 + 1) + k_1 z + k_2$ $\qquad$ ... (iii)

Substituting values from (i) and (ii) in (iii)

$$\Rightarrow i = k_1 i + k_2 \text{ and } i + 1 = -k_1 i + k_2$$

$$\Rightarrow k_2 = i + \frac{1}{2} \text{ and } k_1 = \frac{i}{2}$$

$$\therefore \text{ Remainder} = \frac{iz}{2} + i + \frac{1}{2}$$

Sol 29: (a)

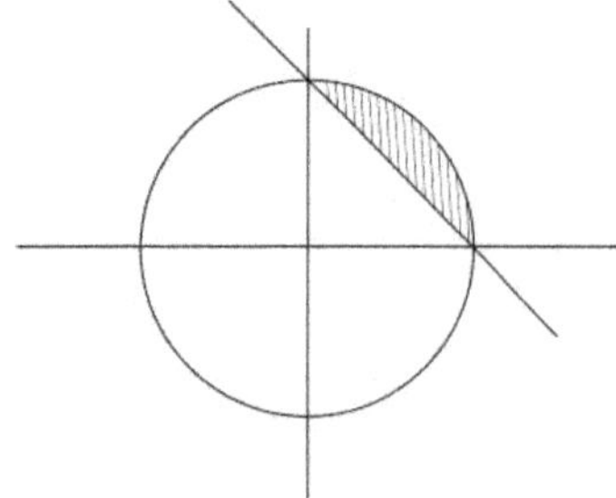

$A = \{z|\ |z| \leq 2\}$

$B = \{z|\ (1 - i)z + (1 + i)\ \bar{z} \geq 4\}$

Let $z = a + ib$

$\Rightarrow (1 - i)(a + ib) + (1 + i)(a - ib) \geq 4$

$\Rightarrow a + b + a + b \geq 4 \Rightarrow a + b \geq 2$

area $= \dfrac{\pi r^2}{4} - \dfrac{1}{2}r^2 = \left(\dfrac{\pi}{4} - \dfrac{1}{2}\right)r^2 = (\pi - 2)$

(b) $f(x) = \dfrac{1}{x - i} = \dfrac{1}{x - i}\left(\dfrac{x + i}{x + i}\right)$

$= \dfrac{x}{x^2 + 1} + i\left(\dfrac{1}{x^2 + 1}\right)$

$x - \text{coordinate} = \dfrac{k}{k^2 + 1}$

$y - \text{coordinate} = \dfrac{1}{k^2 + 1}$

$\therefore$ Locus of the function will be

$x^2 + y^2 - y = 0$

This is circle with diameter 1

Hence the area of the square $= \dfrac{1}{2}$

Sol 30: (a) $(1 + \omega + \ldots\ldots\omega^n)^m$, $m, n \in N$

$n \in 3k \Rightarrow 1$

$n \in 3k + 1 \Rightarrow (-\omega^2)^m$

$n \in 3k + 2 \Rightarrow 0$

$m = 1 \Rightarrow (-\omega^2)$, $m = 2 \Rightarrow (\omega)$, $m = 3 \Rightarrow (-1)$

$m = 4 \Rightarrow (\omega^2)$, $m = 5 \Rightarrow (-\omega)$, $m = 6 \Rightarrow (1)$

No. of distinct elements are 7(S)

(b) Real coefficient,

Root $\rightarrow 2\omega, 2+3\omega, 2+3\omega^2, 2-\omega-\omega^2$

Roots are $2\omega, 2 + 3\omega, 2 + 3\omega^2, 3$

Other root will be $2\bar{\omega}$

Total no. of roots are 5 (q)

(c) $\alpha = 6 + 4i\beta = 2 + 4i$

$\text{amp}\left(\dfrac{z - \alpha}{z - \beta}\right) = \dfrac{\pi}{6}$

$\tan^{-1}\left[\dfrac{(x - 6) + (y - 4)i\ (x - 2) - i(y - 4)}{(x - 2) + i(y - 4)\ (x - 2) - i(y - 4)}\right] = \dfrac{\pi}{6}$

$\text{Re}(z) = (x - 6)(x - 2) + (y - 4)^2$

$\text{Im}(z) = (x - 2)(y - 4) + (x - 6)(y - 4)$

$\Rightarrow \dfrac{(x - 2)(y - 4) + (x - 6)(4 - y)}{(x - 6)(x - 2) + (y - 4)^2} = \dfrac{1}{\sqrt{3}}$

$\Rightarrow \dfrac{-4x - 2y + 8 + 4x + 6y - 24}{x^2 + y^2 - 8x - 8y + 12 + 16} = \dfrac{1}{\sqrt{3}}$

$\Rightarrow x^2 + y^2 - 8x - 8y + 28 = 4\sqrt{3}\,y - 16\sqrt{3}$

$\Rightarrow x^2 + y^2 - 8x + y(-8 - 4\sqrt{3}) + 28 + 16\sqrt{3} = 0$

$\Rightarrow r = \sqrt{16 + (4 + 2\sqrt{3})^2 - 28 - 16\sqrt{3}}$

$= \sqrt{16 + 16 + 12 - 28 + 16\sqrt{3} - 16\sqrt{3}}$

$= \sqrt{16} = 4\text{(p)}$

Exercise 2

Single Correct Choice Type

Sol 1: (C) $z^2 + z + 1$ is real and +ve

$x^2 - y^2 + x + 1 + i(2xy + y)$

$x^2 - y^2 + x + 1 > 0$ and $2xy + y = 0$

$\Rightarrow y(2x + 1) = 0 \Rightarrow y = 0$ or $x = \dfrac{-1}{2}$

If $y = 0\ x^2 + x + 1 > 0 \Rightarrow z$ represents real axis

If $x = \dfrac{-1}{2}\ \dfrac{3}{4} - y^2 > 0$, Line segment joining

$\left[-\dfrac{1}{2}, \dfrac{-\sqrt{3}}{2}\right]$ to $\left[-\dfrac{1}{2}, \dfrac{\sqrt{3}}{2}\right]$

Sol 2: (B) $z\bar{z}^3 + z^3\bar{z} = 350$

$|z|^2\ (z^2 + \bar{z}^2) = 350$

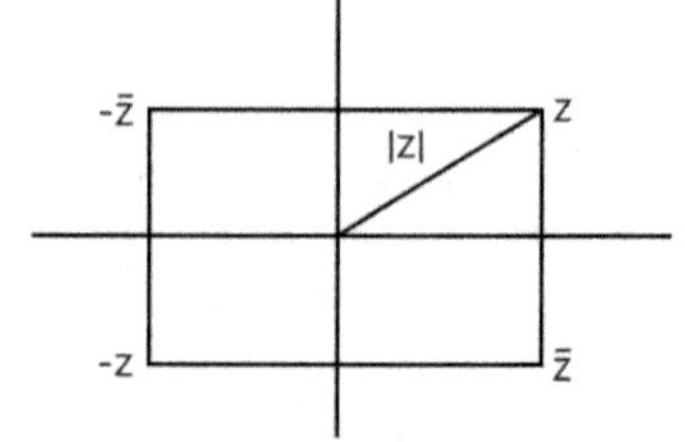

Length of diagonal $2|z|$

$|z|^2 (x^2 - y^2 + 2ixy + x^2 - y^2 - 2ixy) = 350$

$(x^2 + y^2)(x^2 - y^2) = \dfrac{350}{2} = 175$

As x and y are integers

$x = \pm 4;\ y = \pm 3$

$\Rightarrow |z| = 5$

Length of diagonal is 10

Sol 3: (A) $z_1^2 - 2z_1 z_2 + 2z_2^2 = 0$

$\left(z_1 - z_2\right)^2 + z_2^2 = 0$

$\Rightarrow \left(z_1 - z_2\right)^2 = -z_2^2 = |z_1 - z_2| = |z_2|$ and $\left(z_1 - z_2\right) = iz_2$

and

$\left(\dfrac{z_1}{z_2}\right)^2 - 2\dfrac{z_1}{z_2} + 2 = 0$

$\Rightarrow \dfrac{z_1}{z_2} = 1+i$

$\Rightarrow \left|\dfrac{z_1}{z_2}\right| = \sqrt{2}$

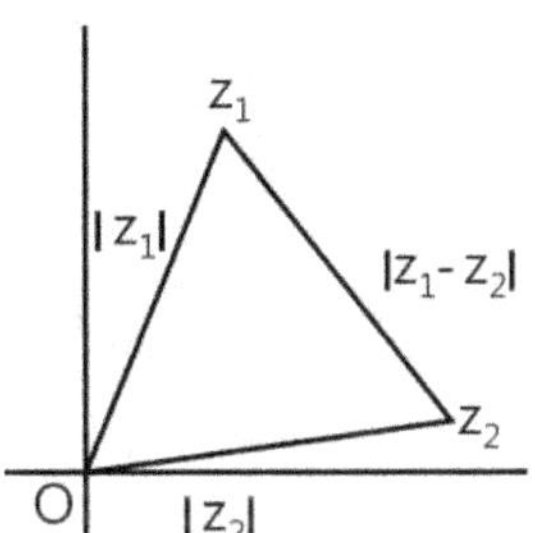

Now,

$\dfrac{z_1 - 0}{z_2 - 0} = \left|\dfrac{z_1}{z_2}\right| e^{i\angle z_1 o z_2}$

$e^{i\angle z_1 o z_2} = \dfrac{1+i}{\sqrt{2}} = e^{i\pi/4}$

$\Rightarrow \angle z_1 o z_2 = 45^\circ = \angle o z_1 z_2$

Also,

$\dfrac{z_2 - z_1}{z_2} = \left|\dfrac{z_2 - z_1}{z_2}\right| e^{i\angle o z_2 z_1}$

$\Rightarrow e^{i\angle o z_2 z_1} = i \Rightarrow \angle o z_2 z_1 = 90^\circ$

Sol 4: (B)

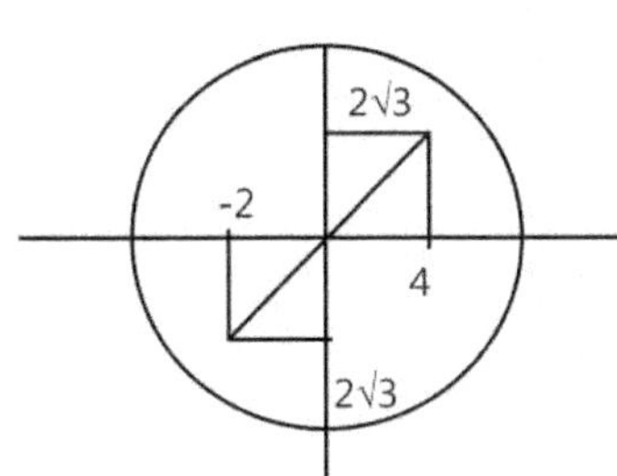

$|Z| \le 4$ and $\text{Arg}(z) = \dfrac{\pi}{3}$

$x^2 + y^2 \le 16$ and $y = \sqrt{3}\,x$

$\Rightarrow x \in \left[-2, 2\right]$ & $y \in \left[-2\sqrt{3}, 2\sqrt{3}\right]$

Set of points lie on radius of circle

Sol 5: (C)

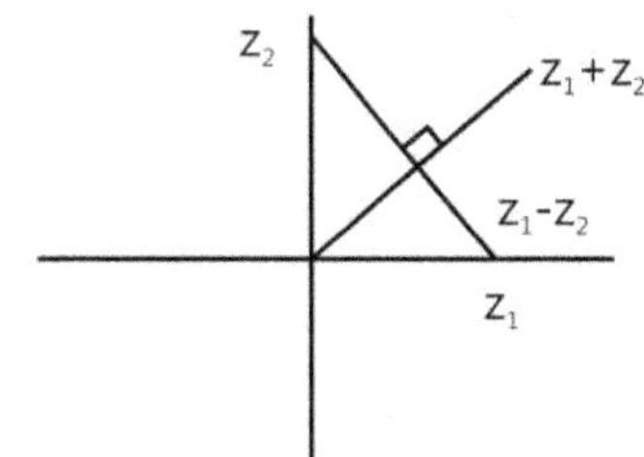

$\arg \dfrac{z_1 + z_2}{z_1 - z_2} = \dfrac{\pi}{2}$

$\tan^{-1}\left(\dfrac{z_1 + z_2}{z_1 - z_2}\right) = \dfrac{\pi}{2}$

$|z_1 + z_2| \ne |z_1 - z_2|$

Diagonals are perpendicular but not equal ie figure represented is rhombus but not a square.

Sol 6: (A) Condition for equilateral

$\Delta : z_1^2 + z_2^2 + z_3^2 = z_1 z_2 + z_2 z_3 + z_3 z_1$

$z^3 - 3\alpha z^2 + 3\beta z + x = 0$

z_1, z_2, z_3

$\Rightarrow z_1 + z_2 + z_3 = 3\alpha$

$\Rightarrow z_1 z_2 + z_2 z_3 + z_3 z_1 = 3\beta$

$\Rightarrow z_1, z_2, z_3 = -x$

$(z_1 + z_2 + z_3)^2 = z_1^2 + z_2^2 + z_3^2 + 2(3\beta)$

$(3\alpha)^2 = 9\beta$

$\alpha^2 = \beta$

Sol 7: (D)

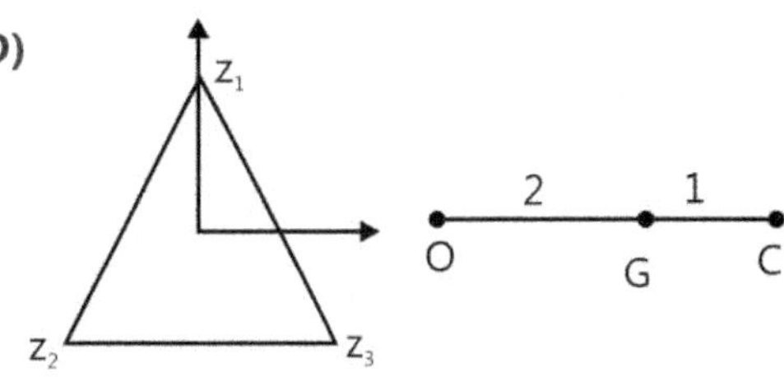

$\dfrac{z_1 + z_2 + z_3}{3} = \text{Centroid},\ z_c = 0$

$\dfrac{z_1 + z_2 + z_3}{3} = \dfrac{2z_c + z_o}{3} \Rightarrow z_o = z_1 + z_2 + z_3$

Sol 8: (C)

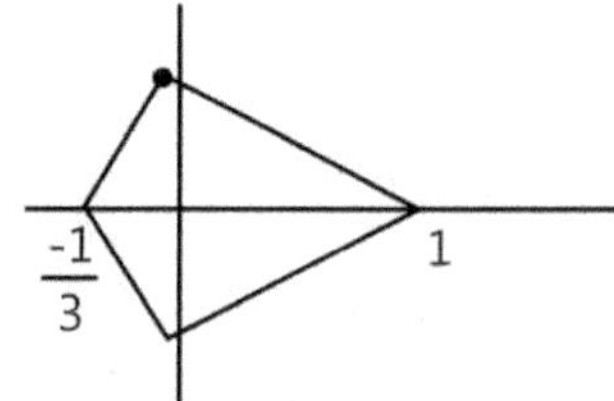

$(z + 1)^4 = 16z^4$

$\Rightarrow z + 1 = 2z,\ 2iz,\ -2z,\ -2iz$

$\Rightarrow z = 1,\ \dfrac{1}{2i-1},\ -\dfrac{1}{3},\ \dfrac{1}{-1-2i} = -\dfrac{1}{3},\ 1,\ \dfrac{-1-2i}{5},\ \dfrac{-1+2i}{5}$

Point equidistant is $z = \dfrac{z_1 + z_2}{2}$

Where $z_1 = (1, 0)$, $z_2 = \left(\dfrac{-1}{3}, 0\right)$

$z = \dfrac{1 - \dfrac{1}{3}}{2} = \dfrac{1}{3}$

Sol 9: (B) $|z - 2| = 3$

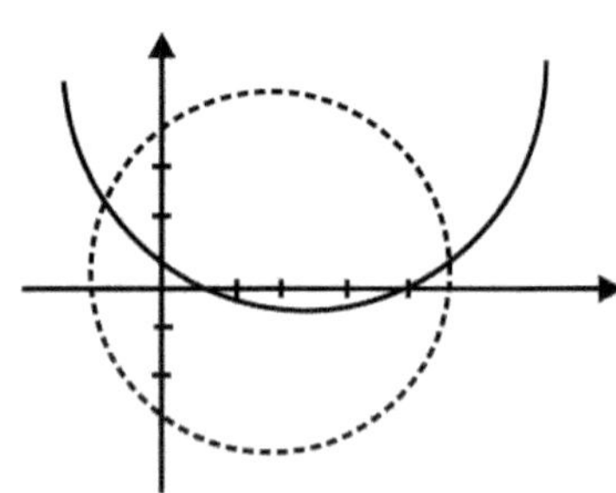

$|z - 2 - 3i| = 4$

$\Rightarrow S_1 = (x - 2)^2 + y^2 = 9$

$\Rightarrow S_2 = (x - 2)^2 + (y - 3)^2 = 16$

Both circles are intersecting. So, radical axis will be

$S_1 - S_2 = 0 \Rightarrow 9 - 6y = 7 \Rightarrow 3y - 1 = 0$

Sol 10: (A) $z^3 + iz - 1 = 0$

Let $z = k$ (real)

$\Rightarrow k^3 + ik - 1 = 0$

$\quad k^3 + ik - 1 = 0$ and $k = 0$

If $k = 0$, then $0 - 1 = 0$

Not possible

Therefore $z \ne k$ (real)

Hence, no real solution

Sol 11: (D) $|z - 4 - 3i| = 2$

z lies on a circle shown in the figure $|z| = |z - 0|$ is nothing but distance from origin

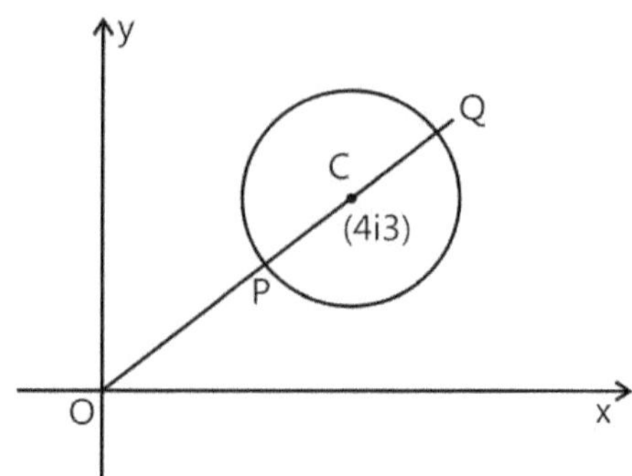

Minimum = OP = OC − CP

$\qquad = \sqrt{4^2 + 3^2} - 2 = 5 - 2 = 3$

Maximum = OC + CQ $= 5 + 9 = 7$

Sol 12: (A) $z = 1 - \sin\alpha + i\cos\alpha$

$|z| = \sqrt{1 + \sin^2\alpha - 2\sin\alpha + \cos^2\alpha}$

$= \sqrt{2 - 2\sin\alpha} = \sqrt{2(1 - \sin\alpha)}$

$\arg z = \tan^{-1}\dfrac{\cos\alpha}{1 - \sin\alpha} = \tan^{-1}\left(\dfrac{1}{\sec\alpha - \tan\alpha}\right)$

$= \tan^{-1}\dfrac{1 + \sin\alpha}{\cos\alpha} = \tan^{-1}\dfrac{\left(\sin\dfrac{\alpha}{2} + \cos\dfrac{\alpha}{2}\right)^2}{\left(\cos^2\dfrac{\alpha}{2} - \sin^2\dfrac{\alpha}{2}\right)^2}$

$= \tan^{-1}\left(\dfrac{\sin\dfrac{\alpha}{2} + \cos\dfrac{\alpha}{2}}{\cos\dfrac{\alpha}{2} - \sin\dfrac{\alpha}{2}}\right) = \tan^{-1}\left(\dfrac{1 + \tan\dfrac{\alpha}{2}}{1 - \tan\dfrac{\alpha}{2}}\right)$

$= \tan^{-1}\left[\tan\left(\dfrac{\pi}{4} + \dfrac{\alpha}{2}\right)\right] = \dfrac{\pi}{4} + \dfrac{\alpha}{2}$

Sol 13: (D) $\left|\dfrac{z_1 + z_2}{2} + \sqrt{z_1 z_2}\right| + \left|\dfrac{z_1 + z_2}{2} - \sqrt{z_1 z_2}\right|$

$= \dfrac{\left|z_1 + z_2 + 2\sqrt{z_1 z_2}\right| + \left|z_1 + z_2 - 2\sqrt{z_1}\sqrt{z_2}\right|}{2}$

$= \dfrac{\left(\sqrt{z_1} + \sqrt{z_2}\right)^2 + \left(\sqrt{z_1} - \sqrt{z_2}\right)^2}{2}$

$= \dfrac{2(z_1)^2 + 2(z_2)^2}{2} = \left|\sqrt{z_1}\right|^2 + \left|\sqrt{z_2}\right|^2$

Sol 14: (A) $u^2 - 2u + 2 = 0$

$\Rightarrow (u-1)^2 = -1$

$\Rightarrow u - 1 = \pm i$

$\Rightarrow u = 1 + i,\ 1 - i = (\alpha,\ \beta)$

$\cot\theta = x + 1$

$\dfrac{(x+\alpha)^n - (x+\beta)^n}{\alpha - \beta} = \dfrac{(\cot\theta + i)^n - (\cot\theta - i)^n}{2i}$

$= \dfrac{(\cos\theta + i\sin\theta)^n - (\cos\theta - i\sin\theta)^n}{2i\sin^n\theta}$

$= \dfrac{e^{i\theta n} - e^{-i\theta n}}{2i\sin^n\theta} = \dfrac{\sin n\theta}{\sin^n\theta}$

Sol 15: (B) $A_1 \ldots\ldots A_n$ vertices of regular polygon in a circle of radius R

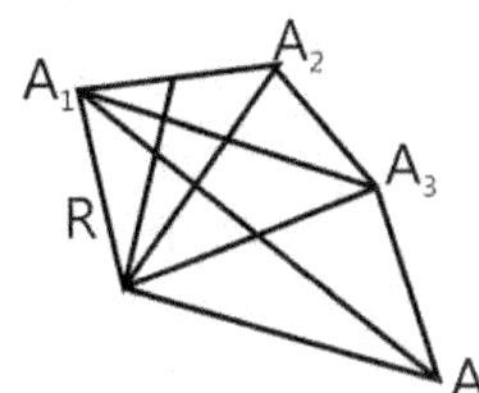

$(A_1A_2)^2 + (A_1A_3)^2 \ldots\ldots (A_1A_n)^2$

$A_1A_2 = 2R\sin\dfrac{\pi}{n}$

$A_1A_3 = 2R\sin\dfrac{2\pi}{n}$

$A_1A_2 = 2R\dfrac{3\pi}{n}$

$4R^2\left[\sin^2\dfrac{\pi}{n} + \sin^2\dfrac{2\pi}{n} + \sin^2\dfrac{3\pi}{n}\ldots\sin^2\dfrac{(n-1)\pi}{n}\right]$

$= -2R^2\left[1 - n + \cos\dfrac{2\pi}{n} + \cos\dfrac{4\pi}{n} + \ldots\cos(2n-1)\dfrac{\pi}{n}\right]$

$= -2R^2\left[1 - n + \dfrac{\sin\pi k}{\sin n\left(\dfrac{\pi}{n}\right)} - \cos\dfrac{2n\pi}{n}\right] = 2nR^2$

Sol 16: (B) $z^4 + a_1z^3 + a_2z^2 + a_3z + a_4 = 0$

$z = ib$

$b^4 - ib^3a_1 - b^2a_2 + iba_3 + a_4 = 0$

$b^4 - b^2a_2 + a_4 = 0 \Rightarrow b^4a_2 + a_4a_2 = b^2a_2^2$

$\Rightarrow ba_3 - b^3a_1 = 0 \Rightarrow a_3 = b^2a_1$

$\dfrac{a_3}{a_1a_2} + \dfrac{a_1a_4}{a_2a_3} = \dfrac{b^2}{a_2} + \dfrac{a_4}{a_2b^2} = \dfrac{b^4a_2 + a_4a_2}{a_2^2b^2} = \dfrac{b^2a_2^2}{b^2a_2^2} = 1$

Sol 17: (D) $(1+z)^6 + z^6 = 0$

$\Rightarrow (1+z)^6 = -z^6$

Now,

$\Rightarrow \left(\dfrac{1+z}{z}\right)^6 = -1$

$\Rightarrow 1 + \dfrac{1}{z} = (-1)^{1/6}$

$(4, 3)$

$\Rightarrow \dfrac{1}{z} = \cos\dfrac{2K\pi + \pi}{6} - 1$

$\Rightarrow \dfrac{1}{z} = -\left\{2\sin\dfrac{2K\pi + \pi}{12}\cos\left(\dfrac{2K\pi + \pi}{12} - \dfrac{\pi}{2}\right)\right\}$

$\Rightarrow z = -\dfrac{1}{2\sin\dfrac{2K\pi + \pi}{12}}\left\{\cos\left(\dfrac{\pi}{2} - \dfrac{2K\pi + \pi}{12}\right)\right\}$

$z = -\dfrac{1}{2}\left\{1 + i\cot\dfrac{2K\pi + \pi}{12}\right\}$

$\Rightarrow z$ lies on line $x = -\dfrac{1}{2}$

$\Rightarrow$ All roots are collinear.

Sol 18: (A)

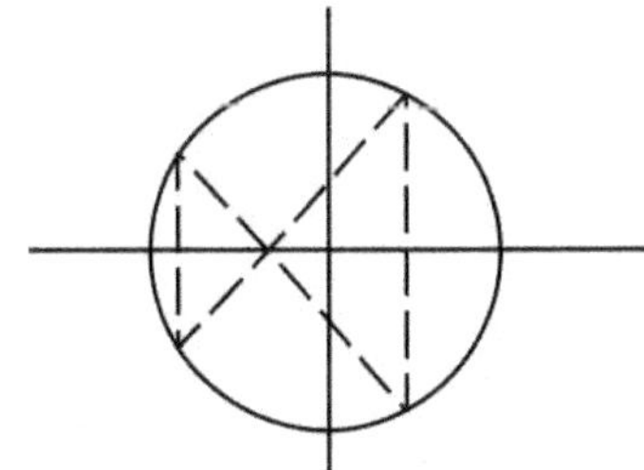

$z^{10} - z^5 = 992$

$z^5(z^5 - 1) = 32(31)$

$z^5 = t$

$\Rightarrow t^2 - t = 992 \Rightarrow t^2 - 32t + t - 992 = 0$

$\Rightarrow t = 32,\ -31 \Rightarrow z^5 = 32,\ -31$

$\Rightarrow z^5 = 32$ has 2 roots with $-$ve real Part

and $z^5 = -31$ has 3 roots with $-$ve real part

Sol 19: (A) $a|z_1| = b|z_2|$

$$\frac{a}{b} = \frac{|z_2|}{|z_1|} \quad T = \frac{az_1}{bz_2} + \frac{bz_2}{az_1} \quad \text{let } \frac{az_1}{bz_2} = z$$

$$T = z + \frac{1}{z}\frac{\bar{z}}{\bar{z}} = z + \frac{\bar{z}}{|z|^2}$$

$|z| = 1 \Rightarrow T = z + \bar{z} = 2\text{Re}(z)$

$\text{Re}(z) \in (-1, 1)$

$\Rightarrow T \in [-2, 2]$

Sol 20: (B) $(p+q)^3 + (p\omega+q\omega^2)^3 + (p\omega^2+q\omega)^3$

$= p^3+q^2+3p^2q+3pq^2+p^3+q^3+3pq^2\omega^5$

$+3p^2q\omega^4+p^3+q^3+3p^2q\omega^5+3pq^2\omega^4$

$= 3(p^3+q^3)$

Sol 21: (A) $\left(1+i\sqrt{3}\right)^x = 2^x \Rightarrow \left(\dfrac{1+i\sqrt{3}}{2}\right)^x = 1$

$\Rightarrow (-\omega)^x = 1$

$\Rightarrow x = 6, 12, 18, \ldots\ldots$

It forms an AP

Multiple Correct Choice Type

Sol 22: (B, C) $x^2+(p+iq)x + 3i = 0$

$a^2 + b^2 = 8$

$(\alpha + \beta) = -(p + iq)$

$a\beta = 3i$

$\Rightarrow (\alpha + \beta)^2 = 8 + 6i = p^2 - q^2 + 2ipq$

$p^2 - q^2 = 8$

$2pq = 6$

$\Rightarrow p = 3, q = +1$ or $p = -3, q = -1$

Sol 23: (A, D) $|z_1| = |z_2| \Rightarrow a^2 + b^2 = c^2 + d^2$

$z_1 = a + ib, a > 0$

$z_2 = c + id, d < 0$

$$\frac{z_1 + z_2}{z_1 - z_2} = \frac{(a+c) + i(b+d)}{(a-c) + i(b-d)}$$

$(a + c)$ & $(b + d)$ can be zero, so value can be zero

$(a - c)$ & $(b - d)$ can be simultaneously zero $\Rightarrow$ purely imaginary.

Sol 24: (A, B, C, D) (a) $a, b, x\, y$ is $\in R$

$$\frac{a + ib}{x + iy} = a - ib$$

$$\Rightarrow x + iy = \frac{a^2 - b^2 + 2abi}{a^2 + b^2}$$

$$\Rightarrow x = \frac{a^2 - b^2}{a^2 + b^2}, y = \frac{2ab}{a^2 + b^2}$$

$\Rightarrow x^2 + y^2 = 1$ A is correct

(b) $\dfrac{1 - ix}{1 + ix} = a - ib$

$$\Rightarrow \frac{1 - x^2 - 2ix}{1 + x^2} = a - ib$$

$$\Rightarrow a = \frac{1 - x^2}{1 + x^2}, b = \frac{2x}{1 + x^2}$$

$\Rightarrow a^2 + b^2 = 1$ B is correct

(c) $\dfrac{a + ib}{a - ib} = x - iy$

$$\Rightarrow \frac{(a + ib)^2}{a^2 + b^2} = x - iy$$

$$\Rightarrow x + iy = \frac{a^2 - b^2 + 2iab}{a^2 + b^2}$$

$$\Rightarrow x + iy = \frac{a^2 - b^2 - 2iab}{a^2 + b^2} \Rightarrow |x + iy| = 1$$

(d) $\dfrac{y - ix}{a + ib} = y + ix$

$$\Rightarrow \frac{y^2 - x^2 - 2ixy}{y^2 + x^2} = a + ib$$

$$\Rightarrow a - ib = \frac{y^2 - x^2 + 2ixy}{x^2 + y^2} \Rightarrow |a - ib| = 1$$

Sol 25: (A, D) $z = x + iy = r(\cos\theta + i\sin\theta)$

$$\sqrt{z} = \sqrt{x + iy} = \sqrt{r(\cos\theta + i\sin\theta)}$$

$$= \sqrt{r}\, e^{i\theta/2} = \sqrt{r}\left[\cos\frac{\theta}{2} + i\sin\frac{\theta}{2}\right] \text{ for } y > 0$$

$$= \sqrt{r}\left[\sqrt{\frac{1 + \cos\theta}{2}} + i\sqrt{\frac{1 - \cos\theta}{2}}\right]$$

$$= \frac{\sqrt{r}}{\sqrt{2}}\left[\frac{\sqrt{r + x} + i\sqrt{r - x}}{\sqrt{r}}\right] = \frac{\sqrt{r + x} + i\sqrt{r - x}}{\sqrt{2}}\ \text{[A]}$$

Similarly

$$= \frac{1}{\sqrt{2}}\left[\sqrt{r + x} - i\sqrt{r - x}\right] \text{ for } y < 0$$

Sol 26: (A, D)

$$\left(az_1 + b\bar{z}_1\right)\left(cz_2 + d\bar{z}_2\right) = \left(cz_1 + d\bar{z}_1\right)\left(az_2 + b\bar{z}_2\right)$$

$$adz_1\bar{z}_2 + bc\,\bar{z}_1 z_2 = bcz_1\bar{z}_2 + ad\,\bar{z}_1 z_2$$

$$\Rightarrow ad = bc$$

$$\frac{a}{b} = \frac{c}{d} \quad \text{(A is correct)}$$

$$\Rightarrow \arg z_1 = \tan^{-1}\frac{b}{a}, \ \arg z_2 = \tan^{-1}\frac{d}{c}$$

$$\Rightarrow \arg z_1 = \arg z_2$$

$$\text{or } z_1\bar{z}_2 = \bar{z}_1 z_2 \Rightarrow \frac{z_1}{\bar{z}_1} = \frac{z_2}{\bar{z}_2}$$

$$\Rightarrow \frac{a+ib}{a-ib} = \frac{c+id}{c-id} \Rightarrow -ad + bc = 0$$

Sol 27: (A, B, C)

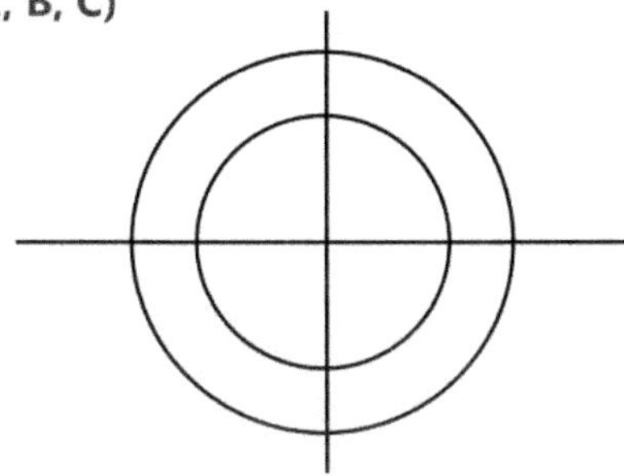

$|z_1| = 1$ and $|z_2| = 2$

Max $|2z_1 + z_2|$

$|2z_1 + z_2| \le 2|z_1| + |z_2| = 4$ (A is correct)

Min $|z_1 - z_2| \ge |z_1| - |z_2|$

$|z_1 - z_2| \ge 1$ (B is correct)

$$\left|z_2 + \frac{1}{|z_1|}\right| \le |z_2| + \frac{1}{|z_1|} \le 2 \quad \text{(C is correct)}$$

Sol 28: (A, B, C, D) $\left|\dfrac{\alpha - \beta}{1 - \bar{\alpha}\beta}\right| = 1$

$$|\alpha - \beta|^2 = |1 - \bar{\alpha}\beta|^2$$

$$\Rightarrow (\alpha - \beta)(\bar{\alpha} - \bar{\beta}) = (1 - \bar{\alpha}\beta)(1 - \alpha\bar{\beta})$$

$$|\alpha|^2 + |\beta|^2 - \alpha\bar{\beta} - \beta\bar{\alpha} = 1 - \alpha\bar{\beta} - \beta\bar{\alpha} + |\alpha|^2|\beta|^2$$

$$|\alpha|^2(\alpha - |\beta|^2) = (1 - |\beta|^2) \Rightarrow (|\alpha|^2 - 1)(1 - |\beta|^2) = 0$$

$\Rightarrow |\alpha| = 1 \qquad$ A is correct

$\Rightarrow |\beta| = 1 \qquad$ B is correct

$\alpha = e^{i0} = \beta$ [C and D are correct]

Sol 29: (A, B, D) $\alpha = 3z - 2$

$\beta = -3z - 2$

$|z| = 1$

$|\alpha + 2| = 3 \rightarrow$ A is correct

$|\beta + 2| = 3 \rightarrow$ B is correct

$\alpha - \beta = 6Z$ D is correct

$\Rightarrow \alpha - \beta$ and z, both move on same circle

Sol 30: (B, D) $(i^n + i^{-n})$ is

(A) $\dfrac{2^{2n} + 2^{2n}}{2^n(1-i)^{2n}} = \dfrac{2\cdot 2^n}{(1-i)^{2n}} = \dfrac{2^{n+1}}{2^{2n}}(1+i)^{2n} =$

$\dfrac{(2i)^n}{2^{n-1}} = 2i^n = i^n + i^n$

(B) $\dfrac{(1+i)^{2n} + (1-i)^{2n}}{2^n} = \dfrac{(2i)^n + (-2i)^n}{2^n} = i^n + (-i)^n$

$= i^n + i^{-n}$

(C) $\dfrac{-(2i)^n}{2^n} + \dfrac{2^n}{(-2i)^n} = -i^n + i^n = 0$

(D) $\dfrac{2^n}{(2i)^n} + \dfrac{2^n}{(-2i)^n} = i^{-n} + i^n$

Sol 31: (A, B, D)

$$\log_{14}\left(13 + \left|z^2 - 4i\right|\right) + \log_{196}\left(\frac{1}{\left(13 + \left|z^2 + 4i\right|\right)^2}\right) = 0$$

$$\log_{14}\frac{13 + \left|z^2 - 4i\right|}{13 + \left|z^2 + 4i\right|} = 0$$

$\Rightarrow 13 + |z^2 - 4i| = 13 + |z^2 + 4i|$

$\Rightarrow |z^2 - 4i| = |z^2 + 4i|$

$\Rightarrow |x^2 - y^2 + 2ixy - 4i| = |x^2 - y^2 + 2ixy + 4i|$

$\Rightarrow (x^2 - y^2)^2 + (2xy - 4)^2 = (x^2 - y^2)^2 + (2xy + 4)^2$

$\Rightarrow -16xy = 16xy \Rightarrow$ either $x = 0$ or $y = 0$

If $y = 0$ can be purely real

$x = 0$ can be purely imaginary

Must be real or purely imaginary.

Sol 32: (A, B) $1 - \log_2 \left[\dfrac{|x+1+2i| - 2}{\sqrt{2} - 1} \right] \geq 0$

$\Rightarrow \log_2 \dfrac{|x+1+2i| - 2}{\sqrt{2} - 1} \leq 1$

$\Rightarrow |x + 1 + 2i| - 2 \leq 2(\sqrt{2} - 1)$

$\Rightarrow |x + 1 + 2i| \leq 2\sqrt{2} \Rightarrow (x+1)^2 + 4 \leq 8$

$\Rightarrow (x+1)^2 \leq 4 \Rightarrow -2 \leq x + 1 \leq 2$

$\Rightarrow -3 \leq x \leq 1 \Rightarrow x \in [-3, 1]$

Sol 33: (D) $x = \cos \alpha$

$y = \cos \beta$

$z = \cos \gamma$

$\Sigma x = \cos\alpha + \cos\beta + \cos\gamma$

$= \cos\alpha + \cos\beta + \cos\gamma + i(\sin\alpha + \sin\beta + \sin\gamma)$

$\Pi x = \cos\alpha \cos\beta \cos\gamma = e^{i\alpha}\, e^{i\beta}\, e^{i\gamma} = e^{i(\alpha + \beta + \gamma)}$

Sol 34: (A, D) $x_r = \cos\left(\dfrac{\pi}{2^r} \right)$

$\underset{n\to\infty}{\text{Lim}} \displaystyle\prod_{r=1}^{n} x_r = e^{i\left(\frac{\pi}{2^r} + \frac{\pi}{2^{r+1}} \cdots \right)} = e^{i\left(\frac{\pi}{2} + \frac{\pi}{4} + \cdots \right)}$

$= e^{i \frac{\pi}{2}\left(1 - \frac{1}{2}\right)} = e^{i\pi}$

$= \cos\pi + i\sin\pi = -1 + 0i$

Real part is -1

Imaginary part is 0

Sol 35: (A, B, D) $\displaystyle\prod_{r=1}^{n-1} (\omega - z_r)$

$(\omega - z_1)(\omega - z_2)(\omega - z_3) \ldots (\omega - z_{n-1}) = \dfrac{\omega^n - 1}{\omega - 1}$

If n is a multiple of 3, value can be zero as z_1 can be ω or ω^2

If $n = 3k + 1$ value is 1

If $n = 3k + 2$ value is $1 + \omega$

Sol 36: (A, B, C) $\bar{z} = -4z$

$x - iy = -4x - 4iy$

$5x + i(3y) = 0$

$x = y = 0$ (A is true)

$\bar{z} = z$ implies z is purely real (B is correct)

$\bar{z} = -z$ implies z is purely imaginary (C is correct)

Sol 37: (A, C)

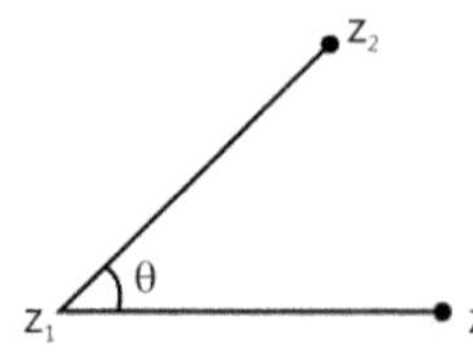

$z_3 = (1 - z_0)z_1 + z_0 z_2$

$z_3' = (1 - z_0)z_1' + z_0 z_2'$

$z_0 = \dfrac{z_3 - z_1}{z_2 - z_1} = \dfrac{z_3' - z_1'}{z_2' - z_1'} \Rightarrow \begin{vmatrix} z_1 & z_1 & 1 \\ z_2 & z_2 & 1 \\ z_3 & z_3 & 1 \end{vmatrix} = 0$

angle is same

The two triangles are similar

Sol 38: (A, B) Z multiplicative inverse is same as additive inverse

$\dfrac{1}{a + ib} = -a - ib$

$\dfrac{a - ib}{a^2 + b^2} = -a - ib$

$\Rightarrow a = -a^3 - ab^2 \Rightarrow a(1 + a^2 + b^2) = 0$

$\Rightarrow a = 0$ and

$b = a^2 b + b^3 \Rightarrow b(1 - a^2 - b^2) = 0$

$b \neq 0$ so $a^2 + b^2 = 1 \Rightarrow b = \pm 1$

$z = 0 \pm i$

Sol 39: (A, B, C, D) $Z = a + bi = \dfrac{(1 - ix)(1 - ix)}{(1 + ix)(1 - ix)}$

$a + bi = \dfrac{1 - x^2 - 2ix}{1 + x^2}$

$\Rightarrow |z| = 1$

$\text{Arg } z = \tan^{-1}\left(\dfrac{2x}{x^2 - 1} \right)$

$\text{Arg } (z) \equiv (-\pi, \pi]$

Previous Years' Questions

Sol 1: Given, $|z - 3 - 2i| \le 2$ .... (i)

To find minimum of $|2z - 6 + 5i|$

Or $2\left|z - 3 + \dfrac{5}{2}i\right|$,

By using triangle inequality

i.e., $\big||z_1| - |z_2|\big| \le |z_1 + \bar{z}_2|$

$\therefore \ \left|z - 3 + \dfrac{5}{2}i\right|$

$= \left|z - 3 - 2i + 2i + \dfrac{5}{2}i\right| = \left|(z - 3 - 2i) + \dfrac{9}{2}i\right|$

$\ge \left||z - 3 - 2i| - \dfrac{9}{2}\right|$

$\ge \left|2 - \dfrac{9}{2}\right| \ge \dfrac{5}{2}$

$\Rightarrow \left|z - 3 + \dfrac{5}{2}i\right| \ge \dfrac{5}{2}$ or $|2z - 6 + 5i| \ge 5$

Sol 2: $\omega = e^{i\frac{2\pi}{3}}$

Then, $\dfrac{|x|^2 + |y|^2 + |z|^2}{|a|^2 + |b|^2 + |c|^2} = 3$

Note: Here, $\omega = e^{i2\pi/3}$, then only integer solution exists.

Sol 3: (B)

$A = \{z : \mathrm{Im}(z) \ge 1\}$

$B = \{z : |z - 2 - i| = 3\}$

$C = \left\{z : \mathrm{Re}\left[(1 - i)z + \sqrt{2}\right]\right\}$

Taking $|z - 2 - i| = 3$

Let $z = z + iy$

$|x + iy - 2 - i| = 3$

$\Rightarrow |(x - 2) + i(y - 1)|^2 = (3)^2$

$\Rightarrow (x - 2)^2 + (y - 1)^2 = 9$ ...(i)

And

$\mathrm{Re}\left[(1 - i)z\right] = \sqrt{2}$

$\mathrm{Re}\left[(1 - i)(x + iy)\right] = \sqrt{2}$

$\Rightarrow x + y = \sqrt{2}$ ....(ii)

And

$\mathrm{Im}(z) \ge 1$

$\Rightarrow \mathrm{Im}(x + iy) \ge 1$

$\Rightarrow y \ge 1$ ....(iii)

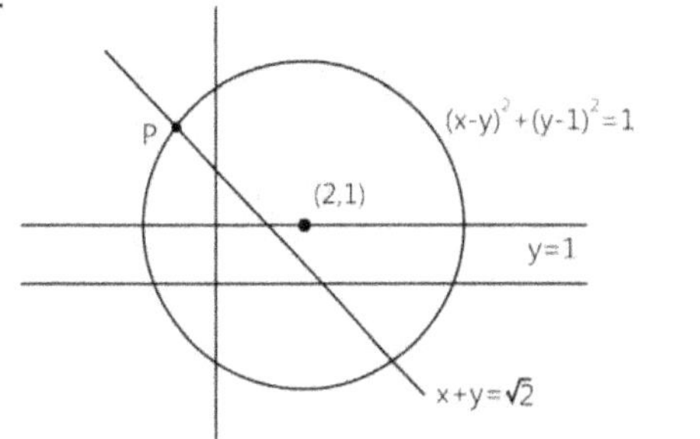

From (i), (ii) and (iii) $A \cap B \cap C$ has only one point P shown in the figure.

Sol 4: (C) $|z + 1 - i|^2 + |z - 5 - i|^2$

$= |z - (-1 + i)|^2 + |z - (5 + i)|^2$

The point A (-1, 1) and dB (5, 1) are end points of one of the diameter

In right angle Δ APB

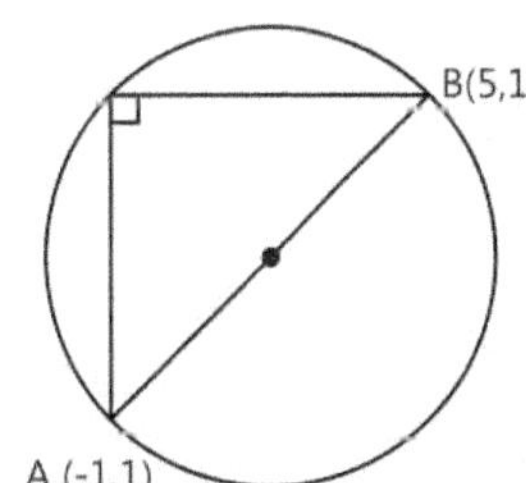

$AP^2 + BP^2 = (AB)^2$

$\Rightarrow |z + 1 - i|^2 + |z - 5 - i|^2$

$= (6)^2 = 36$

Sol 5: (D) $|w - 2 - i| < 3$

From triangle in equality

$|z - w| > \big||z| - |w|\big|$

$\Rightarrow -|z - w| < |z| - |w| < |z - w|$ ...(i)

Given $|w - 2 - i| < 3$, which means that w lies inside the given circle. Since z is presented by point P and W has to be the other end of diameter

$|z - w| = $ length of diameter

From (i), we get

$$-6 < |z| - |w| < 6$$

$$\Rightarrow -6 + 3 < |z| - |w| + 3 < 6 + 3$$

$$\Rightarrow -3 < |z| - |w| + 3 < 9$$

Sol 6: (B) Given, $z = \left(\dfrac{\sqrt{3}}{2} + \dfrac{i}{2}\right)^5 + \left(\dfrac{\sqrt{3}}{2} - \dfrac{i}{2}\right)^5$

$$\left(\because \omega = \dfrac{-1 + i\sqrt{3}}{2} \text{ and } \omega^2 = \dfrac{-1 - i\sqrt{3}}{2}\right)$$

Now, $\dfrac{\sqrt{3} + i}{2} = -i\left(\dfrac{-1 + i\sqrt{3}}{2}\right) = -i\omega$

And $\dfrac{\sqrt{3} - i}{2} = i\left(\dfrac{-1 - i\sqrt{3}}{2}\right) = i\omega^2$

$\therefore z = (-i\omega)^5 + (i\omega^2)^5 = -i\omega^2 + i\omega$

$= i(\omega - \omega^2) = i\left(i\sqrt{3}\right) = -\sqrt{3}$

$\Rightarrow$ Re (z) < 0 and Im (z) = 0

Alternate solution:

We known $z + \bar{z} = 2\text{Re}(z)$

If $z = \left(\dfrac{\sqrt{3}}{2} + \dfrac{i}{2}\right)^5 + \left(\dfrac{\sqrt{3}}{2} - \dfrac{i}{2}\right)^5$, then z is purely real,

i.e. Im (z) = 0

Sol 7: (D) Given, $|z - 4| < |z - 2|$

Since, $|z - z_1| > |z - z_2|$ represents the region on right side of perpendicular bisector of z_1 and z_2

$\therefore |z - 2| > |z - 4|$

$\Rightarrow$ Re(z) > 3 and Im (z) $\in$ R

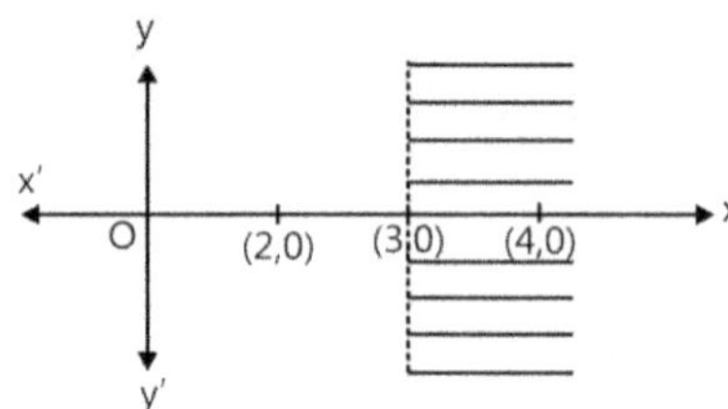

Sol 8: (B) Since a, b, c and u, v, w are the vertices of two triangles.

Also, c = (1 − r) a + rb and w = (1 − r) u + rv   ... (i)

Consider $\begin{vmatrix} a & u & 1 \\ b & v & 1 \\ c & w & 1 \end{vmatrix}$

Applying $R_3 \to R_3 - \{(1 - r) R_1 + rR_2\}$

$$= \begin{vmatrix} a & u & 1 \\ b & v & 1 \\ c - (1-r)a - rb & w - (1-r)u - rv & 1 - (1-r) - r \end{vmatrix}$$

$$= \begin{vmatrix} a & u & 1 \\ b & v & 1 \\ 0 & 0 & 0 \end{vmatrix} = 0 \text{ (from eq. (i))}$$

Hence, two triangles are similar.

Sol 9: (D) $\displaystyle\sum_{k=1}^{6}\left(\sin\dfrac{2k\pi}{7} - i\cos\dfrac{2k\pi}{7}\right)$

$$\sum_{k=1}^{6} i\left(\cos\dfrac{2k\pi}{7} + i\sin\dfrac{2k\pi}{7}\right) = -i\left\{\sum_{k=1}^{6} e^{\frac{i2k\pi}{7}}\right\}$$

$$\left\{\sum_{k=0}^{6} e^{\frac{i2k\pi}{7}} = 0\right\} = -i\left\{\sum_{k=0}^{6} e^{\frac{i2k\pi}{7}} - 1\right\} = -i\left[0 - 1\right] = i$$

Sol 10: (C) Given, $|z + iw| = |z - i\bar{w}| 2 = 2$

$\Rightarrow$ $|z - (-iw)| = |z - (i\bar{w})| = 2$

$\Rightarrow$ $|z - (-iw)| = |z - (-i\bar{w})|$.

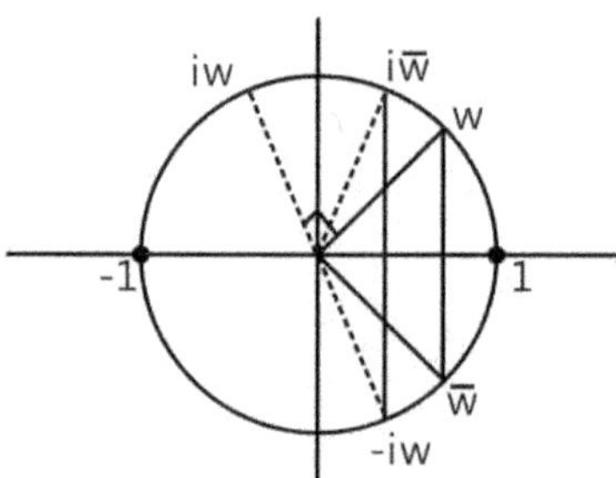

$\therefore$ z may take values given in (c).

Alternate solution

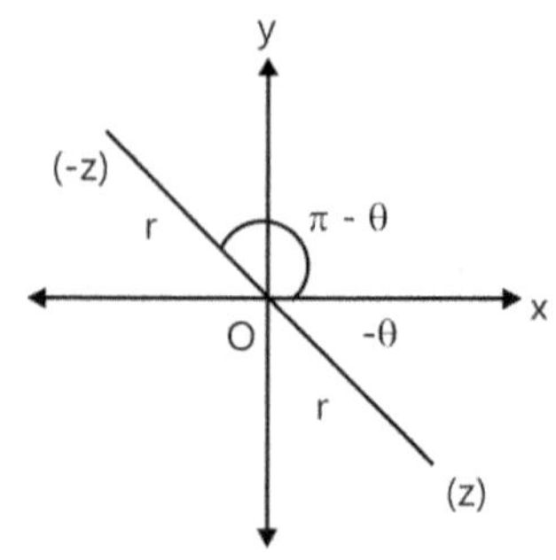

$|z+iw| \leq |z| + |iw| = |z| + |w| \leq 1+1 = 2$

$\therefore \qquad |z+iw| \leq 2$

$\Rightarrow \qquad |z+iw| = 2$ holds when

Arg z – arg $iw = 0$

Similarly, when $|z - \overline{iw}| = 2$

Then $\dfrac{z}{w}$ is purely imaginary

Now, given relation

$|z+iw| = |z - \overline{iw}| = 2$

Put $w = i$, we get

$$|z+i^2| = z+i^2 = 2$$
$$\Rightarrow \qquad |z-1| = 2$$
$$\Rightarrow \qquad z = -1 \qquad (\because |z| \leq 1)$$

Put $w = -i$, we get

$$|z-i^2| = |z-i^2| = 2$$
$$\Rightarrow \qquad |z+1| = 2$$
$$\Rightarrow \qquad z = 1 \qquad (\because |z| \leq 1)$$

$\therefore$ $z = 1$ or -1 is the one correct option given.

Sol 11: (D)

$(1+i)^{n_1} + (1-i)^{n_1} + (1+i)^{n_2} + (1-i)^{n_2}$

$= [{}^{n_1}C_0 + {}^{n_1}C_1 i + {}^{n_1}C_2 i^2 + {}^{n_1}C_3 i^3 + \ldots]$

$\quad + [{}^{n_1}C_0 - {}^{n_1}C_1 i + {}^{n_1}C_2 i^2 - {}^{n_1}C_3 i^3 + \ldots]$

$\quad\quad + [{}^{n_2}C_0 + {}^{n_2}C_1 i + {}^{n_2}C_2 i^2 + {}^{n_2}C_3 i + \ldots]$

$\quad\quad\quad + [{}^{n_2}C_0 - {}^{n_2}C_1 i + {}^{n_2}C_2 i^2 - {}^{n_2}C_3 i^3 + \ldots]$

$= 2[{}^{n_1}C_0 + {}^{n_1}C_2 i^2 + {}^{n_1}C_4 i^4 + \ldots]$

$\quad\quad + 2[{}^{n_2}C_0 + {}^{n_2}C_2 i^2 + {}^{n_2}C_4 i^4 + \ldots]$

$= 2[{}^{n_1}C_0 - {}^{n_1}C_2 + {}^{n_1}C_4 = \ldots]$

$\quad\quad + 2[{}^{n_2}C_0 - {}^{n_2}C_2 + {}^{n_2}C_4 - \ldots]$

This is a real number irrespective of the values of n_1 and n_2

Alternate solution

$\{(1+i)^{n_1} + (1-i)^{n_1}\} + \{(1+i)^{n_2} + (1-i)^{n_2}\}$

$\Rightarrow$ a real number for all n_1 and $n_2 \in R$.

$[\because z + \overline{z} = 2\,\mathrm{Re}(z) \Rightarrow (1+i)^{n_1} + (1-i)^{n_1}$
is real number for all $n \in R]$

Sol 12: (C) If in a complex number a + ib, the ratio a: b is $1 : \sqrt{3}$, then always convert the complex number in the form of ω.

Since, $\qquad \omega = -\dfrac{1}{2} + \dfrac{\sqrt{3}}{2} i$

$\therefore 4 + 5\left(-\dfrac{1}{2} + \dfrac{i\sqrt{3}}{2}\right)^{334} + 3\left(-\dfrac{1}{2} + \dfrac{i\sqrt{3}}{2}\right)^{365}$

$= 4 + 5\omega^{334} + 3\omega^{365}$

$= 4 + 5 \cdot (\omega^3)^{111} \cdot \omega + 3 \cdot (\omega^3)^{121} \cdot \omega^2$

$= 4 + 5\omega + 3\omega^2 \qquad\qquad (\because \omega^3 = 1)$

$= 1 + 3 + 2\omega + 3\omega + 3\omega^2$

$= 1 + 2\omega + 3(1 + \omega + \omega^2) = 1 + 2\omega + 3 \times 0$

$= 1 + (-1 + \sqrt{3}i) = \sqrt{3}i.$

Sol 13: (A) Since, arg $(z) < 0$

$\Rightarrow \qquad \arg(z) = -\theta$

$\Rightarrow \qquad z = r\cos(-\theta) + i\sin(-\theta) = r(\cos\theta - i\sin\theta)$

and $-z = -r[\cos\theta - i\sin\theta]$

$= r[\cos(\pi - \theta) + i\sin(\pi - \theta)]$

$\therefore \qquad\qquad \arg(-z) = \pi - \theta$

Thus, arg $(-z)$ –arg (z)

$= \pi - \theta - (-\theta) = \pi$

Alternate solution:

Reason: arg $(-z) -$ arg $z = \arg\left(\dfrac{-z}{z}\right) = \arg(-1) = \pi$

And also arg $z -$ arg $(-z) = \arg\left(\dfrac{z}{-z}\right) = \arg(-1) = \pi$

Sol 14: (A, B,C) Since, $z_1 = a + ib$ and $z_2 = c + id$

$\Rightarrow |z_1|^2 = a^2 + b^2 = 1$ and $|z_2|^2 = c^2 + d^2 = 1 \quad \ldots(i)$

$(\because |z_1| = |z_2| = 1)$

Also, Re $(z_1 \overline{z_2}) = 0 \Rightarrow \quad ac + bd = 0$

$\Rightarrow \qquad \dfrac{a}{b} = -\dfrac{d}{c} = \lambda \qquad\qquad \text{(say)}\ldots(ii)$

From Eqs. (i) and (ii), $b^2\lambda^2 + b^2 = c^2 + \lambda^2 c^2$

$\Rightarrow \qquad b^2 = c^2$ and $a^2 = d^2$

Also, given $w_1 = a + ic$ and $w_2 = b + id$

Now, $\quad |w_1| = \sqrt{a^2 + c^2} = \sqrt{a^2 + b^2} = 1$

$\qquad |w_2| = \sqrt{b^2 + d^2} = \sqrt{a^2 + b^2} = 1$

and $\text{Re}(w_1 \overline{w}_2) = ab + cd = (b\lambda)b + c(-\lambda c)$

$= \lambda(b^2 - c^2) = 0$

Sol 15: (A, C, D) Given, $z = \dfrac{(1-t)z_1 + tz_2}{(1-t)+t}$

A — P — B
z_1 — z — z_2

$t : (1-t)$

Clearly, z divides z_1 and z_2 in the ratio of t:

(1-t), 0 < t < 1

$\Rightarrow$ AP + BP = AB

ie, $|z - z_1| + |z - z_2| = |z_1 - z_2|$

$\Rightarrow$ Option (a) is true.

And $\arg(z - z_1) = \arg(z_2 - z) = \arg(z_2 - z_1)$

$\Rightarrow$ (b) is false and (d) is true.

Also, $\arg(z - z_1) = \arg(z_2 - z_1)$

$\Rightarrow \arg\left(\dfrac{z - z_1}{z_2 - z_1}\right) = 0$

$\therefore \dfrac{z - z_1}{z_2 - z_1}$ is purely real.

$\Rightarrow \dfrac{z - z_1}{z_2 - z_1} = \dfrac{\overline{z} - \overline{z_1}}{\overline{z_2} - \overline{z_1}}$ or $\begin{vmatrix} z - z_1 & \overline{z} - \overline{z_1} \\ z_2 - z_1 & \overline{z_2} - \overline{z_1} \end{vmatrix} = 0$

Sol 16: (A) $|z - i|z|| = |z + i|z||$

$\Rightarrow |z - i|z||^2 = |z + i|z||^2$

$\Rightarrow (z - i|z|)(\overline{z} + i|z|) = (z + i|z|)(\overline{z} - i|z|)$

$\Rightarrow z\overline{z} + iz|z| - i|z|\overline{z} + |z|^2$

$= z\overline{z} - iz|z| + i|z|\overline{z} + |z|^2$

$\Rightarrow 2iz|z| = 2i|z|\overline{z}$

$\Rightarrow 2i|z|(z - \overline{z}) = 0$

$\Rightarrow |z| = 0$ or $z - \overline{z} = 0$

$\Rightarrow \text{Im}(z) = 0$

Also $|\text{Im}(z)| \leq 1$

$A \rightarrow q, r$

(B) $|z + 4| + |z - 4| = 0$

Its an equation of ellipse having

aq = 4 and 2a = 10

$\Rightarrow e = \dfrac{4}{5}$

$B \rightarrow p$

(C) $z = w - \dfrac{1}{w}$

$\Rightarrow z = w - \dfrac{1}{w\overline{w}} \times \overline{w}$

$= w - \dfrac{\overline{w}}{|w|^2}$

$= w - \dfrac{\overline{w}}{4}$

Let w = p + I q, then

$z = \dfrac{3P}{4} + i\dfrac{5q}{4}$

Now, let z = x + iy

$\Rightarrow x = \dfrac{3P}{4} \Rightarrow P = \dfrac{4x}{3}$

$\Rightarrow y = \dfrac{5q}{4} \Rightarrow q = \dfrac{4y}{5}$

$|w|^2 = P^2 + q^2 = 4$

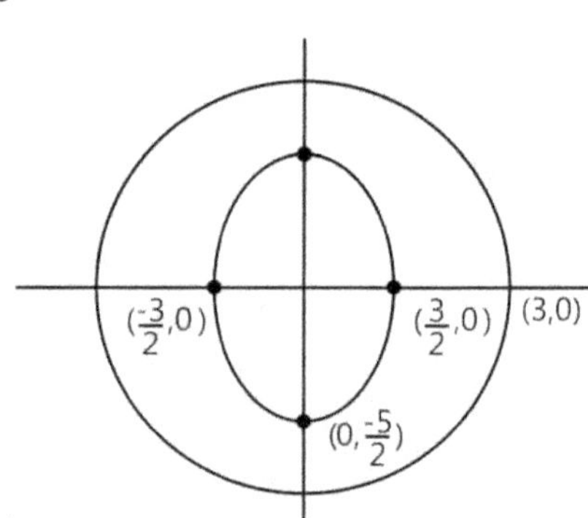

$\Rightarrow \dfrac{16 x^2}{9} + \dfrac{16 y^2}{25} = 4$

$\Rightarrow \dfrac{x^2}{9/4} + \dfrac{y^2}{25/4} = 1$, its an ellipse

$e^2 = 1 - \dfrac{b^2}{a^2}$

$= 1 - \dfrac{9}{25} = \dfrac{16}{25}$

$\Rightarrow e = \dfrac{4}{5}$

From figure, we can conclude that

$\left|\text{Re}(z)\right| \leq 2$

And $|z| \leq 3$

$C \rightarrow p, s, t$

(D) $z = w + \dfrac{1}{w} = w + \dfrac{\overline{w}}{w\,\overline{w}} = w + \dfrac{\overline{w}}{|w|^2} = w + \overline{w}$

Let $w = a + i\,b$

$z = a + i\,b + a - i\,b = 2a$

Now, $|z| = 2\,|a|$ and $\text{Im}(z) = 0$, $\text{Im}(z) \le 1$

$\Rightarrow \text{Re}(z) \le 2$ and $|z| \le 3$

$D \to q, r, s$ and t

Sol 17: (B)

$$\begin{vmatrix} z+1 & \omega & \omega^2 \\ \omega & z+\omega^2 & 1 \\ \omega^2 & 1 & z+\omega \end{vmatrix} = 0$$

$C_1 \to C_1 + C_2 + C_3$

$$\begin{vmatrix} z & \omega & \omega^2 \\ z & z+\omega^2 & 1 \\ z & 1 & z+\omega \end{vmatrix} = 0$$

$$\left\{ \omega = \cos\dfrac{2\pi}{3} + i\sin\dfrac{2\pi}{3} \right\}$$

$$\begin{vmatrix} 1 & \omega & \omega^2 \\ 1 & z+\omega^2 & 1 \\ 1 & 1 & z+\omega \end{vmatrix} = 0$$

Expanding the determinant, we get

$$z\left[\left(z+\omega^2\right)\left(z+\omega\right) - 1 - \omega\left(z+\omega-1\right) + \omega^2\left(1-z-\omega^2\right) \right] = 0$$

$$\Rightarrow z\left[z^2 + z\omega + z\omega^2 + \omega^2 - 1 - z\omega - \omega^2 + \omega + \omega^2 - \omega^2 z - \omega^4 \right] = 0$$

$$\Rightarrow z\left[z^2 \right] = 0$$

$$\Rightarrow z = 0$$

Sol 18: (A, C, D) Given $z = (1-t)z_1 + t z_2$

This equation represents line segment between z_1 and z_2

From fig.

$$|z - z_1| + |z - z_2| = |z_1 - z_2|$$

$$\text{Arg}\left(z - z_1\right) = \text{Arg}\left(z_2 - z_1\right)$$

$$\dfrac{z - z_1}{\overline{z} - \overline{z}_1} = \dfrac{z_2 - z_1}{\overline{z}_2 - \overline{z}_1}$$

$$\Rightarrow \begin{vmatrix} z - z_1 & \overline{z} - \overline{z}_1 \\ z_2 - z_1 & \overline{z}_2 - \overline{z}_1 \end{vmatrix} = 0$$

Sol 19: $|2z - 6 - 5i|$

$$= 2\left| z - 3 + \dfrac{5}{2} i \right|$$

Here $\left| z - \left(3 - \dfrac{5}{2}\, i \right) \right|$ is nothing but the distance between

any point on the circle $|z - 3 - 2i| \le 2$ and point $\left(3, \dfrac{-5}{2} \right)$

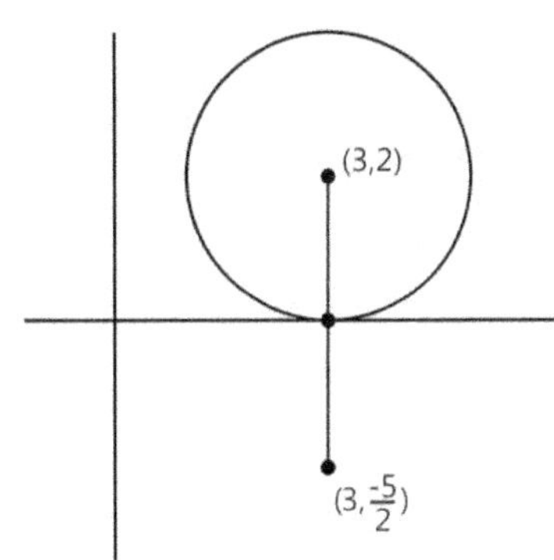

$\therefore$ Minimum value $= \dfrac{5}{2}$

$$|2z - 6 + 5i|_{\min} = 2 \times \dfrac{5}{2} = 5$$

Sol 20: Given: $a + b + c = x$

$a + b\omega + c\omega^2 = y$

$a + b\omega^2 + c\omega = z$

$$\dfrac{|x|^2 + |y|^2 + |z|^2}{|a|^2 + |b|^2 + |c|^2} = \dfrac{x\overline{x} + y\overline{y} + z\overline{z}}{|a|^2 + |b|^2 + |c|^2}$$

$$= \dfrac{\left(a+b+c\right)\left(\overline{a}+\overline{b}+\overline{c}\right) + \left(a+b\omega+c\omega^2\right) + \left(\overline{a}+\overline{b}\,\overline{\omega}+\overline{c}\,\overline{\omega}^2\right)}{|a|^2 + |b|^2 + |c|^2}$$

$$+ \dfrac{\left(a+b\omega^2+c\omega\right)\left(\overline{a}+\overline{b}\,\overline{\omega}^2+\overline{c}\,\overline{\omega}\right)}{|a|^2 + |b|^2 + |c|^2}$$

$$\begin{cases} \omega = e^{i\pi/3} \\ \overline{\omega} = e^{-i\pi/3} \\ \omega^2 = e^{2i\frac{\pi}{3}} = -\overline{\omega} \end{cases}$$

$$= \dfrac{\left(a+b+c\right)\left(\overline{a}+\overline{b}+\overline{c}\right) + \left(a+b\omega-\overline{\omega}c\right)\left(\overline{a}+\overline{b}\,\overline{\omega}+\overline{c}\,\overline{\omega}^2\right)}{}$$

$$+ \dfrac{\left(a-b\overline{\omega}+c\omega\right)\left(\overline{a}+\overline{b}\,\overline{\omega}^2+\overline{c}\,\overline{\omega}\right)}{|a|^2 + |b|^2 + |c|^2}$$

$$= \frac{3\left(|a|^2 + |b|^2 + |c|^2\right)}{|a|^2 + |b|^2 + |c|^2} = 3$$

Sol 21: (A) Given $\begin{bmatrix} 1 & a & b \\ \omega & 1 & c \\ \omega^2 & \omega & 1 \end{bmatrix}$

Determinant $(D) = \begin{bmatrix} 1 & a & b \\ \omega & 1 & c \\ \omega^2 & \omega & 1 \end{bmatrix}$

$|D| = 1\left(1 - c\omega\right) - a\left(\omega - c\omega^2\right) + b\left(\omega^2 - \omega^2\right)$

$\Rightarrow |D| = 1 - (a + c)\omega + ac\omega^2$

$|D| \neq 0$, only when $a = \omega$ and $c = \omega$

$\therefore (a, b, c) \equiv (\omega, \omega, \omega)$ or $(\omega, \omega^2, \omega)$

$\therefore$ Two non-singular matrices are possible.

Sol 22: (D) $a = z^2 + z + 1 = 0$

$\Rightarrow z^2 - z + 1 - a = 0$

Discriminant < 0 {Since imaginary part of z us not zero}

$1 - 4(1 - a) < 0$

$\Rightarrow 1 - 4 + 4a < 0$

$\Rightarrow 4a < 3$

$\Rightarrow a < \dfrac{3}{4}$

Sol 23: (C) $\left(x - x_0\right)^2 + \left(y - y_0\right)^2 = r$

$\Rightarrow |z - z_0| = r$

α lies on it, then

$|\alpha - z_0| = r$

$\Rightarrow \left(\alpha - z_0\right)\left(\bar{\alpha} - \bar{z}_0\right) = r^2$

$= |\alpha|^2 - \bar{z}_0\,\alpha - z_0\,\bar{\alpha} + |z_0|^2 = r^2$ ....(i)

Similarly,

$\left(x - x_0\right)^2 + \left(y - y_0\right)^2 = 2r$

$\Rightarrow |z - z_0| = 2r$

$\dfrac{1}{\alpha}$ lies on it, then

$\left|\dfrac{1}{\alpha} - z_0\right| = 2r$

$\Rightarrow \left(\dfrac{1}{\alpha} - z_0\right)\left(\dfrac{1}{\bar{\alpha}} - \bar{z}_0\right) = 4r^2$

$\Rightarrow \left(\dfrac{1}{\bar{\alpha}\,\alpha} - \dfrac{\bar{z}_0}{\bar{\alpha}} - \dfrac{z_0}{\alpha} + |z_0|^2\right) = 4r^2$

$\left\{\alpha\,\bar{\alpha} = |\alpha|^2 \Rightarrow \bar{\alpha} = \dfrac{|\alpha|^2}{\alpha}\right\}$

$\Rightarrow \dfrac{1}{|\alpha|^2} - \dfrac{\bar{z}_0\,\alpha}{|\alpha|^2} - \dfrac{z_0\,\bar{\alpha}}{|\alpha|^2} + |z_0|^2 = 4r^2$

$\Rightarrow 1 - z_0\,\bar{\alpha} - \bar{z}_0\,\alpha + |\alpha|^2|z_0|^2 = 4r^2\,|z|^2$ ...(ii)

Subtracting (ii) from (i), we get

$1 - |\alpha|^2 + |\alpha|^2|z_0|^2 - |z_0|^2 = r^2\left(4|\alpha|^2 - 1\right)$

$1 - |\alpha|^2 + \dfrac{r^2 + 2}{2}\left(|\alpha|^2 - 1\right) = r^2\left(4|\alpha|^2 - 1\right)$

$\Rightarrow \left(|\alpha|^2 - 1\right)\left[\dfrac{r^2 + 2}{2} - 1\right] = r^2\left(4|\alpha|^2 - 1\right)$

$\Rightarrow \left(|\alpha|^2 - 1\right)\left[\dfrac{r^2}{2}\right] = r^2\left(4|\alpha|^2 - 1\right)$

$\Rightarrow |\alpha|^2 - 1 = 8|\alpha|^2 - 2$

$\Rightarrow 7|\alpha|^2 = 1$

$\Rightarrow |\alpha| = \dfrac{1}{\sqrt{7}}$

Sol 24: (C, D) $P = \left\{w^n : n = 1, 2, 3, \ldots\right\}$

$w = \dfrac{\sqrt{3}}{2} + \dfrac{1}{2} \Rightarrow |w| = 1$

$\Rightarrow$ All the complex number belong to set P lie on circle of unit radius, centre at origin.

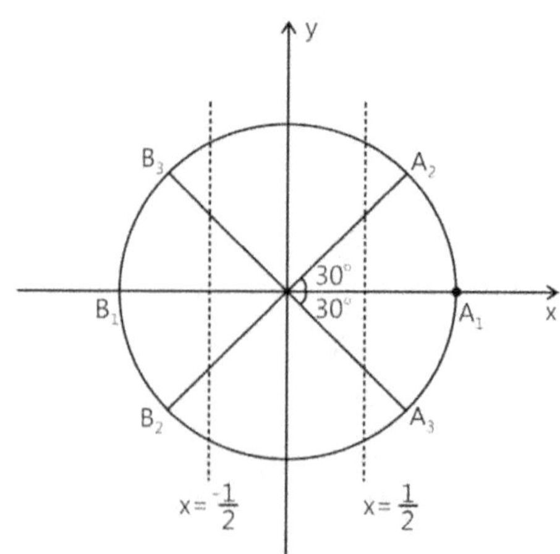

All the complex number belong to PnH_1 lie right of the line $x = \dfrac{1}{2}$ on circle. Possible positions are shown in the figure as A_1, A_2, A_3.

Similarly, all the complex number belong to PnH_2 lie left of the line $x = \dfrac{-1}{2}$ on circle. Possible positions are shown in the figure as $B_1, B_2, B_2 < z_1 \; 0 \; z_2 =$ angle between A_2 and $B_3 = \dfrac{2\pi}{3}$ or angle between A_1 and $B_3 = \dfrac{5\pi}{6}$

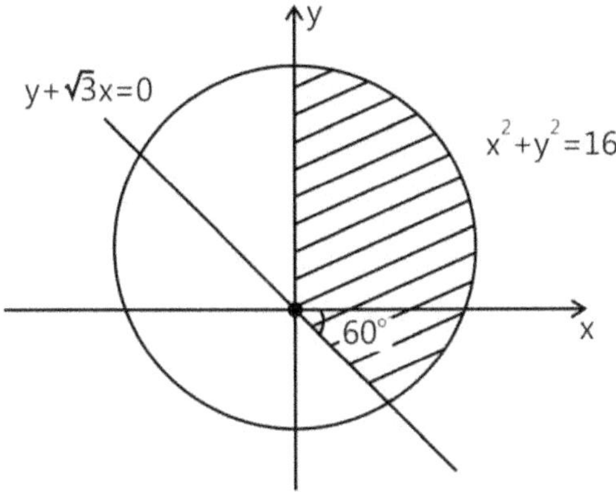

Sol 25: (B, C, D) $P = \left[p_{ij} \right]_{n \times n}$ $\qquad p_{ij} = \omega^{i+j}$

$$P - \begin{bmatrix} \omega^2 & \omega^3 & \omega^4 & \cdots & \omega^{n+1} \\ \omega^3 & \omega^4 & \omega^5 & \cdots & \omega^{n+2} \\ \vdots & \vdots & \vdots & \vdots & \vdots \\ \omega^{2n+1} & \omega^{n+2} & \omega^{n+3} & \cdots & \omega^{n+n} \end{bmatrix}$$

$$\Rightarrow P^2 = \begin{bmatrix} \omega^4 + \omega^6 + \omega^8 + \dots & \omega^5 + \omega^7 + \omega^9 & \cdots & \cdots \\ \vdots & & \vdots & \cdots \\ \vdots & & \vdots & \vdots \\ \omega^{n+3} + \omega^{n+5} + \dots & & \cdots & \cdots \end{bmatrix}$$

Lets take element $P_{11} = \omega^4 + \omega^6 + \omega^8 + \dots n$ terms

$$= \dfrac{\omega^4 \left(\omega^{2n} - 1 \right)}{\omega - 1}$$

If n is multiple of 3, then this element will vanish. Which is the case for every element.

$\therefore$ n can not be multiple of 3, for $P^2 \neq 0$

$\therefore$ Possible values of n are 55, 58, 56

Sol 26: (B) $S_1 = \left\{ z \in C : |z| < 4 \right\}$

z lies inside of a circle given by $|z| = 4$

$$S_2 = \left\{ z \in C : \text{Im} \left[\dfrac{z - 1 + \sqrt{3}i}{1 - \sqrt{3}i} \right] > 0 \right\}$$

$$\text{Im} \left[\dfrac{\left(z - 1 + \sqrt{3}i \right)\left(1 + \sqrt{3}i \right)}{4} \right] > 0$$

Let $z = x + iy$

$$\Rightarrow \text{Im} \left[\dfrac{\left(x + iy \right)\left(1 + \sqrt{3}i \right) - \left(1 - \sqrt{3}i \right)\left(1 + \sqrt{3}i \right)}{4} \right] > 0$$

$$\Rightarrow \text{Im} \left[\dfrac{\left(x + \sqrt{3}y \right) + i\left(\sqrt{3}x + y \right) - 4}{4} \right] > 0$$

$$\Rightarrow \dfrac{\sqrt{3}x + y}{4} > 0 \Rightarrow \sqrt{3}x + y > 0$$

$S_3 \left\{ z \in C : \text{Re}(z) > 0 \right\}$

$\Rightarrow$ z lies in either first or fourth quadrant.

Now, points of intersection of circle $x^2 + y^2 = 16$ and $\sqrt{3}x + y = 0$ is $x^2 + 3x^2 = 16$

$\Rightarrow 4x^2 = 16 \Rightarrow x = \pm 2$

$\Rightarrow y = \pm 2\sqrt{3}$

Area $= \dfrac{1}{2} (4)^2 \times \dfrac{5\pi}{6}$

$= \dfrac{1}{2} \times 16 \times \dfrac{5\pi}{6} = \dfrac{20\pi}{3}$ .sq. unit^2

Sol 27: (C) $|1 - 3i - z|$

$= |z - 1 + 3i| = |z - (1 - 3i)|$

$\text{Min} \left(|z - (1 - 3i)| \right) =$ distance of point $(1, -3)$ from line $\sqrt{3}\,x + y = 0$

$= \left| \dfrac{\sqrt{3} \times 1 - 3}{\sqrt{3} + 1} \right| = \left| \dfrac{\sqrt{3} - 3}{2} \right|$

$= \dfrac{3 - \sqrt{3}}{2}$

Sol 28: (C) $z_k = \cos \dfrac{2\pi x}{10} + i \sin \dfrac{2k\pi}{10}$, $k = 1, 2, \dots 9$

(p) $z_k = e^{i \frac{2\pi k}{10}}$

$$\Rightarrow z_k \cdot z_j = e^{i\frac{2\pi}{10}(k+j)} = 1$$

$$\Rightarrow \cos\frac{2\pi}{10}(k+j) + i\sin\frac{2\pi}{10}(k+j) = 1$$

If $k + j = 10\,m$ (multiple of 10), then above equation is True.

(q) $z_1 \cdot z = z_k$

$$\Rightarrow z = \frac{z_k}{z_1} = \frac{e^{i\frac{2\pi k}{10}}}{e^{i\frac{2\pi}{10}}} = e^{i\frac{(k-1)}{10}2\pi}$$

Clearly, this equation has many solutions

$\Rightarrow$ Q is False

(r) Consider, $z^{10} = 1$

$$z^{10} - 1 = (z-1)(z-z_1)(z-z_2)(z-z_3)....(z-z_9)$$

Where z_k represents the roots of equation $z^{10} = 10$

$$= 1 + z + z^2 + ... + z^9$$

$$\Rightarrow \frac{z^{10}-1}{z-1} = (z-z_1)(z-z_2)......(z-z_9)$$

$$= 1 + z + z^2 + z^3 + ... + z^9$$

$$\frac{z^{10}-1}{z-1} = (z-z_1)(z-z_2)....(z-z_9)$$

$$= 1 + 1 + 1 ...... 10$$

$$= 10$$

(s) We know that sum of roots of $z^{10} - 1 = 0$ is zero

$$\Rightarrow 1 + \sum_{k=1}^{9}\cos\frac{2k\pi}{10} = 0 \Rightarrow \sum_{k=1}^{9}\cos\frac{2k\pi}{10} = -1$$

$$\Rightarrow 1 - \sum_{k=1}^{9}\cos\frac{2k\pi}{10} = 2$$

Sol 29: Given $\alpha_k = \cos\frac{k\pi}{7} + i\sin\frac{k\pi}{7} = e^{i\frac{k\pi}{7}}$

$$\frac{\displaystyle\sum_{k=1}^{12}|\alpha_{k+1} - \alpha_k|}{\displaystyle\sum_{k=1}^{3}|\alpha_{4k-1} - \alpha_{4k-2}|} = \frac{\displaystyle\sum_{k=1}^{12}\left|e^{i\frac{(k+1)\pi}{7}} - e^{i\frac{k\pi}{7}}\right|}{\displaystyle\sum_{k=1}^{3}\left|e^{i(4k-1)\frac{\pi}{7}} - e^{i\frac{(4k-2)}{7}}\right|}$$

$$= \frac{\displaystyle\sum_{k=1}^{12}\left|e^{ik\frac{\pi}{7}}\right|\left|e^{i\frac{\pi}{7}} - 1\right|}{\displaystyle\sum_{k=1}^{3}\left|e^{i\frac{(4k-2)}{7}}\right|\left|e^{i\frac{\pi}{7}} - 1\right|} = \frac{\displaystyle\sum_{k=1}^{12}\left|e^{i\frac{\pi}{7}} - 1\right|}{\displaystyle\sum_{k=1}^{3}\left|e^{i\frac{\pi}{7}} - 1\right|}$$

$$= \frac{\displaystyle\sum_{k=1}^{12} 2\sin\frac{\pi}{14}}{\displaystyle\sum_{k=1}^{3} 2\sin\frac{\pi}{14}} = \frac{12 \times 2\sin\frac{\pi}{4}}{3 \times 2\sin\frac{\pi}{4}} = 4$$

Sol 30: $P = \begin{bmatrix} (-z)^r & z^{2s} \\ z^{2s} & z^r \end{bmatrix}$

$$P^2 = \begin{bmatrix} (-z)^{2r} + (z^{4s}) & (-z)^r z^{2s} + z^r z^{2s} \\ (-z)^r z^{2s} + z^r z^{2s} & z^{2r} + z^{4s} \end{bmatrix}$$

$$= \begin{bmatrix} z^{2r} + z^{4s} & z^{2s}\left[(-z)^r + z^r\right] \\ z^{2s}\left((-z)^r + z^r\right) & z^{2r} + z^{4s} \end{bmatrix}$$

Given that $P^2 - I = \begin{bmatrix} 1 & 0 \\ 0 & 1 \end{bmatrix} \Rightarrow z^{2r} + z^{4s} = -1$

And $z^{2s}\left[(-z)^r + z^r\right] = 0$

Now, we have $z = \dfrac{-1 + i\sqrt{3}}{2} = \omega$

$$\Rightarrow \omega^{2r} + \omega^{4s} = -1 \text{ and } \omega^{2s}\left[(-\omega)^r + \omega^r\right] = 0$$

Only $(r, s) \equiv (1, 1)$ satisfies both the equation.

Only one pair exists.

Sol 31: (A, C, D) $S = \left\{ z \in C : z = \dfrac{1}{a+ibt}, t \in R, t \neq 0 \right\}$

$$z = \frac{1}{a+ibt} \times \frac{a-ibt}{a-ibt}$$

$$\Rightarrow z = \frac{a-ibt}{a^2 + b^2 t^2} = x + iy \quad \text{(Let)}$$

$$\Rightarrow x = \frac{a}{a^2 + b^2 t^2} \text{ and } y = \frac{bt}{a^2 + b^2 t^2}$$

$\Rightarrow \dfrac{x}{y} = \dfrac{a}{bt} \Rightarrow t = \dfrac{ay}{bx}$

Substituting 't' in $x = \dfrac{a}{a^2 + b^2 t^2}$

$x = \dfrac{a}{a^2 + b^2 \dfrac{a^2 y^2}{b^2 x^2}}$

$\Rightarrow x = \dfrac{ax^2}{a^2 x^2 + a^2 y^2} \Rightarrow a^2 x^2 + a^2 y^2 - ax = 0$

$\Rightarrow x^2 + y^2 - \dfrac{x}{a} = 0$

$\left(x - \dfrac{1}{2a}\right)^2 + y^2 = \left(\dfrac{1}{2a}\right)^2$

Centre $\equiv \left(\dfrac{1}{2a}, 0\right)$

Radius $= \dfrac{1}{2a}$, when $a > 0$, $b \neq 0$

If $b = 0$, $a \neq 0$

$y = 0 \Rightarrow x - \text{axis}$

If $a = 0$, $b \neq 0$

$x = 0 \Rightarrow y - \text{axis}$

3. BINOMIAL THEOREM

MATHEMATICAL INDUCTION

The technique of Induction is used to prove mathematical theorems. A variety of statements can be proved using this method. Mathematically, if we show that a statement is true for some integer value, say n = 0, and then we prove that the statement is true for some integer k+1 if it is true for the integer k (k is greater than or equal to 0), then we can conclude that it is true for all integers greater than or equal to 0.

The solution in mathematical induction consists of the following steps:

Step 1: Write the statement to be proved as $P(n)$ where n is the variable.

Step 2: Show that $P(n)$ is true for the starting value of n equal to 0(say).

Step 3: Assuming that $P(k)$ is true for some k greater than the starting value of n, prove that $P(k+1)$ is also true.

Step 4: Once $P(k+1)$ has been proved to be true, we say that the statement is true for all values of the variable.

The following illustrations will help to understand the technique better.

Illustration 1: Prove that $1+2+3+\ldots+n=n(n+1)/2$ for all n, n is natural. **(JEE MAIN)**

Sol: Clearly, the statement $P(n)$ is true for n = 1. Assuming $P(k)$ to be true, add $(k+1)$ on both sides of the statement.

$P(n):1+2+3+\ldots+n=n(n+1)/2$

Clearly, $P(1)$ is true as $1=1.2/2$.

Let $P(k)$ be true. That is, let $1+2+3+\ldots+k$ be equal to $k(k+1)/2$

Now, we have to show that $P(k+1)$ is true, or that

$1+2+3+\ldots+(k+1)=(k+1)(k+2)/2.$

$L.H.S = 1+2+3+\ldots+(k+1)$

$= 1+2+3+\ldots+k+(k+1) = k(k+1)/2 + (k+1)$ (As $P(k)$ is true)

$= (k+1)\,(k/2+1) = (k+1)(k+2)/2$

$= R.H.S$

Illustration 2: Prove that $(n+1)! > 2^n$ for all n>1. **(JEE MAIN)**

Sol: For n = 2, the given statement is true. Now assume the statement to be true for n = m and multiply $(m+2)$ on both sides.

Let $(n+1)! > 2^n$ ... (i)

Putting n=2 in eq. (i), we get,

$3! > 2^2$

$3! > 4$

Since this is true,

Therefore the equation holds true for n=2.

Assume that equation holds true for n=m,

$(m+1)! > 2^m$ ... (ii)

Now, we have to prove that this equation holds true for n=m+1, i.e. $(m+2)! > 2^{m+1}$.

From equation 2, $(m+1)! > 2^m$.

Multiply above equation by m+2

$\qquad (m+2)! > 2^m (m+2)$

$\qquad\qquad > 2^{m+1} + 2^m.m$

$\qquad\qquad > 2^{m+1}$.

$\qquad$ Hence proved.

Illustration 3: Prove that $n^2 + n$ is even for all natural numbers n.   **(JEE MAIN)**

Sol: Consider $P(n) = n^2 + n$. It can written as a product of two consecutive natural numbers. Use this fact to prove the question.

Consider that $P(n)$ $n^2 + n$ is even, $P(1)$ is true as $1^2 + 1 = 2$ is an even number.

Consider $P(k)$ be true,

To prove : $P(k + 1)$ is true.

$P(k + 1)$ states that $(k + 1)^2 + (k + 1)$ is even.

Now, $(k + 1)^2 + (k + 1) = k^2 + 2k + 1 + k + 1 = k^2 + k + 2k + 2$

As $P(k)$ is true, hence $k^2 + k$ is an even number and can be written as 2λ, where λ is sum of natural number.

$\therefore$ $2\lambda + 2k + 2 \Rightarrow 2(\lambda + k + 1)$ = a multiple of 2.

Thus, $(k + 1)^2 + (k + 1)$ is an even number.

Hence, $P(n)$ is true for all n, where n is a natural number.

Illustration 4: Prove that exactly one among n+10, n+12 and n+14 is divisible by 3, considering n is always an natural number.   **(JEE MAIN)**

Sol: We can observe here that

For n = 1,

n+10 = 11

n+12 = 13

n+14 = 15

Exactly one i.e 15 is divisible by 3.

Let us assume that that for n = m exactly one out of n+10, n+12, n+14 is divisible by 3

Without the loss of generality consider for n=m, m+10 was divisible by 3

Therefore, m+10 = 3k

m+12 = 3k+2

m+14 = 3k+4

We need to prove that for n=m+1 , exactly one among them is divisible by 3. Putting m+1 in place of n, we get

(m+1)+10 = m+11 = 3k + 1 (not divisible by 3)

(m+1)+12 = m+13 = 3k+3 = 3(k+1) (divisible by 3)

(m+1)+14 = m+15 = 3k+5 (not divisible by 3)

Therefore, for n=m+1 also exactly one among the three, n+10, n+12 and n+14 is divisible by 3.

Similarly we can prove that exactly one among three of these is divisible by 3 by considering cases when n+12 = 3k and n+14 = 3k.

BINOMIAL THEOREM

1. INTRODUCTION TO BINOMIAL THEOREM

1.1 Introduction

Consider two numbers a and b, then

$$\left(a+b\right)^2 = a^2 + 2ab + b^2$$

$$\left(a+b\right)^3 = \left(a+b\right)\left(a+b\right)^2 \qquad = \left(a+b\right)\left(a^2 + 2ab + b^2\right) \qquad = a^3 + 3a^2b + 3ab^2 + b^3$$

$$\left(a+b\right)^4 = \left(a+b\right)^2\left(a+b\right)^2 \qquad = \left(a^2 + 2ab + b^2\right)\left(a^2 + 2ab + b^2\right) \qquad = a^4 + 4a^3b + 6a^2b^2 + 4ab^3 + b^4$$

As the power increases, the expansion becomes lengthy, difficult to remember and tedious to calculate. A binomial expression that has been raised to a very large power (or degree), can be easily calculated with the help of Binomial Theorem.

1.2 Binomial Expression

A binomial expression is an algebraic expression which contains two dissimilar terms.

For example: $x + y,\ a^2 + b^2,\ 3 - x,\ \sqrt{x^2 + 1} + \dfrac{1}{\sqrt[3]{x^3 + 1}}$ etc.

1.3 Binomial Theorem

Let n be any natural number and x, a be any real number, then

$$\left(x+a\right)^n = {}^nC_0\, x^n a^0 + {}^nC_1\, x^{n-1}a^1 + {}^nC_2\, x^{n-2}a^2 + \ldots\ldots + {}^nC_r\, x^{n-r}a^r + \ldots\ldots + {}^nC_{n-1}\, x^1 a^{n-1} + {}^nC_n\, x^0 a^n$$

i.e. $\left(x+a\right)^n = \displaystyle\sum_{0}^{n} {}^nC_r x^{n-r}a^r$ where ${}^nC_r = \dfrac{n!}{r!\left(n-r\right)!}$

and the co-efficients ${}^nC_0,\ {}^nC_1,\ {}^nC_2,\ \ldots\ldots\ldots$ and ${}^nC_n$ are known as binomial coefficient.

(a) The total number of terms in the expansion of $(x + a)^n = \sum_{r=0}^{n} {}^nC_r x^{n-r} a^r$, is $(n + 1)$.

(b) The sum of the indices of x and a in each term is n.

(c) ${}^nC_0, {}^nC_1, {}^nC_2, \ldots, {}^nC_n$ are called binomial coefficients and also represented by C_0, C_1, C_2 and so on.

(i) $\quad {}^nC_x = {}^nC_y \Rightarrow x = y$ or $x + y = n$ $\qquad$ (ii) $\quad {}^nC_r = {}^nC_{n-r}$

(iii) $\quad {}^nC_r + {}^nC_{r-1} = {}^{n+1}C_r$ $\qquad\qquad$ (iv) $\quad {}^nC_r = n/(n-r) \cdot {}^{n-1}C_r$

Illustration 5: Expand the following binomials

(i) $(x - 2)^5$ $\qquad\qquad$ (ii) $\left(1 - \dfrac{3x^3}{2}\right)^4$ $\hfill$ **(JEE MAIN)**

Sol: By using formula of binomial expansion.

(i) $\quad \left(x - 2\right)^5 = {}^5C_0 x^5 + {}^5C_1 x^4 \left(-2\right)^1 + {}^5C_2 x^3 \left(-2\right)^2 + {}^5C_3 x^2 \left(-2\right)^3 + {}^5C_4 x \left(-2\right)^4 + {}^5C_5 \left(-2\right)^5$

$\qquad = x^5 - 10x^4 + 40x^3 - 80x^2 + 80x - 32$

(ii) $\left(1 - \dfrac{3x^3}{2}\right)^4 = {}^4C_0 + {}^4C_1 \left(-\dfrac{3x^3}{2}\right) + {}^4C_2 \left(-\dfrac{3x^3}{2}\right)^2 + {}^4C_3 \left(-\dfrac{3x^3}{2}\right)^3 + {}^4C_4 \left(-\dfrac{3x^3}{2}\right)^4$

$\qquad = 1 - 6x^3 + \dfrac{27}{2}x^6 - \dfrac{27}{2}x^9 + \dfrac{81}{16}x^{12}$

2. DEDUCTIONS FROM BINOMIAL THEOREM

2.1 Results of Binomial Theorem

D-1 On replacing a by $-a$, in the expansion of $(x + a)^n$, we get

$$\left(x - a\right)^n = {}^nC_0 x^n a^0 - {}^nC_1 x^{n-1} a^1 + {}^nC_2 x^{n-2} \cdot a^2 - \ldots + \left(-1\right)^r {}^nC_r x^{n-r} a^r + \ldots + \left(-1\right)^n {}^nC_n x^0 a^n$$

i.e. $\left(x - a\right)^n = \sum_{r=0}^{n} \left(-1\right)^r {}^nC_r x^{n-r} a^r$

Therefore, the terms in $(x - a)^n$ are alternatively positive and negative, and the sign of the last term is positive or negative depending on whether n is even or odd.

D-2 Putting $x = 1$ and $a = x$ in the expansion of $(x + a)^n$, we get

$$\left(1 + x\right)^n = {}^nC_0 + {}^nC_1 x + {}^nC_2 x^2 + \ldots + {}^nC_r x^r + \ldots + {}^nC_n x^n$$

$$\Rightarrow \left(1 + x\right)^n = \sum_{r=0}^{n} {}^nC_r x^r$$

This is the expansion of $(1 + x)^n$ in ascending powers of x.

D-3 Putting a = 1 in the expansion of $(x + a)^n$, we get

$$(x+1)^n = {}^nC_0 x^n + {}^nC_1 x^{n-1} + {}^nC_2 x^{n-2} + ... + {}^nC_r x^{n-r} + ... + {}^nC_{n-1}x + {}^nC_n \Rightarrow (1+x)^n = \sum_{r=0}^{n} {}^nC_r x^{n-r}$$

This is the expansion of $(1 + x)^n$ in descending powers of x.

D-4 Putting x = 1 and a = −x in the expansion of $(x + a)^n$, we get

$$\left(1-x\right)^n = {}^nC_0 - {}^nC_1 x + {}^nC_2 x^2 - {}^nC_3 x^3 + ... + \left(-1\right)^r {}^nC_r x^r + ... + \left(-1\right)^n {}^nC_n x^n$$

D-5 From the above expansions, we can also deduce the following

$$\left(x+a\right)^n + \left(x-a\right)^n = 2\left[{}^nC_0 x^n a^0 + {}^nC_2 x^{n-2} a^2 + ... \right]$$

and $\left(x+a\right)^n - \left(x-a\right)^n = 2\left[{}^nC_1 x^{n-1} a^1 + {}^nC_3 x^{n-3} a^3 + ... \right]$

If n is odd then $\left\{\left(x+a\right)^n + \left(x-a\right)^n\right\}$ and $\left\{\left(x+a\right)^n - \left(x-a\right)^n\right\}$ both have the same number of terms equal to $\left(\dfrac{n+1}{2}\right)$ where as if n is even, then $\left\{\left(x+a\right)^n + \left(x-a\right)^n\right\}$ has $\left(\dfrac{n}{2}+1\right)$ terms.

2.2 Properties of Binomial Coefficients

Using binomial expansion, we have

$$\left(1+x\right)^n = {}^nC_0 + {}^nC_1 x + {}^nC_2 x^2 + ... + {}^nC_r x^r + ... + {}^nC_n x^n$$

Also, $\left(1+x\right)^n = {}^nC_0 x^n + {}^nC_1 x^{n-1} + {}^nC_2 x^{n-2} + ... + {}^nC_r x^{n-r} + ... + {}^nC_{n-1}x + {}^nC_n$

Let us represent the binomial coefficients ${}^nC_0, {}^nC_1, {}^nC_2, ..., {}^nC_{n-1}, {}^nC_n$ by $C_0, C_1, C_2, ..., C_{n-1}, C_n$ respectively. Then the above expansions become

$$\left(1+x\right)^n = C_0 + C_1 x + C_2 x^2 + ... + C_n x^n \text{ i.e. } \left(1+x\right)^n = \sum_{r=0}^{n} C_r x^r$$

Also, $\left(1+x\right)^n = C_0 x^n + C_1 x^{n-1} + C_2 x^{n-2} + ... + C_r x^{n-r} + ... + C_{n-1}x + C_n$ i.e. $\left(1+x\right)^n = \sum_{r=0}^{n} C_r x^{n-r}$

The binomial coefficients $C_0, C_1, C_2, ...C_{n-1},$ and C_n posses the following properties:

Property-I In the expansion of $(1 + x)^n$, the coefficients of terms equidistant from the beginning and the end are equal.

Property-II The sum of the binomial coefficients in the expansion of $(1 + x)^n$ is 2^n.

i.e. $C_0 + C_1 + C_2 + ... + C_n = 2^n$ or, $\sum_{r=0}^{n} {}^nC_r = 2^n$.

Property-III The sum of the coefficient of the odd terms in the expansion of $(1 + x)^n$ is equal to the sum of the coefficient of the even terms and each is equal to 2^{n-1}.

i.e. $C_0 + C_2 + C_4 + \ldots\ldots = C_1 + C_3 + C_5 + \ldots\ldots = 2^{n-1}$

Property-IV $^nC_r = \dfrac{n}{r} \cdot {}^{n-1}C_{r-1} = \dfrac{n}{r} \cdot \dfrac{n-1}{r-1} \cdot {}^{n-2}C_{r-2}$ and so on.

Property-V $C_0 - C_1 + C_2 - C_3 + C_4 - \ldots\ldots + (-1)^n C_n = 0$

i.e. $\displaystyle\sum_{r=0}^{n} (-1)^r \, {}^nC_r = 0$

NOMORECLASS CONCEPTS

(a) $^{(n+1)}C_r = {}^nC_r + {}^nC_{r-1}$ (b) $r\,{}^nC_r = n^{n-1}C_{r-1}$ (c) $\dfrac{{}^nC_r}{r+1} = \dfrac{{}^{n+1}C_{r+1}}{n+1}$

(d) When n is even, $(x+a)^n + (x-a)^n = 2\left(x^n + {}^nC_2 x^{n-2}a^2 + {}^nC_4 x^{n-4}a^4 + \ldots + {}^nC_n a_n\right)$

When n is odd, $(x+a)^n + (x-a)^n = 2\left(x^n + {}^nC_2 x^{n-2}a^2 + \ldots + {}^nC_{n-1} x a^{n-1}\right)$

When n is even $(x+a)^n - (x-a)^n = 2\left({}^nC_1 x^{n-1}a + {}^nC_3 x^{n-3}a^3 + \ldots + {}^nC_{n-1} x a^{n-1}\right)$

When n is odd $(x+a)^n - (x-a)^n = 2\left({}^nC_1 x^{n-1}a + {}^nC_3 x^{n-3}a^3 + \ldots + {}^nC_n a^n\right)$

Illustration 6: If $(1+x)^n = C_0 + C_1 x + C_2 x^2 + \ldots\ldots + C_n x^n$, then show that $\hfill$ **(JEE MAIN)**

(i) $C_0 + 4C_1 + 4^2 C_2 + \ldots\ldots + 4^n C_n = 5^n$

(ii) $C_0 + 2C_1 + 3C_2 + \ldots\ldots + (n+1)C_n = 2^{n-1}(n+2)$

(iii) $C_0 - \dfrac{C_1}{2} + \dfrac{C_2}{3} - \dfrac{C_3}{4} + \ldots\ldots + (-1)^n \dfrac{C_n}{n+1} = \dfrac{1}{n+1}$

Sol: By using properties of binomial coefficients and methods of summation, differentiation, and integration we can easily prove given equations.

(i) $(1+x)^n = C_0 + C_1 x + C_2 x^2 + \ldots\ldots + C_n x^n$

Putting x=4, we have $C_0 + 4C_1 + 4^2 C_2 + \ldots\ldots + 4^n C_n = 5^n$

(ii) $C_0 + 2C_1 + 3C_2 + \ldots\ldots + (n+1)C_n = 2^{n-1}(n+2)$

Method 1: By Summations

r^{th} term in the series is given by $(r+1).{}^nC_r$

Therefore, L.H.S $= {}^nC_0 + 2.{}^nC_1 + 3.{}^nC_2 + \ldots\ldots + (n+1).{}^nC_n = \displaystyle\sum_{r=0}^{n}(r+1).{}^nC_r$

$$= \sum_{r=0}^{n} r \cdot {}^{n}C_r + \sum_{r=0}^{n} {}^{n}C_r = n \sum_{r=0}^{n} {}^{n-1}C_{r-1} + \sum_{r=0}^{n} {}^{n}C_r = n.2^{n-1} + 2^n = 2^{n-1}(n+2) = \text{R.H.S}$$

Method 2: By Differentiation

$$(1+x)^n = C_0 + C_1 x + C_2 x^2 + \dots\dots + C_n x^n$$

Multiplying x on both sides, $x(1+x)^n = C_0 x + C_1 x^2 + C_2 x^3 + \dots\dots + C_n x^{n+1}$

On differentiating, we have $(1+x)^n + xn(1+x)^{n-1} = C_0 + 2.C_1 x + 3.C_2 x^2 + \dots\dots + (n+1)C_n x^n$

Putting x = 1, we get $C_0 + 2.C_1 + 3.C_2 + \dots\dots + (n+1)C_n = 2^n + n.2^{n-1}$

$$C_0 + 2.C_1 + 3.C_2 + \dots\dots + (n+1)C_n = 2^{n-1}(n+2)$$

(iii) $C_0 - \dfrac{C_1}{2} + \dfrac{C_2}{3} - \dfrac{C_3}{4} + \dots\dots + (-1)^n \dfrac{C_n}{n+1} = \dfrac{1}{n+1}$

Method 1: By Summations

r^{th} term in the series is given by $(-1)^r \cdot \dfrac{{}^{n}C_r}{r+1}$

Therefore, L.H.S. $= C_0 - \dfrac{C_1}{2} + \dfrac{C_2}{3} - \dfrac{C_3}{4} + \dots\dots + (-1)^n \cdot \dfrac{C_n}{n+1} = \sum_{r=0}^{n} (-1)^r \cdot \dfrac{{}^{n}C_r}{r+1}$

$$= \dfrac{1}{n+1} \sum_{r=0}^{n} (-1)^r \, {}^{n+1}C_{r+1} \quad \left\{ \text{using } \dfrac{n+1}{r+1} \cdot {}^{n}C_r = {}^{n+1}C_{r+1} \right\} = \dfrac{1}{n+1} \left[{}^{n+1}C_1 - {}^{n+1}C_2 + {}^{n+1}C_3 - \dots\dots + (-1)^n \cdot {}^{n+1}C_{n+1} \right]$$

Adding and subtracting the term ${}^{n+1}C_0$, we have

$$= \dfrac{1}{n+1} \left[-{}^{n+1}C_0 + {}^{n+1}C_1 - {}^{n+1}C_2 + \dots\dots + (-1)^n \cdot {}^{n+1}C_{n+1} + {}^{n+1}C_0 \right]$$

$$= \dfrac{1}{n+1} \qquad \text{as} \left[-{}^{n+1}C_0 + {}^{n+1}C_1 - {}^{n+1}C_2 + \dots\dots + (-1)^n \cdot {}^{n+1}C_{n+1} = 0 \right] = \text{R.H.S.}$$

Method 2: By Integration

$(1+x)^n = C_0 + C_1 x + C_2 x^2 + \dots\dots + C_n x^n.$

On integrating both sides within the limits −1 to 0, we have

$$\int_{-1}^{0} (1+x)^n \, dx = \int_{-1}^{0} \left(C_0 + C_1 x + C_2 x^2 + \dots\dots + C_n x^n \right) dx$$

$$\Rightarrow \left[\dfrac{(1+x)^{n+1}}{n+1} \right]_{-1}^{0} = \left[C_0 x + C_1 \dfrac{x^2}{2} + C_2 \dfrac{x^3}{3} + \dots\dots + C_n \dfrac{x^{n+1}}{n+1} \right]_{-1}^{0}$$

$$\Rightarrow \dfrac{1}{n+1} - 0 = 0 - \left[-C_0 + \dfrac{C_1}{2} - \dfrac{C_2}{3} + \dots\dots + (-1)^{n+1} \dfrac{C_n}{n+1} \right] \Rightarrow C_0 - \dfrac{C_1}{2} + \dfrac{C_2}{3} + \dots\dots + (-1)^n \dfrac{C_n}{n+1} = \dfrac{1}{n+1}$$

Illustration 7: If $(1+x)^n = C_0 + C_1 x + C_2 x^2 + \dots\dots + C_n x^n$, then prove that

(i) $\quad C_0^2 + C_1^2 + C_2^2 + \dots\dots + C_n^2 = {}^{2n}C_n$

(ii) $\quad C_0C_2 + C_1C_3 + C_2C_4 + \ldots\ldots + C_{n-2}C_n = {}^{2n}C_{n-2}$ or ${}^{2n}C_{n+2}$

(iii) $\quad 1 \cdot C_0^2 + 3 \cdot C_1^2 + 5 \cdot C_2^2 + \ldots\ldots + (2n+1) \cdot C_n^2 = 2n \cdot {}^{2n-1}C_n + {}^{2n}C_n$ **(JEE ADVANCED)**

Sol: In the expansion of $(1+x)^{2n}$, (i) and (ii) can be proved by comparing the coefficients of x^n and x^{n-2} respectively. The third equation can be proved by two methods - the method of summation and the methods of differentiation.

(i) $(1+x)^n = C_0 + C_1 x + C_2 x^2 + \ldots\ldots + C_n x^n$ $\hfill \ldots(i)$

Also, $(x+1)^n = C_0 x^n + C_1 x^{n-1} + C_2 x^{n-2} + \ldots\ldots + C_n x^0$ $\hfill \ldots(ii)$

Multiplying equation (i) and (ii)

$(1+x)^{2n} = \left(C_0 + C_1 x + C_2 x^2 + \ldots\ldots + C_n x^n\right)\left(C_0 x^n + C_1 x^{n-1} + \ldots\ldots + C_n x^0\right)$ $\hfill \ldots(iii)$

On comparing the coefficients of x^n both sides, we have

$\Rightarrow {}^{2n}C_n = C_0^2 + C_1^2 + C_2^2 + \ldots\ldots + C_n^{\,2}$ $\qquad$ Hence, Proved.

(ii) From (iii), on comparing the coefficients of x^{n-2} or x^{n+2}, we have

$C_0C_1 + C_1C_3 + C_2C_4 + \ldots\ldots + C_{n-2}C_n = {}^{2n}C_{n-2}$ or ${}^{2n}C_{n+2}$

(iii) $1 \cdot C_0^2 + 3 \cdot C_1^2 + 5 \cdot C_2^2 + \ldots\ldots + (2n+1) \cdot C_n^2 = 2n \cdot {}^{2n-1}C_n + {}^{2n}C_n$

Method 1: By Summation

r^{th} term in the series is given by $(2r+1) {}^nC_r^2$

L.H.S. $= 1.C_0^2 + 3.C_1^2 + 5.C_2^2 + \ldots\ldots + (2n+1)C_n^2 \qquad = \sum_{r=0}^{n} (2r+1) {}^nC_r^2$

$= \sum_{r=0}^{n} 2.r.\left({}^nC_r\right)^2 + \sum_{r=0}^{n} \left({}^nC_r\right)^2 = 2\sum_{r=1}^{n} .n.{}^{n-1}C_{r-1}\, {}^nC_r + {}^{2n}C_n$

$(1+x)^n = {}^nC_0 + {}^nC_1 x + {}^nC_2 x^2 + \ldots\ldots + {}^nC_n x^n$ $\hfill \ldots(i)$

$(x+1)^{n-1} = {}^{n-1}C_0 x^{n-1} + {}^{n-1}C_1 x^{n-2} + \ldots\ldots + {}^{n-1}C_{n-1} x^0$ $\hfill \ldots(ii)$

Multiplying (i) and (ii) and comparing coefficients of x^n, we have

${}^{2n-1}C_n = {}^{n-1}C_0.{}^nC_1 + {}^{n-1}C_1.{}^nC_2 + \ldots\ldots + {}^{n-1}C_{n-1}.{}^nC_n$

i.e. $\sum_{r=1}^{n} {}^{n-1}C_{r-1}\, {}^nC_r = {}^{2n-1}C_n$

Hence, required summation is $2n.\,{}^{2n-1}C_n + {}^{2n}C_n$

Method 2: By Differentiation

$\left(1+x^2\right)^n = C_0 + C_1 x^2 + C_2 x^4 + C_3 x^6 + \ldots\ldots + C_n x^{2n}$

Multiplying x on both sides

$x\left(1+x^2\right)^n = C_0 x + C_1 x^3 + C_2 x^5 + \ldots\ldots + C_n x^{2n+1}$

3.8

Differentiating both sides

$$x.n\left(1+x^2\right)^{n-1}.2x+\left(1+x^2\right)^n = C_0 + 3.C_1 x^2 + 5.C_2 x^4 + \ldots\ldots + \left(2n+1\right)C_n x^{2n} \qquad \text{.... (i)}$$

$$\left(x^2+1\right)^n = C_0 x^{2n} + C_1 x^{2n-2} + C_2 x^{2n-4} + \ldots\ldots + C_n \qquad \text{.... (ii)}$$

On multiplying (i) and (ii), we have

$$2nx^2\left(1+x^2\right)^{2n-1} + \left(1+x^2\right)^{2n} = \left(C_0 + 3C_1 x^2 + 5C_2 x^4 + \ldots\ldots + \left(2n+1\right)C_n x^{2n}\right)\left(C_0 x^{2n} + C_1 x^{2n-2} + \ldots\ldots + C_n\right)$$

Comparing coefficient of x^{2n},

$$2n.^{2n-1}C_{n-1} + {}^{2n}C_n = C_0^2 + 3C_1^2 + 5C_2^2 + \ldots\ldots + \left(2n+1\right)C_n^2$$

$$\therefore\; C_0^2 + 3C_1^2 + 5C_2^2 + \ldots\ldots + \left(2n+1\right)C_n^2 = 2n.^{2n-1}C_n + {}^{2n}C_n$$

Illustration 8: If $\left(1+x\right)^n = C_0 + C_1 x + C_2 x^2 + \ldots\ldots + C_n x^n$,

Prove that $C_0 C_r + C_1 C_{r+1} + C_2 C_{r+2} + \ldots\ldots + C_{n-r}C_n = \dfrac{2n!}{\left(n-r\right)!\left(n+r\right)!}$ **(JEE MAIN)**

Sol: Clearly the differences of lower suffixes of binomial coefficients in each term is r.

By using properties of binomial coefficients we can easily prove given equations.

Given $\left(1+x\right)^n = C_0 + C_1 x + C_2 x^2 + \ldots\ldots + C_{n-r}x^{n-r} + \ldots\ldots + C_n x^n$ .... (i)

Now $\left(x+1\right)^n = C_0 x^n + C_1 x^{n-1} + C_2 x^{n-2} + \ldots\ldots + C_r x^{n-r} + C_{r+1}x^{n-r-1} + \ldots\ldots + C_n$ ....(ii)

Multiplying (i) and (ii), we get

$$\left(x+1\right)^{2n} = \left(C_0 + C_1 x + C_2 x^2 + \ldots\ldots + C_{n-r}x^{n-r} + \ldots\ldots + C_n x^n\right)$$

$$\times\left(C_0 x^n + C_1 x^{n-1} + C_2 x^{n-2} + \ldots\ldots + C_r x^{n-r} + C_{r+1}x^{n-r-1} + C_{r+2}x^{n-r-2} + \ldots\ldots + C_n\right) \qquad \text{.... (iii)}$$

Now coefficient of x^{n-r} on L.H.S. of (iii) $= {}^{2n}C_{n-r} = \dfrac{2n!}{\left(n-r\right)!\left(n+r\right)!}$

and coefficient of x^{n-r} on R.H.S. of (iii) $= C_0 C_r + C_1 C_{r+1} + C_2 C_{r+2} + \ldots\ldots + C_{n-r}C_n$

But (iii) is an identity, therefore, of x^{n-r} in R.H.S. = Coefficient of x^{n-r} in L.H.S.

$$\Rightarrow C_0 C_r + C_1 C_{r+1} + C_2 C_{r+2} + \ldots\ldots + C_{n-r}C_n = \dfrac{2n!}{\left(n-r\right)!\left(n+r\right)!}$$

Hence, Proved.

Illustration 9: Prove that ${}^nC_0 \cdot {}^{2n}C_r - {}^nC_1 \cdot {}^{2n-2}C_r + \ldots\ldots = 2^{2n-r}.{}^nC_{r-n}$ if $r > n$ and 0 if $r < n$.   **(JEE MAIN)**

Sol: By comparing coefficient of x^r in L.H.S. and R.H.S. in the expansion of $\left[\left(1+x\right)^2 - 1\right]^n$ we can prove it.

$$\left[\left(1+x\right)^2 - 1\right]^n = {}^nC_0\left(1+x\right)^{2n} - {}^nC_1\left(1+x\right)^{2n-2} + {}^nC_2\left(1+x\right)^{2n-4} + \ldots \qquad \text{.... (i)}$$

Coefficient of x^r in R.H.S. $= {}^nC_0.{}^{2n}C_r - {}^nC_1.{}^{2n-2}C_r + \ldots\ldots\ldots$ ... (ii)

L.H.S. $= [(1 + x)^2 - 1]^n = [2x + x^2]^n = x^n (2 + x)^n$

$\therefore$ Coefficient of x^r in $x^n (2 + x)^n$

$\quad$ = Coefficient of x^{r-n} in $(2 + x)^n = {}^nC_{r-n} 2^{2n-r}$ if $r > n$

$\quad$ = 0 if $r < n$ $\qquad$ (Since lower suffix cannot be negative)

But (i) is an identity, therefore coefficient of x^r in R.H.S. = coefficient of x^r in L.H.S.

Hence ${}^nC_0 . {}^{2n}C_r - {}^nC_1 . 2^{n-2}C_r + \ldots\ldots = {}^nC_{r-n} 2^{2n-r}$ $\quad$ if $r > n$

$\qquad\qquad\qquad\qquad\qquad\qquad = 0$ $\qquad\qquad$ if $r < n$.

Illustration 10: Show that $C_0 . {}^{2n}C_n - C_1 . {}^{2n-1}C_n + C_2 . {}^{2n-2}C_n - C_3 . {}^{2n-3}C_n + \ldots\ldots + (-1)^n C_n . {}^nC_n = 1$ $\qquad$ **(JEE ADVANCED)**

Sol: Observe the pattern in the terms on the LHS. The first term $C_0 . {}^{2n}C_n$ is the co-efficient of x^n in the expansion of $C_0 (1+x)^{2n}$. Similarly, $C_1 . {}^{2n-1}C_n$ is the co-efficient of x^n in $C_0 (1+x)^{2n}$ and so on. On adding all the coefficients of x^n we can prove the given equation.

Note that $C_0 . {}^{2n}C_n - C_1 . {}^{2n-1}C_n + C_2 . {}^{2n-2}C_n - C_3 . {}^{2n-3}C_n + \ldots\ldots + (-1)^n {}^nC_n . {}^nC_n$

= Coefficient of x^n in $\left[C_0 (1+x)^{2n} - C_1 (1+x)^{2n-1} + C_2 (1+x)^{2n-2} - C_3 (1+x)^{2n-3} + \ldots (-1)^n C_n (1+x)^n \right]$

= Coefficient of x^n in $(1+x)^n \left[C_0 (1+x)^n - C_1 (1+x)^{n-1} + C_2 (1+x)^{n-2} - C_3 (1+x)^{n-3} + \ldots (-1)^n C_n \right]$

= Coefficient of x^n in $(1+x)^n \left[(1+x) - 1 \right]^n$

= Coefficient of x^n in $(1+x)^n (x)^n$

= Coefficient of the constant terms in $(1 + x)^n = 1$

3. TERMS IN BINOMIAL EXPANSION

3.1 General Term in Binomial Expansion

We have, $(x + a)^n = {}^nC_0 x^n a^0 + {}^nC_1 x^{n-1} a^1 + {}^nC_2 x^{n-2} x^2 + \ldots\ldots + {}^nC_r x^{n-r} a^r + \ldots\ldots + {}^nC_n x^0 a^n$

$(r+1)^{th}$ term is given by ${}^nC_r x^{n-r} a^r$

Thus, if T_{r+1} denotes the $(r+1)^{th}$ term, then $T_{r+1} = {}^nC_r x^{n-r} a^r$

This is called the general term of the binomial expansion.

(a) The general term in the expansion of $(x - a)^n$, is given by $T_{r+1} = (-1)^r . {}^nC_r x^{n-r} a^r$

(b) The general term in the expansion of $(1 + x)^n$, is given by $T_{r+1} = {}^nC_r x^r$

(c) The general term in the expansion of $(1 - x)^n$, is given by $T_{r+1} = (-1)^r {}^nC_r x^r$

(d) In the binomial expansion of $(x + a)^n$, the rth term from the end is $((n + 1) - r + 1)^{th}$ term i.e. $(n - r + 2)^{th}$ term from the beginning.

Illustration 11: The number of dissimilar terms in the expansion of $(1 - 3x + 3x^2 - x^3)^{20}$ is $\qquad$ **(JEE MAIN)**

Sol: As we know that number of dissimilar terms in the expansion of $(1 - x)^n$ is $n+1$. Rewrite the given expression in the form of $(1 - x)^n$.

$(1 - 3x + 3x^2 - x)^{20} = [(1 - x)^3]^{20} = (1 - x)^{60}$

Therefore number of dissimilar terms in the expansion of $(1 - 3x + 3x^2 - x^3)^{20}$ is 61.

Illustration 12: Find (i) 28th term of $(5x + 8y)^{30}$ (ii) 7th term of $\left(\dfrac{4x}{5} - \dfrac{5}{2x}\right)^9$ **(JEE MAIN)**

Sol: Here in this problem, by using $T_{r+1} = {}^nC_r x^{n-r} a^r$ we can easily obtain $(r+1)^{th}$ term of given expansion.

(i) 28th term of $(5x + 8y)^{30}$

$$T_{28} = T_{27+1} = {}^{30}C_{27}(5x)^{30-27}(8y)^{27} = \frac{30!}{3!.27!}(5x)^3 .(8y)^{27}$$

(ii) 7th term of $\left(\dfrac{4x}{5} - \dfrac{5}{2x}\right)^9$

$$T_7 = T_{6+1} = {}^9C_6\left(\frac{4x}{5}\right)^{9-6}\left(-\frac{5}{2x}\right)^6 = \frac{9!}{3!6!}\left(\frac{4x}{5}\right)^3\left(\frac{5}{2x}\right)^6 = \frac{10500}{x^3}$$

Illustration 13: Find the number of rational terms in the expansion of $(9^{1/4} + 8^{1/6})^{1000}$. **(JEE ADVANCED)**

Sol: In this problem, by using $T_{r+1} = {}^nC_r x^{n-r} a^r$ we can easily obtain $(r+1)^{th}$ term of given expansion and after that by using the conditions of rational number we can obtain number of rational terms.

The general term in the expansion of $(9^{1/4} + 8^{1/6})^{1000}$ is

$$T_{r+1} = {}^{1000}C_r\left(9^{\frac{1}{4}}\right)^{1000-r}\left(8^{1/6}\right)^r = {}^{1000}C_r\, 3^{\frac{1000-r}{2}}\, 2^{\frac{r}{2}}$$

T_{r+1} will be rational if the power of 3 and 2 are integers. It means $\dfrac{1000-r}{2}$ and $\dfrac{r}{2}$ must be integers.

Therefore the possible set of values of r is {0, 2, 4... ... 1000}. Hence, number of rational terms is 501.

3.2 Middle Term in Binomial Expansion

(a) If n is even, then the number of terms in the expansion i.e. (n + 1) is odd, therefore, there will be only one middle term which is $\left(\dfrac{n+2}{2}\right)^{th}$ term i.e. $\left(\dfrac{n}{2}+1\right)^{th}$ term.

So middle term $= \left(\dfrac{n}{2}+1\right)^{th}$ term i.e. $T_{\left(\frac{n}{2}+1\right)} = {}^nC_{\frac{n}{2}} x^{\frac{n}{2}} a^{\frac{n}{2}}$

(b) If n is odd, then the number of terms in the expansion i.e. (n+1) is even, therefore there will be two middle terms which are

$$= \left(\frac{n+1}{2}\right)^{th} \text{ and} \left(\frac{n+3}{2}\right)^{th} \text{ term i.e. } T_{\left(\frac{n+1}{2}\right)} = {}^nC_{\left(\frac{n-1}{2}\right)} x^{\frac{n+1}{2}} a^{\frac{n-1}{2}} \text{ and } T_{\left(\frac{n+3}{2}\right)} = {}^nC_{\left(\frac{n+1}{2}\right)} x^{\frac{n-1}{2}} a^{\frac{n+1}{2}}$$

- When there are two middle terms in the expansion then their binomial coefficients are equal.
- Binomial coefficient of middle term is the greatest Binomial coefficient.

Illustration 14: Find the middle term(s) in the expansion of (i) $\left(1-\dfrac{x^2}{2}\right)^{14}$ (ii) $\left(3a-\dfrac{a^3}{6}\right)^{9}$ **(JEE MAIN)**

Sol: By using appropriate formula of finding middle term(s) i.e. $\left(\dfrac{n}{2}+1\right)^{th}$ when n is even and $\left(\dfrac{n+1}{2}\right)^{th}$ and $\left(\dfrac{n+3}{2}\right)^{th}$ when n is odd, we can obtain the middle terms of given expansion.

(i) $\left(1-\dfrac{x^2}{2}\right)^{14}$ Since, n is even, therefore middle term is $\left(\dfrac{14}{2}+1\right)^{th}$ term.

$$\therefore T_8 = {}^{14}C_7\left(-\dfrac{x^2}{2}\right)^{7} = -\dfrac{429}{16}x^{14}$$

(ii) $\left(3a-\dfrac{a^3}{6}\right)^{9}$

Since, n is odd therefore, the middle terms are $\left(\dfrac{9+1}{2}\right)^{th}$ and $\left(\dfrac{9+1}{2}+1\right)^{th}$.

$$\therefore T_5 = {}^{9}C_4\left(3a\right)^{9-4}\left(-\dfrac{a^3}{6}\right)^{4} = \dfrac{189}{8}a^{17} \quad \text{and} \quad T_6 = {}^{9}C_5\left(3a\right)^{9-5}\left(-\dfrac{a^3}{6}\right)^{5} = -\dfrac{21}{16}a^{19}.$$

3.3 Determining a Particular Term

In the expansion of $\left(x^\alpha \pm \dfrac{1}{x^\beta}\right)^{n}$, if x^m occurs in T_{r+1}, then r is given by

$$n\alpha - r\left(\alpha+\beta\right) = m \qquad \Rightarrow r = \dfrac{n\alpha-m}{\alpha+\beta}$$

Thus in above expansion if constant term i.e. the term independent of x, occurs in T_{r+1} then r is determined by

$$n\alpha - r\left(\alpha+\beta\right) = 0 \qquad \Rightarrow r = \dfrac{n\alpha}{\alpha+\beta}$$

Illustration 15: The term independent of x in the expansion of $\left(\dfrac{4}{3}x^2 - \dfrac{3}{2x}\right)^{9}$ is **(JEE MAIN)**

Sol: By using the result proved above i.e. $r = \dfrac{n\alpha}{\alpha+\beta}$, we can obtain the term independent of x. Here, α and β are obtained by comparing given expansion to $\left(x^\alpha \pm \dfrac{1}{x^\beta}\right)^{n}$.

On comparing $\left(\dfrac{4}{3}x^2 - \dfrac{3}{2x}\right)^{9}$ with $\left(x^\alpha \pm \dfrac{1}{x^\beta}\right)^{n}$, we get $\alpha = 2, \beta = 1, n = 9$

i.e. $r = \dfrac{9(2)}{2+1} = 6$ $\therefore (6 + 1) = 7^{\text{th}}$ term is independent of x.

Illustration 16: The ratio of the coefficient of x^{15} to the term independent of x in $\left(x^2 + \dfrac{2}{x}\right)^{15}$ is **(JEE MAIN)**

Sol: Here in this problem, by using standard formulas of finding general term and term independent of x we can obtain the required ratio.

General term in the expansion is $T_{r+1} = {}^{15}C_r \left(x^2\right)^{15-r} \left(\dfrac{2}{x}\right)^r$ i.e., ${}^{15}C_r x^{30-3r}.2^r$

For x^{15}, $30 - 3r = 15 \Rightarrow 3r = 15 \qquad \Rightarrow r = 5$

$\therefore T_6 = T_{5+1} = {}^{15}C_5 \left(x^2\right)^{15-5} \left(\dfrac{2}{x}\right)^5$ i.e., ${}^{15}C_5 x^{15}.2^5$

$\therefore$ Coefficient of x^{15} is ${}^{15}C_5 2^5$ $(r = 5)$

For the constant term $30 - 3r = 0 \Rightarrow r = 10$.

$\therefore T_{11} = T_{10+1} = {}^{15}C_{10} \left(x^2\right)^{15-10} \left(\dfrac{2}{x}\right)^{10}$ i.e., ${}^{15}C_{10} 2^{10}$

$\therefore$ Coefficient of constant term is ${}^{15}C_{10}2^{10}$.

Hence, the required ratio is $1 : 32$.

Illustration 17: The term independent of x in the expansion of $\left(\sqrt[6]{x} - \dfrac{1}{\sqrt[3]{x}}\right)^9$ is equal to **(JEE MAIN)**

Sol: By using the formula $T_{r+1} = {}^nC_r x^{n-r} a^r$ we can solve it.

$T_{r+1} = {}^9C_r \left(\sqrt[6]{x}\right)^{9-r} \left(-\dfrac{1}{\sqrt[3]{x}}\right)^r \qquad = {}^9C_r (-1)^r x^{\frac{9-r}{6} - \frac{r}{3}} \qquad = {}^9C_r (-1)^r x^{\left(\frac{9-3r}{6}\right)}$

$\Rightarrow \dfrac{9-3r}{6} = 0 \qquad \Rightarrow r = 3 \qquad \therefore T_4 = T_{3+1} = -{}^9C_3$

Illustration 18: If the second, third and fourth terms in the expansion of $(b+a)^n$ are 135, 30 and 10/3 respectively, then n is equal to **(JEE MAIN)**

Sol: In this problem, by using the formula of finding general term we will get the equation of given terms and by taking ratios of these terms we can get the value of n.

$T_2 = {}^nC_1 ab^{n-1} = 135$ ...(i)

$T_3 = {}^nC_2 a^2b^{n-2} = 30$ ...(ii)

$T_4 = {}^nC_3 a^3b^{n-3} = \dfrac{10}{3}$ ...(iii)

On dividing (i) by (ii), we get

$\dfrac{{}^nC_1 ab^{n-1}}{{}^nC_2 a^2b^{n-2}} = \dfrac{135}{30} \qquad \Rightarrow \dfrac{n}{\dfrac{n(n-1)}{2}} \dfrac{b}{a} = \dfrac{9}{2}$ ...(iv)

$\therefore \dfrac{b}{a} = \dfrac{9}{4}(n-1)$ ...(v)

Dividing (ii) and (iii), we get

$$\dfrac{\dfrac{n(n-1)}{2}}{\dfrac{n(n-1)(n-2)}{3.2}}\cdot\dfrac{b}{a}=\dfrac{30\times 3}{10}=9 \qquad\Rightarrow\quad \dfrac{3}{(n-2)}\dfrac{b}{a}=9 \qquad\qquad\qquad \text{...(vi)}$$

Eliminating a and b from (v) and (vi) $\Rightarrow$ n = 5

Illustration 19: If a, b, c and d are the coefficients of any four consecutive terms in the expansion of $(1+x)^n$, n being positive integer, show that $\dfrac{a}{a+b}+\dfrac{c}{c+d}=\dfrac{2b}{b+c}$ **(JEE MAIN)**

Sol: Consider four consecutive terms and use $^nC_{r-1}+{}^nC_r={}^{n+1}C_r$.

The $(r+1)^{th}$ term is $T_{r+1}={}^nC_r x^r$

$\therefore$ The coefficient of term $T_{r+1}={}^nC_r$

$\therefore$ Now take four consecutive terms as $(r-1)$th, rth, $(r+1)$th and $(r+2)$th

$\therefore$ We get $a={}^nC_{r-2},\, b={}^nC_{r-1},\, c={}^nC_r,\, d={}^nC_{r+1}$

$a+b={}^nC_{r-2}+{}^nC_{r-1}={}^{n+1}C_{r-1}$

$b+c={}^nC_{r-1}+{}^nC_r={}^{n+1}C_r$

$c+d={}^nC_r+{}^nC_{r+1}={}^{n+1}C_{r+1}$

$$\therefore\ \dfrac{a}{a+b}=\dfrac{^nC_{r-2}}{^{n+1}C_{r-1}}=\dfrac{n!}{(r-2)!(n-r+2)!}\times\dfrac{(r-1)!(n-r+2)!}{(n+1)!}=\dfrac{r-1}{n+1}$$

$$\dfrac{b}{b+c}=\dfrac{^nC_{r-1}}{^{n+1}C_r}=\dfrac{n!}{(r-1)!(n-r+2)!}\times\dfrac{r!(n-r+1)!}{(n+1)!}=\dfrac{r}{n+1}$$

$$\dfrac{c}{c+d}=\dfrac{^nC_r}{^{n+1}C_{r+1}}=\dfrac{n!}{n!(n-r)!}\times\dfrac{(r+1)!(n-r)!}{(n+1)!}=\dfrac{r+1}{n+1}$$

$$\therefore\ \dfrac{a}{a+b}+\dfrac{c}{c+d}=\dfrac{r-1}{n+1}+\dfrac{r+1}{n+1}=\dfrac{2r}{n+1}=2\left(\dfrac{r}{n+1}\right)=\dfrac{2b}{b+c}$$

3.4 Finding a Term from the End of Expansion

In the expansion of $(x+a)^n$, $(r+1)^{th}$ term from end $=(n-r+1)^{th}$ term from beginning i.e. $T_{r+1}(E)=T_{n-r+1}(B)$

$$\therefore T_r(E)=T_{n-r+2}(B)$$

Illustration 20: The 4th term from the end in the expansion of $(2x-1/x^2)^{10}$ is **(JEE MAIN)**

Sol: By using $T_r(E)=T_{n-r+2}(B)$ we will get the fourth term from the end in the given expansion.

Required term $=T_{10-4+2}=T_8={}^{10}C_7(2x)^3\left(-\dfrac{1}{x^2}\right)^7=-960\,x^{-11}$

3.5 Greatest Term in the Expansion

Let T_{r+1} and T_r be $(r+1)$th and rth terms respectively in the expansion of $(x+a)^n$. Then, $T_{r+1} = {}^nC_r x^{n-r}a^r$ and $T_r = {}^nC_{r-1}x^{n-r+1}a^{r-1}$.

$$\therefore \frac{T_{r+1}}{T_r} = \frac{{}^nC_r x^{n-r}a^r}{{}^nC_{r-1}x^{n-r+1}a^{r-1}} = \frac{n!}{(n-r)!r!} \times \frac{(r-1)!(n-r+1)!}{n!} \cdot \frac{a}{x} = \frac{n-r+1}{r} \cdot \frac{a}{x}$$

Now, $T_{r+1} >=< T_r \quad \Rightarrow \frac{T_{r+1}}{T_r} >=< 1 \quad \Rightarrow \frac{n-r+1}{r} \cdot \frac{a}{x} >=< 1 \quad \Rightarrow \left\{ \left(\frac{n+1}{r}\right) - 1 \right\} \frac{a}{x} >=< 1$

$$\Rightarrow \frac{n+1}{r} - 1 >=< \frac{x}{a} \quad \Rightarrow \frac{n+1}{r} >=< \left(1 + \frac{x}{a}\right) \quad \Rightarrow \frac{n+1}{1 + \frac{x}{a}} >=< r$$

Thus, $T_{r+1} >=< T_r$ according as $\left(\dfrac{n+1}{1 + \frac{x}{a}}\right) >=< r$ ...(i)

Now, two cases arise

Case-I: When $\dfrac{n+1}{1 + \frac{x}{a}}$ is an integer Let $\dfrac{n+1}{1 + \frac{x}{a}} = m$, Then, from (i), we have

$T_{r+1} > T_r$, for $r = 1, 2, 3, \ldots(m-1)$ ....(ii)

$T_{r+1} = T_r$, for $r = m$ ...(iii)

and, $T_{r+1} < T_r$, for $r = m+1, \ldots n$ ....(iv)

$\therefore T_2 > T_1, T_3 > T_2, T_4 > T_3, \ldots T_m > T_{m-1}$ [From (ii)]

$T_{m+1} = T_m$ [From (iii)]

and, $T_{m+2} < T_{m+1}, T_{m+3} < T_{m+2}, T_{n+1} < T_n$ [From (iv)]

$\Rightarrow T_1 < T_2 < \ldots < T_{m-1} < T_m = T_{m+1} > T_{m+2} \ldots > T_n$

This shows that m^{th} and $(m+1)^{th}$ terms are greatest terms.

Case-II: When $\left[\dfrac{n+1}{1 + \frac{x}{a}}\right] = m$. Then, from (i), we have

$T_{r+1} > T_r$ for $r = 1, 2, \ldots, m$ ....(v)

and $T_{r+1} < T_r$ for $r = m+1, m+2, \ldots, n$ ....(vi)

$\therefore T_2 > T_1, T_3 > T_2, \ldots, T_{m+1} > T_m$ [From (v)]

and, $T_{m+2} < T_{m+1}, T_{m+3} < T_{m+2}, \ldots, T_{n+1} < T_n$ [From (vi)]

$\Rightarrow T_1 < T_2 < T_3 < \ldots < T_m < T_{m+1} > T_{m+2} > T_{m+3} \ldots < T_{n+1}$

$\Rightarrow (m + 1)^{th}$ term is the greatest term.

Following algorithm may be used to find the greatest term in a binomial expansion.

3.6 Algorithm to Find Greatest Term

Step I: From the given expansion, get T_{r+1} and T_r

Step II: Find $\dfrac{T_{r+1}}{T_r}$

Step III: Put $\dfrac{T_{r+1}}{T_r} > 1$

Step IV: Simplify the inequality obtained in step III, and write it in the form of either $r < m$ or $r > m$.

Step V: If m is an integer, then m^{th} and $(m+1)^{th}$ terms are the greatest terms and they are equal.

If m is not an integer, then $([m]+1)^{th}$ term is the greatest term, where [m] means the integral part of m.

3.7 Greatest Coefficient

Case-I When n is even, we have

$$\frac{{}^{n}C_r}{{}^{n}C_{r+1}} = \frac{n!}{(n-r)!\,r!} \times \frac{(r+1)!(n-r-1)!}{n!} = \frac{r+1}{n-r} \qquad \qquad \text{....(i)}$$

Now, for $0 \le r \le \dfrac{n}{2} - 1 \qquad \qquad \Rightarrow 1 \le r+1 \le \dfrac{n}{2}$ and $\dfrac{n}{2}+1 < n-r \le n$

$$\Rightarrow \frac{r+1}{n-r} < 1 \qquad \text{[Using (i)]} \Rightarrow \frac{{}^{n}C_r}{{}^{n}C_{r+1}} < 1 \Rightarrow {}^{n}C_r < {}^{n}C_{r+1}$$

Putting $r = 0,1,2,\ldots\ldots,\left(\dfrac{n}{2}-1\right)$, we get ${}^{n}C_0 < {}^{n}C_1, {}^{n}C_1 < {}^{n}C_2, {}^{n}C_2 < {}^{n}C_3 \ldots\ldots < {}^{n}C_{\frac{n}{2}-1} < {}^{n}C_{\frac{n}{2}}$

$$\Rightarrow {}^{n}C_0 < {}^{n}C_1 < {}^{n}C_2 < \ldots\ldots < {}^{n}C_{\frac{n}{2}-1} < {}^{n}C_{\frac{n}{2}} \qquad \qquad \text{...(ii)}$$

Since ${}^{n}C_{n-r} = {}^{n}C_r$

$$\therefore {}^{n}C_0 = {}^{n}C_n, {}^{n}C_1 = {}^{n}C_{n-1}, {}^{n}C_2 = {}^{n}C_{n-2} \ldots\ldots, {}^{n}C_{\frac{n}{2}-1} < {}^{n}C_{\frac{n}{2}}$$

Substituting these values in (ii), we get

$$\qquad \qquad \text{...(iii)}$$

$${}^{n}C_n < {}^{n}C_{n-1} < {}^{n}C_{n-2} < \ldots\ldots < {}^{n}C_{\frac{n}{2}+1} < {}^{n}C_{\frac{n}{2}}$$

From (ii) and (iii), we refer that the maximum value of ${}^{n}C_r$ is ${}^{n}C_{n/2}$.

Case-II When n is odd

We have, $\dfrac{{}^{n}C_r}{{}^{n}C_{r+1}} = \dfrac{r+1}{n-r} \qquad \qquad \text{...(i)}$

Now, $0 \le r < \dfrac{n-3}{2} \qquad \qquad \Rightarrow 0 < r+1 < \dfrac{n-1}{2}$ and $\dfrac{n-1}{2} \le n-r \le n$

$\Rightarrow \dfrac{r+1}{n-1} < 1 \Rightarrow \dfrac{{}^nC_r}{{}^nC_{r+1}} < 1 \quad [\text{Using (i)}] \Rightarrow {}^nC_r < {}^nC_{r+1}$

Putting $r = 0, 1, 2, \ldots\ldots, \dfrac{n-3}{2}$

We get ${}^nC_0 < {}^nC_1, \; {}^nC_1 < {}^nC_2, \; {}^nC_2 < {}^nC_3, \ldots\ldots, < {}^nC_{\frac{n-3}{2}} < {}^nC_{\frac{n-1}{2}} = {}^nC_{\frac{n+1}{2}}$

$\Rightarrow {}^nC_0 < {}^nC_1 < {}^nC_2 < {}^nC_3 < \ldots\ldots < {}^nC_{\frac{n-3}{2}} < {}^nC_{\frac{n-1}{2}} = {}^nC_{\frac{n-1}{2}}$
$\qquad\qquad\qquad\qquad\qquad\qquad\qquad\qquad\qquad\qquad\qquad\qquad\qquad\qquad$ ...(ii)

Since ${}^nC_{n-r} = {}^nC_r$, Therefore,

$\therefore \; {}^nC_0 = {}^nC_n, \; {}^nC_1 = {}^nC_{n-1}, \; {}^nC_2 = {}^nC_{n-2}, \ldots\ldots, {}^nC_{\frac{n-1}{2}} = {}^nC_{\frac{n+1}{2}}$
$\qquad\qquad\qquad\qquad\qquad\qquad\qquad\qquad\qquad\qquad\qquad\qquad\qquad\qquad$ ...(iii)

From (ii) and (iii), it follows that the maximum value of ${}^nC_r$ is ${}^nC_{\frac{n-1}{2}} = {}^nC_{\frac{n+1}{2}}$

Illustration 21: Find the numerically greatest term in the expansion of $(3 - 4x)^{15}$, when $x = \dfrac{1}{4}$.
 (JEE MAIN)

Sol: Follow the algorithm mentioned above.

Let r^{th} and $(r + 1)^{th}$ be two consecutive terms in the expansion of $(3 - 4x)^{15}$
$T_{r+1} > T_r$

$${}^{15}C_r \, 3^{15-r} \left(\left|-4x\right|\right)^r > {}^{15}C_{r-1} \, 3^{15-(r-1)} \left(\left|-4x\right|\right)^{r-1}$$

$$\dfrac{(15)!}{(15-r)!\,r!}\left|-4x\right| > \dfrac{3.(15!)}{(16-r)!(r-1)!} \qquad \Rightarrow 5.\dfrac{1}{5}(16-r) > 3r \quad \Rightarrow 16 - r > 3r$$

$\Rightarrow 4r < 16 \qquad \Rightarrow r < 4$

Hence, we have $T_1 < T_2 < T_3 < T_4$.

Similarly, if we simplify $T_{r+1} = T_r$, we get $r = 4$.

Therefore the numerically greatest term is T_4 and T_5.

4. APPLICATION OF BINOMIAL THEOREM

4.1 Divisibility Test

Illustration 22: Show that $7^{2n} + 7$ is divisible by 8, where n is a positive integer.
 (JEE MAIN)

Sol: Write $7^{2n} + 7$ in the form of $8\lambda + c$, where c is a constant. If $c = 0$ then we can conclude that $7^{2n} + 7$ is divisible by 8.

$7^{2n} + 7 = (8 - 1)^{2n} + 7 \qquad = {}^{2n}C_0 8^{2n} - {}^{2n}C_1.8^{2n-1} + {}^{2n}C_2.8^{2n-2} - \ldots\ldots + {}^{2n}C_{2n} + 7$

$\qquad = 8^{2n}.{}^{2n}C_0 - 8^{2n-1}.{}^{2n}C_1 + \ldots\ldots - 8.{}^{2n}C_{2n-1} + 8 \qquad = 8\lambda$ where λ is a positive integer

Hence, $7^{2n} + 7$ is divisible by 8.

Illustration 23: Prove that $13^{99} - 19^{57}$ is divisible by 162. **(JEE ADVANCED)**

Sol: Reduce $13^{99} - 19^{57}$ into the form of $162\lambda + C$ using binomial expansion and If $C = 0$ then $13^{99} - 19^{57}$ is divisible by 162.

Let the given number be called S. Hence, $S = 13^{99} - 19^{57} = (1 + 3 \times 4)^{99} - (1 + 9 \times 2)^{57}$

$$S = \left\{ 1 + {}^{99}C_1.(3 \times 4) + {}^{99}C_2.(3 \times 4)^2 + {}^{99}C_3.(3 \times 4)^3 + \ldots\ldots + {}^{99}C_{99}.(3 \times 4)^{99} \right\}$$

$$- \left\{ 1 + {}^{57}C_1.(9 \times 2) + {}^{57}C_2.(9 \times 2)^2 + {}^{57}C_3.(9 \times 2)^3 + \ldots\ldots + {}^{57}C_{57}.(9 \times 2)^{57} \right\}$$

$$S = \left\{ 1 + {}^{99}C_1.(3 \times 4) + \left(3^4 \times 2\right)k_1 \right\} - \left\{ 1 + {}^{57}C_1.(9 \times 2) + \left(3^4 \times 2\right)k_2 \right\}$$

All terms like $\left\{ {}^{99}C_1.(3 \times 4)^2, {}^{99}C_2.(3 \times 4)^3, \ldots\ldots, {}^{99}C_{99}.(3 \times 4)^{99} \right\}$ and

$\left\{ {}^{57}C_2.(9 \times 2)^2, {}^{57}C_3.(9 \times 2)^3, \ldots\ldots, {}^{57}C_{57}.(9 \times 2)^{57} \right\}$ have a common factor of $\left(3^4.2 = 162\right)$.

Hence they can be written as $(3^4.2)\, k_1$ and $(3^4.2)\, k_2$ respectively, where k_1 and k_2 are integers.

Therefore, $S = 1 + {}^{99}C_0.(3 \times 4) - 1 - {}^{57}C_1.(9 \times 2) + (162)(k_1 - k_2)$

$= (1188 - 1026) + \left\{ 162 \times (k_1 - k_2) \right\} = (162 \times \text{some integer})$

Hence the given number S is exactly divisible by 162.

4.2 Finding Remainder

Illustration 24: What is the remainder when 5^{2015} is divisible by 13. **(JEE MAIN)**

Sol: In this problem, we can obtain required remainder by reducing 5^{2015} into the form of $13\lambda + a$, where λ and a are integers.

$5^{2015} = 5.5^{2014} = 5.(25)^{1007}$

$$= 5\left(26 - 1\right)^{1007} = 5\left[{}^{1007}C_0 \left(26\right)^{1007} - {}^{1007}C_1 \left(26\right)^{1006} + \ldots\ldots + {}^{1007}C_{1006} \left(26\right)^1 - {}^{1007}C_{1007} \left(26\right)^0 \right]$$

$$= 5\left[{}^{1007}C_0 \left(26\right)^{1007} - {}^{1007}C_1 \left(26\right)^{1006} + \ldots\ldots + {}^{1007}C_{1006} \left(26\right)^1 - 1 \right]$$

$$= 5\left[{}^{1007}C_0 \left(26\right)^{1007} - {}^{1007}C_1 \left(26\right)^{1006} + \ldots\ldots + {}^{1007}C_{1006} \left(26\right)^1 - 13 \right] + 60$$

$$= 13\left(k\right) + 52 + 8 = 13 \times \text{(some integer)} + 8.$$

4.3 Finding Digits of a Number

Illustration 25: Find the last two digits of the number $(13)^{10}$. **(JEE MAIN)**

Sol: Write $(13)^{10}$ in the form of $\left(x - 1\right)^n$, such that x is a multiple of 10. Then using expansion formula we will get last two digits.

$$(13)^{10} = (169)^5 = \left(170 - 1\right)^5 = {}^5C_0 \left(170\right)^5 - {}^5C_1.\left(170\right)^4 + \ldots\ldots + {}^5C_4 \left(170\right)^1 - {}^5C_5 \left(170\right)^0$$

$$= {}^{5}C_0 (170)^5 - {}^{5}C_1 \cdot (170)^4 + \ldots + {}^{5}C_3 (170)^2 + 5 \times 170 - 1 = \text{A multiple of } 100 + 849$$

Therefore, the last two digits are 49

Illustration 26: Find the last three digits of 13^{256}. **(JEE MAIN)**

Sol: Similar to above problem..

We have $13^2 = 169 = 170 - 1$

Now, $13^2 = \left(13^2\right)^{128} = (170 - 1)^{128}$

$$= {}^{128}C_0 (170)^{128} - {}^{128}C_1 \cdot (170)^{127} + {}^{128}C_2 \cdot (170)^{126} - \ldots + {}^{128}C_{126} (170)^2 - {}^{128}C_{127} (170) + 1$$

$= 1000\, m + (128)\,(170)\,(10794) + 1 \quad$ (where m is a positive integer)

$= 1000\, m + 234877440 + 1 = 1000m + 234877441$

Thus, the last three digits of 13^{256} are 441.

4.4 Relation between Two Numbers

Illustration 27: Which number is smaller $(1.01)^{1000000}$ or 10,000 **(JEE MAIN)**

Sol: By reducing $(1.01)^{1000000}$ into the form of $\left(1 + 0.01\right)^n$ and solve it by using expansion formula we can obtain the value of $(1.01)^{1000000}$.

$(1.01)^{1000000} = (1 + 0.01)^{1000000}$

$= 1 + {}^{1000000}C_1 \left(0.01\right) + {}^{1000000}C_2 \left(0.01\right)^2 + {}^{1000000}C_3 \left(0.01\right)^3 + \ldots$

$= 1 + 1000000 \times (0.01) + \text{ some positive terms}$

$= 1 + 10000 + \text{ some positive terms}$

Hence $10{,}000 < (1.01)^{1000000}$.

5. MULTINOMIAL THEOREM

Using binomial theorem, we have

$$\left(x + a\right)^n = \sum_{r=0}^{n} {}^{n}C_r x^{n-r} a^r, \quad n \in N$$

$$= \sum_{r=0}^{n} \frac{n!}{(n-r)!\, r!} x^{n-r} a^r \qquad = \sum_{r+s=n} \frac{n!}{r!\, s!} x^s a^r, \quad \text{where } s = n - r$$

Let us now consider the expansion of $\left(x_1 + x_2 + x_3\right)^n$

$$\left(x_1 + x_2 + x_3\right)^n = \sum_{k=0}^{n} {}^{n}C_k x_1^{n-k} \left(x_2 + x_3\right)^k = \sum_{k=0}^{n} \frac{n!}{(n-k)!\, k!} x_1^{n-k} \left(\sum_{p=0}^{k} \frac{k!}{(k-p)!\, p!} x_2^{k-p} x_3^p \right)$$

$$= \sum_{k=0}^{n} \sum_{p=0}^{k} \frac{n!}{(n-k)!\, (k-p)!\, p!} x_1^{n-k} x_2^{k-p} x_3^p \qquad = \sum_{p+q+r=n} \frac{n!}{r!\, q!\, p!} x_1^r x_2^q x_3^p \quad \text{where, } k - p = q,\ n - k = r.$$

And so on, if we want to generalize for n terms, we get

$$\left(x_1 + x_2 + ... + x_k\right)^n = \sum_{r_1+r_2+...+r_k=n} \frac{n!}{r_1!\,r_2!.....r_k!} x_1^{r_1} x_2^{r_2} .... x_k^{r_k}$$

Therefore, general term in the expansion of $\left(x_1 + x_2 + ... + x_k\right)^n$ is $\dfrac{n!}{r_1!\,r_2!\,r_3!.....r_k!} x_1^{r_1} x_2^{r_2} x_3^{r_3} .... x_k^{r_k}$

The number of terms is equal to the number of non-negative integral solution of the equation $r_1 + r_2 +..........+ r_k = n$, because each solution of this equation gives a term in the above expansion. The number of such solutions is $^{n+k-1}C_{k-1}$.

Number of terms for the following expansions

(a) $\left(x+y+z\right)^n = \sum_{r+s+t=n} \dfrac{n!}{r!s!t!} x^r y^s z^t$ The above expansion has $^{n+3-1}C_{3-1} = \,^{n+2}C_2$ terms.

(b) $\left(x+y+z+u\right)^n = \sum_{p+q+r+s=n} \dfrac{n!}{p!q!r!s!} x^p y^q z^r u^s$. There are $^{n+4-1}C_{4-1} = \,^{n+3}C_3$ term in the above.

The greatest coefficient in the expansion of $\left(x_1 + x_2 + ... + x_m^n\right)$ is $\dfrac{n!}{\left(q!\right)^{m-r}\left[(q+1)!\right]^r}$, where q and r are the quotient and remainder respectively when n is divided by m.

6. BINOMIAL THEOREM FOR ANY INDEX

Let n be a rational number and x be a real number such that $|x| < 1$, then

$$\left(1+x\right)^n = 1 + nx + \frac{n(n-1)}{2!}x^2 + ... + \frac{n(n-1)(n-2)...(n-r+1)}{r!}x^r + ... + \text{terms upto } \infty$$

The general term in the expansion of $(1 + x)^n$ is $\dfrac{n(n-1)(n-2)..........(n-r+1)}{r!} x^r$ and is represented by T_{r+1}.

The above result is also true for complex x, n.

Illustration 28: If x is very large and n is a negative integer or a proper fraction, then an approximate value of $\left(\dfrac{1+x}{x}\right)^n$ is equal to__________ **(JEE MAIN)**

Sol: Since x is very large therefore $\dfrac{1}{x}$ will be very small. Neglect the terms containing three and higher powers of $\dfrac{1}{x}$ in the expansion to obtain the approximate value of $\left(\dfrac{1+x}{x}\right)^n$.

$$\left(1+\frac{1}{x}\right)^n = 1+\frac{n}{x}+\frac{n(n-1)}{1.2}\left(\frac{1}{x}\right)^2 +\ldots\ldots$$ Since x is very large, we can ignore terms after the 2nd term.

Illustration 29: If $\dfrac{(1-3x)^{1/2}+(1-x)^{5/3}}{\sqrt{4-x}}$ is approximately equal to a + bx for small values of x, then (a, b) is equals to. **(JEE MAIN)**

Sol: Calculate the value of $\dfrac{(1-3x)^{1/2}+(1-x)^{5/3}}{\sqrt{4-x}}$ and equate it to a + bx.

Using the binomial expansion for any rational index, we have

$$\frac{(1-3x)^{1/2}+(1-x)^{5/3}}{2\left[1-\dfrac{x}{4}\right]^{1/2}} = \frac{\left[1+\dfrac{1}{2}(-3x)+\dfrac{1}{2}\left(-\dfrac{1}{2}\right)\dfrac{1}{2}(-3x)^2 +\ldots\ldots\right]+\left[1+\dfrac{5}{3}(-x)+\dfrac{5}{3}\dfrac{2}{3}\dfrac{1}{2}(-x)^2 +\ldots\ldots\right]}{2\left[1+\dfrac{1}{2}\left(-\dfrac{x}{4}\right)+\dfrac{1}{2}\left(-\dfrac{1}{2}\right)\dfrac{1}{2}\left(-\dfrac{x}{4}\right)^2 +\ldots\ldots\right]}$$

$$= \frac{\left[1-\dfrac{19}{12}x+\dfrac{53}{144}x^2 -\ldots\ldots\right]}{\left[1-\dfrac{x}{8}-\dfrac{1}{8}x^2 -\ldots\ldots\right]} = 1-\frac{35}{24}x+\ldots\ldots$$

Neglecting the higher powers of x, $\Rightarrow a+bx = 1-\dfrac{35}{24}x \Rightarrow a=1, b=-\dfrac{35}{24}$

Illustration 30: Find the coefficient of $a^3b^2c^4d$ in the expansion of $(a-b-c+d)^{10}$ **(JEE ADVANCED)**

Sol: Expand $(a-b-c+d)^{10}$ using multinomial theorem and by using coefficient property we can obtain the required result.

Using multinomial theorem, we have

$$(a-b-c+d)^{10} = \sum_{r_1+r_2+r_3+r_4=10} \frac{(10)!}{r_1!\,r_2!\,r_3!\,r_4!}(a)^{r_1}(-b)^{r_2}(-c)^{r_3}(d)^{r_4}$$

We want to get coefficient of $a^3b^2c^4$, this implies that $r_1 = 3, r_2 = 2, r_3 = 4, r_4 = 1$

$\therefore$ Coefficient of $a^3b^2c^4d$ is $\dfrac{(10)!}{3!2!4!}(-1)^2(-1)^4 = 12600$

Illustration 31: In the expansion of $\left(1+x+\dfrac{5}{x}\right)^{11}$ find the term independent of x. **(JEE ADVANCED)**

Sol: By expanding $\left(1+x+\dfrac{5}{x}\right)^{11}$ using multinomial theorem and obtaining the coefficient of x^0 we will get the term independent of x.

$$\left(1+x+\frac{5}{x}\right)^{11} = \sum_{r_1+r_2+r_3=11} \frac{(11)!}{r_1!\,r_2!\,r_3!}(1)^{r_1}(x)^{r_2}\left(\frac{5}{x}\right)^{r_3}$$

The exponent 11 is to be divided in such a way that we get x^0. Therefore, possible set of values of (r_1, r_2, r_3) are (11, 0, 0), (9, 1, 1), (7, 2, 2) (5, 3, 3), (3, 4, 4), (1, 5, 5) Hence the required term is

$$\frac{(11)!}{(11)!}\left(5^0\right)+\frac{(11)!}{9!1!1!}5^1+\frac{(11)!}{7!2!2!}5^2+\frac{(11)!}{5!3!3!}5^3+\frac{(11)!}{3!4!4!}5^4+\frac{(11)!}{1!5!5!}5^5$$

$$=1+\frac{(11)!}{9!2!}\cdot\frac{2!}{1!1!}5^1+\frac{(11)!}{7!4!}\cdot\frac{4!}{2!2!}5^2+\frac{(11)!}{5!6!}\cdot\frac{6!}{3!3!}5^3+\frac{(11)!}{3!8!}\cdot\frac{8!}{4!4!}5^4+\frac{(11)!}{1!10!}\cdot\frac{(10)!}{5!5!}5^5$$

$$=1+{}^{11}C_2\times{}^2C_1\times5^1+{}^{11}C_4\times{}^4C_2\times5^2+{}^{11}C_6\times{}^6C_3\times5^3+{}^{11}C_8\times{}^8C_4\times5^4+{}^{11}C_{10}\times{}^{10}C_5\times5^5$$

$$=1+\sum_{r=1}^{5}{}^{11}C_{2r}\cdot{}^{2r}C_r\times5^r$$

PROBLEM-SOLVING TACTICS

Summation of series involving binomial coefficients

For $(1+x)^n={}^nC_0+{}^nC_1x+{}^nC_2x^2+\ldots+{}^nC_nx^n$, the binomial coefficients are ${}^nC_0,{}^nC_1,{}^nC_2,\ldots,{}^nC_n$. A number of series may be formed with these coefficients figuring in the terms of a series.

Some standard series of the binomial coefficients are as follows:

(a) By putting x = 1, we get $\qquad {}^nC_0+{}^nC_1+{}^nC_2+\ldots+{}^nC_n=2^n$ $\qquad$ ...(i)

(b) By putting x =-1, we get $\qquad {}^nC_0-{}^nC_1+{}^nC_2-\ldots+(-1)^n\cdot{}^nC_n=0$ $\qquad$ ...(ii)

(c) On adding (i) and (ii), we get $\qquad {}^nC_0+{}^nC_2+{}^nC_4+\ldots=2^{n-1}$ $\qquad$ ...(iii)

(d) On subtracting (ii) from (i), we get $\quad {}^nC_1+{}^nC_3+{}^nC_5+\ldots=2^{n-1}$ $\qquad$ ...(iv)

(e) $\quad {}^{2n}C_0+{}^{2n}C_1+{}^{2n}C_2+\ldots+{}^{2n}C_{n-1}+{}^{2n}C_n=2^{2n-1}$

Proof: From the expansion of $(1+x)^{2n}$, we get $\;{}^{2n}C_0+{}^{2n}C_1+{}^{2n}C_2+\ldots+{}^{2n}C_{2n-1}+{}^{2n}C_{2n}=2^{2n}$

$$\Rightarrow 2\left({}^{2n}C_0+{}^{2n}C_1+{}^{2n}C_2+\ldots+{}^{2n}C_{n-1}\right)+{}^{2n}C_n=2^{2n}\;[\because\;{}^{2n}C_0={}^{2n}C_{2n},\;{}^{2n}C_1={}^{2n}C_{2n-1}\;\text{and so on. }]$$

(f) $\quad {}^{2n+1}C_0+{}^{2n+1}C_1+{}^{2n+1}C_2+\ldots+{}^{2n+1}C_n=2^{2n}$

Proof: (as above)

(g) Sum of the first half of ${}^nC_0+{}^nC_1+\ldots+{}^nC_n=$ Sum of the last half of ${}^nC_0+{}^nC_1+\ldots+{}^nC_n=2^{n-1}$

(h) **Bino-geometric series:** ${}^nC_0+{}^nC_1x+{}^nC_2x^2+\ldots+{}^nC_nx^n=(1+x)^n$

(i) **Bino-arithmetic series:** $a{}^nC_0+(a+d){}^nC_1+(a+2d){}^nC_2+\ldots+(a+nd){}^nC_n$

Consider an AP - a, (a+d), (a+2d), ... , (a+nd)

Sequence of Binomial Co-efficient - ${}^nC_0,{}^nC_1,{}^nC_2,\ldots,{}^nC_n$

A **bino-arithmetic** series is nothing but the sum of the products of corresponding terms of the sequences. It can be added in two ways.

(i) By elimination of r in the multiplier of binomial coefficient from the $(r+1)^{th}$ term of the series

$$\left(\text{By using } r\cdot{}^nC_r=n{}^{n-1}C_{r-1}\right)$$

(ii) By differentiating the expansion of $x^d\left(1+x^d\right)^n$.

(j) **Bino-harmonic series:** $\dfrac{{}^nC_0}{a} + \dfrac{{}^nC_1}{a+d} + \dfrac{{}^nC_2}{a+2d} + \dots\dots + \dfrac{{}^nC_n}{a+nd}$

Consider an HP - $\dfrac{1}{a}, \dfrac{1}{a+d}, \dfrac{1}{a+2d}, \dots, \dfrac{1}{a+nd}$

Sequence of Binomial Co-efficient - ${}^nC_0, {}^nC_1, {}^nC_2, \dots\dots, {}^nC_n$

It is obtained by the sum of the products of corresponding terms of the sequences. Such series are calculated in two ways :

(i) By elimination of r in the multiplier of binomial coefficient from the $(r + 1)^{th}$ term of the series

$$\left(\text{By using} \dfrac{1}{r+1}\, {}^nC_r = \dfrac{1}{n+1}\, {}^{n+1}C_{r+1} \right)$$

(ii) By integrating suitable expansion.

For explanation see illustration 2

(k) **Bino-binomial series:** ${}^nC_0.{}^nC_r + {}^nC_1.{}^nC_{r+1} + {}^nC_2.{}^nC_{r+2} + \dots\dots + {}^nC_{n-r}.{}^nC_r$

or, ${}^mC_0.{}^nC_r + {}^mC_1.{}^nC_{r-1} + {}^mC_2.{}^nC_{r-2} + \dots\dots + {}^mC_r.{}^nC_0$

As the name suggests such series are obtained by multiplying two binomial expansion, one involving the first factors as coefficient and the other involving the second factors as coefficient. They can be calculated by equating coefficients of a suitable power on both sides.

For explanation see illustration 4

FORMULAE SHEET

Binomial theorem for any positive integral index:

$$(x+a)^n = {}^nC_0x^n + {}^nC_1x^{n-1}a + {}^nC_2x^{n-2}a^2 + \dots\dots + {}^nC_rx^{n-r}a^r + \dots\dots + {}^nC_na^n = \sum_{r=0}^{n} {}^nC_rx^{n-r}a^r$$

(a) General term $- T_{r+1} = {}^nC_rx^{n-r}a^r$ is the $(r + 1)^{th}$ term from beginning.

(b) $(m + 1)^{th}$ term from the end $= (n - m + 1)^{th}$ from beginning $= T_{n-m+1}$

(c) Middle term

(i) If n is even then middle term $= \left(\dfrac{n}{2}+1\right)^{th}$ term

(ii) If n is odd then middle term $= \left(\dfrac{n+1}{2}\right)^{th}$ and $\left(\dfrac{n+3}{2}\right)^{th}$

Binomial coefficient of middle term is the greatest binomial coefficient.

To determine a particular term in the given expansion:

Let the given expansion be $\left(x^\alpha \pm \dfrac{1}{x^\beta}\right)^{th}$, if x^n occurs in T_{r+1} $(r + 1)^{th\ term}$ then r is given by $n\alpha - r(\alpha+\beta) = m$ and for $x^0, n\alpha - r(\alpha+\beta) = 0$

Properties of Binomial coefficients:

For the sake of convenience the coefficients ${}^nC_0, {}^nC_1, {}^nC_2 \ldots {}^nC_r \ldots {}^nC_n$ are usually denoted by $C_0, C_1, \ldots C_r \ldots C_n$ respectively.

$$C_0 + C_1 + C_2 + \ldots + C_n = 2^n$$

$$C_0 - C_1 + C_2 - C_3 + \ldots + C_n = 0$$

$$C_0 + C_2 + C_4 + \ldots = C_1 + C_3 + C_5 + \ldots = 2^{n-1}$$

$${}^nC_r = \frac{n}{r}{}^{n-1}C_{r-1} = \frac{n}{r} \cdot \frac{n-1}{r-1}{}^{n-2}C_{r-2} \text{ and so on}\ldots$$

$${}^{2n}C_{n+r} = \frac{2n!}{(n-r)!(n+r)!}$$

$${}^nC_r + {}^nC_{r-1} = {}^{n+1}C_r$$

$$C_1 + 2C_2 + 3C_3 + \ldots + {}^nC_n = n.2^{n-1}$$

$$C_1 - 2C_2 + 3C_3 \ldots = 0$$

$$C_0 + 2C_1 + 3C_2 + \ldots + (n+1)C_n = (n+2)2^{n-1}$$

$$C_0^2 + C_1^2 + C_2^2 + \ldots + C_n^2 = \frac{(2n)!}{(n!)^2} = {}^{2n}C_n$$

$$C_0^2 - C_1^2 + C_2^2 - C_3^2 + \ldots = \begin{cases} 0, & \text{if } n \text{ is odd} \\ (-1)^{n/2}\,{}^nC_{n/2}, & \text{if } n \text{ is even} \end{cases}$$

Note: $\quad {}^{2n+1}C_0 + {}^{2n+1}C_1 + \ldots + {}^{2n+1}C_n = {}^{2n+1}C_{n+1} + {}^{2n+1}C_{n+2} + \ldots + {}^{2n+1}C_{2n+1} = 2^{2n}$

$$C_0 + \frac{C_1}{2} + \frac{C_2}{3} + \ldots + \frac{C_n}{n+1} = \frac{2^{n+1}-1}{n+1}; \quad C_0 - \frac{C_1}{2} + \frac{C_2}{3} - \frac{C_3}{4} \ldots + \frac{(-1)^n C_n}{n+1} = \frac{1}{n+1}$$

(a) Greatest term:

(i)   If $\dfrac{(n+1)a}{x+a} \in Z$ (integer) then the expansion has two greatest terms. These are k^{th} and $(k+1)^{th}$ where x and a are +ve real numbers.

(ii)   If $\dfrac{(n+1)a}{x+a} \notin Z$ then the expansion has only one greatest term. This is $(K+1)^{th}$ term $k = \left[\dfrac{(n+1)a}{x+a}\right]$ denotes greatest integer less than or equal to x}

(b) Multinomial theorem:

Generalized $\left(x_1 + x_2 + \ldots + x_k\right)^n = \displaystyle\sum_{r_1 + r_2 + \ldots r_k = n} \frac{n!}{r_1!\,r_2!\ldots r_k!} x_1^{r_1} x_2^{r_2} \ldots x_k^{r_k}$

(c) Total no. of terms in the expansion $\left(x_1 + x_2 + \ldots x_n\right)^m$ is ${}^{m+n-1}C_{n-1}$

JEE Main/Boards

Example 1: Find the coefficient of $\dfrac{1}{y^2}$ in $\left(\dfrac{c^3}{y^2}+y\right)^{10}$

Sol: By using formula of finding general term we can easily get coefficient of $\dfrac{1}{y^2}$.

In the binomial expansion, $(r+1)^{th}$ term is

$$T_{r+1} = {}^nC_r\,(y)^r\left(\dfrac{c^3}{y^2}\right)^{n-r} : n=10$$

$$\Rightarrow T_{r+1} = {}^{10}C_r\,(y)^r\,(c^3)^{10-r}\left(\dfrac{1}{y^2}\right)^{10-r}$$

$$= {}^{10}C_r\,c^{30-3r}\,y^{3r-20} \qquad\qquad \text{...(i)}$$

$$\therefore\ 3r - 20 = -2; r = 6$$

$\therefore$ 7th term will contain y^{-2} and from (i) the coefficient of y^{-2} is $= 210\,c^{12}$

Example 2: Use Binomial theorem to find the value of $(10.1)^5$.

Sol: After reducing $(10.1)^5$ into the form of $(10 + 0.1)^n$ we can use binomial expansion to get required result.

$(10.1)^5 = (10 + 0.1)^5$

$$= (10)^5 + {}^5C_1(10)^4(0.1) + {}^5C_2(10)^3(0.1)^2$$

$$+ {}^5C_3(10)^2(0.1)^3 + {}^5C_4 10(0.1)^4 + (0.1)^5$$

$$= (10)^5 + 5(10^3) + 10(10)^3(0.01) + 10(10)^2$$

$$(0.001) + 5(10)(0.0001) + (0.00001)$$

$$= 100000 + 5000 + 100 + 1 + 0.005 + 0.00001$$

$$= 105101.00501$$

Example 3: Find the middle term(s) in the expansion of $\left(2x^2 - \dfrac{1}{x}\right)^7$.

Sol: Since $n = 7$ is a odd number. Therefore, find the $\dfrac{n+1}{2}^{th}$ and $\dfrac{n+3}{2}^{th}$ term.

The total number of terms in the expansion are 8.

Therefore $\dfrac{7+1}{2}^{th}$ and $\dfrac{7+3}{2}^{th}$ i.e. 4th and 5th terms are the two middle terms. 4th term $= {}^7C_3\left(2x^2\right)^{7-3}\left(-\dfrac{1}{x}\right)^3$

$$= -\dfrac{7!}{3!\,4!}16x^{8-3} = -560x^5$$

and 5th term $= {}^7C_4\left(2x^2\right)^{7-4}\left(\dfrac{-1}{x}\right)^4 = 280x^2$

Hence the two middle terms are $-560x^5$ and $280x^2$.

Example 4: The coefficient of $(r-1)^{th}$, r^{th} and $(r+1)^{th}$ term in the expansion of $(x+1)^n$ are in the ratio 1:3:5. Find n and r.

Sol: In this problem, by using the formula of general term we will get the equation of given terms and by taking ratios of these terms we can get the value of n and r.

Coefficient of $(r-1)^{th}$ term is ${}^nC_{r-2}$

Coefficient of rth term is ${}^nC_{r-1}$

Coefficient of $(r+1)^{th}$ term is ${}^nC_r$

Coefficient are in ratio of $1:3:5$

$$\dfrac{{}^nC_{r-2}}{{}^nC_{r-1}} = \dfrac{1}{3} \quad\text{and}\quad \dfrac{{}^nC_{r-1}}{{}^nC_r} = \dfrac{3}{5}$$

$$\text{or}\ \dfrac{r-1}{n-r+2} = \dfrac{1}{3} \quad\text{and}\quad \dfrac{r}{n-r+1} = \dfrac{3}{5}$$

i.e. $n - 4r + 5 = 0$ and $3n - 8r + 3 = 0$

Solving both we get $n = 7$ & $r = 3$

Example 5: Find the remainder when $27^{10} + 7^{51}$ is divided by 10

Sol: We can obtain the remainder by reducing $27^{10} + 7^{51}$ into the form of $10\lambda + a$, where λ is any integer and a is an integer less than 10.

We have $27^{10} = 3^{30} = 9^{15} = (10-1)^{15}$

$7^{51} = 7.7^{50} = 7.(49)^{25} = 7\,(50-1)^{25}$

$27^{10} = 10m_1 \qquad\qquad\qquad\qquad \text{...(i)}$

$7^{51} = 7(50-1)^{25} = 10m_2 - 7 \qquad\quad \text{...(ii)}$

Adding (i) and (ii)

$27^{10} + 7^{51} = (10m_1 - 1) + (10m_2 - 7) = 10m_1 + 10m_2 - 8$

$$= 10m_1 + 10m_2 - 10 + 2$$

Thus, the remainder is 2 when $27^{10} + 7^{51}$ is divided by 10.

Example 6: If A be the sum of odd numbered terms and B the sum of even numbered terms in the expansion of $(x + a)^n$ prove that $A^2 - B^2 = (x^2 - a^2)^n$

Sol: Do it yourself.

$$\left(x+a\right)^n = {}^nC_0 x^n + {}^nC_1 x^{n-1}a$$

$$x + {}^nC_2 x^{n-2}a^2 + ... + {}^nC_n a^n = A + B$$

When $A = {}^nC_0 x^n + {}^nC_2 x^{n-2}a^2 + {}^nC_4 x^{n-4}a^4 + ...$

$$B = {}^nC_1 x^{n-1}a + {}^nC_3 x^{n-3}a^3 + {}^nC_5 x^{n-5}a^5 + ...$$

$$\therefore \left(x-a\right)^n = A - B, \ A^2 - B^2 = \left(A-B\right)\left(A+B\right)$$

$$= \left(x-a\right)^n \left(x+a\right)^n = \left(x^2 - a^2\right)^n$$

Example 7: If C_r denotes the binomial coefficient ${}^nC_r$, prove that :

$$C_0^2 + C_1^2 + ... + C_n^2 = \frac{2n!}{\left(n!\right)^2}.$$

Sol: Multiply the expansion of $(x+1)^n$ and $(1+x)^n$ and compare the coefficients of x^n on both sides.

We know that $\left(1+x\right)^n = {}^nC_0 + {}^nC_1 x$

$$+ {}^nC_2 x^2 + ... + {}^nC_{n-1} x^{n-1} + {}^nC_n x^n$$

$$\left(x+1\right)^n = {}^nC_0 x^n + {}^nC_1 x^{n-1}$$

$$+ {}^nC_2 x^{n-2} + ... + {}^nC_{n-1} x + {}^nC_n$$

Multiplying these equations side by side, we get

$$\left(1+x\right)^n \left(x+1\right)^n = \left(C_0 + C_1 x + C_2 x^2 + ... + C_{n-1} x^{n-1} + C_n x^n\right)$$

$$\times \left(C_0 x^n + C_1 x^{n-1} + C_2 x^{n-2} + ... + C_{n-1} x + {}^nC_n\right)$$

Coefficient of x^n on R.H.S. is equal to

$$C_0^2 + C_1^2 + C_2^2 + ... + C_{n-1}^2 + C_n^2$$

Coefficient of x^n in L.H.S. is ${}^{2n}C_n = \dfrac{2n!}{n!n!}.$

This proves the required identity.

Example 8: If $\left(1 + x + x^2\right)^n = a_0 + a_1 x + a_2 x^2 + ... + a_{2n} x^{2n}$ show that

(i) $a_0 + a_1 + a_2 + ... + a_{2n} = 3^n$

(ii) $a_0 - a_1 + a_2 - a_3 + ... + a_{2n} = 1$

(iii) $a_0 + a_3 + a_6 + ... = 3^{n-1}$

Sol: By putting x = 1, -1, and ω, ω^2

Respectively in the expansion of $(1 + x + x^2)^n$ we will get the result.

Given $(1 + x + x^2)^n$

$$= a_0 + a_1 x + a_2 x^2 + ... + a_{2n} x^{2n} \qquad ....(i)$$

(i) Putting x = 1, we get

$$3^n = a_0 + a_1 + a_2 + ... + a_{2n} \quad .....(A)$$

(ii) Putting x = −1 in (i), we get

$$1 = a_0 - a_1 + a_2 - a_3 ..... + a_{2n}$$

(iii) Putting $x = \omega, \omega^2$ successively in (i), we get
$$0 = a_0 + a_1\omega + a_2\omega^2 + a_3$$

$$+ a_4\omega + a_5\omega^2 + ... + a_{2n}\omega^{2n} \ .....(B) \ \ 0 = a_0 + a_1\omega^2 + a_2\omega + a_3$$

$$+ a_4\omega^2 + a_5\omega + a_6 + ... + a_{2n}\omega^{4n} \qquad ...(C)$$

Adding (A), (B) and (C) we have

$$3^n = 3\left(a_0 + a_3 + a_6 + ...\right)$$

$$\therefore a_0 + a_3 + a_6 + ... = 3^{n-1}$$

Example 9: If $\left(1+x\right)^n = C_0 + C_1 x +$

$$C_2 x^2 + C_3 x^3 + ... + C_n x^n$$

then prove that $C_1^2 + 2C_2^2 + 3C_3^2 + ... + nC_n^2 = \dfrac{(2n-1)!}{\left((n-1)!\right)^2}$

Sol: Expanding $\left(1+x\right)^n$ and $\left(x+1\right)^n$ and multiplying these two expansion and comparing the coefficient of x^{n-1} we will prove above equation.

Given $\left(1+x\right)^n = C_0 + C_1 x +$

$$C_2 x^2 + C_3 x^3 + ... + C_n x^n$$

Differentiating both sides w. r. t. to x, we get

$$n\left(1+x\right)^{n-1} = 0 + C_1 + 2C_2 x + 3C_3 x^2 + ... + nC_n x^{n-1}$$

$$\Rightarrow n\left(1+x\right)^{n-1} = C_1 + 2C_2 x$$

$$+ 3C_2 x^2 + ... + nC_n x^{n-1} \qquad ....(i)$$

and $\left(x+1\right)^n = C_0 x^n + C_1 x^{n-1} + C_2 x^{n-2}$

$$+ C_3 x^{n-3} + C_4 x^{n-4} + ... + C_n \qquad ....(ii)$$

Multiplying (i) and (ii), we get

$$n(1+x)^{2n-1} = \left(C_1 + 2C_2 x + 3C_3 x^2 + \ldots + nC_n x^{n-1}\right)$$

$$\times \left(C_0 x^n + C_1 x^{n-1} + C_2 x^{n-2} + C_3 x^{n-3} + \ldots + C_n\right) \quad \ldots\text{(iii)}$$

Now, coefficient of x^{n-1} on R.H.S.

$$= C_1^2 + 2C_2^2 + 3C_3^2 + \ldots + nC_n^2 \text{ and coefficient of } x^{n-1} \text{ on}$$

L.H.S. $= n.^{2n-1}C_{n-1}$

$$= n\frac{(2n-1)!}{(n-1)!n!} = \frac{(2n-1)!}{(n-1)!(n-1)!} = \frac{(2n-1)!}{\left[\left((n-1)!\right)^2\right]}$$

But (iii) is an identity, therefore the coefficient of x^{n-1} in R.H.S. = coefficient of x^{n-1} in R.H.S.

$$\Rightarrow C_1^2 + 2C_2^2 + 3C_3^2 + \ldots + nC_n^2 = \frac{(2n-1)!}{\left((n-1)!\right)^2}$$

Example 10: Find the numerically greatest term in the expansion of $(3 - 5x)^{15}$ when $x = 1/5$.

Sol: Follow the algorithm for the greatest term.

Using standard notations w.r.t. $(x + a)^n$

$$\frac{n+1}{1+\left|\dfrac{x}{a}\right|} = \frac{16}{1+\left|\dfrac{3}{(-1)}\right|} = 4$$

T_4 and T_5 are numerically equal to each other and are greater than any other term.

Example 11: If $(1 + x + x^2) = a_0 + a_1 x + a_2 x^2 + a_3 x^3 \ldots + a_{2n} x^{2n}$

Then show that

$$a_0 + a_3 + a_6 + \ldots = a_1 + a_4 + a_7 + \ldots = 3^{n-1}.$$

Sol: By Putting $x = 1$ ω, ω^2 respectively in the given equation and adding these values we can prove it.

$$3^n = a_0 + a_1 + a_2 + a_3 + a_4 + \ldots \quad \ldots\text{(i)}$$

$$0 = a_0 + a_1\omega + a_2\omega^2 + a_3\omega^3 + a_4\omega^4 + \ldots \quad \ldots\text{(ii)}$$

Because $1 + \omega + \omega^2 = 0$

$$0 = a_0 + a_1\omega^2 + a_2\omega^4 + a_3\omega^6 + a_4\omega^8 + \ldots \quad \ldots\text{(iii)}$$

Adding these

$$3^n = 3(a_0) + a_1\left(1 + \omega + \omega^2\right) + a_2$$

$$\left(1 + \omega^2 + \omega^4\right) + a_3\left(1 + \omega^3 + \omega^6\right)$$

$$+\ldots = 3\left(a_0 + a_3 + a_6 + \ldots\right)$$

$$\therefore a_0 + a_3 + a_6 + \ldots = 3^{n-1}$$

From (i) + (ii) $\times \omega^2$ (iii) $\times \omega$, + we get,

$$3^n + 0 \times \omega^2 + 0 \times \omega$$

$$= a_0\left(1 + \omega^2 + \omega\right) + a_1\left(1 + \omega^3 + \omega^3\right)$$

$$+a_2\left(1 + \omega^4 + \omega^5\right) + a_3\left(1 + \omega^5 + \omega^7\right)$$

$$+a_4\left(1 + \omega^6 + \omega^9\right) + \ldots$$

$$\therefore 3^n = 3\left(a_1 + a_4 + a_7 + \ldots\right)$$

Because coefficient of each is

$$1 + \omega + \omega^2 = 0, \text{ using } \omega^3 = 1$$

$$\therefore a_1 + a_4 + a_7 + \ldots = 3^{n-1}$$

Again, from (i) + (ii) ω + (iii) $\times \omega^3$, we get

$$= 3^n = a_0\left(1 + \omega + \omega^2\right) + a_1\left(1 + \omega^2 + \omega^4\right)$$

$$+a_2\left(1 + \omega^3 + \omega^3\right) + \ldots = 3\left(a_2 + a_5 + a_8 + \ldots\right)$$

Example 12: Sum the series

$$C_0 + \frac{C_1}{2} + \frac{C_2}{3} + \ldots + \frac{C_n}{n+1}$$

Sol: Expanding $(1 + x)^n$ integrating it from 0 to 1 or by using summation method we will get result.

$$\text{Sum} = \sum_{r=1}^{n+1} \frac{{}^{n+1}C_{r-1}}{r} = \sum_{r=1}^{n+1} \frac{1}{n+1}{}^{n+1}C_r$$

$$= \frac{1}{n+1}\left({}^{n+1}C_0 + {}^{n+1}C_1 + \ldots + {}^{n+1}C_{r+1} - {}^{n+1}C_0\right)$$

$$= \frac{1}{n+1}\left(2^{n+1} - 1\right)$$

Alternative method

$$(1 + x)^n = C_0 + C_1 + C_2 x^2 + \ldots + C_n x^n$$

Integrating both sides w.r.t. x from 0 to 1

$$\int_0^1 (1 + x)^n \, dx = \int_0^1 \left(C_0 + C_1 x + \ldots + C_n x^n\right) dx$$

$$\frac{2^{n+1} - 1}{n+1} = C_0 + \frac{C_1}{2} + \frac{C_2}{3} + \ldots + \frac{C_n}{n+1}$$

Example 13: Find the last three digits of 27^{26}.

Sol: By reducing 27^{26} into the form $(730-1)^n$ and using simple binomial expansion we will get required digits.

We have $27^2 = 729$.

Now $27^{26} = (729)^{13} = (730-1)^{13}$

$$= {}^{13}C_0 (730)^{13} - {}^{13}C_1 (730)^{12} + {}^{13}C_2 (730)^{11}$$

$$-..... - {}^{13}C_{10} (730)^3 - {}^{13}C_{12} (730)^2$$

$$- {}^{13}C_{12} (730) + 1$$

$$= 1000m + \frac{(13)(12)}{2}(14)^2 - (13)(730) + 1$$

Where m is a positive integer

$= 1000\,m + 15288 - 9490 + 1$

$= 1000\,m + 5799$

Thus, the last three digits of 17^{256} are 799.

JEE Advanced/Boards

Example 1: Find the coefficient of x^4 in the expansion of

(i) $(1 + x + x^2 + x^3)^{11}$

(ii) $(2 - x + 3x^2)^6$

Sol: By expanding given equation using expansion formula we can get the coefficient x^4 .

(i) $1 + x + x^2 + x^3 = (1 + x) + x^2(1 + x) = (1 + x)(1 + x^2)$

$$\therefore \left(1 + x + x^2 + x^3\right)^{11} = (1+x)^{11}\left(1+x^2\right)^{11}$$

$$= \left(1 + {}^{11}C_1 x + {}^{11}C_2 x^2 + {}^{11}C_3 x^3 + {}^{11}C_4 x^4 + ...\right)$$

$$\left(1 + {}^{11}C_1 x^2 + {}^{11}C_2 x^4 + ...\right)$$

To find term in x^4 from the product of two brackets on the right-hand-side, consider the following products terms as

$$1 \times {}^{11}C_2 x^4 + {}^{11}C_2 x^2 \times {}^{11}C_1 x^2 + {}^{11}C_4 x^4$$

$$= \left[{}^{11}C_2 + {}^{11}C_2 \times {}^{11}C_1 + {}^{11}C_4\right] x^4$$

$$\left[55 + 605 + 330\right] x^4 = 990 x^4$$

$\therefore$ The coefficient of x^4 is 990.

(ii) $\left(2 - x + 3x^2\right)^6 = \left[2 - x(1 - 3x)\right]^6$

$$= [\,2^6 - {}^6C_1 \times 2^5 \times x(1 - 3x) + {}^6C_2 2^4$$

$$\times x^2 (1 - 3x)^2 - {}^6C_3 2^3 \times x^3 (1 - 3x)^3$$

$$+ {}^6C_4 2^2 \times x^4 (1 - 3x)^4 - 2 \times {}^6C_5$$

$$\times x^5 (1 - 3x)^5 + {}^6C_6 \times x^6 (1 - 3x)^6\,]$$

The term in x^4 will come only from the three terms, viz.

(a) ${}^6C_2 \times 2^4 \times x^2 (1 - 3x)^2 = 15 \times 16x^2 (1 - 6x + 9x^2)$

$\therefore$ The term in x^4 is $(15)\,(16)\,(9x^4)$

(b) $-{}^6C_3 2^3 \times x^3 (1 - 3x)^3$

$$= -20 \times 8 \times x^3 \left[1 - 9x + 27x^2 - 27x^3\right]$$

$\therefore$ The term in x^4 is $-20 \times (-9) \times (8)x^4$

(c) ${}^6C_4 2^2 x^4 (1 - 3x)^4 = 15 \times 4x^4 (1 - 4 \times 3x + ...)$

$\therefore$ The term in x^4 is $15 \times 4 \times x^4$

$\therefore$ The total term in x^4 is

$$\left[15 \times 16 \times 9 + 20 \times 8 \times 9 + 15 \times 4\right] \times x^4$$

$$= \left[2160 + 1440 + 60\right] x^4 = 3660 x^4$$

$\therefore$ The coefficient of x^4 is 3660.

Example 2: Show that $\displaystyle\sum_{r=0}^{n} r(n-r) C_r^2 = n^2 \cdot {}^{2n-2}C_n$

Sol: By expanding and differentiating $(1+x)^n$ and $(x+1)^n$ and then multiplying these expansion we can prove given equations by comparing coefficient of x^{n-2} on both side.

We have

$$(1 + x)^n = C_0 + C_1 x + C_2 x + ... + C_n x^n \qquad ....(i)$$

Differentiating both side w.r.t x, we get

$$n(1 + x)^{n-1} = C_1 + 2C_2 x + 3C_3 x^2 + ... + {}^nC_n x^n \qquad ...(ii)$$

(i) can also the be written as

$$(1 + x)^n = (x + 1)^n = C_0 x^n$$

$$+ C_1 x^{n-1} + C_2 x^{n-2} + ... + C_{n-1} x + C_n \qquad ....(iii)$$

Differentiating both sides w.r.t. x, we get

$$n(1 + x)^{n-1} = nC_0 x^{n-1} + (n-1)$$

$$C_1 x^{n-2} + (n-2)C_2 x^{n-3} + ... + C_{n-1} \qquad ....(iv)$$

Multiplying (ii) and (iv), we have

$$n^2(1+x)^{n-1}(x+1)^{n-1} = n^2(1+x)^{2n-2}$$

$$= [C_1 + 2C_2 + 3C_3x^2 + ... + {}^rC_nx^{n-1}]$$

$$x\ [nC_0x^{n-1} + (n-1)C_1x^{n-2} + (n-2)$$

$$C_2x^{n-3} + ... + C_{n-2}x + C_{n-1}] \qquad ....(v)$$

The coefficient of x^{n-2} on the LHS of (v) is

$$n^2.{}^{2n-2}C_{n-2} = n^2.{}^{2n-2}C_n$$

The coefficient of x^{n-2} on the RHS of (v) is

$$1.(n-1)C_1^2 + 2.(n-2)C_2^2 + ... + (n-1).1C_{n-1}^2$$

$$= \sum_{r=0}^{n-1} r(n-r)C_r^2 = \sum_{r=0}^{n} r(n-r)C_r^2$$

Hence, $\displaystyle\sum_{r=0}^{n} r(n-r)C_r^2 = n^2\left({}^{2n-2}C_n\right)$

Example 3: Prove that

(i) $\quad C_0 + \dfrac{C_1}{2} + \dfrac{C_2}{3} + ... + \dfrac{C_n}{n+1} = \dfrac{2^{n+1}-1}{n+1}$

(ii) $\quad 2.C_0 + 2^2.\dfrac{C_1}{2} + 2^3.\dfrac{C_2}{3} + ...$

$\qquad +2^{n+1}.\dfrac{C_n}{n+1} = \dfrac{3^{n+1}-1}{n+1}$

(iii) $\quad C_0 - \dfrac{1}{2}C_1 + \dfrac{1}{3}C_2 - \dfrac{1}{4}C_3 + ...$

$\qquad +(-1)^n\dfrac{C_n}{n+1} = \dfrac{1}{n+1}$

(iv) $\quad \dfrac{C_0}{1.2} + \dfrac{C_1}{2.3} + \dfrac{C_2}{3.4} + ...$

$\qquad +\dfrac{C_n}{(n+1).(n+2)} = \dfrac{2^{n+2}-n-3}{(n+1)(n+2)}$

(v) $\quad C_0 + \dfrac{C_2}{3} + \dfrac{C_4}{5} + ... = \dfrac{2^n}{n+1}$

Sol: Expand $(1+x)^n$ and integrate it within the limit 0 to 1, 0 to 2, -1 to 0 and -1 to 1 respectively to prove these equations

$$(1+x)^n = C_0 + C_1x + C_2x^2 + C_3x^3 + ... + C_nx^n \qquad ....(i)$$

(i) Integrating both sides of equation (i) within limits 0 to 1, we get

$$\int_0^1 (1+x)^n dx =$$

$$\int_0^1 (C_0 + C_1x + C_2x + C_2x^2 + C_3x^3 + ...C_nx^n)\ dx$$

$$\left(\dfrac{(1+x)^{n+1}}{n+1}\right)\Bigg|_0^1 = C_0x + C_1\dfrac{x^2}{2} +$$

$$C_2\dfrac{x^3}{3} + ... + C_n\dfrac{x^{n+1}}{n+1}\Bigg]_0^1$$

$$\dfrac{2^{n+1}-1}{n+1} = C_0 + \dfrac{C_1}{2} + \dfrac{C_2}{3} + ... + \dfrac{C_n}{n+1}$$

(ii) Integrating both sides of equation (i) within limits 0 to 2.

$$\int_0^2 (1+x)^n = \int_0^2 (C_0 + C_1x + C_2x^2 + C_3x^3 + ...C_nx^n)dx$$

or $\left(\dfrac{(1+x)^{n+1}}{n+1}\right)^2\Bigg|_0 = \left[C_0x + C_1\dfrac{x^2}{2} + C_2\dfrac{x^3}{3} + ... + C_n\dfrac{x^{n+1}}{n+1}\right]_0^2$

or $\dfrac{3^{n+1}-1}{n+1} = C_0.2 + 2^2. + \dfrac{C_1}{2} + 2^3.\dfrac{C_2}{3} + ... + 2^{n+1}.\dfrac{C_n}{n+1}$

(iii) Integrating both sides of equation (i) within limits -1 to 0,

$$\int_{-1}^0 (1+x)^n dx = \int_{-1}^0 (C_0 + C_1x + C_2x^2 + C_3x^3 + ...C_nx^n)dx$$

$$\left(\dfrac{(1+x)^{n+1}}{n+1}\right)^0\Bigg|_{-1} = C_0x + C_1\dfrac{x^2}{2} + C_2\dfrac{x^3}{3} + ... + C_n\dfrac{x^{n+1}}{n+1}\Big|_{-1}^0$$

$$\dfrac{1}{n+1} - 0 = 0 - \left[-C_0 + \dfrac{C_1}{2} - \dfrac{C_2}{3} + ... + (-1)^{n+1}\dfrac{C_n}{n+1}\right]$$

$$\dfrac{1}{n+1} = C_0 - \dfrac{C_1}{2} + \dfrac{C_2}{3} + ... + (-1)^n\dfrac{C_n}{n+1}$$

(iv) General term of L.H.S $= \dfrac{{}^nC_k}{(k+1)(k+2)}$

$$= \dfrac{{}^{n+1}C_{k+1}}{(n+1)(k+2)} = \left[\because \dfrac{{}^nC_r}{n} = \dfrac{{}^{n-1}C_{r-1}}{r}\right] = \dfrac{{}^{n+2}C_{k+2}}{(n+1)(n+2)}$$

$\therefore$ The sum of terms on L.H.S.

$$= \sum_{k=0}^{n} \dfrac{{}^{n+2}C_{k+2}}{(n+1)(n+2)} = \dfrac{1}{(n+1)(n+2)} . \sum_{k=0}^{n} {}^{n+2}C_{k+2}$$

$$= \frac{1}{(n+1)(n+2)}\left[2^{n+2} - {}^{n+2}C_0 - {}^{n+2}C_1\right]$$

$$= \frac{1}{(n+1)(n+2)}\left[2^{n+2} - 1 - (n+2)\right] = \frac{2^{n+2} - n - 3}{(n+1)(n+2)}$$

(v) Integrating both sides of equation (i) within limits −1 to 1, we get

$$\int_{-1}^{1} (1+x)^n \, dx =$$

$$\int_{-1}^{1} \left(C_0 + C_1 x + C_2 x^2 + C_3 x^3 + \dots\dots + C_n x^n\right) dx$$

$$\left(\frac{(1+x)^{n+1}}{n+1}\right)_{-1}^{1} = C_0 x + C_1 \frac{x^2}{2} + C_2 \frac{x^3}{3} + \dots\dots + C_n \frac{x^{n+1}}{n+1}\Big|_{-1}^{1}$$

$$\frac{2^{n+1} - 0}{n+1} = \left[C_0 + \frac{C_1}{2} + \frac{C_2}{3} + \dots\dots + \frac{C_n}{n+1}\right] - \left[-C_0 + \frac{C_1}{2} - \frac{C_2}{3} + \dots\dots\right]$$

$$\frac{2^{n+1}}{n+1} = 2\left[C_0 + \frac{C_2}{3} + \frac{C_4}{5} + \dots\dots\right]$$

$$\Rightarrow \frac{2^n}{n+1} = C_0 + \frac{C_2}{3} + \frac{C_4}{5} + \dots\dots$$

Example 4: Prove, by binomial expansion, that

(i) $\displaystyle\sum_{k=1}^{n} k^2 \cdot {}^nC_k = n(n+1)2^{n-2}$

(ii) $\displaystyle\prod_{k=1}^{n}(C_{k-1} + C_k) = \frac{C_0 C_1 \dots\dots C_{n-1}(n+1)^n}{n!}$

Sol: Expanding $(1+x)^n$ and differentiating it twice we will prove given equation (i) and by multiplying and dividing by $C_0 C_1 C_2 \dots\dots C_{n-1}$ in L.H.S. of equation (ii) we can prove it.

(i) Now $(1+x)^n = C_0 + C_1 x + C_2 x^2 + C_3 x^3 + \dots\dots + C_n x^n$

Differentiating twice w.r.t. x, we get

$$n(n-1)(1+x)^{n-2} = 2.C_2 + 3.2.C_3 x$$

$$+ 4.3 C_4 x^2 + \dots\dots + n(n-1)C_n x^{n-2}$$

Substituting x = 1, we get

$$n(n-1)2^{n-2} = \sum_{k=1}^{n} k(k-1)(C_k)$$

$$\therefore \sum_{k=1}^{n} \left(k^2\right)\left({}^nC_k\right) = n(n-1)2^{n-2} + n.2^{n-1}$$

$$\left[\because k.{}^nC_k = n.{}^{n-1}C_{k+1}\right]$$

$$= 2^{n-2}\left[n^2 - n + 2n\right]$$

$$\therefore \sum_{k=1}^{n} k^2 \, {}^nC_k = n(n+1)2^{n-2}$$

(ii) To prove $\left(C_0 + C_1\right)\left(C_1 + C_2\right)\left(C_2 + C_3\right)\dots\dots$

$$\left(C_{n-1} + C_n\right) = \frac{C_0 C_1 \dots\dots C_{n-1}(n+1)^n}{n!}$$

Multiply and divide L.H.S. by $C_0 C_1 C_2 \dots\dots C_{n-1}$; then,

$$\text{L.H.S.} = C_0 C_1 C_2 \dots\dots C_{n-1}\left(1 + \frac{C_1}{C_0}\right)$$

$$\left(1 + \frac{C_2}{C_1}\right)\dots\dots\left(1 + \frac{C_n}{C_{n-1}}\right)$$

On using $\dfrac{{}^nC_r}{{}^nC_{r-1}} = \dfrac{n-r+1}{r}$ we have,

$$\text{L.H.S.} = C_0 C_1 C_2 \dots\dots C_{n-1}$$

$$\left(1 + \frac{C_1}{C_0}\right)\left(1 + \frac{C_2}{C_1}\right)\dots\dots\left(1 + \frac{C_n}{C_{n-1}}\right)$$

$$= C_0 C_1 C_2 \dots C_{n-1}(1+n)\left(\frac{1+n}{2}\right)\left(\frac{1+n}{3}\right)\dots\left(\frac{n+1}{n}\right)$$

$$= \frac{C_0 C_1 \dots C_{n-1}(1+n)^n}{1.2.3\dots\dots n} = \frac{C_0 C_1 \dots\dots C_{n-1}(n+1)^n}{n!} = \text{R.H.S.}$$

Example 5: If $(1+x)^n = C_0 + C_1 x + C_2 x^2$
$$+ C_3 x^3 + \dots\dots + C_n x^n$$

Then find the value of $\displaystyle\sum_{0 \le i \, < j \le n} \sum \left(C_i + C_j\right)^2$

Sol: By using summation and coefficients properties we can prove given equations.

$$\sum_{0 \le i \, < j \le n} \sum \left(C_i + C_j\right)^2$$

$$= \left(C_0 + C_1\right)^2 + \left(C_0 + C_2\right)^2 + \dots\dots +$$

$$\left(C_0 + C_n\right)^2 + \left(C_1 + C_2\right)^2 + \dots\dots + \dots\dots +$$

$$\left(C_1 + C_n\right)^2 + \left(C_2 + C_3\right)^2 + \dots + \dots + \left(C_2 + C_n\right)^2$$

$$+ \ldots\ldots + \left(C_{n-1} + C_n\right)^2$$

$$= n\left(C_0^2 + C_1^2 + C_2^2 + \ldots\ldots + C_n^2\right) + 2\sum_{0 \le i \, < j \le n} \sum C_i \cdot C_j$$

The square of the sum of n terms is given by

$$= \left(C_0 + C_1 + C_2 + C_3 + \ldots\ldots + C_n\right)^2$$

$$= \left(C_0^2 + C_1^2 + C_2^2 + \ldots\ldots + C_n^2\right) + 2\sum_{0 \le i \, < j \le n} \sum C_i \cdot C_j$$

$$\therefore 2\sum_{0 \le i \, < j \le n} \sum C_i \cdot C_j$$

$$= \left[\left(C_0 + C_1 + C_2 + C_3 + \ldots\ldots + \right)^2 - \left(C_0^2 + C_1^2 + \ldots\ldots + C_n^2\right)\right]$$

$$= \left(2^n\right)^2 - {}^{2n}C_n$$

$$\therefore \sum_{0 \le i \, < j \le n} \sum \left(C_i + C_j\right)^2 = \left[n.\,{}^{2n}C_n\right] + \left[2^{2n} - {}^{2n}C_n\right],$$

$$= \left(n-1\right){}^{2n}C_n + 2^{2n}$$

Example 6: Show that

$$\frac{C_0}{1} - \frac{C_1}{4} + \frac{C_2}{7} - \frac{C_3}{10} + \ldots\ldots + 3\frac{(-1)^n C_n}{2n+1} = \frac{3^n n!}{1.4.7\ldots\ldots(3n+1)}$$

Sol: By expanding $\left(1-x^3\right)^n$ using binomial expansion and integrating it within a limit 0 to 1 we will prove given equation.

$$\left(1-x^3\right)^n = C_0 - C_1 x^3 + C_2 x^6$$

$$- C_3 x^9 + C_4 x^{12} + \ldots\ldots + \left(-1\right)^n C_n x^{3n}$$

Integrating both sides between limits 0 and 1, we get

$$\int_0^1 \left(1-x^3\right)^n dx = C_0 - \frac{C_1}{4} + \frac{C_2}{7} - \frac{C_3}{10} + \ldots\ldots + \frac{(-1)^n C_n}{3n+1} \qquad \ldots\text{(i)}$$

Also $I_n = \displaystyle\int_0^1 \left(1-x^3\right)^n dx$

$$= \left[x\left(1-x^3\right)^n\right]_0^1 - \int_0^1 n\left(1-x^3\right)^{n-1} ; \qquad \left(-3x^2\right).x\,dx$$

$$= 3n\int_0^1 x^3 \left(1-x^3\right)^{n-1} dx$$

$$= 3n\int_0^1 \left(x^3 - 1 + 1\right)\left(1-x^3\right)^{n-1} dx$$

$$= 3nI_{n-1} - 3nI_n ; \quad \left(1+3n\right)I_n = 3nI_{n-1} \quad \therefore I_n = \frac{3n}{3n+1}I_{n-1}$$

Replacing n by 1, 2, 3, 4, ... n–1 successively in the above reduction formula, we get

$$I_n = \frac{3n}{3n+1} \cdot \frac{3(n-1)}{3n-2} \cdot \frac{3(n-2)}{3n-5} \ldots \frac{3}{4}I_0 \qquad \ldots\text{(ii)}$$

But $I_0 = \displaystyle\int_0^1 \left(1-x^3\right)^0 dx = \int_0^1 dx = 1$

Hence, from (ii),

$$I_n = \frac{3^n n!}{\left(3n+1\right)\left(3n-2\right)\left(3n-5\right)\ldots\ldots 7.4}$$

Using (i)

$$\frac{C_0}{1} - \frac{C_1}{4} + \frac{C_2}{7} - \frac{C_3}{10} + \ldots\ldots + \frac{(-1)^n C_n}{3n+1} = \frac{3^n n!}{1.4.7\ldots\ldots(3n+1)}$$

Example 7: Prove that

$$\frac{1}{m!}C_0 + \frac{n}{\left(m+1\right)!}C_1 + \frac{n(n-1)}{\left(m-2\right)!}C_2 + \ldots\ldots + \frac{n(n-1)\ldots\ldots 2.1}{\left(m+n\right)!}C_n$$

$$= \frac{\left(m+n+1\right)\left(m+n+2\right)\ldots\ldots\left(m+2n\right)}{\left(m+n\right)!}$$

Sol: As $\left(1+x\right)^{m+n}.\left(1+x\right)^n = \left(1+x\right)^{m+2n}$ and expanding this by using expansion formula and equating the coefficient of x^n we can prove given equation.

$$\Rightarrow \left({}^{m+n}C_0 + {}^{m+n}C_1 x + {}^{m+n}C_2 x^2 + \ldots. + {}^{m+n}C_{m+n} x^{m+n}\right)$$

$$\times \left(C_0 + C_1 x + C_2 x^2 + \ldots\ldots + C_n x^n\right) = \left(1+x\right)^{m+2n}$$

Equating the coefficients of x^n on both sides, we find

$${}^{m+n}C_n.C_0 + {}^{m+n}C_{n-1}.C_1 + {}^{m+n}C_{n-2}.C_2$$

$$+ \ldots\ldots + {}^{m+n}C_0.C_n = {}^{m+2n}C_n$$

$$\Rightarrow \frac{\left(m+n\right)!}{m!n!}C_0 + \frac{\left(m+n\right)!}{\left(n-1\right)!\left(m+1\right)!}C_1$$

$$+ \frac{\left(m+n\right)!}{\left(n-2\right)!\left(m+2\right)!}C_2 + \ldots. + \frac{\left(m+n\right)!}{\left(m+n\right)!}C_n = \frac{\left(m+2n\right)!}{\left(m+n\right)!n!}$$

Dividing both sides by (m + n)!/n! we find

$$\frac{1}{m!}C_0 + \frac{n}{\left(m+1\right)!}C_1 + \frac{n(n-1)}{\left(m+2\right)!}C_2 + \ldots\ldots + \frac{n(n-1)\ldots\ldots 2.1}{\left(m+n\right)!}C_n$$

$$= \frac{(m+2n)!}{(m+n)!(m+n)!} = \frac{(m+n+1)(m+n+2).....(m+2n)}{(m+n)!}$$

Example 8: Find the sum of the following series

$$S = C_1^2 + \frac{1+2}{2}C_2^2 + \frac{1+2+3}{3}C_3^2 + ...\text{Upto n term}$$

Sol: In this problem, first obtain the rth term and then by using binomial expansion and coefficient property we can get required sum.

The rth term of the given series

$$= \frac{1+2+......+r}{r}C_r^2 = \frac{r(r+1)}{2r}C_r^2 = \frac{1}{2}(r+1)C_r^2$$

$$\therefore S = \frac{1}{2}(1+1)C_1^2 + \frac{1}{2}(2+1)C_2^2 + \frac{1}{2}$$

$$(3+1)C_2^3 + ... + \frac{1}{2}(n+1)C_n^2$$

We know that

$$C_0 + C_1 x + C_2 x^2 + ... + C_n x^n = (1+x)^n$$

$$\Rightarrow C_0 x + C_1 x^2 + C_2 x^3 + ... + C_n x^{n+1} = x(1+x)^n$$

Differentiating both sides w.r.t. x we get

$$C_0 + 2C_1 x + 3C_2 x^2 + 4C_3 x^3$$

$$+.... + (n+1)C_n x^n = (1+x)^n$$

$$+ nx(1+x)^{n-1} \qquad ...(i)$$

Also

$$C_0 + C_1\left(\frac{1}{x}\right) + C_2\left(\frac{1}{x}\right)^2 + C_3\left(\frac{1}{x}\right)^3$$

$$+.... + C_n\left(\frac{1}{x}\right)^n = \left(1+\frac{1}{x}\right)^n \qquad ...(ii)$$

Now, $C_0^2 + 2C_1^2 + 3C_2^2 + 4C_3^2 + ... + (n+1)C_n^2$

= Coefficient of constant term in

$$\left[C_0 + 2C_1 x + 3C_2 x^2 + 4C_3 x^3 + ... + (n+1)C_n x^n\right] \times$$

$$\left[C_0 + C_1\left(\frac{1}{x}\right) + C_2\left(\frac{1}{x}\right)^2 + ... + C_n\left(\frac{1}{x}\right)^n\right]$$

= Coefficient of constant term in

$$\left[(1+x)^n + nx(1+x)^{n-1}\right](1+1/x)^n$$

= Coefficient of x^n in

$$\left[(1+x)^n + nx(1+x)^{n-1}\right](x+1)^n$$

= Coefficient of x^n in

$$\left[(1+x)^{2n} + nx(1+x)^{2n-1}\right] = {}^{2n}C_n + n.{}^{2n-1}C_{n-1}$$

$$= \frac{(2n)!}{n!n!}\left(1+\frac{n}{2}\right) = {}^{2n}C_n\left(1+\frac{n}{2}\right)$$

$$\Rightarrow 2C_1^2 + 3C_2^2 + 4C_3^2 + ... + (n+1)C_n^2$$

$$= {}^{2n}C_n\left(1+\frac{n}{2}\right) - 1 \qquad \left[\because C_0 = 1\right]$$

$$\Rightarrow S = \frac{1}{2}\left[{}^{2n}C_n\left(1+\frac{n}{2}\right) - 1\right]$$

Example 9: If n be a positive integer, then prove that the integral part I of $\left(5+2\sqrt{6}\right)^n$ is an odd integer. If f be the fractional part of $\left(5+2\sqrt{6}\right)^n$ prove that $I = \frac{1}{1-f} - f$.

Sol: By using expansion formula we can expand the given binomial and separating its integral and fractional part we can prove given equations.

Let $P = \left(5+2\sqrt{6}\right)^n = I+f$

Or $I + f = 5^n + C_1 5^{n-1}\left(2\sqrt{6}\right)$

$$+ C_2 5^{n-2}\left(2\sqrt{6}\right)^2 + ... + C_n\left(2\sqrt{6}\right)^n \qquad ...(i)$$

$$0 < 5 - 2\sqrt{6} < 1 \Rightarrow 0 < \left(5 - 2\sqrt{6}\right)^n < 1$$

Let $\left(5 - 2\sqrt{6}\right)^n = f'$, where $0 < f' < 1$.

$$f' = 5^n - C_1 5^{n-1}\left(2\sqrt{6}\right)$$

$$+ C_2 5^{n-2}\left(2\sqrt{6}\right)^2 - C_3 5^{n-3}\left(2\sqrt{6}\right)^3 + ... \qquad ...(ii)$$

Adding (i) and (ii) $\quad I + f + f' =$

$$2\left[5^n + {}^nC_2 5^{n-2}\left(2\sqrt{6}\right)^2 + {}^nC_4 5^{n-1}\left(2\sqrt{6}\right)^4 ....\right]$$

Or $I + f + f' =$ even integer

Now $0 \leq f < 1$ and $0 < f' < 1$.

$$\therefore 0 < f + f' < 2$$

$\therefore\ f + f' = 1$ and $\therefore\ I$ is an odd integer

Now $I + f = \left(5 + 2\sqrt{6}\right)^n$,

$\left(5 - 2\sqrt{6}\right)^n = f' = 1 - f \Rightarrow (I + f)(1 - f) = 1$

$\therefore\ (I + f) = \dfrac{1}{1 - f}$ $\qquad \therefore\ I = \dfrac{1}{1 - f} - f$

Example 10: If $\left(1 + x + x^2\right) = a_0 + a_1 x$

$\qquad\qquad + a_2 x^2 + a_3 x^3 \ldots + a_{2n} x^{2n}$

Then show that

$a_0 + a_3 + a_6 + \ldots\ldots = a_1 + a_4 + a_7 + \ldots\ldots = 3^{n-1}$.

Sol: By using properties of binomial coefficients and cube root unity $1, \omega, \omega^2$ we can prove given problem.

The rth term of the given series

Putting $x = 1, \omega, \omega^2$, where ω is a non real cube root of unity.

$3^n = a_0 + a_1 + a_2 + a_3 + a_4 + \ldots\ldots$ $\qquad$ ...(i)

$0 = a_0 + a_1\omega + a_2\omega^2 + a_3\omega^3 + a_4\omega^4 + \ldots\ldots$ $\qquad$ ...(ii)

Because $1 + \omega + \omega^2 = 0$

$0 = a_0 + a_1\omega^2 + a_2\omega^4 + a_3\omega^6 + a_4\omega^8 + \ldots\ldots$ $\qquad$ ...(iii)

Adding these

$3^n = 3(a_0) + a_1\left(1 + \omega + \omega^2\right) + a_2$

$\left(1 + \omega^2 + \omega^4\right) + a_3\left(1 + \omega^3 + \omega^6\right)$

$+ \ldots = 3\left(a_0 + a_3 + a_6 + \ldots\right)$

$\therefore\ a_0 + a_3 + a_6 + \ldots\ldots = 3^{n-1}$

From (i) + (ii) × ω^2 + (iii) × ω,

we get, $\qquad 3^n + 0 \times \omega^2 + 0 \times \omega$

$= a_0\left(1 + \omega^2 + \omega\right) + a_1\left(1 + \omega^2 + \omega^3\right)$

$+ a_2\left(1 + \omega^4 + \omega^5\right) + a_3\left(1 + \omega^5 + \omega^7\right)$

$+ a_4\left(1 + \omega^6 + \omega^9\right) + \ldots$

$\therefore\ 3^n = 3\left(a_1 + a_4 + a_7 + \ldots\ldots\right)$

Because coefficient of each is

$1 + \omega + \omega^2 = 0$, using $\omega^2 = 1$

$\therefore\ a_1 + a_4 + a_7 + \ldots\ldots\ldots = 3^{n-1}$

Again, from (i) + (ii) ω + (iii) × ω^3, we get

$= 3^n = a_0\left(1 + \omega + \omega^2\right) + a_1\left(1 + \omega^2 + \omega^4\right)$

$+ a_2\left(1 + \omega^3 + \omega^3\right) + \ldots\ldots = 3\left(a_2 + a_5 + a_8 + \ldots\ldots\right)$

Example 11: Find the

(i) Last digit

(ii) Last two digits and

(iii) Last three digits of 17^{256}.

Sol: By reducing 17^{256} into the form $(x - 1)^n$ and using simple binomial expansion we will get required digits.

Since

$17^{256} = \left(17^2\right)^{128} = (289)^{128} = (290 - 1)^{128}$

$\therefore\ 17^{256} = {}^{128}C_0 (190)^{128} - {}^{128}C_1 (290)^{127}$

$+ {}^{128}C_2 (290)^{126} - \ldots - {}^{128}C_{125} (290)^3$

$+ {}^{128}C_{126} (290)^2 - {}^{128}C_{127} (290) + 1$

$\left[{}^{128}C_0 (290)^{128} - {}^{128}C_1 (290)^{127} \right.$

$+ {}^{128}C_2 (290)^{126} - \ldots\ldots - {}^{128}C_{125} (290)^3 \left.\right]$

$+ {}^{128}C_{126} (290)^2 - {}^{128}C_{127} (290) + 1$

$= 1000m + {}^{128}C_2 (290)^2 - {}^{128}C_1 (290) + 1\ \left(m \in I_+\right)$

$= 1000m + \dfrac{(128)(127)}{2}(290)^2 - 128 \times 290 + 1$

$= 1000\,m + (128)(127)(290)(145) - 128 \times 290 + 1$

$= 1000\,m + (128)(290)(127 \times 145 - 1) + 1$

$= 1000\,m + (128)(290)(18414) + 1$

$= 1000\,(m + 683527) + 681$

Hence last three digits of 17^{256} must be 681. As result last two digits of 17^{256} or 81 and last digit of 17^{256} is 1.

Example 12: If $32^{32^{32}}$ is divided by 7, then find the remainder

Sol: Here in this problem, we can obtain required remainder by reducing $32^{32^{32}}$ into the form of $7\lambda + a$, where λ is any integer and a is an integer less than 7.

We have $32 = 2^5$

$\therefore (32)^{32} = (2^5)^{32} = 2^{160}$; $(32)^{32} = (3-1)^{160}$

$= {}^{160}C_0 3^{160} - {}^{160}C_1 3^{159} + ... + {}^{160}C_{159} 3 + {}^{160}C_{160} + 1$

$= 3\left(3^{159} - {}^{160}C_1 3^{158} + ... - {}^{160}C_{159}\right) + 1$

$= 3m + 1, \quad m \in I^+$

Now, $32^{32^{32}} = 32^{3m+1} = 2^{5(3m+1)} = 2^{15m+5}$

$\therefore 32^{32^{32}} = 2^{3(5m+1)}.2^2 = 4.(8)^{5m+1}$

$= 4.(7+1)^{5m+1}$

$= 4.({}^{5m+1}C_0 (7)^{5m+1} + {}^{5m+1}C_1 (7)^{5m}$

$+ {}^{5m+1}C_2 (7)^{5m-1} + ... +$

${}^{5m+1}C_{5m} 7 + {}^{5m+1}C_{5m+1})$

$= 4[7\{{}^{5m+1}C_0 - 7^{5m} + {}^{5m+1}C_1 7^{5m-1}$

$+ {}^{5m+1}C_2 7^{5m-2} + ... + {}^{5m+1}C_{5m}\} + 1]$

$= 4\left[7n+1\right], \qquad n \in I_+ \qquad = 28n + 4$

This show that where $32^{32^{32}}$ is divided by 7, then remainder is 4.

JEE Main/Boards

Exercise 1

Q.1 Expand $\left(x^2 + 2a\right)^5$ by binomial theorem.

Q.2 Expand $\left(a+b\right)^6 - \left(a-b\right)^6$. Hence find the value of $\left(\sqrt{2}+1\right)^6 - \left(\sqrt{2}-1\right)^6$.

Q.3 Show that $\left(101\right)^{50} > \left(100\right)^{50} + \left(99\right)^{50}$

Q.4 If $x > 1$ and the third term in the expansion of $\left(\dfrac{1}{x} + x^{\log_{10} x}\right)^5$ is 1000, find the value of x.

Q.5 Find the sum of rational terms in the expansion of $\left(\sqrt{2} + 3^{1/5}\right)^{10}$.

Q.6 Find the middle term in the expansion of $\left(2x^2 - \dfrac{1}{x}\right)^7$

Q.7 Find the middle term in the expansion of $\left(1 - 2x + x^2\right)^n$.

Q.8 Show that the greatest coefficient in the expansion of $\left(x + \dfrac{1}{x}\right)^{2n}$ is $\dfrac{1.3.5....(2n-1).2^n}{n!}$.

Q.9 Given that the 4th term in the expansion of $\left(px + \dfrac{1}{x}\right)^n$ is $\dfrac{5}{2}$, find n and p.

Q. 10 If in the expansion of $(1 + x)^m (1 - x)^n$ the coefficient of x and x^2 are 3 and −6 respectively then find m.

Q.11 If the coefficients of a^{r-1}, a^r, a^{r+1} in the binomial expansion of $(1 + a)^n$ are in A.P., prove that $n^2 - n(4r+1) + 4r^2 - 2 = 0$.

Q.12 If n be a positive integer, then prove that $6^{2n} - 35n - 1$ is divisible by 1225.

Q.13 Using binomial theorem, show that $3^{4n+1} + 16n - 3$ is divisible by 256 if n is a positive integer.

Q.14 If a_1, a_2, a_3 and a_4 be any four consecutive coefficients in the expansion of $(1 + x)^n$, prove that
$$\frac{a_1}{a_1 + a_2} + \frac{a_3}{a_3 + a_4} = \frac{2a_2}{a_2 + a_3}$$

Q.15 If 3 consecutive coefficients in the expansion of $(1 + x)^n$ are in the ratio 6 : 33 : 110, find n and r.

Q.16 If a, b, c be the three consecutive coefficients in the expansion of a power of $(1 + x)$, prove that the index of the power is $\dfrac{2ac + b(a+c)}{b^2 - ac}$

Q.17 Expand $\left(x - \dfrac{1}{y}\right)^{11}$, $y \neq 0$

Q.18 Expand $(1 - x + x^2)^4$

Q.19 Which number is larger, $(1.2)^{4000}$ or 800?

Q.20 If in the expansion of $(1 + x)^n$, the coefficients of 14th, 15th and 16th terms are in A.P., find n.

Q.21 If three consecutive coefficient in the expansion of $(1 + x)^n$ be 165, 330 and 462, find n and the position of the coefficient.

Q.22 Find the greatest term in the expansion of;

$(7 - 5x)^{11}$, where $x = \dfrac{2}{3}$

Q.23 Find the coefficient of x^{-1} in $(1 + 3x^2 + x^4)\left(1 + \dfrac{1}{x}\right)^8$

Q.24 Find the value of k so that the term

independent of x in $\left(\sqrt{x} + \dfrac{k}{x^2}\right)^{10}$ of 405.

Q.25 If A be the sum of odd terms and B the sum of even terms in the expansion of $(x + a)^n$, prove that

$$2\left(A^2 + B^2\right) = \left(x + a\right)^{2n} + \left(x - a\right)^{2n}$$

Q.26 Find the coefficient of x^{40} in the expansion of

$\left(1 + 2x + x^2\right)^{27}$

Q.27 Find the term independent of x in $\left(\dfrac{3}{2}x^2 - \dfrac{1}{3x}\right)^9$.

Q.28 If $(1 + ax)^n = 1 + 8x + 24x^2 + \ldots$ Find a and n.

Exercise 2

Single Correct Choice Type

Q.1 Given that the term of the expansion $\left(x^{1/3} - x^{-1/2}\right)^{15}$ which does not contain x is 5 m where $m \in N$, then m =

(A) 1100   (B) 1010   (C) 1001   (D) None

Q.2 If the coefficients of x^7 & x^8 in the expansion of $\left[2 + \dfrac{x}{3}\right]^n$ are equal, then the value of n is:

(A) 15   (B) 45   (C) 55   (D) 56

Q.3 The coefficient of x^{49} in the expansion of $(x - 1)$

$\left(x - \dfrac{1}{2}\right)\left(x - \dfrac{1}{2^2}\right) \cdots \left(x - \dfrac{1}{2^{49}}\right)$ is equal to

(A) $-2\left(1 - \dfrac{1}{2^{50}}\right)$ (B) + ve coefficient of x

(C) – ve coefficient of x   (D) $-2\left(1 - \dfrac{1}{2^{49}}\right)$

Q.4 The last digit of $(3^P + 2)$ is

(A) 1   (B) 2   (C) 4   (D) 5

Where $P = 3^{4n}$ and $n \in N$

Q.5 The sum of the binomial coefficient of $\left[2x + \dfrac{1}{x}\right]^n$ is equal to 256. The constant term in the expansion is :

(A) 1120   (B) 2110   (C) 1210   (D) None

Q.6 The coefficient of x^4 in $\left[\dfrac{x}{2} - \dfrac{3}{x^2}\right]^{10}$ is

(A) $\dfrac{405}{256}$ (B) $\dfrac{504}{259}$ (C) $\dfrac{450}{263}$ (D) $\dfrac{405}{512}$

Q.7 If $(11)^{27} + (21)^{27}$ when divided by 16 leaves the remainder

(A) 0   (B) 1   (C) 2   (D) 14

Q.8 Last three digits of the number $N = 7^{100} - 3^{100}$ are

(A) 100   (B) 300   (C) 500   (D) 000

Q.9 The last two digits of the number 3^{400} are:

(A) 81   (B) 43   (C) 29   (D) 01

Q.10 If $\left(1 + x + x^2\right)^{25} = a_0 + a_1x + a_2x^2 + \ldots + a_{50}.x^{50}$

then $a_0 + a_2 + a_4 + \ldots + a_{50}$ is:

(A) Even

(B) Odd and of the form 3n

(C) Odd and of the form (3n –1)

(D) Odd and of the form (3n+1)

Q.11 The sum of the series

$$\left(1^2+1\right).1!+\left(2^2+1\right).2!+\left(3^2+1\right).3!+....+\left(n^2+1\right).n!$$ is

(A) $(n+1).(n+2)!$ (B) $n.(n+1)!$

(C) $(n+1).(n+1)!$ (D) None of these

Q.12 Let P_m stand for ^nP_m. Then the expression $1.P_1+2.P_2+3.P_3+.....+n.P_n=$

(A) $(n+1)!-1$ (B) $(n+1)!+1$

(C) $(n+1)!$ (D) None of these

Q.13 The expression

$$\frac{1}{\sqrt{4x+1}}\left[\left[\frac{1+\sqrt{4x+1}}{2}\right]^7-\left[\frac{1-\sqrt{4x+1}}{2}\right]^7\right]$$

is a polynomial in x of degree

(A) 7   (B) 5   (C) 4   (D) 3

Q.14 If the second term of the expansion $\left[a^{1/13}+\dfrac{a}{\sqrt{a^{-1}}}\right]^n$ is $14a^{5/2}$ then the value of $\dfrac{^nC_3}{^nC_2}$ is

(A) 4   (B) 3   (C) 12   (D) 6

Q.15 If $\left(1+x\right)\left(1+x+x^2\right)$

$$\left(1+x+x^2+x^3\right).....\left(1+x+x^2+x^3+....+x^n\right)$$

$$=a_0+a_1x+a_2x^2+a_3x^3+....+a_mx^m$$

Then $\displaystyle\sum_{r=0}^{m}a_r$ has the value equal to

(A) $n!$ (B) $(n+1)!$

(C) $(n-1)!$ (D) None of these

Q.16 In the expansion of $(1+x)^{43}$ if the coefficient of the $(2r+1)^{th}$ and the $(r+2)^{th}$ terms are equal, the value of r is :

(A) 12   (B) 13   (C) 14   (D) 15

Q.17 The positive value of a so that the coefficient of x^5 is equal to that of x^{15} in the expansion of $\left(x^2+\dfrac{a}{x^3}\right)^{10}$ is

(A) $\dfrac{1}{2\sqrt{3}}$ (B) $\dfrac{1}{\sqrt{3}}$ (C) 1   (D) $2\sqrt{3}$

Q.18 In the expansion of $x^2+\left(\dfrac{9}{43}\right)^{10}$ the term which does not contain x is :

(A) $^{10}C_0$ (B) $^{10}C_7$ (C) $^{10}C_4$ (D) None of these

Q.19 If the 6th term in the expansion of the binomial $\left[\dfrac{1}{x^{8/3}}+x^2\log_{10}x\right]^8$ is 5600, then x equals to

(A) 5   (B) 8   (C) 10   (D) 100

Q.20 $\left(1+x\right)\left(1+x+x^2\right)\left(1+x+x^2+x^3\right).....$

$\left(1+x+x^2+....+x^{100}\right)$ when written in the ascending power of x then the highest exponent of x is_____.

(A) 4950   (B) 5050   (C) 5150   (D) None of these

Q.21 Let $\left(5+2\sqrt{6}\right)^n=p+f$ where $n\in N$ and $p\in N$ and $0<f<1$ then the value of, $f^2-f+pf-p$ is

(A) A natural number   (B) A negative integer

(C) A prime number   (D) Are irrational number

Q.22 Number of rational terms in the expansion of $\left(\sqrt{2}+\sqrt[4]{3}\right)^{100}$ is-

(A) 25   (B) 26   (C) 27   (D) 28

Q.23 The greatest value of the term independent of x in the expansion of $\left(x\sin\theta+\dfrac{\cos\theta}{x}\right)^{10}$ is

(A) $^{10}C_5$ (B) 2^5 (C) $2^5.\,^{10}C_5$ (D) $\dfrac{^{10}C_5}{2^5}$

Q.24 If $\left(1+x-3x^2\right)^{2145}=a_0+a_1x+a_2x^2+....$ then $a_0-a_1+a_2-a_3+....$ end with

(A) 1   (B) 3   (C) 7   (D) 9

Q.25 Coefficient of x^6 in the binomial expansion $\left(\dfrac{4x^2}{3}-\dfrac{3}{2x}\right)^9$ is

(A) 2438   (B) 2688   (C) 2868   (D) None

Q.26 The expression

$$\left[x+\left(x^3-1\right)^{1/2}\right]^5+\left[x-\left(x^3-1\right)^{1/2}\right]^5$$

is a polynomial of degree

(A) 5   (B) 6   (C) 7   (D) 8

Q.27 Given $\left(1-2x+5x^2-10x^3\right)\left(1+x\right)^n = 1 + a_1 x + a_2 x^2 + \dots$ and that $a_1^2 = 2a_2$ then the value of n is

(A) 6   (B) 2   (C) 5   (D) 3

Q.28 The sum of the series

$$aC_0 + \left(a+b\right)C_1 + \left(a+2b\right)C_2 + \dots + \left(a+nb\right)C_n$$

is where C_r denotes combinatorial coefficient in the expansion of $(1 + x)^n$, $n \in N$

(A) $(a + 2nb)2^n$ (B) $(2a + nb)2^n$

(C) $(a + nb)2^{n-1}$ (D) $(2a + nb)2^{n-1}$

Previous Years' Questions

Q.1 Given positive integers $r > 1$, $n > 2$ and the coefficient of $(3r)^{th}$ and $(r + 2)^{th}$ terms in the binomial expansion of $(1 + x)^{2n}$ are equal. Then **(1980)**

(A) n = 2r   (B) n = 2r + 1

(C) n = 3r   (D) None of these

Q.2 If C_r stands for ^nC_r, then the sum of the series

$$\frac{2\left(\frac{n}{2}\right)!\left(\frac{n}{2}\right)!}{n!} \cdot \left[C_0^2 - 2C_1^2 + 3C_2^2 - \dots + \left(-1\right)^n \left(n+1\right)C_n^2\right]$$

Where n is an even positive integer, is equal to **(1986)**

(A) $\left(-1\right)^{n/2}\left(n+2\right)$ (B) $\left(-1\right)^n\left(n+1\right)$

(C) $\left(-1\right)^{n/2}\left(n+1\right)$ (D) None of these

Q.3 The expression

$$\left[x+\left(x^3-1\right)^{1/2}\right]^5+\left[x-\left(x^3-1\right)^{1/2}\right]^5$$

is a polynomial of degree **(1992)**

(A) 5   (B) 6   (C) 7   (D) 8

Q.4 For $2 \le r \le n$, $^nC_r + 2\,^nC_{r-1} + \,^nC_{r-2}$

Is equal to **(2000)**

(A) $^{n+1}C_{r-1}$ (B) $2^{n+1}C_{r+1}$

(C) $2^{n+2}C_r$ (D) $^{n+2}C_r$

Q.5 Let T_n denotes the number of triangles which can be formed using the vertices of a regular polygon of n sides. If $T_{n+1} - T_n = 21$, then n equals **(2001)**

(A) 5   (B) 7   (C) 6   (D) 4

Q.6 If $^{n-1}C_r = \left(k^2-3\right)\,^nC_{r+1}$, then k belongs to **(2004)**

(A) $\left(-\infty,-2\right]$ (B) $\left[-2,-\sqrt{3}\right)\cup\left(\sqrt{3},2\right]$

(C) $\left[-\sqrt{3},\sqrt{3}\right]$ (D) $\left(\sqrt{3},\infty\right]$

Q.7 $^{30}C_0\,^{30}C_{10} - \,^{30}C_1\,^{30}C_{11} + \dots \,^{30}C_{20}\,^{30}C_{30}$ is equal to **(2005)**

(A) $^{30}C_{11}$ (B) $^{60}C_{10}$ (C) $^{30}C_{10}$ (D) $^{65}C_{55}$

Q.8 For $r = 0, 1, \dots$, let A_r, B_r and C_r denote, respectively, the coefficient of x^r in the expansions of $(1 + x)^{10}$, $(1 + x)^{20}$ and $(1 + x)^{30}$. Then $\displaystyle\sum_{r=1}^{10} A_r \left(B_{10}B_r - C_{10}A_r\right)$ is equal to **(2010)**

(A) $B_{10} - C_{10}$ (B) $A_{10}\left(B_{10}^2 - C_{10}A_{10}\right)$

(C) 0   (D) $C_{10} - B_{10}$

Q.9 If the coefficients of x^3 and x^4 in the expansion of $\left(1+ax+bx^2\right)\left(1-2x\right)^{18}$ in powers of x are both zero, then (a, b) is equal to: **(2014)**

(A) $\left(16,\dfrac{251}{3}\right)$ (B) $\left(14,\dfrac{251}{3}\right)$

(C) $\left(14,\dfrac{272}{3}\right)$ (D) $\left(16,\dfrac{272}{3}\right)$

Q.10 The sum of coefficients of integral powers of x in

the binomial expansion of $\left(1 - 2\sqrt{x}\right)^{50}$ is: **(2015)**

(A) $\dfrac{1}{2}\left(3^{50} + 1\right)$

(B) $\dfrac{1}{2}\left(3^{50}\right)$

(C) $\dfrac{1}{2}\left(3^{50} - 1\right)$

(D) $\dfrac{1}{2}\left(2^{50} + 1\right)$

Q.11 If the number of terms in the expansion of $\left(1 - \dfrac{2}{x} + \dfrac{4}{x^2}\right)^n, x \neq 0$, is 28, then the sum of the coefficients of all the terms in this expansion, is: **(2016)**

(A) 64
(B) 2187
(C) 243
(D) 729

JEE Advanced/Boards

Exercise 1

Q.1 Let $f(x) = 1 - x + x^2 - x^3 \ldots + x^{16} - x^{17}$

$= a_0 + a_1(1 + x) + a_2(1 + x)^2 + \ldots + a_{17}(1 + x)^{17}$,

Find the value of a_2.

Q.2 (a) Find the term independent of x in the expansion of

(i) $\left[\sqrt{\dfrac{x}{3}} + \dfrac{\sqrt{3}}{2x^2}\right]^{10}$

(ii) $\left[\dfrac{1}{2}x^{1/3} + x^{-1/5}\right]^8$

(b) Find the value of x for which the fourth term in the expansion,

$$\left(5^{\frac{2}{5}\log_5\sqrt{4^x + 44}} + \dfrac{1}{5^{\log_5\sqrt[3]{2^{x-1}+7}}}\right)^8 \text{ is } 336.$$

Q.3 Find the coefficients:

(i) x^7 in $\left(ax^2 + \dfrac{1}{bx}\right)^{11}$

(ii) x^{-7} in $\left(ax - \dfrac{1}{bx^2}\right)^{11}$

(iii) Find the relation between a and b, so that these coefficients are equal.

Q.4 (a) If the coefficients of the rth, $(r + 1)^{th}$ & $(r + 2)^{th}$ terms in the expansion of $(1 + x)^{14}$ are in AP, find r.

(b) If the coefficients of 2nd, 3rd & 4th terms in the expansion of $(1 + x)^{2n}$ are in AP, show that $2n^2 - 9n + 7 = 0$.

Q.5 Let a and b be the coefficient of x^3 in $(1 + x + 2x^2 + 3x^3)^4$ and $(1 + x + 2x^2 + 3x^3 + 4x^4)^4$ respectively. Find the value of $(a - b)$.

Q.6 Prove that the ratio of the coefficient of x^{10} in $(1 - x^2)^{10}$ & the term independent of x in $\left(x - \dfrac{2}{x}\right)^{10}$ is $1 : 32$.

Q.7 Find the coefficient of

(a) $x^2 y^3 z^4$ in the expansion of $(ax - by + cz)^9$.

(h) $a^2 b^3 c^4 d$ in the expansion of $(a - b - c + d)^{10}$.

Q.8 Given $S_n = 1 + \dfrac{q+1}{2} + \left(\dfrac{q+1}{2}\right)^2 + \ldots + \left(\dfrac{q+1}{2}\right)^n$,

$q \neq 1$, prove that $^{n+1}C_1 + {}^{n+1}C_2.S_1 + {}^{n+1}C_3.S_2$

$+ \ldots + {}^{n+1}C_{n+1}.S_n = 2^n.S_n.$

Q.9 Find numerically the greatest term in the expansion of

(i) $(2 + 3x)^9$ when $x = \dfrac{3}{2}$

(ii) $(3 - 5x)^{15}$ when $x = \dfrac{1}{5}$

Q.10 Given that

$$\left(1 + x + x^2\right)^n = a_0 + a_1 x + a_2 x^2 + \ldots + a_{2n} x^{2n},$$

Find the values of :

(i) $a_0 + a_1 + a_2 + \ldots + a_{2n}$;

(ii) $a_0 - a_1 + a_2 - a_3 \ldots + a_{2n}$;

(iii) $a_0^2 - a_1^2 + a_2^2 - a_3^2 + \ldots + a_{2n}^2$

Q.11 For which positive values of x is the fourth term in the expansion of $(5 + 3x)^{10}$ is the greatest.

Q.12 Find the index n of the binomial $\left(\dfrac{x}{5} + \dfrac{2}{5}\right)^n$ if the 9th term of the expansion has numerically the greatest coefficient $(n \in N)$.

Q.13 Find the number of divisors of the number
$$N = {}^{2000}C_1 + 2.\,{}^{2000}C_2 + 3.\,{}^{2000}C_3 + ... + 2000.\,{}^{2000}C_{2000}$$

Q.14 Find number of different dissimilar terms in the sum
$$(1+x)^{2012} + (1+x^2)^{2011} + (1+x^3)^{2010}$$

Q.15 Find the term independent of x in the expansion of $\left(1 + x + 2x^3\right)\left(\dfrac{3x^2}{2} - \dfrac{1}{3x}\right)^9$.

Q.16 Let $f(n) = \displaystyle\sum_{r=0}^{n}\sum_{k=r}^{n}\binom{k}{r}$. Find the total number of divisors of f (11).

Q.17 Find the sum $\displaystyle\sum_{j=0}^{11}\sum_{i=j}^{11}\binom{i}{j}$.

[Note : $\binom{n}{r} = {}^nC_r$]

Q.18 Let $\left(1+x^2\right)^2.\left(1+x\right)^n = \displaystyle\sum_{k=0}^{n+4} a_k.x^k$. If a_1, a_2 and a_3 are in AP, find n.

Q.19 Prove that $\displaystyle\sum_{K=0}^{n} {}^nC_k \sin kx.\cos(n-k)x = 2^{n-1}\sin nx$.

Q.20 Find the sum of the roots (real or complex) of the equation $x^{2001} + \left(\dfrac{1}{2} - x\right)^{2001} = 0$.

Q.21 If for $n \in N$, $\displaystyle\sum_{k=0}^{2n}(-1)^k\,({}^{2n}C_k)^2 = A$, then what will be the value of $\displaystyle\sum_{k=0}^{2n}(-1)^k\,(k - 2n)\,({}^{2n}C_k)^2$?

Paragraph for questions. 22 and 23

A path of length n is a sequence of points $(x_1, y_1), (x_2, y_2), ..., (x_n, y_n)$ with integer coordinates such that for all i between 1 and n − 1 both inclusive, either $x_{i+1} = x_i + 1$ and $y_{i+1} = y_i$ (in which case we say the ith step is rightward) or $x_{i+1} = x_i$ and $y_{i+1} = y_i + 1$ (in which case we say that the ith step is upward).

This path is said to start at (x_1, y_1) and end at (x_n, y_n). Let P (a, b), for a and b non negative integers, denotes the number of paths that start at (0, 0) and end at (a, b)

Q.22 The value of $\displaystyle\sum_{i=0}^{10} P(i, 10 - i)$, is

(A) 1024   (B) 512   (C) 256   (D) 128

Q.23 Number of ordered pairs (i, j) where $i \neq j$ for which P (i, 100 − i) = P(j, 100 − j), is

(A) 50   (B) 99   (C) 100   (D) 101

Q.24 If $\left(6\sqrt{6} + 14\right)^{2n+1}$ = N & F be the fractional part of N, prove that NF = 20^{2n+1} $(n \in N)$.

Q.25 Let P = $\left(2 + \sqrt{3}\right)^5$ and f = P − [P], where [P] denotes the greatest integer function. Find the value of $\left(\dfrac{f^2}{1-f}\right)$.

Q.26 If $C_0, C_1, C_2, ..., C_n$ are the combinatorial coefficients in the expansion of $(1 + x)^n$, $n \in N$ then prove the following:

(a) $C_1 + 2C_2 + 3C_3 + ... + n.C_n = n.2^{n-1}$

(b) $C_0 + 2C_1 + 3C_2 + ... + (n+1)C_n = (n+2)2^{n-1}$

(c) $C_0 + 3C_1 + 5C_2 + ... + (2n+1)C_n = (n+1)2^n$

(d) $\left(C_0 + C_1\right)\left(C_1 + C_2\right)\left(C_2 + C_3\right)....\left(C_{n-1} + C_n\right)$
$$= \dfrac{C_0.C_1.C_2....C_{n-1}\left(n+1\right)^n}{n!}$$

(e) $1.C_0^2 + 3.C_1^2 + 5.C_2^2 + ... + \left(2n+1\right)C_n^2 = \dfrac{(n+1)(2n)!}{n!n!}$

Q.27 Let I denotes the integral part and F the proper fractional part of $\left(3 + \sqrt{5}\right)^n$ where $n \in N$ and if ρ denotes the rational part and σ the irrational part of the same, show that

$\rho = \dfrac{1}{2}(I+1)$ and $\sigma = \dfrac{1}{2}(I + 2F - 1)$

Q.28 Prove that

(a) $\dfrac{C_1}{C_0} + \dfrac{2C_2}{C_1} + \dfrac{3C_3}{C_2} + ... + \dfrac{n.C_n}{C_{n-1}} = \dfrac{n(n+1)}{2}$

(b) $C_0 + \dfrac{C_1}{2} + \dfrac{C_2}{3} + ... + \dfrac{C_n}{n+1} = \dfrac{2^{n+1}-1}{n+1}$

(c) $2.C_0 + \dfrac{2^2.C_1}{2} + \dfrac{2^3.C_2}{3} + \dfrac{2^4.C_3}{4}$

$\quad +.... + \dfrac{2^{n+1}.C_n}{n+1} = \dfrac{3^{n+1}-1}{n+1}$

(d) $C_0 - \dfrac{C_1}{2} + \dfrac{C_2}{3} - .... + (-1)^n \dfrac{C_n}{n+1} = \dfrac{1}{n+1}$

Q.29 Prove the following identities using the theory of permutation where $C_0,\ C_1,\ C_2,\ .... ,\ C_n$ are the combinatorial coefficients in the expansion of $(1+x)^n$, $n \in N$, the prove the following :

(a) $C_0C_1 + C_1C_2 + C_2C_3 + ... +$

$\qquad C_{n-1}C_n = \dfrac{2n!}{(n+1)!(n-1)!}$

(b) $C_0C_r + C_1C_{r+1} + C_2C_{r+2} + ... + C_{n-r}C_n = \dfrac{2n!}{(n-r)!(n+r)!}$

(c) $\displaystyle\sum_{r=0}^{n-2} \left({}^nC_r.{}^nC_{r+2} \right) = \dfrac{(2n)!}{(n-2)!(n+2)!}$

(d) ${}^{100}C_{10} + 5.{}^{100}C_{11} + 10.{}^{100}C_{12} + 10.$

$\qquad {}^{100}C_{13} + 5.{}^{100}C_{14} + {}^{100}C_{15} = {}^{105}C_{90}$

Q.30 If $a_0,\ a_1,\ a_2,.....$ be the coefficients in the expansion of $(1 + x + x^2)^n$ in ascending powers of x, then prove that :

(i) $a_0a_1 - a_1a_2 + a_2a_3 - .... = 0$

(ii) $a_0a_2 - a_1a_3 + a_2a_4 - .... + a_{2n-2}a_{2n} = a_{n+1}$ or a_{n-1}

(iii) $E_1 = E_2 = E_3 = 3_{n-1};$

Where $E_1 = a_0 + a_3 + a_6 + ...; E_2 = a_1 + a_4 + a_7$

$\qquad +.....\ \&\ E_3 = a_2 + a_5 + a_8 + ...$

Q.31 Let $\displaystyle\sum_{r=0}^{100}\sum_{s=0}^{100} \left(C_1^2 + C_s^2 + C_rC_s \right) = m\left({}^{2n}C_n \right) + 2^p$

Where m, n and p are even natural numbers and C_r represents the coefficient of x_r in the expansion of $(1 + x)^{100}$. Find the value of $(m + n + p)$.

Q.32 The expressions $1 + x,\ 1 + x + x^2,\ 1 + x + x^2 + x^3,\ .....$ $1 + x + x^2 + ... + x^n$ are multiplied together and the terms of the product thus obtained are arranged in increasing powers of x in the form of $a_0 + a_1x + a_2x^2 + ...,$ then

(a) How many terms are there in the product.

(b) Show that the coefficients of the terms in the product, equidistant from the beginning and end are equal.

(c) Show that the sum of the odd coefficients = the sum of the even coefficients = $\dfrac{(n+1)!}{2}$

Q.33 Let $S_1 = \displaystyle\sum_{0 \le i < j \le 100}\sum C_iC_j, S_2 =$

$\displaystyle\sum_{0 \le j < i \le 100}\sum C_iC_j$ and $S_3 = \displaystyle\sum_{0 \le i = j \le 100}\sum C_iC_j$

Where C_r represents coefficient of x^r in the binomial expansion of $(1 + x)^{100}$

If $S_1 + S_2 + S_3 = a^b$ where $a,\ b \in N$ then find the least value of $(a + b)$.

Exercise 2

Single Correct Choice Type

Q.1 In the binomial $(2^{1/3} + 3^{-1/3})$, if the ratio of the seventh term from the beginning of the expansion to the seventh term from its end is $1/6$, then $n =$

(A) 6   (B) 9   (C) 12   (D) 15

Q.2 The remainder, when $(15^{23} + 23^{23})$ is divided by 19, is

(A) 4   (B) 15   (C) 0   (D) 18

Q.3 The value of $4\left\{ {}^nC_1 + 4.{}^nC_2 + 4^2.{}^nC_3 + ... + 4^{n-1} \right\}$ is

(A) 0   (B) $5^n + 1$ (C) 5^n (D) $5^n - 1$

Q.4 If n be a positive integer such that $n \ge 3$, then the value of the sum to n terms of the series

$1.n - \dfrac{(n-1)}{1!}(n-1) + \dfrac{(n-1)(n-2)}{2!}$

$(n-2) - \dfrac{(n-1)(n-2)(n-3)}{3!}(n-3) + \ldots$ is

(A) 0

(B) 1

(C) –1

(D) None of these

Q.5 If the 6th term in the expansion of the binomial

$\left[\dfrac{1}{x^{8/3}} + x^2 \log_{10} x\right]^8$ is 5600, then x equals to-

(A) 5 (B) 8 (C) 10 (D) 100

Q.6 Coefficient of α^t in the expansion of,

$(\alpha + p)^{m-1} + (\alpha + p)^{m-2}(\alpha + q) +$

$(\alpha + p)^{m-3}(\alpha + q)^2 + \ldots + (\alpha + q)^{m-1}$

Where $\alpha \neq -q$ and $p \neq q$ is:

(A) $\dfrac{{}^mC_t\left(p^t - q^t\right)}{p - q}$

(B) $\dfrac{{}^mC_t\left(p^{m-t} - q^{m-t}\right)}{p - q}$

(C) $\dfrac{{}^mC_t\left(p^t + q^t\right)}{p - q}$

(D) $\dfrac{{}^mC_t\left(p^{m-t} + q^{m-t}\right)}{p - q}$

Q.7 If $\left(1 + x - 3x^2\right)^{2145} = a_0 + a_1 x + a_2 x^2 + \ldots$

then $a_0 - a_1 + a_2 - a_3 + \ldots$ end with

(A) 1 (B) 3 (C) 7 (D) 9

Q.8 Coefficient of x^6 in the binomial expansion $\left(\dfrac{4x^2}{3} - \dfrac{3}{2x}\right)^9$ is

(A) 2438

(B) 2688

(C) 2868

(D) None of these

Q.9 The term independent of 'x' in the expansion of $\left(9x - \dfrac{1}{3\sqrt{x}}\right)^{18}$, $x > 0$, is α times the corresponding binomial coefficient. Then 'α' is:

(A) 3 (B) $\dfrac{1}{3}$ (C) $-\dfrac{1}{3}$ (D) 1

Q.10 The expression

$$\left[x + \left(x^3 - 1\right)^{1/2}\right]^5 + \left[x - \left(x^3 - 1\right)^{1/2}\right]^5$$

Is a polynomial of degree

(A) 5 (B) 6 (C) 7 (D) 8

Q.11 Value of the expression $C_0^2 + C_1^2 + C_2^2 + \ldots + C_n^2$ is

(A) 2^{2n-1}

(B) $2n\,({}^{2n}C_n)$

(C) ${}^{2n}C_n$

(D) None of these

Q.12 The sum of the series

$aC_0 + (a+b)C_1 + (a+2b)\,C_2 + \ldots + (a+nb)C_n$ is

where C_r denotes combinatorial coefficient in the expansion of $(1 + x)^n$, $n \in N$

(A) $(a + 2nb)2^n$

(B) $(2a + nb)2^n$

(C) $(a + nb)2^{n-1}$

(D) $(2a + nb)2^{n-1}$

Previous Years' Questions

Q.1 Prove that $C_1^2 - 2.C_2^2 + C_3^2 - \ldots - 2n.C_{2n}^2 = (-1)^n n.C_n$

(1979)

Q.2 Given,

$S_n = 1 + q + q^2 + \ldots + q^n$

$S_n = 1 + \dfrac{q+1}{2} + \left(\dfrac{q+1}{2}\right)^2 + \ldots + \left(\dfrac{q+1}{2}\right)^n , q \neq 1$

Prove that ${}^{n+1}C_1 + {}^{n+1}C_2 S_1 + {}^{n+1}C_3 S_2$

$+ \ldots + {}^{n+1}C_{n+1}S_n = 2^n S_n$ *(1984)*

Q.3 Find the sum of the series

$$\sum_{r=0}^{n}(-1)^r {}^nC_r\left[\dfrac{1}{2^r} + \dfrac{3^r}{2^{2r}} + \dfrac{7^r}{2^{3r}} + \dfrac{15^r}{2^{4r}}\ldots\text{upto m terms}\right]$$

(1985)

Q.4 If $\displaystyle\sum_{r=0}^{2n}a_r(x-2)^r = \sum_{r=0}^{2n}b_r(x-3)^r$ and $a_k = 1$ for all $k \geq n$, then show that $b_n = {}^{2n+1}C_{n+1}$ *(1992)*

Q.5 Let n be a positive integer and

$$\left(1+x+x^2\right)^n = a_0 + a_1 x + ... + a_{2n} x^{2n}.$$

Show that $a_0^2 - a_1^2 + ... + a_{2n}^2 = a_n$ ***(1994)***

Q.6 Prove that   ***(2003)***

$$2^k \, {}^nC_0 {}^nC_k - 2^{k-1} \, {}^nC_1 {}^{n-1}C_{k-1} + 2^{k-2}$$
$${}^nC_2 {}^{n-2}C_{k-2} - ...... + (-1)^k \, {}^nC_k {}^{n-k}C_0 = {}^nC_k$$

Q.7 For $r = 0, 1, ..., 10$, let A_r, B_r and C_r denote, respectively, the coefficient of x^r in the expansions of $\left(1+x\right)^{10}, \left(1+x\right)^{20}$ and $\left(1+x\right)^{30}$. Then $\sum_{r=1}^{10}\left(B_{10}B_r - C_{10}A_r\right)$ is equal to   ***(2010)***

(A) $B_{10} - C_{10}$ (B) $A10\left(B_{10}^2 - C_{10}A_{10}\right)$

(C) 0   (D) $C_{10} - B_{10}$

Q.8 The coefficients of three consecutive terms of $\left(1+x\right)^{n+5}$ are in the ratio $5 : 10 : 14$. Then n =__ ***(2013)***

Q.9 Coefficient of x^{11} in the expansion of $\left(1+x^2\right)^4 \left(1+x^3\right)^7 \left(1+x^4\right)^{12}$ is   ***(2014)***

(A) 1051   (B) 1106   (C) 1113   (D) 1120

Q.10 The coefficient of x^9 in the expansion of $\left(1+x\right)\left(1+x^2\right)\left(1+x^3\right)....\left(1+x^{100}\right)$ is   ***(2015)***

Q.11 Let $z = \dfrac{-1+\sqrt{3}i}{2}$, where $i = \sqrt{-1}$, and $r, s \in \{1,2,3\}$.

Let $P = \begin{bmatrix} (-z)^r & z^{2s} \\ z^{2s} & z^r \end{bmatrix}$ and I be the identity matrix of order 2. Then the total number of ordered pairs (r, s) for which $P^2 = -I$ is   ***(2016)***

Important Questions

JEE Main/Boards

Exercise 1

Q. 3	Q. 16	Q. 19	Q. 23
Q. 28	Q. 32	Q. 34	

Exercise 2

Q. 7	Q. 13	Q. 15	Q. 21
Q. 22	Q. 25	Q. 29	

Previous Years' Questions

Q. 2	Q. 3	Q. 5	Q. 6
Q. 8			

JEE Advanced/Boards

Exercise 1

Q. 14	Q. 23	Q. 26	Q. 31
Q. 34	Q. 35		

Exercise 2

Q. 2	Q. 4	Q. 12

Previous Years' Questions

Q. 3	Q. 4

JEE Main/Boards

Exercise 1

Q.1 $x^{10} + 10x^8a + 40x^6a^2 + 80x^4a^3 + 80x^2a^4 + 32a^5$

Q.2 $4ab\left[3a^4 + 10a^2b^2 + 3b^4\right]$, $140\sqrt{2}$

Q.4 100

Q.5 41

Q.6 $280x^2$

Q.7 $\dfrac{2n!}{n!n!}(-1)^n x^n$

Q.9 $n = 6$, $p = \dfrac{1}{2}$

Q.10 $m = 12$

Q.15 $n = 12$, $r = 1$

Q.20 34, 23

Q.21 11, T_{3+1}, T_{3+2}, T_{3+3}

Q.23 232

Q.24 $k = \pm 3$

Q.26 $^{54}C_{14}$

Q.27 $\dfrac{7}{18}$

Q.28 $n = 4$, $a = 2$

Exercise 2

Single Correct Choice Type

Q.1 C	**Q.2** C	**Q.3** A	**Q.4** D	**Q.5** A	**Q.6** A
Q.7 A	**Q.8** D	**Q.9** D	**Q.10** A	**Q.11** B	**Q.12** A
Q.13 D	**Q.14** A	**Q.15** B	**Q.16** C	**Q.17** A	**Q.18** C
Q.19 C	**Q.20** B	**Q.21** B	**Q.22** B	**Q.23** D	**Q.24** B
Q.25 B	**Q.26** C	**Q.27** A	**Q.28** D		

Previous Years' Questions

Q.1 A	**Q.2** A	**Q.3** C	**Q.4** D	**Q.5** B	**Q.6** B
Q.7 C	**Q.8** D	**Q.9** D	**Q.10** A	**Q.11** D	

JEE Advanced/Boards

Exercise 1

Q.1 816

Q.2 (a) (i) $\dfrac{5}{12}$ (ii) $T_6 = 7$, (b) $x = 0$ or 1

Q.3 (i) $^{11}C_5 \dfrac{a^6}{b^5}$ (ii) $^{11}C_6 \dfrac{a^5}{b^6}$ (iii) $ab = 1$

Q.4 (a) $r = 5$ or 9

Q.5 0

Q.7 (a) $-1260\, a^2b^3c^4$; (b) -12600

Q.9 (i) $T_7 = \dfrac{7.3^{13}}{3}$ (ii) 455×3^{12}

Q.10 (i) 3^n (ii) 1, (iii) a_n

Q.11 $\dfrac{5}{8} < x < \dfrac{20}{21}$ **Q.12** $n = 12$ **Q.13** 8016   **Q.14** 4023

Q.15 $\dfrac{17}{54}$ **Q.16** 24   **Q.17** 4095   **Q.18** $n = 2$ or 3 or 4

Q.20 500   **Q.22** A   **Q.23** C   **Q.25** 722

Q.31 502   **Q.32** (a) $\dfrac{n^2+n+2}{2}$, (b) $a_0 = a\dfrac{n(n+1)}{2}$, (c) $\dfrac{(n+1)!}{2}$ **Q.33** 66

Exercise 2

Single Correct Choice Type

Q.1 B   **Q.2** C   **Q.3** D   **Q.4** A   **Q.5** C   **Q.6** B

Q.7 B   **Q.8** B   **Q.9** D   **Q.10** C   **Q.11** C   **Q.12** D

Previous Years' Questions

Q.3 $\dfrac{2^{mn}-1}{2^{mn}\left(2^n-1\right)}$ **Q.9** C

Solutions

JEE Main/Boards

Exercise 1

Sol 1: $(x^2 + 2a)^5$

$= {}^5C_0(x^2)^5 + {}^5C_1(x^2)^{5-1}(2a)^1 + {}^5C_2(x^2)^{5-2}$

$(2a)^3 + {}^5C_3(x^2)^{5-3}(2a)^3 + {}^5C_4(x^2)^{5-4}$

$(2a)^4 + {}^5C_5(x^2)^{5-5}(2a)^5$

$= x^{10} + 5x^8(2a) + 10x^6(2a)^2 + 10x^4(2a)^3 + 5x^2(2a)^4 + (2a)^5$

$= x^{10} + 10x^8a + 40x^6a^2 + 80x^4a^3 + 80x^2a^4 + 32a^5$

Sol 2: $(a+b)^6 - (a-b)^6$

${}^6C_0a^6 + {}^6C_1a^5b + {}^6C_2a^4b^2 + {}^6C_3a^3b^3$

$+ {}^6C_4a^2b^4 + {}^6C_5ab^5 + {}^6C_6b^6$

$-({}^6C_0a^6 - {}^6C_1a^5b + {}^6C_2a^4b^2 - {}^6C_3a^3b^3$

$+ {}^6C_4a^2b^4 - {}^6C_5a^4b^5 + {}^6C_0b^6)$

$= 2[6a^5b + 20a^3b^3 + 6ab^5] = 4ab[3a^4 + 10a^2b^2 + 3b^4]$

For finding the value, put $a = \sqrt{2}$ $b = 1$

$\therefore \sqrt{2}\,(12 + 20 + 3)$

$\Rightarrow 140\sqrt{2}$

Sol 3: $(101)^{50} > (100)^{50} + (99)^{50}$

$(100+1)^{50} > (100)^{50} + (100-1)^{50}$

$= (100+1)^{50} - (100-1)^{50} > 100^{50}$

Both binomial will cancel every odd terms of each others rest of the even terms are.

$= 2[{}^{50}C_1(100)^{49} + {}^{50}C_3(100)^{47}] + {}^{50}C_5(100)^{45} +$

$......+ {}^{50}C_{49}100]$

$= 100(100)^{49} + 2[{}^{50}C_3(100)^{47} + ... + {}^{50}C_{49}(100)]$

$$= (100)^{50} + 2[\,^{50}C_3(100)^{47} + ... + \,^{50}C_{47}(100)] > 100^{50}$$

Which is always true

So $(101)^{50} - (99)^{50} > (100)^{50}$

$$= (101)^{50} > 100^{50} - (99)^{50}$$

Sol 4: $x > 1$

$$\left(\frac{1}{x} + x^{\log_{10} x}\right)^5 \text{ and } T4 = \,^7C_3\,(2x^2)^3 \cdot \left(\frac{1}{x}\right)^4 = 280y^2$$

$$T_3 = T_{2+1} = \,^5C_2\left(\frac{1}{x}\right)^{5-2}\left(x^{\log_{10} x}\right)^2 = 1000$$

$$= 10\left(\frac{1}{x^3}\right)\left(x^{2\log_{10} x}\right) = 1000 \Rightarrow x^{\log_{10} x^2} = 100x^3$$

Assume $x = 10^y$

$$\Rightarrow 10^{y\log_{10}(10^y)^2} = 100(10^y)^3 = 10^{2+3y}$$

$$\Rightarrow 10^{2y(\log_{10} 10^y)} = 10^{2y^2} = 10^{2+3y}$$

$$\Rightarrow 2y^2 = 2+3y \Rightarrow 2y^2 - 3y - 2 = 0$$

$$\Rightarrow (y-2)(2y+1) = 0$$

$$\Rightarrow y = 2 \text{ or } y = -\frac{1}{2}$$

$$\Rightarrow x = 10^2 \text{ or } x = 10^{-1/2} \Rightarrow x = 100 \text{ or } x = \frac{1}{\sqrt{10}}$$

But $x > 1$ so $x = 100$

Sol 5: $(\sqrt{2} + 3^{1/5})^{10}$

For rational number

$$(\sqrt{2})^y \to y = 2n, n \in N$$

$$(3^{1/5})^z \to z = 5n, n \in N$$

Rational terms

$$^{10}C_0(\sqrt{2})^{10} + \,^{10}C_{10}(\sqrt{2})^0(3^{1/5})^{10} = 2^5 + 3^2 = 32 + 9 = 41$$

Sol 6: $\left(2x^2 - \frac{1}{x}\right)^7$

Middle terms are $T_4 = T_{3+1}$ and $T_5 = T_{4+1}$

$$T_{4+1} = \,^7C_4(2x^2)^{7-4}\left(-\frac{1}{x}\right)^4 = \frac{7\times6\times5}{1.2.3}(2x^2)^3\frac{1}{x^4}$$

$$= 35\times8\times\frac{x^6}{x^4} = 280x^2$$

Sol 7: $(1-2x+x^2)^n = (1-2x+x^2)^n$

$$= (-1+x)^{2n} = (x-1)^{2n}$$

Middle term $= T_{n+1} = \,^{2n}C_n(x)^{2n-n}(-1)^n$

$$= \frac{2n!}{n!n!}x^n(-1)^n$$

Sol 8: $\left(x + \frac{1}{x}\right)^{2n}$

Greatest coefficient $= \,^{2n}C_n$

$$= \frac{2n!}{n!(2n-n)!} = \frac{2n!}{n!n!}$$

$$= \frac{2n(2n-1)(2n-2)(2n-3)(2n-4).....3.2.1}{n!(n(n-1)(n-2)(n-3)..........3.2.1)}$$

$$= \frac{2^n[n(n-1)(n-2)(n-3)\cdots1]1.3.5.7........(2n-1)}{n!(n(n-1)......3.2.1)}$$

$$= \frac{2^n1.3.5.......(2n-1)}{n!}$$

Sol 9: $= \left(px + \frac{1}{x}\right)^n$

Given $= 4^{\text{th}}$ term $= \frac{5}{2}$

$$T_4 = T_{3+1} = \,^nC_3(px)^{n-3}\left(\frac{1}{x}\right)^3 = \frac{5}{2}$$

$$\Rightarrow \,^nC_3 p^{n-3}x^{n-3+3(-1)} = \frac{5}{2}x^0$$

$$\Rightarrow n = 6 \Rightarrow \,^6C_3 p^{6-3}x^0 = \frac{5}{2}$$

$$\Rightarrow \frac{6\times4\times5}{1.2.3}\,p^3 = \frac{5}{2}$$

$$\Rightarrow p^3 = \frac{1}{8} = \left(\frac{1}{2}\right)^3 \Rightarrow p = \frac{1}{2} \text{ and } n = 6$$

Sol 10: $(1+x)^m (1-x)^n$

$$= (\,^mC_0 x^0 + \,^mC_1 x^1 + \,^mC_2 x^2 + ... + \,^mC_m x^m)$$

$$(\,^nC_0 + \,^nC_1(-x) + \,^nC_2(-x)^2 + ... \,^nC_2(-x)^n)$$

terms of $x = \,^mC_0\,^nC_1\,(-1) + \,^nC_0\,^mC_1$

$$= (-n) + m = m - n = 3 \text{ (given)} \qquad\qquad .... \text{(i)}$$

terms of

$$x^2 = \,^mC_0\,^nC_2(-1)^2 + \,^mC_2\,^nC_0 + \,^mC_1\,^nC_1(-1)$$

$$= 1 \cdot \frac{n \times (n-1)}{1 \cdot 2} 1 + \frac{m(m-1)}{1 \cdot 2} 1 + m(-n)$$

$$= \frac{n^2 - n}{2} + \frac{m^2 - m}{2} - mn = -6 \qquad \dots \text{(ii)}$$

In equation (i) $m - n = 3 \Rightarrow n = (m-3)$

Put the values of n in eq. (ii)

$$\frac{(m-3)^2 - (m-3)}{2} + \frac{m^2 - m}{2} - m(m-3) = -6$$

$m^2 + 3^2 - 3(2)(m) - m + 3 + m^2 - m - 2m^2 + 6m = 12$

$2m^2 - 6m + 9 + 3 - 2m - 2m^2 + 6m = -12 \quad 12 - 2m = -12$

$\Rightarrow 2m = 24 \Rightarrow m = 12$ and $n = 9$

Sol 11: Coefficient of a^{r-1}, a^r, a^{r+1} in the binomial expansion of $(1+a)^n$ are in A. P. so

Terms of $a^{r-1} = T_r = {}^nC_{r-1}(a)^{r-1}$

Terms of $a^r = T_{r+1} = {}^nC_r a^r$

Terms of $a^{r+1} = T_{r+2} = {}^nC_{r+1}a^{r+1}$

Coefficients of T_r, T_{r+1}, T_{r+2} are in A. P. so

$${}^nC_{r-1} + {}^nC_{r+1} = 2\,{}^nC_r$$

$$\frac{n!}{(r-1)!(n-r+1)!} + \frac{n!}{(r+1)!(n-r-1)!} = 2\frac{n!}{r!(n-r)!}$$

$$\Rightarrow \frac{1}{(r-1)!(n-r+1)(n-r)(n-r-1)!}$$

$$+ \frac{1}{(r+1)r(r-1)!(n-r-1)!} = \frac{2}{r(r-1)!(n-r)(n-r-1)!}$$

$$\Rightarrow \frac{1}{(n-r)(n-r+1)} + \frac{1}{r(r+1)} = \frac{2}{r(n-r)}$$

$$\Rightarrow \frac{r(r+1) + (n-r)(n-r+1)}{r(r+1)(n-r)(n-r+1)} = \frac{2}{r(n-r)}$$

$\Rightarrow r^2 + r + n^2 - nr + n - nr + r^2 - r = 2(r+1)(n-r+1)$

$\Rightarrow 2r^2 + n^2 - 2nr + n = 2rn - 2r^2 + 2r + 2n - 2r + 2$

$\Rightarrow n^2 + 4r^2 - 4rn - n - 2 = 0$

$\Rightarrow n^2 - n(4r+1) + 4r^2 - 2 = 0$

Sol 12: n is a positive integer

$\Rightarrow 6^{2n} - 35n - 1 = (6^2)^n - 35n - 1 = (36)^n - 35n - 1$

$= (35+1)^n - 35n - 1$

$= {}^nC_0 35^n + {}^nC_1 35^{n-1} + \dots + {}^nC_{n-2}35^2$

$+ {}^nC_{n-1}35 + {}^nC_n 35^0 - 35n - 1$

And $1225 = 35^2$

so each term is a multiple of 35^2 and is divisible by 1225

Sol 13: $3^{4n+1} + 16n - 3$ is divisible by 256

$256 = 2^8 = 4^4$

$= 3^{4n+1} + 16n - 3$

$= 3 \cdot 3^{4n} + 16n - 3$

$= 3[4-1]^{4n} + 16n - 3$

$= 3[\,{}^{4n}C_0 4^{4n} + {}^{4n}C_1 4^{4n-1}(-1) + \dots +$

$\quad {}^{4n}C_{4n-2}(4)^{4n-4n+2} - {}^{4n}C_{4n-1}(4)^{4n-4n+1} + {}^{4n}C_{4n}] + 16n - 3$

= all terms which is multiple of 4^4 is divisible by 256. So rest of the terms

$= 3[-{}^{4n}C_{4n-2}(4)^2 - {}^{4n}C_{4n-1}(4)^1 + 1] + 16n - 3$

$3[\frac{4n(4n-1)}{1 \cdot 2} \times 4^2 - 4n \times 4 + 1] + 16n - 3$

$= 128n^2 \cdot 3 - 128n = 128(3n^2 - 1)$

$= 128n(3n-1)$ and $(3n-1)$ is always even

so $128n(3n-1) = 128 \times 2^{x+1}$ (assume), $x \in N$

$= 256 \times 2^x$, which is divisible by 256

Sol 14: a_1, a_2, a_3 and a_4 are any four consecutive coefficients in the expansion of $(1+x)^n$

$a_1 = {}^nC_r, \qquad a_2 = {}^nC_{r+1}$

$a_3 = {}^nC_{r+2}, \qquad a_4 = {}^nC_{r+3}$

L. H. S.

$$= \frac{a_1}{a_1 + a_2} + \frac{a_3}{a_3 + a_4}$$

$$= \frac{{}^nC_r}{{}^nC_r + {}^nC_{r+1}} + \frac{{}^nC_{r+2}}{{}^nC_{r+2} + {}^nC_{r+3}}$$

$$- \frac{{}^nC_r}{{}^{n+1}C_{r+1}} + \frac{{}^nC_{r+2}}{{}^{n+1}C_{r+3}}$$

$$= \frac{n!(r+1)!(n-r)!}{(n+1)!(r)!(n-r)!} + \frac{n!(r+3)!(n-r-2)!}{(n+1)!(r+2)!(n-r-2)!}$$

$$= \frac{(r+1)}{(n+1)} + \frac{(r+3)}{(n+1)} = \frac{2r+4}{n+1} = \frac{2(r+2)}{n+1}$$

$$= 2\frac{n!(r+2)!(n-r-1)!}{(n+1)!(r+1)!(n-r-1)!} = 2\frac{{}^{n}C_{r+1}}{{}^{n+1}C_{r+2}}$$

$$= 2\frac{{}^{n}C_{r+1}}{{}^{n}C_{r+2} + {}^{n}C_{r+1}} = \frac{2a_2}{a_2 + a_3}$$

Sol 15: 3 consecutive coefficients in the expansion of $(1+x)^n$ are in the ratio $6 : 33 : 110$

$$= T_{r+1} : T_{r+2} : T_{r+3}$$

$$= {}^{n}C_{r} : {}^{n}C_{r+1} : {}^{n}C_{r+2}$$

$$= \frac{n!}{r!n-r!} : \frac{n!}{r+1!n-r-1!} : \frac{n!}{r+2!n-r-2!}$$

$$= \frac{1}{(n-r)(n-r-1)} : \frac{1}{(r+1)(n-r-1)} : \frac{1}{(r+1)(r+2)}$$

$$= 6 : 33 : 110$$

$$\Rightarrow \frac{(r+1)(n-r-1)}{(n-r)(n-r-1)} = \frac{6}{33} = \frac{2}{11} \Rightarrow \frac{r+1}{n-r} = \frac{2}{11}$$

$$\Rightarrow 11r + 11 = 2n - 2r \Rightarrow 2n - 13r - 11 = 0 \qquad \text{.... (i)}$$

And $\dfrac{(r+1)(r+2)}{(r+1)(n-r-1)} = \dfrac{33}{110} = \dfrac{3}{10} \Rightarrow \dfrac{r+2}{n-r-1} = \dfrac{3}{10}$

$$\Rightarrow 10r + 20 = 3n - 3r - 3 \Rightarrow 3n - 13r - 23 = 0 \qquad \text{.... (ii)}$$

Subtracting equation (i) from (ii), we get $n = 12$

Putting $n = 12$ in equation (i)

$13r = 2n - 11 = 2(12) - 11 = 24 - 11 = 13 \Rightarrow r = 1$

So terms are $T_{r+1}, T_{r+2}, T_{r+3}$

Sol 16: a, b, c are three consecutive coefficients in the expansion of power (say n) of $(1+x)^n$

So $a = {}^{n}C_{r}$, $b = {}^{n}C_{r+1}$, $c = {}^{n}C_{r+2}$

$$\frac{a}{b} = \frac{(r+1)}{(n-r)} \Rightarrow an - ar = br + b$$

$$\Rightarrow r = \frac{an - b}{a + b} \qquad \text{......(i)}$$

$$\frac{b}{c} = \frac{(r+2)}{(n-r-1)}$$

$$\Rightarrow bn - br - b = cr + 2c$$

$$\Rightarrow r = \frac{bn - b - 2c}{b + c} \qquad \text{....... (ii)}$$

From (i) and (ii)

$$\frac{an - b}{a + b} = \frac{bn - b - b - 2c}{b + c}$$

$$\Rightarrow abn - b^2 + acn - bc = abn - 2ab - 2ac + b^2n - 2b^2 - 2bc$$

$$\Rightarrow n(ac - b^2) = -ab - 2ac - bc$$

$$\Rightarrow n = \frac{ab + 2ac + bc}{b^2 - ac} = \frac{2ac + b(a+c)}{b^2 - ac}$$

Sol 17: $\left(x - \dfrac{1}{y}\right)^{11}$, $y \neq 0$

$$= {}^{11}C_0 x^{11} + {}^{11}C_1 x^{11-1}\left(-\frac{1}{y}\right) + {}^{11}C_2 x^{11-2}\left(-\frac{1}{y}\right)^2$$

$$+ {}^{11}C_3 x^{11-3}\left(-\frac{1}{y}\right)^3 + \cdots + {}^{11}C_{11}\left(-\frac{1}{y}\right)^{11}$$

Sol 18: $(1-x+x^2)^4 = ((1-x)+x^2)^4$

$$= {}^{4}C_0(1-x)^4 + {}^{4}C_1(1-x)^3 x^2 + {}^{4}C_2(1-x)^2(x^2)^2$$

$$+ {}^{4}C_3(1-x)(x^2)^3 + {}^{4}C_4(x^2)^4$$

$$= (1-x)^4 + 4x^2(1-x)^3 + 6(1+x^2-2x)x^4 + 4x^6(1-x) + x^8$$

$$= ({}^{4}C_0 - {}^{4}C_1 x + {}^{4}C_2 x^2 - {}^{4}C_3 x^3 + {}^{4}C_4 x^4)$$

$$+ 4x^2\,({}^{3}C_0 - {}^{3}C_1 x + {}^{3}C_2 x^2 - {}^{3}C_3 x^3)$$

$$+ (6 + 6x^2 - 12x)x^4 + 4x^6 - 4x^7 + x^8$$

$$= 1 - 4x + 6x^2 - 4x^3 + x^4 + 4x^2 - 12x^3$$

$$+ 12x^4 - 4x^5 + 6x^4 + 6x^6 - 12x^5 + 4x^6 - 4x^7 + x^8$$

$$= 1 - 4x + 10x^2 - 16x^3 + 19x^4 - 16x^5 + 10x^6 - 4x^7 + x^8$$

Sol 19: $(1.2)^{4000} = (1 + 0.2)^{4000}$

$$= {}^{4000}C_0(0.2)^0 + {}^{4000}C_1(0.2)^1 + {}^{4000}C_2(0.2)^2 + \ldots\ldots$$

$$= 1 + 4000(0.2) + \ldots\ldots\ldots = 1 + 800 + \ldots\ldots . = 801 + \ldots\ldots$$

So $(1.2)^{4000}$ is greater than 800

Sol 20: For $(1+x)^n$

T_{14}, T_{15} and T_{16} are in A. P.

$$\Rightarrow T_{14} + T_{16} = 2T_{15} \Rightarrow {}^{n}C_{13} + {}^{n}C_{15} = 2\,{}^{n}C_{14}$$

$$\Rightarrow {}^{n}C_{13} + {}^{n}C_{15} = 2\,{}^{n}C_{14}$$

$$\frac{n!}{13!(n-13)!} + \frac{n!}{15!(n-15)!} = 2\frac{n!}{14!(n-14)!}$$

$$\frac{1}{(n-13)(n-14)} + \frac{1}{15\times 14} = \frac{2}{14(n-14)}$$

$$\frac{15\times 14 + (n-13)(n-14)}{15\times 14(n-13)(n-14)} = \frac{2}{14(n-14)}$$

$210 + n^2 + 182 - n(13+14) = 2\times 15(n-13)$

$n^2 + 392 - 27n = 30n - 390$

$n - 57n + 782 = 0$

$$n = \frac{57 \pm \sqrt{57^2 - 4(1)(782)}}{2(1)} = \frac{57 \pm \sqrt{121}}{2} = \frac{57 \pm 11}{2}$$

$n = 34,\ 23$

Sol 21: Given

$^nC_r = 165;\ ^nC_{r+1} = 330;\ ^nC_{r+2} = 462$

$$\frac{^nC_r}{^nC_{r+1}} = \frac{165}{330} = \frac{1}{2}$$

$$\Rightarrow \frac{n!(r+1)!(n-r-1)!}{r!(n-r)!n!} = \frac{1}{2}$$

$$\Rightarrow \frac{(r+1)}{(n-r)} = \frac{1}{2}$$

$\Rightarrow 2r+2 = n-r$

$\Rightarrow 3r = n-2$ .... (i)

$$\frac{^nC_{r+2}}{^nC_{r+1}} = \frac{462}{330} = \frac{7}{5}$$

$$\frac{(r+1)!n!(n-r-1)!}{n!(r+2)!(n-r-2)!} = \frac{n-r-1}{r+2} = \frac{7}{5}$$

$\Rightarrow 5n - 5r - 5 = 7r + 14$

$\Rightarrow 12r = 5n - 19$ .... (ii)

From eq (i) and (ii)

$4(n2) = 5n19 \qquad \Rightarrow n = 11$

So $3r = 11 - 2 = 9 \quad \Rightarrow r = 9/3 = 3$

Position of coefficients are $T_{3+1}, T_{3+2}, T_{3+3}$

Sol 22: $(7-5x)^{11},\ x = \dfrac{2}{3}$

$$\frac{n+1}{1+\left|\frac{x}{a}\right|} = \frac{11+1}{1+\left(\frac{7\times 3}{5\times 2}\right)} = \frac{12}{1+\frac{21}{10}} = \frac{12}{3.1} = 3.87$$

So greatest is 4^{+n}.

$$|T_4| = |T_{3+1}| = {}^{11}C_3 (7)^{11-3}\left(5\times \frac{2}{3}\right)^3$$

$$= \frac{11\times 10\times 9}{1.2.3}\times 7^8\ 5^3 \times \frac{2^3}{3^3} = \frac{11}{9}2^3 5^4 7^8$$

Sol 23: $(1 + 3x^2 + x^4)\left(1 + \dfrac{1}{x}\right)^8$

For Coefficient of x^{-1}

$$= (1)\ {}^8C_1\left(\frac{1}{x}\right) + 3x^2\ {}^8C_3\frac{1}{x^3} + x^4\ {}^8C_5\frac{1}{x^5}$$

$$= \frac{8}{x} + \frac{3\times 8\times 7\times 6}{1\times 2\times 3}\times x^{-1} + \frac{8\times 7\times 6}{1.2.3}x^{-1}$$

$$= \frac{1}{x}[8 + 168 + 56] = \frac{232}{x}$$

Coefficient of $x^{-1} = 232$

Sol 24: $\left(\sqrt{x} + \dfrac{k}{x^2}\right)^{10}$

$$T_{r+1} = {}^{10}C_r(\sqrt{x})^{10-r}\left(\frac{k}{x^2}\right)^r = {}^{10}C_r(x)^{\frac{10-r}{2}}k^r x^{2r}$$

T_{r+1} is independent of x

So $\dfrac{10-r}{2} - 2r = 0 \Rightarrow 10 - 5r = 0 \Rightarrow r = \dfrac{10}{5} = 2$

So Coefficient is $= {}^{10}C_2 k^2$

$$= \frac{10\times 9}{1\times 2}\times k^2 = 405 \Rightarrow k^2 = 9 \Rightarrow k = \pm 3$$

Sol 25: $(x + a)^n$

A = Sum of odd terms

B = Sum of even terms

(ii) $2(A^2 + B^2) = (x+a)^{2n} + (x-a)^{2n}$

$A = {}^nC_0 x^n + {}^nC_2 x^{n-2}a^2 + \dots\ {}^nC_n x^0 a^n$

$B = {}^nC_1 x^{x-1}a + {}^nC_3 x^{n-3}a^3 + \dots + {}^nC_{n-1} + xa^{n-1}$

$2(A^2 + B^2) = (A+B)^2 + (A-B)^2 = (x+a)^{2n} + (x-a)^{2n}$

L. H. S. = R. H. S.

Sol 26: $(1+2x+x^2)^{27} = ((1+x)^2)^{27} = (1+x)^{54}$

$T_{r+1} = {}^{54}C_r x^r$

Coefficient of $x^{40} \Rightarrow r=40$

Coefficient $= {}^{54}C_{40} = {}^{54}C_{54-40} = {}^{54}C_{14}$

Sol 27: $\left(\dfrac{3}{2}x^2 - \dfrac{1}{3x}\right)^9$

$T_{r+1} = {}^9C_r\left(\dfrac{3}{2}x^2\right)^{9-r}\left(-\dfrac{1}{3x}\right)^r$

For independence of x

$2(9-r)-r = 18-2r-r = 18-3r = 0$

Coefficient $r = 6$

$T_{r+1} = T_{6+1} = {}^9C_6\left(\dfrac{3}{2}\right)^{9-6}\left(-\dfrac{1}{3}\right)^6$

$= {}^9C_3 \times \dfrac{3^3}{2^3}\dfrac{1}{3^6} = \dfrac{9\times8\times7}{1.2.3}\times\dfrac{(1)}{2^33^3} = \dfrac{7\times3}{2.3^3} = \dfrac{7}{18}$

Sol 28: $(1+ax)^n = 1+8x+24x^2+\ldots\ldots$

${}^nC_0 + {}^nC_1 ax + {}^nC_2(ax)^2 + \ldots. = 1 + 8x + 24x^2 + \ldots\ldots$

So ${}^nC_1 a = 8$ and ${}^nC_2 a^2 = 24$

$na = 8$ and $\dfrac{n(n-1)}{2}a^2 = 24$

$a^2n^2 - na^2 = 48 \Rightarrow (8)^2 - 8a = 48 \Rightarrow 64 - 8a = 48$

$\Rightarrow a = 2 \Rightarrow n = 4$

Exercise 2

Single Correct Choice Type

Sol 1: (C) $(x^{1/3} - x^{-1/2})^{15}$

$T_{r+1} = {}^{15}C_r (x^{1/3})^{15-r}(-x^{-1/2})^r$

Power of $x = \dfrac{15-r}{3} - \dfrac{r}{2} = 0$ for x^0

$2(15-r) - 3r = 0$

$30 - 2r - 3r = 0 \Rightarrow 5r = 30 \Rightarrow r = 30/5 = 6$

Coefficient $T_{r+1} = {}^{15}C_6 \times 1 = 5005$

$5m = 5005 \Rightarrow m = 1001$

Sol 2: (C) In the expansion $\left(2+\dfrac{x}{3}\right)^n$ the coefficients of x^7 & x^8 are equal

$${}^nC_7(2)^{n-7}\left(\dfrac{1}{3}\right)^7 = {}^nC_8(2)^{n-8}\left(\dfrac{1}{3}\right)^8$$

$\dfrac{6}{(n-7)} = \dfrac{1}{8} \Rightarrow n-7 = 48 \Rightarrow n = 48+7 = 55$

Sol 3: (A) $(x-1)\left(x-\dfrac{1}{2}\right)\left(x-\dfrac{1}{2^2}\right)\ldots\ldots\left(x-\dfrac{1}{2^{49}}\right)$

Max power of $x = 50$

Coefficient of x^{49}

$= -1 - \dfrac{1}{2} - \dfrac{1}{2^2} - \dfrac{1}{2^3}\cdots-\dfrac{1}{2^{49}}$

$= \left[\dfrac{1-\left(\dfrac{1}{2}\right)^{50}}{1-\dfrac{1}{2}}\right] = -2\left[1-\dfrac{1}{2^{50}}\right]$

Sol 4: (D) (3^P+2)

$P = 3^{4n}, n\in N = 3^{3^{4n}} + 2$

$3^0 = 1, 3^1 = 3,\ 3^2 = 9,\ 3^3 = 27, 3^4 = 81$

Last digit $= 1,3,9,7$

Last digit of 3^x repeat after every power of 4 so 3^{4n} last digit $= 1$

$3^1 = 3$

$3^1 + 2 = 5$

So last digit of $3^{3^{4n}} + 2$ is 5

Sol 5: (A) $\left(2x+\dfrac{1}{x}\right)^n$

Sum of binomial coefficient $= 2^n = 256$

$2^n = 2^8 \Rightarrow n = 8$

Constant term $=$

${}^8C_4(2x)^4\cdot\left(\dfrac{1}{x}\right)^4 = \dfrac{8\times7\times6\times5}{1.2.3.4}\times 2^4 x^{4-4} = 1120$

Sol 6: (A) $\left(\dfrac{x}{2} - \dfrac{3}{x^2}\right)^{10}$

$T_{r+1} = {}^{10}C_r\left(\dfrac{x}{2}\right)^{10-r}\left(-\dfrac{3}{x^2}\right)^r$

Power of $x = 10 - r - 2(r) = 4$ (given)

$\Rightarrow 10 - 3r = 4 \Rightarrow 3r = 10 - 4 = 6 \Rightarrow r = 6/3 = 2$

Coefficient $T_{r+1} = {}^{10}C_2 \left(\dfrac{1}{2}\right)^{10-2} (-3)^2$

$= \dfrac{10 \times 9}{1.2} 2^{-8} 3^2 = \dfrac{5 \times 9 \times 9}{2^8} = \dfrac{405}{256}$

Sol 7: (A) $11^{27} + 21^{27}$

$= (16-5)^{27} + (16+5)^{27}$

$= 2\left[{}^{27}C_0 16^{27} + \dots + {}^{27}C_{26} 16 \right]$

$= 2.\, 16k = 32k$

Always divisible by 16

Sol 8: (D) $N = 7^{00} - 3^{100}$

$N = (5+2)^{100} - (5-2)^{200}$

$N = 2\left[{}^{100}C_1 5^{99}.2 + \dots + {}^{100}C_{99} 5.2^{99} \right]$

$N = [{}^{100}C_1 5^{97}.100 + 10^3 \, {}^{100}C_3 5^{94}$

$\qquad \dots + {}^{100}C_{99} 10.2^{99}]$

$N = 1000.[10.5^{97} + \dots + 2^{99}]$

Integer

Last three digits $= 000$

Sol 9: (D) $3^{400} = (3^2)^{200}$

$(9)^{200} = (10-1)^{200}$

$= {}^{200}C_0 10^{200} - {}^{200}C_1 10^{199} + \dots {}^{200}C_{199} 10^1 + {}^{200}C_{200}.1$

$= 10m + 1 \qquad (m \in N)$

Last 2 digits are 01

Sol 10: (A) $(1+x+x^2)^{25} = a_0 + a_1 x + \dots + a_{50} x^{50}$

$x = 1$

$3^{25} = a_0 + a_1 + a_2 + \dots + a_{50}$ \qquad ...(i)

$x = -1$

$(1 - 1 + 1)^{25} = 1$

$= a_0 - a_1 + a_2 - a_3 + \dots + a_{50}$ \qquad .... (ii)

Sum of both eq^n.

$3^{25} + 1 = 2(a_0 + a_2 + a_4 + \dots + a_{50})$

$a_0 + a_2 + a_4 + \dots + a_{50} = \dfrac{1}{2}(3^{25} + 1)$

$= \dfrac{1}{2}\left[(4-1)^{25} + 1\right]$

$= \dfrac{1}{2}\left[\left({}^{25}C_0 4^{25} - {}^{25}C_1 4^{24} + \dots + {}^{25}C_{24} 4 - 1 \right) + 1 \right]$

$= 2$ m always even. ($\because$ divisible by 2)

Sol 11: (B) $(1^2+1)1! + (2^2+1)2! + (3^2+1)3! + \dots + (n^2+1)n!$

$T_n = (n^2 + 1) n! = n(n + 1)! - (n - 1) n!$

$S_n = n(n+1)!$

Sol 12: (A) $P_m \to {}^{n}P_m$

$1P_1 + 2P_2 + 3P_3 + \dots + n.\, P_n$

$= 1.\, n + 2.\, n(n - 1) + 3n(n - 1)(n - 2)$

$\qquad + 4n(n - 1)(n - 2)(n - 3) + \dots + n.\, n!$

Add ($+1$ and -1)

$= 1 + {}^{n}C_1 + 2\,{}^{n}C_2 2! + 3\,{}^{n}C_3 3! + 4\,{}^{n}C_4 4! + \dots n\,{}^{n}C_n n! - 1$

$= -1 + 1 + \displaystyle\sum_{i=0}^{n} i\,{}^{n}C_i (i)!$

$= -1 + 1 \displaystyle\sum_{i=0}^{n} iP_i$

When $1 + 1P_1 + 2P_2 + 3P_3 + \dots + nP_n = (n+1)!$

$= -1 + 1(n+1)! - 1$

$= (n+1)! - 1$

Sol 13: (D)

$$\dfrac{1}{\sqrt{4x+1}} \left[\left[\dfrac{1 + \sqrt{4x+1}}{2}\right]^7 - \left[\dfrac{1 - \sqrt{4x+1}}{2}\right]^7 \right]$$

$= \dfrac{1.2}{2^7 \sqrt{4x+1}} \left[{}^{7}C_1 \sqrt{4x+1} + {}^{7}C_3 (\sqrt{4x+1})^3 \right.$

$\qquad\qquad \left. + \dots + {}^{7}C_7 (\sqrt{4x+1})^7 \right]$

$= 2^{-6} \left[{}^{7}C_1 + {}^{7}C_3 (4x+1) + {}^{7}C_5 \right.$

$\left. (4x+1)^2 + {}^{7}C_7 (4x+1)^3 \right]$

$\Rightarrow$ Max. power of $x = 3$

Sol 14: (A)

$$\left(a^{1/13} + \frac{a}{\sqrt{a^{-1}}}\right)^n = \left(a^{1/13} + a^{1+1/2}\right)^n = \left(a^{1/13} + a^{3/2}\right)^n$$

$$T_2 = {}^nC_1 (a^{1/13})^{n-1} + (a^{3/2})^1 = 14a^{5/2}$$

$$\Rightarrow n \; a^{\frac{n-1}{13}+\frac{3}{2}} = 14a^{5/2} \Rightarrow n = 14$$

$$\frac{{}^{14}C_3}{{}^{14}C_2} = \frac{14-3+1}{3} = \frac{12}{3} = 4$$

Sol 15: (B) $(1+x)(1+x+x^2)(1+x+x^2+x^3)$

..... $(1+x+..... +x^n)$

$$= a_0 + a_1 x + a_2 x^2 + ... + a_m x^m$$

$$\sum_{r=0}^{m} ar = a_0 + a_1 + a_2 + ... + a_m$$

At $x = 1$

$$= 2.3.4.5.6.....(n+1) = (n+1)!$$

Sol 16: (C) $(1+x)^{43}$

Given $T_{2r+1} = T_{r+2}$

$${}^{43}C_{2r} = {}^{43}C_{r+1} = {}^{43}C_{43-(r+1)}$$

$$\Rightarrow 2r = 43 - r - 1 = 42 - r \Rightarrow 3r = 42 \Rightarrow r = \frac{42}{3} = 14$$

Sol 17: (A) $\left(x^2 + \frac{a}{x^3}\right)^{10}$

Coefficient of x^5 is equal to that of x^{15}

$$T_{r+1} = {}^{10}C_r (x^2)^{10-r}\left(\frac{a}{x^3}\right)^r$$

Power of $x = 2(10-r) - 3r = 20-5r$

$$20 - 5r = 5 \Rightarrow r = 3$$

$$20 - 5r = 15 \Rightarrow r = 1$$

$$T_{3+1} = T_{1+1}$$

$${}^{10}C_3 a^3 = {}^{10}C_1 a$$

$$\frac{10 \times 9 \times 8}{1.2.3} a^2 = 10$$

$$a^2 = \frac{1}{12} \Rightarrow a = \frac{1}{\sqrt{12}} = \frac{1}{2\sqrt{3}}$$

Sol 18: (C) $\left(x^2 + \frac{a}{x^3}\right)^{10}$

Power of x for term

$$T_{r+1} = {}^{10}C_r (x^2)^{10-r}\left(\frac{a}{x^3}\right)^r$$

Power of $x = 2(10-r) - 3r = 20 - 5r = 0$

$$\Rightarrow 5r = 20 = r = 20/5 = 4$$

$$T_{4+1} = {}^{10}C_4 \text{ binomial coefficient}$$

Sol 19: (C) $\left(\frac{1}{x^{8/3}} + x^2 \log_{10} x\right)^8$

$$T_6 = T_{5+1} = {}^8C_5 \left(\frac{1}{x^{8/3}}\right)^{8-5} \left(x^2 \log_{10} x\right)^5 = 5600$$

$$\Rightarrow \frac{8 \times 7 \times 6}{1.2.3} \times \left(x^{-8/3}\right)^3 x^{10} (\log_{10} x)^5 = 5600$$

$$x^{-8+10}(\log_{10} x)^5 = 100 \Rightarrow x^2 (\log_{10} x)^5 = 100$$

Assume $x = 10^y$

So $10^{2y}(\log_{10} 10^y)^5 = 10^2 \Rightarrow 10^{2y-2} \; y^5 = 1$

$$\Rightarrow y = 1 \Rightarrow x = 10$$

Sol 20: (B) $(1+x)(1+x+x^2)(1+x+x^2+x^3)$

..... $(1+x+..... x^{100})$

Highest power of $x = 1+2+3+..... +100$

$$= \frac{100(100+1)}{2} = 50 \times 101 = 5050$$

Sol 21: (B) $(5 + 2\sqrt{6})^n = p + f$

$$p = [(5 + 2\sqrt{6})^n] - f$$

$$f^2 - f + pf - p = f(f-1) + p(f-1) = (f-1)(f+p)$$

Assume $F = (5 - 2\sqrt{6})^n = \left(\frac{1}{5 + 2\sqrt{6}}\right)^n$

$$0 < f < 1, \; 0 < F < 1$$

$$F + f + p = (5 + 2\sqrt{6})^n + (5 - 2\sqrt{6})^n = \text{integer} = 2I$$

$$F + f = 2I - p = \text{Integer}$$

$$0 < F + f < 2 \Rightarrow F + f = 1 \Rightarrow F = 1 - f$$

$$(F)(f+p) = (5 - 2\sqrt{6})^n (5 + 2\sqrt{6})^n = -1$$

Sol 22: (B) $(\sqrt{2} + \sqrt[4]{3})^{100} = (2^{1/2} + 3^{1/4})^{100}$

L. C. M. of 2 and 4 =4

Total terms $= n+1 = 100+1 = 101$

T rational $= {}^{100}C_{4n}\left(2^{1/2}\right)^{100-4n}\left(3^{1/4}\right)^{4n}$

$\Rightarrow 0 \le 100 - 4n \le 100$

$\Rightarrow 0 \le n \le 25 \; n \in N$

$n = \{0,1,2,3.....25\}$

Total number for $n = 26$

Sol 23: (D) $\left(x\sin\theta + \dfrac{\cos\theta}{x}\right)^{10}$

$T_{r+1} = {}^{10}C_r (x\sin\theta)^{10-r}\left(\dfrac{\cos\theta}{x}\right)^r$

Power of $x = 10 - r + r(-1) = 10 - 2r = 0$ (given)

$\Rightarrow r = 5$

$T_{r+1} = {}^{10}C_5 (\sin\theta)^5 (\cos\theta)^5 = {}^{10}C_5 (\sin\theta\cos\theta)^5$

$= {}^{10}C_5 \left(\dfrac{\sin 2\theta}{2}\right)^5$. Max value when $\sin 2\theta = 1$

$\therefore$ Max. value $= \dfrac{{}^{10}C_5}{2^5}$

Sol 24: (B) $(1+x-3x^2)^{2145} = a_0 + a_1 x + a_2 x^2 +$

At $x = -1$

$(1 - 1 - 3)^{2145} = -(3)^{2145}$

$= a_0 - a_1 + a_2 - a_3 + ...$

L. H. S. $= 3^{2145} = 3.3^{2144} = 3[9]^{1072}$

Even power of 9 ends with 1. Hence 3^{2145} ends with 3.

Sol 25: (B) $\left(\dfrac{4x^2}{3} - \dfrac{3}{2x}\right)^9$

$T_{r+1} = {}^9C_r\left(\dfrac{4x^2}{3}\right)^{9-r}\left(-\dfrac{3}{2x}\right)^r$

Power of $x = 2(9-r) + (-1)r = 18 - 3r = 6$

$\Rightarrow 3r = 18 - 6 = 12 \Rightarrow r = 4$

Coefficient $\;{}^9C_4\left(\dfrac{4}{3}\right)^{9-4}\left(\dfrac{-3}{2}\right)^4 = \dfrac{9\times 8\times 7\times 6}{1.2.3.4}\left(\dfrac{4}{3}\right)^5\left(\dfrac{3}{2}\right)^4$

$= 9\times 2\times 7\times \dfrac{2^{10}\times 3^4}{3^5 \times 2^4} = 21\times 2^7 = 2688$

Sol 26: (C) $[x+(x^3-1)^{1/2}]^5 + [x(x^3-1)^{1/2}]^5$

$= 2[{}^5C_0 x^5 + {}^5C_2 x^{5-2}(x^3-1) + {}^5C_4 x^{5-4}(x^3-1)^2]$

Max power of $x = 7$

Sol 27: (A) $(1-2x+5x^2-10x^3)(1+x)^n$

$= 1+a_1 x - 1a_2 x^2 + ...$ and $a_1^2 = 2a_2$

Coefficient of $x = a_1 = {}^nC_1 - 2 = n-2$

Co-efficient of $x^2 = a_2 = 5 + {}^nC_2 - 2{}^nC_1 = 5 + \dfrac{n(n-1)}{2} - 2n$

$= 5 + \dfrac{n^2 - n - 4n}{2} = 5 + \dfrac{n^2 - 5n}{2}$

$a_1^2 = 2a_2 \Rightarrow (n-2)^2 = 2\left[\dfrac{10 + n^2 - 5n}{2}\right]$

$\Rightarrow n^2 + 4 - 4n = 10 + n^2 - 5n \Rightarrow n = 6$

Sol 28: (D) $aC_0 + (a+b)C_1 + (a+2b)C_2 + ... + (a+nb)C_n$

$a(C_0 + C_1 + C_2 + ... + C_n)$

$+ b(C_1 + 2C_2 + ... + nC_n)$

$= a2^n + b[n2^{n-1}] = 2^{n-1}[2a + nb]$

Previous Years' Questions

Sol 1: (A) In the expansion

$(1+x)^{2n}, t_{3r} = {}^{2n}C_{3r-1}\left(x\right)^{3r-1}$

$t_{r+2} = {}^{2n}C_{r+1}\left(x\right)^{r+1}$

Since, binomial coefficient of t_{3r} and t_{r+2} are equal.

$\Rightarrow {}^{2n}C_{3r-1} = {}^{2n}C_{r+1}$

$\Rightarrow 3r - 1 = r + 1$ or $2n = (3r-1) + (r+1)$

$\Rightarrow 2r = 2$ or $2n = 4r$

$\Rightarrow r = 1$ or $n = 2r$

But $r > 1$,

$\therefore$ We take $n = 2r$

Sol 2: (A) We have $C_n^2 - 2C_1^2 + 3C_2^2 - 4C_3^2$

$+ .... + (-1)^n (n+1) C_n^2$

$= \{C_0^2 - C_1^2 + C_2^2 - C_3^2 + ... + (-1)^n C_n^2\}$

$-\{C_1^2 - 2C_2^2 + 3C_3^2 - .... + (-1)^n n C_n^2\}$

$= (-1)^{n/2} \cdot \dfrac{n!}{\left(\dfrac{n}{2}\right)!\left(\dfrac{n}{2}\right)!} - (-1)^{\frac{n}{2}-1} \dfrac{n}{2} \dfrac{n!}{\left(\dfrac{n}{2}\right)!\left(\dfrac{n}{2}\right)!}$

$= (-1)^{n/2} \dfrac{n!}{\left(\dfrac{n}{2}\right)!\left(\dfrac{n}{2}\right)!}\left(1 + \dfrac{n}{2}\right)$

$\therefore \dfrac{2\left(\dfrac{n}{2}\right)!\left(\dfrac{n}{2}\right)!}{n!}$

$\{C_0^2 - 2C_1^2 + 3C_2^2 - .... + (-1)^r (n+1) C_n^2\}$

$= \dfrac{2\left(\dfrac{n}{2}\right)!\left(\dfrac{n}{2}\right)!}{n!}(-1)^{n/2} \dfrac{n!}{\left(\dfrac{n}{2}\right)!\left(\dfrac{n}{2}\right)!}\dfrac{(n+2)}{2} = (-1)^{n/2}(n+2)$

Sol 3: (C) We know that $(a+b)^5 + (a-b)^5$

$= {}^5C_0 a^5 + {}^5C_1 a^4 b + {}^5C_2 a^3 b^2$

$+ {}^5C_3 a^2 b^3 + {}^5C_4 ab^4 + {}^5C_5 b^5 + {}^5C_0 a^5 - {}^5C_1 a^4 b$

$+ {}^5C_2 a^3 b^2 - {}^5C_3 a^2 b^3 + {}^5C_4 ab^4 - {}^5C_5 b^5$

$= 2\left[a^5 + 10a^3 b^2 + 5ab^4\right]$

$\therefore \left[x + \left(x^3 - 1\right)^{1/2}\right]^5 + \left[x - \left(x^3 - 1\right)^{1/2}\right]^5$

$= 2\left[x^5 + 10x^3 \left(x^3 - 1\right) + 5x\left(x^3 - 1\right)^2\right]$

Therefore, the given expression is a polynomial of degree 7.

Sol 4: (D) ${}^nC_r + 2\,{}^nC_{r-1} + {}^nC_{r-2}$

$= \left({}^nC_r + {}^nC_{r-1}\right) + \left({}^nC_{r-1} + {}^nC_{r-2}\right)$

We know that

$${}^nC_r + {}^nC_{r-1} = {}^{n+1}C_r$$

$$\therefore {}^{n+1}C_r + {}^{n+1}C_{r-1} = {}^{n+2}C_r$$

Sol 5: (B) $\dbinom{n}{r} + 2\dbinom{n}{r-1} + \dbinom{n}{r-2}$

$= \left[\dbinom{n}{r} + \dbinom{n}{r-1}\right] + \left[\dbinom{n}{r-1} + \dbinom{n}{r-2}\right]$

$= \dbinom{n+1}{r} + \dbinom{n+1}{r-1} = \dbinom{n+2}{r}$

According to given condition, $T_n = {}^nC_3$

and $T_{n+1} - T_n = 21$

$\Rightarrow {}^{n+1}C_3 - {}^nC_3 = 21$

$\Rightarrow \dfrac{1}{6}(n+1)(n)(n-1) - \dfrac{1}{6}n(n-1)(n-2) = 21$

$\Rightarrow \dfrac{n(n-1)}{6}\left[(n+1)-(n-2)\right] = 21$

$\Rightarrow \dfrac{n(n-1)}{6} = 21 \Rightarrow n(n-1) = 42$

$\Rightarrow n = 7$

Sol 6: (B) Given, ${}^{n-1}C_r = \left(k^2 - 3\right){}^nC_{r+1}$

$\Rightarrow {}^{n-1}C_r = \left(k^2 - 3\right)\dfrac{n}{r+1}\,{}^{n-1}C_r$

$\Rightarrow k^2 - 3 = \dfrac{r+1}{n}$

(Since, $n \geq r \Rightarrow \dfrac{r+1}{n} \leq 1$ and $n, r > 0$)

$\Rightarrow 0 < k^2 - 3 \leq 1$

$\Rightarrow 3 < k^2 \leq 4$

$\Rightarrow k \in \left[-2, -\sqrt{3}\right) \cup \left(\sqrt{3}, 2\right]$

Sol 7: (C) Let $\dbinom{30}{0}\dbinom{30}{10} - \dbinom{30}{1}\dbinom{30}{11}$

$+ \dbinom{30}{2}\dbinom{30}{12} - .... + \dbinom{30}{20}\dbinom{30}{30}$

$\therefore A = {}^{30}C_0 \cdot {}^{30}C_{10} - {}^{30}C_1 \cdot {}^{30}C_{11}$

$+ {}^{30}C_2 \cdot {}^{30}C_{12} - .... + {}^{30}C_{20} \cdot {}^{30}C_{30}$

= Coefficient of x^{20} in $(1 + x)^{30}.(1 - x)^{30}$

= Coefficient of x^{20} in $(1 - x^2)^{30}$

= Coefficient of x^{20} in $\sum_{r=0}^{30}(-1)^r\ {}^{30}C_r\left(x^2\right)^r$

$\therefore$ For coefficient of x^{20} clearly $2r = 20 \Rightarrow r = 10$

Put $(r = 10) = {}^{30}C_{10}$

Sol 8: (D) A_r = Coefficient of x^r in $(1 + x)^{10} = {}^{10}C_r$

B_r = Coefficient of x^r in $(1 + x)^{2n} = {}^{20}C_r$

C_r = Coefficient of x^r in $(1 + x)^{30} = {}^{30}C_r$

$\therefore \sum_{r=1}^{10}A_r\left(B_{10}B_r - C_{10}A_r\right) = \sum_{r=1}^{10}A_rB_{10}B_r - \sum_{r=1}^{10}A_rC_{10}A_r$

$= \sum_{r=1}^{10}{}^{10}C_r\ {}^{20}C_{10}\ {}^{20}C_r - \sum_{r=1}^{10}{}^{10}C_r\ {}^{30}C_{10}\ {}^{10}C_r$

$= \sum_{r=1}^{10}{}^{10}C_{10-r}\cdot{}^{20}C_{10}\ {}^{20}C_r - \sum_{r=1}^{10}{}^{10}C_{10-r}\ {}^{30}C_{10}\ {}^{10}C_r$

$= {}^{20}C_{10}\sum_{r=1}^{10}{}^{10}C_{10-r}\cdot{}^{20}C_r - {}^{30}C_{10}\sum_{r=1}^{10}{}^{10}C_{10-r}\ {}^{10}C_r$

$= {}^{20}C_{10}\left({}^{30}C_{10} - 1\right) - {}^{30}C_{10}\left({}^{20}C_{10} - 1\right)$

$- {}^{30}C_{10} - {}^{20}C_{10} = C_{10} - B_{10}$

Sol 9: (D) $\left(1 + ax + bx^2\right)$

$\left[1 - {}^{18}C_1 2x + {}^{18}C_2\left(2x\right)^2 - {}^{18}C_3\left(2x\right)^3 + {}^{18}C_4\left(2x\right)^4\right]$

Coefficient of x^3 is

$-{}^{18}C_3\left(2^3\right) + a\left({}^{18}C_2 \times 4\right) - b\left({}^{18}C_1 \times 2\right) = 0$ ...(i)

Coefficient of x^4 is

${}^{18}C_4\left(2^4\right) + a\left(-{}^{18}C_3 \times 2^3\right) + {}^{18}C_2 b 2^2 = 0$ ....(ii)

or solving both these equation

$a = 16$ and $b = 272/3$.

Sol 10: (A)

$\left(1 - 2\sqrt{x}\right)^{50} = {}^{50}C_0 - {}^{50}C_1\left(2\sqrt{x}\right)^1 + {}^{50}C_2\left(2\sqrt{x}\right)^2$

$- {}^{50}C_3\left(2\sqrt{x}\right)^3 + {}^{50}C_4\left(2\sqrt{x}\right)^4$

So, sum of coefficient Integral powers of x

$S = {}^{50}C_0 + {}^{50}C_2.2^2 + {}^{50}C_4.2^4 + ... + {}^{50}C_{50}.2^{50}$

Now,

$\left(1 + x\right)^{50} = 1 + {}^{50}C_1 x + {}^{50}C_2 x^2 + {}^{50}C_3 x^3 + {}^{50}C_4 x^4$

$+ ..... + {}^{50}C_{50} x^{50}$

Put $x = 2, -2$

$3^{50} = 1 + {}^{50}C_1.2 + {}^{50}C_2.2^2 + {}^{50}C_3.2^3$

$+ {}^{50}C_4.2^4 + ... + {}^{50}C_{50}.2^{20}$ ....(i)

$1 = 1 - {}^{50}C_1.2 + {}^{50}C_2.2^2 - {}^{50}C_3.2^3$

$+ {}^{50}C_4 2^4 - .... + {}^{50}C_{50}.2^{50}$ ....(ii)

(i) + (ii)

$3^{50} + 1 = 2\left[1 + {}^{50}C_2.2^2 + {}^{50}C_4.2^4 + ... + {}^{50}C_{50}.2^{50}\right]$

$\therefore \dfrac{3^{50} + 1}{2} = 1 + {}^{50}C_2.2^2 + {}^{50}C_4.2^4 + ... + {}^{50}C_{50}.2^{50}$

Sol 11: (D) Number of terms $= \dfrac{\left(n+1\right)\left(n+2\right)}{2} = 28$

$\Rightarrow n = 6$

$\therefore a_0 + \dfrac{a_1}{x} + \dfrac{a_2}{x^2} + ... + \dfrac{a_{2n}}{x^{2n}} = \left(1 - \dfrac{2}{x} + \dfrac{4}{x^2}\right)^n$

Put $x = 1, n = 6,$

$a_0 + a_1 + a_2 + ... + a_{2n} = 3^6 = 729$

JEE Advanced/Boards

Exercise 1

Sol 1: $f(x) = 1 - x + x^2 - x^3 + ... x^{16} - x^{17}$

$= a_0 + a_1(1+x) + a_2(1+x)^2 + ... + a_{17}(1+x)^{17}$

Differentiating both sides

$-1 + 2x ... -17x^{16} = a_1 + 2a_2(1 + x) + ... + 17a_{17}(1 + x)^{16}$

Again differentiating

$2 - 6x + ... = 2a_2 + 6a_3(1 + x) + ...$

Putting $x = -1$

$\Rightarrow 2 + 6 + 12 + 20 + \ldots\ldots + 17 \times 16 = 2a_2$

$2a_2 = 1.\,2 + 2.\,3 + 3.\,4 \ldots\ldots + 16.\,17$

T_n for $1.\,2 + 2.\,3 + 3.\,4$ is $T_n = n(n+1)$

$2a_2 = \sum_{i=1}^{16} T_n = \sum_{i=1}^{16} n^2 + \sum_{i=1}^{16} n$

$= \dfrac{(2(16)+1)16(16+1)}{6} + \dfrac{16(16+1)}{2} = 1632$

$\Rightarrow a_2 = 816$

Sol 2: (a) (i) $\left(\sqrt{\dfrac{x}{3}} + \dfrac{\sqrt{3}}{2x^2} \right)^{10}$

$T_{r+1} = {}^{10}C_r \left(\dfrac{\sqrt{x}}{\sqrt{3}} \right)^{10-r} \left(\dfrac{\sqrt{3}}{2x^2} \right)^{r}$

$= {}^{10}C_r \left(\dfrac{1}{\sqrt{3}} \right)^{10-r} x^{\frac{10-r}{2}-2r} \left(\dfrac{\sqrt{3}}{2} \right)^{r}$

For term independent of x-

$\dfrac{10-r}{2} - 2r = 0 \Rightarrow 10 - r - 4r = 0 \Rightarrow r = 2$

So $T_3 = T_{2+1} = {}^{10}C_2 \left(\dfrac{1}{\sqrt{3}} \right)^{10-2} \left(\dfrac{\sqrt{3}}{2} \right)^{2}$

$= \dfrac{10 \times 9}{2} \times \dfrac{1}{3^4} \times \dfrac{3}{4} = \dfrac{5}{12}$

(ii) $\left[\dfrac{1}{2} x^{1/3} + x^{-1/5} \right]^{8}$

$T_{r+1} = {}^{8}C_r \left(\dfrac{1}{2} x^{1/3} \right)^{8-r} (x^{-1/5})^{r}$

Power of x $= \dfrac{8-r}{3} - \dfrac{r}{5} = 0$ for independence

$\Rightarrow 5(8-r) - 3r = 0 \Rightarrow 40 - 5r - 3r = 0 \Rightarrow r = 5$

$T_{5+1} = T_6 = {}^{8}C_5 \left[\dfrac{1}{2}(x^{1/3}) \right]^{8-5} \cdot (x^{-1/5})^5 = \dfrac{8 \times 7 \times 6}{1.2.3} \times \left(\dfrac{1}{2} \right)^3 = 7$

(b) $\left(5^{\frac{2}{5}\log_5 \sqrt{4^x+44}} + \dfrac{1}{5^{\log_5 \sqrt[3]{2^{x-1}+7}}} \right)^{8}$

$= (a_1 + a_2)^8$ assume

$T_4 = T_{3+1} = {}^{8}C_3 (a_1)^{8-3} (a_2)^3$

$a_1 = 5^{\log_5 (\sqrt{4^x+44})^{2/5}} = ((4^x+44)^{1/2})^{2/5} = (4^x+44)^{1/5}$

$a_2 = \dfrac{1}{5^{\log_5 (2^{x-1}+7)^{1/3}}} = \dfrac{1}{(2^{x-1}+7)^{1/3}} = (2^{x-1}+7)^{-1/3}$

$T_4 = {}^{8}C_3 (4^x+44)^{5/5} (2^{x-1}+7)^{-3/3}$

$= \dfrac{8 \times 7 \times 6}{1.2.3} \times (4^x+44)(2^{x-1}+7)^{-1} = 336$

$\Rightarrow \dfrac{4^x+44}{2^{x-1}+7} = \dfrac{336}{8x7} = 6$

$\Rightarrow 4^x + 44 = 6 \times 2^{x-1} + 6 \times 7 = 3.2^x + 42$

$(2^x)^2 - 3(2)^x + 44 - 42 = 0$

Assume $2^x = y$

$y^2 - 3y + 2 = 0 \Rightarrow (y-2)(y-1) = 0$

$\Rightarrow y = 1$ or $y = 2 \Rightarrow x = 0$ or $x = 1$

Sol 3: $\left(ax^2 + \dfrac{1}{bx} \right)^{11}$

$T_{r+1} = {}^{11}C_r (ax^2)^{11-r} \left(\dfrac{1}{bx} \right)^{r}$

Power of x $= 2(11-r) + r(-1) = 7$ (given)

$\Rightarrow 22 - 2r - r = 7 \Rightarrow r = 5$

Coefficient $T_{5+1} = {}^{11}C_5 (a)^{11-5} \left(\dfrac{1}{b} \right)^{5} = {}^{11}C_5 a^6 b^{-5}$

(ii) $\left(ax - \dfrac{1}{bx^2} \right)^{11}$

$T_{r+1} = {}^{11}C_r (ax)^{11-r} \left(-\dfrac{1}{bx^2} \right)^{r}$

Power of x $= 11 - r - 2r = -7$ (given)

$\Rightarrow r = 6$

Coefficient $T_{r+1} = T_{6+1} = {}^{11}C_6 a^{11-6} \left(-\dfrac{1}{b} \right)^{6} = {}^{11}C_6 a^5 b^{-6}$

(iii) Given that both coefficient are equal

$\Rightarrow {}^{11}C_5 a^6 b^{-5} = {}^{11}C_6 a^5 b^{-6} \Rightarrow ab = 1$

Sol 4: (a) $(1+x)^{14}$

Coefficients $T_r = T_{(r-1)+1} = {}^{14}C_{r-1}$

$T_{r+1} = {}^{14}C_r;\ T_{(r+1)+1} = {}^{14}C_{r+1}$

Its given that they are in A. P. so. $T_r + T_{r+2} = 2T_{r+1}$

${}^{14}C_{r-1} + {}^{14}C_{r+1} = 2\ {}^{14}C_r$

$$\frac{14!}{(r-1)!(14-r+1)!} + \frac{14!}{(r+1)!(14-r-1)!} = 2\frac{14!}{r!(14-r)!}$$

$$\Rightarrow \frac{1}{(15-r)(14-r)} + \frac{1}{(r+1)r} = \frac{2}{r(14-r)}$$

$$\Rightarrow \frac{r(r+1)+(15-r)(14-r)}{r(r+1)(14-r)(15-r)} = \frac{2}{r(14-r)}$$

$\Rightarrow r^2+r+210-14r-15r+r^2=2(r+1)(15-r)$

$\Rightarrow 2r^2-28r+210=30r-2r^2+30-2r$

$\Rightarrow 4r^2-56r+180=0 \Rightarrow r^2-14r+45=0$

$\Rightarrow (r-9)(r-5)=0 \Rightarrow r=9$ or $r=5$

(b) $(1+x)^{2n}$

Coefficients $T_2 = T_{1+1} = {}^{2n}C_1 = 2n$

$T_3 = T_{2+1} = {}^{2n}C_2 = \dfrac{2n(2n-1)}{1.2} = n(2n1)$

$T_4 = {}^{2n}C_3 = \dfrac{2n(2n-1)(2n-2)}{1.2.3} = \dfrac{n(2n-1)(2n-2)}{3}$

They all are in A. P.

So, $T_2 + T_4 = 2T_3$

$2n + \dfrac{n(2n-1)(2n-2)}{3} = 2n(2n-1)$

$\Rightarrow 3+(n-1)(2n-1)=3(2n-1)=6n-3$

$\Rightarrow 3+2n^2-2n-n+1=6n-3$

$\Rightarrow 2n^2-9n+7=0$

Sol 5: a = Coefficient of x^3 in $(1+x+2x^2+3x^3)^4$

b = Coefficient of x^3 in $(1+x+2x^2+3x^3+4x^4)^4$

$4x^4$ has no effect on the coefficient of x^3.

Hence $a = b$

$\therefore a - b = 0$

Sol 6: $(1-x^2)^{10}$

$T_{r+1} = {}^{10}C_r(-x^2)^r$

Given that $2r=10 \Rightarrow r=5$

So coefficient is $= (1)^r\ {}^{10}C_r = {}^{10}C_5$

And in $\left(x - \dfrac{2}{x}\right)^{10}$

$T_{r+1} = {}^{10}C_r(x)^{10-r}\left(-\dfrac{2}{x}\right)^r$

Power of $x = 10-r-r = 0$

$\Rightarrow 10-2r=0 \Rightarrow r=5$

Coefficient $= {}^{10}C_5(-2)^5 = -\ {}^{10}C_5\,2^5$

Ratio of both coefficients $= \dfrac{{}^{10}C_5}{{}^{10}C_5\,2^5} = \dfrac{1}{2^5} = \dfrac{1}{32}$

Sol 7: (a) $(ax - by + cz)^9$

General term $= \dfrac{9!}{r_1!r_2!r_3!}(ax)^{r_1}(-by)^{r_2}(cz)^{r_3}$

$r_1 + r_2 + r_3 = 9$

For coefficient of $x^2y^3z^4$ so $\Rightarrow r_1=2,\ r_2=3,\ r_3=4$

So Coefficient $= \dfrac{9!}{2!3!4!} \times a^2 \cdot b^3\, c^4$

$= -1260\, a^2 \cdot b^3 \cdot c^4$

(b) $(a-b-c+d)^{10}$

General Term $= \dfrac{10!}{r_1!r_2!r_3!r_4!}(a)^{r_1}(-b)^{r_2}(-c)^{r_3}(d)^{r_4}$

$r_1+r_2+r_3+r_4=10$

It given that $r_1=2,\ r_2=3,\ r_3=4,\ r_4=1$

Coefficient $\dfrac{10!}{2!3!4!1!}(-1)^3(-1)^4$

$= -\dfrac{10 \times 9 \times 8 \times 7 \times 6 \times 5}{2!3!} = -12600$

Sol 8: $s_n = 1 + q + q^2 + \ldots + q^n =$

$S_n = 1 + \dfrac{q+1}{2} + \left(\dfrac{(q+1)}{2}\right)^2 + \ldots + \left(\dfrac{q+1}{2}\right)^n, q \neq 1$

$= {}^{n+1}C_1 + {}^{n+1}C_2\, s_1 + {}^{n+1}C_3 s_2$

$+ \ldots + {}^{n+1}C_{n+1}s_n$

Constant term

${}^{n+1}C_1 + {}^{n+1}C_2 + \ldots + {}^{n+1}C_{n+1} = 2^{n+1}-1$

In S_n constant term $= 1 + \dfrac{1}{2} + \left(\dfrac{1}{2}\right)^2 + \ldots \left(\dfrac{1}{2}\right)^n$

$$= \frac{1-\left(\dfrac{1}{2}\right)^{n+1}}{1-\dfrac{1}{2}} = 2\left(1-\dfrac{1}{2}\right)^{n+1} = \frac{(2^{n+1}-1)}{2^n}$$

So $(2^{n+1}-1) = (2^n), \left(\dfrac{(2^{x+1}-1)}{2^n}\right)$

$$^{n+1}C_1 + {}^{n+1}C_2 S_1 + {}^{n+1}C_3 S_n = 2^n S_n$$

We can prove this with other terms also.

Sol 9: (i) $(2+3x)^9, x = \dfrac{3}{2}$. Now we have

$$\frac{n+1}{1+\left|\dfrac{a}{x}\right|} = \frac{9+1}{1+\dfrac{2\times 2}{3\times 3}} = \frac{10}{1+\dfrac{4}{9}}$$

$$= \frac{10}{1+0.44} = \frac{10}{1.44} = 6.944$$

Greatest terms is

$$T_7 = T_{6+1} = {}^9C_6(2)^{9-6}(3x)^6 = \frac{9\times 8\times 7}{1.2.3} 2^3 \times 3^6 \left(\dfrac{3}{2}\right)^6$$

$$= \frac{3^2.7.3^6.3^6}{1.2.3} = \frac{7.3^{13}}{2}$$

(ii) $(3-5x)^{15}$ When $x = \dfrac{1}{5}$

$$\frac{n+1}{1+\left|\dfrac{a}{x}\right|} = \frac{15+1}{1+\left|\dfrac{3\times 5}{5\times 1}\right|} = \frac{16}{1+3} = \frac{16}{4} = 4$$

So T_4 and T_{4+1} are same greatest term. $T_4 = {}^{15}C_4(3)^{15-4}(-5x)^4$

$$= \frac{15\times 14\times 13\times 12}{1.2.3.4} 3^{11}\left(\dfrac{-5}{5}\right)^4 = 455.\, 3^{12}$$

Sol 10: (i) $(1+x+x^2)^n = a_0 + + a_1 x + \ldots + a_{2n} x^{2n}$

(i) at $x=1$

$$a_0 + a_1 + a_2 + a_3 + \ldots + a_{2n} = (1+1+1)^n = 3^n$$

(ii) at $x=-1 \Rightarrow [1-1+(-1)^2]^n = 1$

$$\Rightarrow 1 = a_0 - a_1 + a_2 - a_3 + \ldots + a_{2n}$$

(iii) $(1+x+x^2)^n(x^2-x+1)^{2n}$

$$= (a_0 x^{2n} - a_1 x^{2n-1} + \ldots)(a_0 - a_1 x + \ldots)$$

Compare $x \to -x$ in

$(x^2+x+1) \to (x^2-x+1)\, a_0 - a_1 x + a_2 x^2 - a_3 x^3 + \ldots + a_{2x} x^{2x}$

$[(1+x+x^2)(x^2-x+1)]^n$

$$= a_0^2 x^{2n} - a_1^2 x^{2n} - a_3\, x^{2n} + a_4 \cdot x^{2n} - \ldots + a_{2n} x^{2n}$$

$\therefore$ For $x = 1$

$$a_0^2 - a_1^2 - a_3^2 + \ldots + a_{2n}^2 = 3^n$$

Sol 11: $(5+3x)^{10}$

$$T_4 = {}^{10}C_3(5)^{10-3}(3x)^3$$

$$= {}^{10}C_3 5^7 3^3 x^3 \quad {}^{10}C_3 5^7 3^3 x^3 \text{ is the greatest term}$$

So, $\dfrac{n+1}{1+\left|\dfrac{a}{x}\right|} = \dfrac{10+1}{1+\left|\dfrac{5}{3x}\right|}$

For greatest term to be T_4

$$= 3 < \frac{10+1}{1+\dfrac{5}{3x}} < 4$$

$$3 < \frac{33x}{3x+5} < 4$$

$$3(3x+5) < 33x < 4(3x+5)$$

$$9x+15 < 33x < 12x+20$$

Solving each inequality separately we get

$$9x + 15 < 33x$$

$$\Rightarrow 24x > 15$$

$$\Rightarrow x > \frac{15}{24}$$

$$\Rightarrow x > \frac{5}{8}$$

Also, $12x + 20 > 334$

$$\Rightarrow x < \frac{20}{21}$$

$$\therefore \frac{5}{8} < x < \frac{20}{21}$$

Sol 12: In the expansion of $\left(\dfrac{x}{5} + \dfrac{2}{5}\right)^n$, we have

$$T_9 = {}^nC_8\left(\dfrac{x}{5}\right)^{n-8}\left(\dfrac{2}{5}\right)^8$$

Coefficient = $^nC_8(5)^{8-n}2^85^{-8} = \; ^nC_85^{-n}2^8$

Which is greatest coefficient

$$8 < \frac{n+1}{1+\left|\dfrac{x}{a}\right|} < 9 \text{ assume } x=1 \text{ for find}$$

$$8 < \frac{n+1}{1+\dfrac{1\times5}{5\times2}} < 9 \text{ greatest coefficient}$$

$$8 < \frac{n+1}{1+\dfrac{1}{2}} < 9 \;=\; 8 < \frac{n+1}{\dfrac{3}{2}} < 9$$

$$8\times\frac{3}{2} < n+1 < 9\times\frac{3}{2}$$

$$12 < n+1 < \frac{27}{2}$$

$$17 < n < \frac{22}{2}-1 = \frac{25}{2} = 12.5$$

$$11 < n < 12.5$$

There is only one natural no. in region i.e., 12

Sol 13: $N = {}^{2000}C_1 + 2.\; {}^{2000}C_2 + 3.$

$${}^{2000}C_3 + \ldots\ldots + 2000{}^{2000}C_{2000}$$

$N = n.\; 2^{n-1}$ here $n = 2000$

$$\Rightarrow N = 2000 \times 2^{n-1}$$

$$\Rightarrow N = 2\times(2\times5)^3 \times 2^{n-1} = 2^3 \times 5^3 2^n$$

$$\Rightarrow N = 2\times(2\times5)^3\times 2^{n-1} = 2^3\times5^3 2^n$$

$$N = 2^{n+3}5^3$$

Number of divisors $= (n + 3 + 1)(3 + 1)$

$$= (2000 + 4)(4) = 2004 \times 4 = 8016$$

Sol 14: $(1+x)^{2012} + (1+x^2)^{2011} + (1+x^3)^{2010}$

Number of different dissimilar terms

$= 2012 + 2011 + 2010$ (no. of terms which is common in $(1+x)^{2012}$ and $(1+x^2)^{2011}$

of terms which is similar $(1+x)^{2012}$ and $(1+x^3)^{2010}$)

(no. of terms which is similar in $(1+x^2)^{2011} + (1+x^3)^{2010}$)

$$= 2012 + 2011 + 2010 - \left[\frac{2011}{2}\right] - \left[\frac{2012+1}{3}\right] - \left[\frac{1005}{3}\right] + 1$$

Where $(+1)$ for constant term. And $[x]$ is a singularity function. $[1.\,35] = 1$

$$= 6034 - 1005 - 671 - 335 = 4023$$

Sol 15: $(1+x+2x^3)\left(\dfrac{3x^2}{2} - \dfrac{1}{3x}\right)^9$

For independent terms

Coefficient of x^0 in $\left(\dfrac{3x^2}{2} - \dfrac{1}{3x}\right)^9 = A_0$

Coefficient of x^{-1} in $\left(\dfrac{3x^2}{2} - \dfrac{1}{3x}\right)^9 = A_1$

Coefficient of x^{-3} in $\left(\dfrac{3x^2}{2} - \dfrac{1}{3x}\right)^9 = A_2$

$$T_{r+1} = {}^9C_r\left(\frac{3x^2}{2}\right)^{9-r}\left(-\frac{1}{3x}\right)^r$$

Power of $x = 2(9-r) - r = 18 - 2r - r = 18 - 3r$

$$x^0 \;\Rightarrow\; 18 - 3r = 0 \;\Rightarrow\; r = \frac{18}{3} = 6$$

$$T_6 = T_{5+1} = A_0 = {}^9C_6\left(\frac{3}{2}\right)^{9-5}\left(-\frac{1}{3}\right)^6$$

$$A_0 = \frac{7}{18}$$

For $x^{-1} = 18 - 3r = -1$

$$3r = 18 + 1 = 19$$

$r = 19/3$ not natural no.

So $A_1 = 0$

For x^{-3}

$$18 - 3r = -3 \Rightarrow 3r = 18 + 3 = 21 \Rightarrow r = \frac{21}{3} = 7$$

So $A_2 = {}^9C_7\left(\dfrac{3}{2}\right)^{9-7}\left(-\dfrac{1}{3}\right)^7 = \dfrac{-1}{27}$

So coefficient of x^0 in $(1+x+2x^3)\left(\dfrac{3x^2}{2} - \dfrac{1}{3x}\right)^9$

$$= \frac{7}{18} + 2\times\left(-\frac{1}{27}\right) = \frac{21-2(2)}{54} = \frac{17}{54}$$

Sol 16: $f(n) = \sum\limits_{r=0}^{n} \sum\limits_{k=r}^{n} {}^{k}C_{r}$

For $f(11) = \sum\limits_{r=0}^{11} \sum\limits_{k=r}^{11} {}^{k}C_{r}$

$= ({}^{0}C_{0} + {}^{1}C_{0} + {}^{2}C_{0} + \ldots + {}^{11}C_{0})$

$+ {}^{1}C_{1} + {}^{2}C_{1} + \ldots + {}^{1}C_{1}$

$+ {}^{2}C_{2} + {}^{3}C_{2} + \ldots + {}^{11}C_{2} \;\; \vdots\vdots\; \vdots\vdots\; \vdots\vdots$

${}^{10}C_{10} + {}^{11}C_{10}$

${}^{11}C_{11}$

$= {}^{11}C_{0} + \ldots + {}^{11}C_{11} + {}^{10}C_{0} + \ldots + {}^{10}C_{10} \; \vdots\vdots$

$+ {}^{1}C_{1} + {}^{0}C_{0}$

$= 2^{11} + 2^{10} + 2^{9} + \ldots + 2^{1} + 2^{0}$

$= \dfrac{2^{11+1} - 1}{2 - 1} = 4095 = 4095 = 5^{1}.3^{2}\,7^{1}.\,13^{1}$

No. of divisors $= (1+1).\,(2+1).\,(1+1)(1+1)$

$= 2 \times 3 \times 4 = 24$

Sol 17: $\sum\limits_{j=0}^{11} \sum\limits_{i=j}^{11} {}^{i}C_{j}$

$= {}^{0}C_{0} + \left({}^{1}C_{0} + {}^{1}C_{1}\right) + \left({}^{2}C_{0} + \ldots + {}^{2}C_{2}\right)$

$+ \left({}^{3}C_{0} + \ldots + {}^{3}C_{3}\right) + \ldots + \left({}^{11}C_{0} + {}^{11}C_{1} + \ldots + {}^{11}C_{11}\right)$

$= 2^{0} + 2^{1} + \ldots + 2^{11}$

$= 2^{12} - 1$

Sol 18: $(1 + x^{2}).(1 + x)^{n} = \sum\limits_{k=0}^{n+4} a_{k}.x^{k}$

a_{1}, a_{2} and a_{3} are in A.P

$(1 + x^{4} + 2x^{2})({}^{n}C_{0} + {}^{n}C_{1}x + {}^{n}C_{2}x^{2} + {}^{n}C_{3}x^{3} + \ldots {}^{n}C_{n}x^{n})$

$= a_{0} + a_{1}x + a_{2}x^{2} + a_{3}x^{3} + \ldots$

Compare terms of x^{r} in both side

$x^{1} \Rightarrow$ L. H. S. $= {}^{n}C_{1} = n$

R. H. S. $= a_{1}$

$\Rightarrow n = a_{1}$ .... (i)

$x^{2} \Rightarrow 2{}^{n}C_{0} + {}^{n}C_{2} = a_{2}$

$2 + \dfrac{n(n-1)}{2} = a_{2}$ .... (ii)

$x^{3} \Rightarrow 2\,{}^{n}C_{1} + {}^{n}C_{3} = a_{3}$

$2n + \dfrac{n(n-1)(n-2)}{1.2.3} = a_{3}$ .... (iii)

It's given that a_{1}, a_{2}, a_{3} are in A.P

$2a_{2} = a_{1} + a_{3}$

$4 + n(n-1) = n + 2 + \dfrac{n(n-1)\,(n-2)}{6}$

$= \dfrac{6n + 12n + n(n-1)\,(n-2)}{6}$

$24 + 6n(n-1) = 18n + n(n-1)\,(n-2)$

Solving this we get, $n = 2$ or 3 or 4

Sol 19: $\sum\limits_{k=0}^{n} {}^{n}C_{k}\,\sin kx.\,\cos(n-k)x = 2^{n-1}\,\sin nx$

L. H. S. $= \sum\limits_{k=0}^{n} {}^{n}C_{k}\,\sin kx \cos(n-k)x$

We know that $2\sin A\cos B = \sin(A+B) + \sin(A-B)$

$\because\ A + B = kx + (n-k)x = kx + nx - kx = nx$

$A - B = kx - (n-k)k = kx - nx + kx = 2kx - nx$

So, $\sum\limits_{k=0}^{n} \dfrac{1}{2}\,{}^{n}C_{k}[\sin nx + \sin(2kx - nx)]$

$= \sum\limits_{k=0}^{n} \dfrac{1}{2}\,{}^{n}C_{k}\sin nx + \sum\limits_{k=0}^{n} \dfrac{1}{2}\,{}^{n}C_{k}\sin(2kx - nx)$

$= \dfrac{1}{2}\sin nx \sum\limits_{k=0}^{n} {}^{n}C_{k} + \dfrac{1}{2}$

$[\,{}^{n}C_{0}\sin(-nx) + \ldots + {}^{n}C_{n}\sin(nx)]$

$= \dfrac{1}{2}\sin nx.2^{n} + 0 = (\sin nx)2^{n-1} = 2^{n-1}\sin nx$

Sol 20: $x^{2001} + \left(\dfrac{1}{2} - x\right)^{2001} = 0$

$= x^{2001} + \left[\,{}^{2001}C_{0}\left(\dfrac{1}{2}\right)^{2001} + \ldots + {}^{2001}C_{1999}\right.$

$\left(\dfrac{1}{2}\right)^{2}(-x)^{1999} + {}^{2001}C_{2000}\left(\dfrac{1}{2}\right)(-x)^{2000}\Bigg]$

$= x + {}^{2001}C_{2001}\left(\dfrac{1}{2}\right)^{0}(-x)^{2001}$

$$= x^{2001} + \ldots\ldots + {}^{2001}C_{1999}\frac{(-x)^{1999}}{4}$$

$$+ {}^{2001}C_{2000}\frac{(-x)^{2000}}{2} - x^{2001}$$

= Now maximum pointer of x = 2000

Sum of all solution is = $\dfrac{\text{Coefficient of } x^{2001-1}}{\text{Coefficient of } x^{2000}}$

$$= \frac{{}^{2001}C_{1999}\times\dfrac{1}{4}}{{}^{2001}C_{2000}\times\dfrac{1}{2}} = \frac{2001\times2000\times1}{1.2\times2001\times1}\times\frac{1}{2} = 500$$

Sol 21: Let

$$S = \sum_{k=0}^{2n} (-1)^k \, (k-2n) \, ({}^{2n}C_k)^2 \qquad\ldots(i)$$

$$\Rightarrow \quad S = \sum_{k=0}^{2n} (-1)^{2n-k} \, (2n-k) \, ({}^{2n}C_{2n-k})^2$$

Writing the terms in S in the reverse order, we get

$$S = \sum_{k=0}^{2n} (-1)^k \, k \, ({}^{2n}C_k)^2 \qquad\ldots(ii)$$

Adding (i) and (ii) we get

$$2S = 2n\sum_{k=0}^{2n} (-1)^k \, ({}^{2n}C_k)^2 = -2nA$$

$$\Rightarrow \quad S = -nA$$

Sol 22: (A) $\left(\displaystyle\sum_{i=0}^{10} P(i,10-i)\right)$

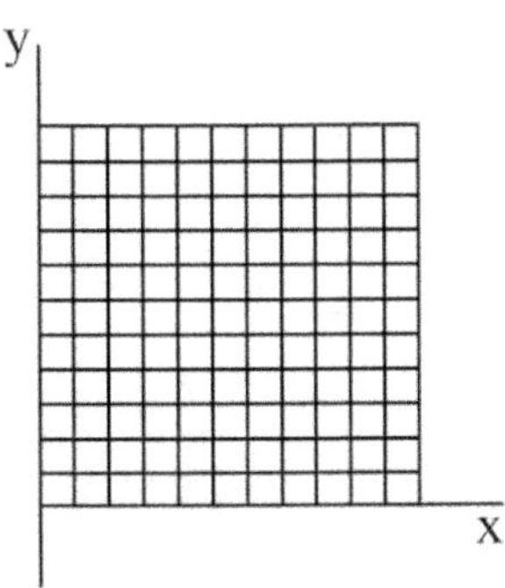

$$P(0,10)+P(1,9)+\ldots\ldots+P(10,0)$$

$$= 1+(9+1)(1)+\frac{10\times9}{2}+\ldots.$$

$$= {}^{10}C_0 - {}^{10}C_1 + {}^{10}C_2 + {}^{10}C_3 + \ldots\ldots + {}^{10}C_{10}$$

$$= 2^{10} = 1024$$

Sol 23: (C) $P(i,100-i) = P(j,100-j)$

$${}^{100}C_i = {}^{100}C_j \text{ and } i \neq j$$

$${}^{100}C_i = {}^{100}C_{100-j}$$

$$I = 100 - j$$

$$i+j = 100$$

$$i,j \in N,0$$

$$(0,100)(1,99)\ldots\ldots(99,1)(100,1)$$

Total no. of ordered pairs

$$(i,j) = 100$$

Sol 24: $(6\sqrt{6}+14)^{2n+1} = I+F$

(assume)

$$(6\sqrt{6}-14) = \frac{(6\sqrt{6})^2-(14)^2}{6\sqrt{6}+14}$$

$$= \frac{20}{6\sqrt{6}+14}$$

$$I = [(6\sqrt{6}+14)^{2n+1}] = 0 < F < 1$$

$$e = (6\sqrt{6}-14)^{2n+1} = 0 < e < 1$$

$$I+F-e = (6\sqrt{6}+14)^{2n+1}$$

$$- (6\sqrt{6}-14)^{2n+1}$$

$$= 2({}^nC_1(6\sqrt{6})^{2n+1}14$$

$$+ {}^nC_3(6\sqrt{6})^{2n-11-3}14^3 + \ldots\ldots)$$

$$= 2K \text{ (K is const. integer)}$$

$$0 \leq F-e < 1$$

$$F - e = 2K-I = \text{Integer}$$

$$F - e = 0 = e = F = F = e = (6\sqrt{6}-14)^{2n+1}$$

$$F = \frac{(20)^{2n+1}}{(6\sqrt{6}+14)^{2n+1}}$$

$$(I+F)F = (6\sqrt{6}+14)^{2n+1}$$

$$\frac{20^{2n+1}}{(6\sqrt{6}+14)^{2n+1}}$$

$$(I+F)F = 20^{2n+1}$$

Sol 25: $P = (2 + \sqrt{3})^5$

$f = P - [P]$

$2 - \sqrt{3} = \dfrac{2^2 - (\sqrt{3})^2}{2 + \sqrt{3}} = \dfrac{1}{2 + \sqrt{3}}$

$\because 0 < 2 - \sqrt{3} < 1$

$\because 0 < f < 1$

$(2 + \sqrt{3})^5 = \left(\dfrac{1}{2 - \sqrt{3}}\right)^5 = \dfrac{1}{F}$

$[P] + f + f = (2 + \sqrt{3})^5 + (2 - \sqrt{3})^5$

$= 2[{}^5C_0 2^5 + {}^5C_2 2^{5-2}(\sqrt{3})^2$

$+ \ldots + {}^5C_4 2^{5-4}(\sqrt{3})^4]$

$2\left[2^5 + \dfrac{5 \times 4}{2} \times 2^3 \times 3 + 5 \times 2 \times 3^2\right]$

$f + f = $ Integer

$0 \le f + f < 2$

$\therefore \ f + f = 1$

$f = 1 - f = (2 - \sqrt{3})^5$

$f = 1 - (2\sqrt{3})^5$

$\dfrac{f^2}{1-f} = \dfrac{f^2 - 1^2 + 1}{1 - f} = \dfrac{(f-1)^{-1}(f+1)}{(1-f)} + \dfrac{1}{(1-f)}$

$= -(f + 1) + \dfrac{1}{f}$

$f = 1 - f = f - 2 = -(f + 1)$

$= f - 2 + \dfrac{1}{f}$

$= (2 + \sqrt{3})^5 - 2 + (2 - \sqrt{3})^5$

$= 2[32 + 15 \times 2^4 + 5 \times 2 \times 3^2] - 2$

$= 724 - 2 = 722$

Sol 26: $(1 + x)^n = {}^nC_0 + C_1 +$

$C_2 x^2 + C_3 x^3 + \ldots + C_n x^n$ ... (i)

(a) Differentiating at both sides

$n(1 + x)^{n-1} = C_1 + 2C_2 x + \ldots + {}^3C_3 x^{2n}$

$x = 1$

Put $n \cdot 2^{n-1} = C_1 + 2C_2 + \ldots + nC_n$ ... (ii)

(b) Sum of eq. (i) and (ii)

$n2^{n-1} + (1 + x)^n = C_1 + {}^2C_2 + \ldots + {}^nC_n$

$+ C_0 + C_1 x + C_2 x^2 + \ldots + C_n x^n$

At $x = 1$

$2^{n-1}(n + 2) = C_0 + 2C_1$

$+ 3C_2 + \ldots + (n + 1)C_n$

(c) Eq. (1) + 2 × eq. (2)

At $x = 1$

$2^n + n2^n = C_0 + C_1(1 \times 2 + 1)$

$+ C_2(2 \times 2 + 1) + \ldots + C_n(2n + 1)$

$2^{n-1}(n + 2) = C_0 + 2C_1$

$+ 7C_3 + \ldots + (2n + 1)C_n$

(d) $(C_0 + C_1)(C_1 + C_2)\ldots(C_{n-1} + C_n) =$

$\dfrac{C_0 C_1 C_2 . C_{n-1}(n+1)^{3n}}{n!}$

Multiply and divide L. H. S. by

$C_0 C_1 C_2 C_3 \ldots C_{n-1}$

$= C_0 C_1 C_2 \ldots C_{n-1}\left(1 + \dfrac{C_1}{C_0}\right)\ldots\left(1 + \dfrac{C_n}{C_{n-1}}\right)$

On using $\dfrac{{}^nC_r}{{}^nC_{r-1}} = \dfrac{n - r + 1}{r} = $ L. H. S.

$C_0 C_1 C_2 \ldots C_{n-1}(1 + n)\left(\dfrac{1+n}{2}\right)\left(\dfrac{1+n}{3}\right)\ldots\left(\dfrac{1+n}{n}\right)$

$= \dfrac{C_0 C_1 C_2 \ldots C_{n-1}(n+1)^n}{n!}$

(e) $1C_0^2 + 3C_1^2 + 5.C_2^2 + \ldots +$

$(2n + 1)C_n^2 = \dfrac{(n+1)(2n)!}{n!n!}$

We know that (part (C))

$C_0 + 3C_1 + 5C_2 + \ldots +$

$(2n + 1)C_n + (n + 1)2^n$

$C_0 + 3C_1 + 5C_2 x^2 + \ldots$

$C_0 + {}^3C_1 + {}^5C_2 x^2 + \ldots$

$+(2n+1)C_n x^n$

$= (n+1)(1+x)^n = (n+1)(1+x)^n$

Multiply with

$C_0 x^n + C_1 x^{n-1} + \ldots + C_n = (x+1)^n$

$=$ and compare x^n and coefficient

$C_0 + 3C_1^2 + 5C_2^2 + \ldots + (2n+1)C_n^2$

$=$ Coefficient of x^n in $(n+1)(1+x)^{n+n}$

$= (n+1) \, {}^{2n}C_n =$

L. H. S. = R. H. S.

Sol 27: $I = [(3\sqrt{5})^n]$

$I + F = (3+\sqrt{5})^n$

P = rational part

σ = irrational part

$3 - \sqrt{5} = \dfrac{9-(\sqrt{5})^2}{3+\sqrt{5}} = \dfrac{4}{3+\sqrt{5}}$

$0 < 3 - \sqrt{5} < 1$

$F = (3-\sqrt{5})^n$

$I + F + F = (3+\sqrt{5})^n + (3-\sqrt{5})^n$

$= 2\left({}^nC_0 3^n + {}^nC_2 3^{n-2}(\sqrt{5})^2 + \ldots \right)$

Rational part

$0 < F + F < 2$

F+F is 1 only integer between 0 and 2

$I + 1 = 2F = P = \dfrac{1}{2}(I+1)$

$I + F - F = (3+\sqrt{5})^n - (3-\sqrt{5})^n$

$I + F + F - F - F = 2({}^nC_1 3^{n-1}$

$(\sqrt{5})^1 + {}^nC_3 3^{n-3}(\sqrt{5})^3 + \ldots)$

$I + 2F - (F+F) = 2\sigma$

$I + 2F - 1 = 2\sigma$

$\sigma = \dfrac{1}{2}(I + 2F - 1)$

Sol 28:

(a) $\dfrac{C_1}{C_0} + \dfrac{2C_2}{C_1} + \dfrac{3C_3}{C_2} + \ldots + \dfrac{nC_n}{C_{n-1}} = \dfrac{n(n+1)}{2}$

We know that $\dfrac{{}^nC_r}{{}^nC_{r-1}} = \dfrac{n-r+1}{r}$

$\Rightarrow \dfrac{r \, {}^nC_r}{{}^nC_{r-1}} = (n-r+1)$

L. H. S. $= (n-1+1) + (n-2+1)$

$+ (n-3+1) + \ldots (n-n+1)$

$n^2 + n - (1+2+3+\ldots n)$

$= n^2 + n - \dfrac{n(n+1)}{2}$

$= \dfrac{n^2+n}{2} = \dfrac{n(n+1)}{2}$

(b) $2C_0 + \dfrac{2^2 C_1}{2} + \dfrac{2^3 C_2}{3}$

$+ \dfrac{2^{n+1} C_n}{n+1} = \dfrac{3^{n+1} - 1}{n+1}$

(c) In equation (i) from above que.

$x = 2$

$2C_0 + \dfrac{2^2 C_1}{2} + \ldots + \dfrac{2^{n+1} C_n}{n+1} = \dfrac{3^{n+1} - 1}{n+1}$

(d) In eq. (i) $x = -1$

$\dfrac{(0)^{n+1} - 1}{n+1} = C_0(-1) + \dfrac{C_1}{2} - \dfrac{C_2}{3}$

$+ \ldots + (-1)^{n+1} \dfrac{C_n}{n+1} = C_0 - \dfrac{C_1}{2} + \dfrac{C_2}{3} + \ldots$

$+ (-1)^n \dfrac{C_n}{n+1} = \dfrac{1}{n+1}$

Sol 29: (a) In equation (ii) compare coefficient of x^{n-1}

$^{2n}C_{n-1} = C_0 C_1 + C_1 C_2 + \ldots + C_{n-1} C_n$

$^{2n}C_{n-1} = \dfrac{2n!}{(n-1)!(n+1)!}$

$\because \; 2n - (n-1) = n+1$

L. H. S. = R. H. S.

(b) In some equ. (ii) compare coefficient of x^{n-r}

$$^{2n}C_{n-r} = C_0 C_r + \dots + C_{n-r}C_n \qquad \dots \text{(ii)}$$

$$^{2n}C_{n-r} = \frac{2n!}{(n+r)!(n-r)!}$$

L. H. S. = R. H. S.

(c) $\displaystyle\sum_{r=0}^{n-2}(^nC_r \ ^nC_{r+2}) = \frac{2n!}{(n-2)!(n+2)!}$

In equ. (iii) if $r = 2$

$$= \ ^{2n}C_{n-2} = C_0 C_2 + C_1 C_3 + \dots + C_{n-2}C_n$$

$$= \ ^{2n}C_{n-2} = \frac{2n!}{(n-2)!(n+2)!} \qquad \dots\text{(iii)}$$

L. H. S. = R. H. S.

(d) $^{100}C_{10} + 5.\,^{100}C_{11} + 10.\,^{100}C_{12}$

$+10.\,^{100}C_{13} + 5.\,^{100}C_{14} + \,^{100}C_{18}$

$$= \ ^{105}C_{90} = \ ^{105}C_{105-90} = \ ^{105}C_{15} = \frac{105!}{90!15!}$$

L. H. S. $= \dfrac{100!}{90!10!} + \dfrac{5100!}{11!89!} + \dfrac{10\times100!}{12!88!}$

$+\dfrac{10\times1001}{13!87!} + \dfrac{5\times100!}{14!80!} + \dfrac{100!}{15!85!}$

$100!\left[\dfrac{15\times14\times13\times12\times11}{90!15!} + \right.$

$\dfrac{5\times90\times15\times14\times13\times12}{15!90} + \dfrac{10\times90\times89\times15\times14\times13}{15!90!}$

$+\dfrac{10\times15\times14\times90\times89\times88}{70!15!} + \dfrac{5.90\times89.8887}{15!90!}$

$\left.+\dfrac{90.89.88.87.86}{90!15!}\right]$

$$= \frac{100\times101\times102\times103\times104\times105}{90!15!} = \frac{105!}{90!15!}$$

$$= \ ^{105}C_{15} = \ ^{105}C_{90}$$

Sol 30: (i) $(1+x+x^2)^n = a_0+a_1x + a_2x^2+\dots+a_{2n}x^{2n}$

$(1+x+x^2)^n = (x^2+x+1)^n$

So $a_0 = a_{2n}$

$a_1 = a_{2n-1}$

$a_{n-1} = a_{n+1}$

So, $a_0a_1 + a_2a_3 + a_4a_5 + \dots +$

$= a_{2n}a_{2n-1} + a_{2n-2}a_{2n-3}$

$+\dots + a_1a_2 + a_3a_4$

$a_0a_1 - a_1a_2 + a_2a_3 \dots = 0$

(ii) $(1-x+x^2)^n$

$= a_0 - a_1x + a_2x^2 - a_3x^3 + \dots$

$(1+x+x^2)^n(1-x+x^2)^n = (a_0 - a_1x + \dots)$

$(a_0x^{2n} + a_1x^{2n-1} + \dots)$

$(1+x^2+x^4)^n = a_0a_2x^{2n-2} - a_1a_3x^{2n-2} + \dots$

Compare x^{2n-2} coefficient

$a_{n+1} = a_{n-1} = a_0a_2 - a_1a_3 + \dots$

$(\because$ in $(1+x+x^2)^n \quad x \to x^2$

So Coefficient of $x^{2(n-1)} = a_{n-1} = a_{n+1})$

(iii) $(1+x+x^2)^n = a_0 + a_1x + \dots + a_{2n}x^{2n}$

Put $x = 1$

$3^n = a_a + a_1 + a_2 + \dots + a_{2n} \quad \dots 1$

$x = \omega$

$0 = a_0 + a_1\omega + a_2\omega^2 + a_3 + \dots a_{2n}\omega^{2n} \,..\, 2$

$x = \omega^2$

$0 = a_0 + a_1\omega^2 + a_2\omega + \dots + a_{2n}\omega^{4n} \,\dots 3$

$A + B + C =$

$3^n = 3(a_0 + a_3 + a_6 + \dots)$

$a_0 + a_3 + a_6 + \dots 3^{n-1} \qquad \dots \text{(i)}$

$x(1+x+x^2)^n = a_0x + a_1x^2 + \dots + a_{2n}x^{2n+1} \quad \dots \text{A}$

$x = \omega = \ 0 = a_0\omega + a_1\omega^2 + a_2 + \dots + a_{2n}\omega^{2n+1} \,\dots \text{B}$

$x = \omega^2 = \ 0 = a_0\omega^2 + a_1\omega + a_2 + \dots + a_{2n}\omega^{4n+2} \,..\, \text{C}$

$A,B,C, = \ 3^n = 3(a_2 + a_5 + a_8 + \dots)$

$a_2 + a_5 + a_8 = 3^{n-1} \qquad \dots \text{(ii)}$

Sum as above

$$x^2(1+x+x^2)^n = a_0 x^2$$

$$x^2(1+x+x^2)^n = a_0 x^2 + a_1 x^3 + \dots + a_{2n}x^{2n+1}$$

$$x = w =$$

$$0 = a_0\omega^2 + a_1\omega^3 + \dots + a_{2n}\omega^{2n+1} \quad \dots\ B_2$$

$$x = \omega^2 = 0 = a_0\omega^4 + a_1 + \dots + a_{2n} \quad \dots\ C_2$$

$$A + B_2 + C_2 \ (a_1 + a_4 + a_7 + \dots)$$

$$= \frac{3^n}{3} = 3^{n-1} \qquad \dots\ (iii)$$

From (i), (ii) and (iii)

$$E = E_2 = E_3 = 3^{n-1}$$

Sol 31: $\displaystyle\sum_{r=0}^{100}\sum_{s=0}^{100}(C_r^2 + C_s^2 + C_r C_s) = m(\,^{2n}C_n) + 2P$

M, n and p are even natural number

$$(1+x)^{100}$$

C_r = coefficients of x^r in $(1+x)^{100}$

$$= \sum_{r=0}^{100}[C_r^2 \times 101 + C_0^2 + C_1^2 + \dots + C_{100}^2$$

$$+ C_r(C_0 + C_1 + \dots + C_{100})]$$

$$= 101\left(\sum C_r^2\right) + 101(\,^{2n}C_n) + \sum C_r(2^n)$$

$$= 101\,^{2n}C_n + 101\,^{2n}C_n + 2^n(2^n)$$

$$= 202\,^{2n}C_n + 2^{100+100}$$

$$= 202\,^{2n}C_n + 2^{200} = m(\,^{2n}C_n) + 2^P$$

n=100, m=202, P=200

Hence, n+m+p = 200+100+202=502

Sol 32: $(1+x)(1+x+x^2)\dots(1+x+x^2+\dots+x^n)$

Max. power of x

$$= 1+2+3+\dots n = \frac{n(n+1)}{2}$$

(a) Total terms $= \dfrac{1+n(n+1)}{2} = \dfrac{n^2+n+2}{2}$

(b) $1+x=x+1$

$1+x+x^2 = x^2+x+7$, So now product is

$= (x+1)(x^1+x+1) = \dots\dots$

$$(x^n + x^{n-1} + \dots + x^2 + x + 1)$$

So, $a_0 = \dfrac{a_{n(n+1)}}{2}$

Or if $x = \dfrac{1}{y}$

$$(y^-)\frac{n(n+1)}{2}$$

$$(y+1)(y^2+y+1)\dots(y^n+y^2+y+1)$$

$$= a_0 y\frac{n(n+1)}{2} + \dots + a\frac{n(n+1)}{2}$$

$$a_0 = a\frac{n(n+1)}{2}$$

(c) Odd coefficient $= a_1 + a_3 + a_5 + \dots$

At $x = 1$

$$2.\,3.\,4.\dots(n+1) = (n+1)!$$

$$= a_0 a_1 + a_2 + \dots + a_{\frac{n(n+1)}{2}}$$

$x = -1 = a_0 - a_1 + a_2 - a_3 = 0$

$$= a_0 + a_2 + a_4 + \dots = a_1 + a_3 + a_5 + \dots$$

Assume P = Q

$$P + Q = 2P = 2Q = (n+i)! = P = Q = \frac{(n+1)!}{2}$$

Sol 33: $S_1 = \displaystyle\sum_{0\le i}< \sum_{j\le 100} C_i C_j$

$$S_2 = \sum_{0\le j}< \sum_{i\le 100} C_i C_j$$

$$S_3 = \sum_{0\le i}= \sum_{j\le 100} C_i C_j$$

$(1+x)^{100} \Rightarrow n = 100$

$$S_1 + S_2 + S_3 = a^b$$

$$S_1 + S_2 + S_3 =$$

$$\sum_{0\le i}< \sum_{j\le 100} C_i C_j + \sum_{0\le j}< \sum_{i\le 100} C_i C_j + \sum_{0\le i}\sum_{j\le 100} C_i C_j$$

$$S_1 = S_2 \ \because \sum_{0\le i}\sum_{j\le 100} C_i C_j < \sum_{0\le i}\sum_{j\le 100} C_j C$$

$$S_3 = S_1 + C_0 C_0 + C_1 C_1 + C_2^2 + \dots + C_{100}^2$$

$$S_3 = S_1 + {}^{2n}C_n$$

$$S_1 + S_2 + S_3 = 2S_1 + 2^n C_n$$

$= [C_0(C_1 + C_2 + ... + C_{100}) + C_1(C_2 + ... + C_{100}) + ...]$

$= {}^{2n}C_0 + {}^{2n}C_1 + {}^{2n}C_2 + ... + {}^{2n}C_{2n}$

When $C_1 + C_2 + ... + C_{100} = 2^n$

$= 2^{2n} = 2^{200} = 4^{100} = 16^{50} = a^b$

$a + b = 16 + 50 = 66$

Exercise 2

Sol 1: (B) Given binomial is $(2^{1/3} + 3^{-1/3})^n$

$\therefore T_7 = T_{6+1} = {}^nC_6(2^{1/3})^{n-6}(3^{-1/3})^6$

T_7' from end $= {}^nC_{n-6}(3^{-1/3})^{n-6}(2^{1/3})^6$

$\Rightarrow \dfrac{T_7}{T_7'} = \dfrac{1}{6} = \dfrac{{}^nC_6\, 2^{n/3}2^{-2}3^{-2}}{{}^nC_6\, 3^{-n/3}3^2 2^2} = \dfrac{(2.3)^{n/3}}{(6)^{2+2}} = (6)^{(n/3)-4}$

$6^{(n/3)-4} = \dfrac{1}{6} \Rightarrow \dfrac{n}{3} - 4 = -1 \Rightarrow n = 9$

Sol 2: (C) We have $15^{23} + 23^{23} = (19-4)^{23} + (19+4)^{23}$

$= 2\left[{}^{23}C_0 19^{23} + {}^{23}C_2 19^{21} + ... + {}^{23}C_{22}19 \right]$

$= 2.19K$ always divisible by 19

So the remainder is zero

Sol 3: (D) $4\{{}^nC_1 + 4\,{}^nC_2 + 4^2.\,{}^nC_3 + ... + 4^{n-1}\}$

$= \{4\,{}^nC_1 + 4^2\,{}^nC_2 + 4^3\,{}^nC_3 + ... + 4^n\,{}^nC_n\}$

$= (1+x)^n = C_0 + C_1 x + C_2 x^2 + ... + x^n\,{}^nC_n$

At $x = 4$

$5^n = 1 + 4C_1 + 4^2 C_2 + ...$

So $4^n C_1 + 4^2\,{}^nC_2 + 4^3\,{}^nC_3 + ... + 4^n\,{}^nC_n = 5^n - 1$

Sol 4: (A) $n \geq 3$

$n - \dfrac{(n-1)}{1!}(n-1) + \dfrac{(n-1)(n-2)}{2!}(n-2)$

$\quad - \dfrac{(n-1)(2-n)(n-3)}{3!}(n-3) + ...$

At $n = 3$

$= \dfrac{1.3 - (3-1)}{1}(3-1) + \dfrac{(3-1)(3-2)}{2!}(3-2) - 0$

$= 3 - 2 \times 2 + \dfrac{2 \times 1}{2!} = 3 + 1 - 4 = 0$

Or

$= n - {}^{n-1}C_1(n-1) + {}^{n-1}C_2(n-2) + ...$

$+ 3\,{}^{n-1}C_{n-3}(-1)^{n-3} + 2\,{}^{n-1}C_{n-2}(-1)^{n-2}$

$= n - {}^{n-1}C_0 + 2\,{}^{n-1}C_1 + 3\,{}^{n-1}C_2 + ...$

$+ (-1)^{n-1}\,{}^{n-1}C_{n-1}$

$= n - ({}^{n-1}C_1 + {}^{n-1}C_0) + 0 = n - (n - 1 + 1) = 0$

Sol 5: (C) t_6 in $\left[x^{-8/3} + x^2 \log_{10}^x \right]^8 = 5600$

$^8C_5\left(x^{-8/3}\right)^3 \left(x^2 \log_{10}^x\right) = 5600$

$\Rightarrow x^2 \left(\log_{10}^x\right)^5 = 100$

$\Rightarrow x = 10$

Sol 6: (B) $(\alpha + p)^{m-1} + (\alpha + p)^{m-2}(\alpha + q)$

$+ (\alpha + p)^3(\alpha + q)^2 + ... + (\alpha + q)^{m-1}$

Coefficient of ^{t}

$= (\alpha + p)^{m-1}\left[1 + \left(\dfrac{\alpha + q}{\alpha + p}\right) + ... + \left(\dfrac{\alpha + q}{\alpha + p}\right)^{m-1}\right]$

$= (\alpha + p)^{m-1}\left[\dfrac{1 - \left(\dfrac{\alpha + q}{\alpha + p}\right)^m}{1 - \left(\dfrac{\alpha + q}{\alpha + p}\right)}\right]$

$= (\alpha + p)^{m-1}\dfrac{\left[1 - \dfrac{(\alpha + q)}{\alpha + p}\right]}{\alpha + p - \alpha - q}(\alpha + p)$

$= (\alpha + p)^m\left[\dfrac{1 - \left(\dfrac{\alpha + q}{\alpha + p}\right)^m}{p - q}\right] = \left[\dfrac{(\alpha + p)^m - (\alpha + q)^m}{p - q}\right]$

Coefficient of $\alpha^t = \dfrac{{}^mC_t[p^{m-t} - q^{m-t}]}{p - q}$

Sol 7: (B) $\left(1 + x - 3x^2\right)^{2145} = a_0 + a_1 x + a_2 x^2 + ...$

Put $x = -1$

$\Rightarrow (1 - 1 - 3)^{2145} = a_0 - a_1 + a_2 - a_3 + ...$

$\Rightarrow a_0 - a_1 + a_2 - a_3 + ... (-3)^{2145}$

Last digit of $(-3)^{2145}$ is 3.

Sol 8: (B) $\left(\dfrac{4x^2}{3} - \dfrac{3}{2x} \right)^9$

$T_{r+1} = {}^9C_r \left(\dfrac{4x^2}{3} \right)^{9-r} \left(-\dfrac{3}{2x} \right)^r$

Power of $x = 2(9 - r) + (-1)r$

$\Rightarrow 18 - 2r - r = 18 - 3r = 6$ (given)

$\Rightarrow 3r = 18 - 6 = 12 \Rightarrow r = 12/3 = 4$

Coefficient $\ {}^9C_4 \left(\dfrac{4}{3} \right)^{9-4} \left(\dfrac{-3}{2} \right)^4 = \dfrac{9 \times 8 \times 7 \times 6}{1.2.3.4} \left(\dfrac{4}{3} \right)^5 \left(\dfrac{3}{2} \right)^4$

$= 9 \times 2 \times 7 \times \dfrac{2^{10} \times 3^4}{3^5 \times 2^4} = 21 \times 2^7 = 2688$

Sol 9: (D) $\left(9x - \dfrac{1}{3\sqrt{x}} \right)^{18}, x > 0$

$T_{r+1} = {}^{18}C_r (9x)^{18-r} \left(\dfrac{1}{\sqrt{9x}} \right)^r$

Power of $x = 18 - r - \dfrac{r}{2} = 18 - \dfrac{3r}{2} = 0$ (given)

$\alpha = 9^{18-r} \left(\dfrac{1}{\sqrt{9}} \right)^r = (9)^{18 - r - \frac{r}{2}} = (9)^{18 - \frac{3r}{2}} = 9^0 = 1$

Sol 10: (C) $\left[x + \sqrt{x^3 - 1} \right]^5 + \left[x - \sqrt{x^3 - 1} \right]^5$

$= 2 \left[{}^5C_0 x^5 + {}^5C_2 x^3 (x^3 - 1) + {}^5C_4 x (x^3 - 1)^2 \right]$

$\Rightarrow$ Highest power is 7.

Sol 11: (C) We have

$C_0^2 + C_1^2 + C_2^2 + ... + C_n^2 = \sum_{r=0}^{n} ({}^nC_r)({}^nC_r)$

$= \sum_{r=0}^{n} ({}^nC_r)({}^nC_{n-r}) \qquad [\because {}^nC_r = nC_{n-r}]$

= Number of ways of choosing n persons out of n men and n women

= Number of ways of choosing n person out of 2n persons

$= {}^{2n}C_n$

Sol 12: (D) $aC_0 + (a+b)C_1 + ... + (a+nb)Cn$

$= \left[C_0 + C_1 + ... + C_n \right]$

$+ b \left[0 \times C_0 + 1 \times C_1 + 2 \times C_2 + ... + nC_n \right]$

$= a2^n + bn2^{n-1}$

$= (2a + nb)2^{n-1}$

Previous Years' Questions

Sol 1: We know, $(1 + x)^{2n} = C_0 + C_1 x + C_2 x^2 + ... + C_{2n}x^{2n}$

On differentiating both sides w.r.t. x, we get

$2n(1 + x)^{2n-1} = C_1 + 2.C_2 x$

$\mid 3.C_3 x^2 + ... + 2nC_{2n}x^{2n-1} \qquad ... (i)$

And

$\left(1 - \dfrac{1}{x} \right)^{2n} = C_0 - C_1 . \dfrac{1}{x} + C_2 . \dfrac{1}{x^2}$

$- C_3 . \dfrac{1}{x^3} + ... + C_{2n} . \dfrac{1}{x^{2n}} \qquad ... (ii)$

On multiplying Eqs. (i) and (ii), we get

$2n(1 + x)^{2n-1} \left(1 - \dfrac{1}{x} \right)^{2n}$

$= \left[C_1 + 2.C_2 x + 3.C_3 x^2 + ... + 2n.C_{2n}x^{2n-1} \right]$

$\times \left[C_0 - C_1 \left(\dfrac{1}{x} \right) + C_2 \left(\dfrac{1}{x^2} \right) - .... + C_{2n} \left(\dfrac{1}{x^{2n}} \right) \right]$

The coefficient of $\left(\dfrac{1}{x} \right)$ on the LHS

$=$ Coefficient of $\dfrac{1}{x}$ in $2n \left(\dfrac{1}{x^{2n}} \right)(1 + x)^{2n-1}(x - 1)^{2n}$

$=$ Coefficient of x^{2n-1} in $2n(1 - x^2)^{2n-1}(1 - x)$

$$= 2n(-1)^{n-1} \cdot (2n-1)\, C_{n-1}$$

$$= (-1)^n (2n) \frac{(2n-1)!}{(n-1)! \, n!}$$

$$= -(-1)^n \, n \cdot \frac{(2n)!}{(n!)^2} \cdot n \qquad \ldots \text{(iii)}$$

$$= -(-1)^n \, n \cdot C_n$$

Again, the coefficient of $\left(\dfrac{1}{x}\right)$ on the RHS

$$= -\left(C_1^2 - 2.C_2^2 + 3.C_3^2 + \ldots + -2nC_{2n}^2\right) \qquad \ldots \text{(iv)}$$

From Eqs. (iii) and (iv), we get

$$C_1^2 - 2.C_2^2 + 3.C_3^2 - \ldots - 2n.C_{2n}^2 = (-1)^n \, n.C_n$$

Sol 2: $^{n+1}C_1 + {}^{n+1}C_2 s_1 + {}^{n+1}C_3 s_2$

$$+ \ldots + {}^{n+1}C_{n+1} s_n = \sum_{r=1}^{n+1} {}^{n+1}C_r s_{r-1}$$

Where $s_n = 1 + q + q^2 + \ldots + q^n = \dfrac{1-q^{n+1}}{1-q}$

$$\therefore \sum_{r=1}^{n+1} {}^{n+1}C_r \left(\frac{1-q^r}{1-q}\right)$$

$$= \frac{1}{1-q}\left(\sum_{r=1}^{n+1} {}^{n+1}C_r - \sum_{r=1}^{n+1} {}^{n+1}C_r q^r\right)$$

$$= \frac{1}{1-q}\left[(1+1)^{n+1} - (1+q)^{n+1}\right]$$

$$= \frac{1}{1-q}\left[2^{n+1} - (1+q)^{n+1}\right] \qquad \ldots \text{(i)}$$

Also, $S_n = 1\left(\dfrac{q+1}{2}\right) + \left(\dfrac{q+1}{2}\right)^2 + \ldots + \left(\dfrac{q+1}{2}\right)^n$

$$= \frac{1 - \left(\dfrac{q+1}{2}\right)^{n+1}}{1 - \left(\dfrac{q+1}{2}\right)} = \frac{2^{n+1} - (q+1)^{n+1}}{2^n(1-q)} \qquad \ldots \text{(ii)}$$

From eqs. (i) and (ii), we get

$$^{n+1}C_r + {}^{n+1}C_2 s_1 + {}^{n+1}C_3 s_2 + \ldots + {}^{n+1}C_{n+1} s_n = 2^n s_n$$

Sol 3: $\displaystyle\sum_{r=0}^{n} (-1)^r \, {}^n C_r$

$$\left[\frac{1}{2^r} + \frac{3^r}{2^{2r}} + \frac{7^r}{2^{3r}} + \frac{15^r}{2^{4r}} + \ldots \text{upto } m \text{ terms}\right]$$

$$\sum_{r=0}^{n} (-1)^r \, {}^n C_r \left(\frac{1}{2}\right)^r +$$

$$\sum_{r=0}^{n} (-1)^r \, {}^n C_r \left(\frac{3}{4}\right)^r + \sum_{r=0}^{n} (-1)^r \, {}^n C_r \left(\frac{7}{8}\right)^r + \ldots$$

Upto m terms

$$\left\{\text{using} \sum_{r=0}^{n} (-1)^r \, {}^n C_r x^r = (1-x)^n\right\}$$

$$= \left(1 - \frac{1}{2}\right)^n + \left(1 - \frac{3}{4}\right)^n + \left(1 - \frac{7}{8}\right)^n + \ldots$$

Upto m terms

$$= \left(\frac{1}{2}\right)^n + \left(\frac{1}{4}\right)^n + \left(\frac{1}{8}\right)^n + \ldots$$

Upto m terms

$$= \left(\frac{1}{2}\right)^n \left[\frac{1 - \left(\dfrac{1}{2^n}\right)^m}{1 - \dfrac{1}{2^n}}\right] = \frac{2^{mn} - 1}{2^{mn}(2^n - 1)}$$

Sol 4: Let $y = (x-a)^m$, where m is a positive integer, $r \le m$,

Now, $\dfrac{dy}{dx} = m(x-a)^{m-1}$

$$\Rightarrow \frac{d^2 y}{dx^2} = m(m-1)(x-a)^{m-2}$$

$$\Rightarrow \frac{d^4 y}{dx^4} = m(m-1)(m-2)(m-3)(x-a)^{m-4}$$

$$\ldots\ldots\ldots\ldots\ldots\ldots\ldots\ldots\ldots\ldots\ldots\ldots$$

On differentiating r times, we get

$$\frac{d^r y}{dx^r} = m(m-1)\ldots(m-r+1)(x-a)^{m-r}$$

$$= \frac{m!}{(m-r)!}(x-a)^{m-r} = r!\left({}^m C_r\right)(x-a)^{m-r}$$

And for $r > m$, $\dfrac{d^r y}{dx^r} = 0$

Now,

$$\sum_{r=0}^{2n} a_r (x-2)^r = \sum_{r=0}^{2n} b_r (x-3)^r \quad \text{(given)}$$

On differentiating both sides n times w.r.t. x, we get

$$\sum_{r=n}^{2n} a_r (n!)^r C_n (x-2)^{r-n}$$

$$= \sum_{r=n}^{2n} b_r (n!)^r C_n (x-3)^{r-n}$$

On putting x = 3, we get

$$\sum_{r=n}^{2n} a_r (n!)^r C_n = (b_n)n!$$

$$\Rightarrow b_r = \frac{1}{n!} \sum_{r=n}^{2n} a_r (n!)^r C_n$$

$$= {}^{2n+1}C_n$$

$$= {}^{2n+1}C_{n+1}$$

Sol 5: $\left(1+x+x^2\right)^n = a_0 + a_1 x + \dots + a_{2n} x^{2n}$ ... (i)

Replacing x by $-1/x$, we get

$$\left(1 - \frac{1}{x} + \frac{1}{x^2}\right)^n$$

$$= a_0 - \frac{a_1}{x} + \frac{a_2}{x^2} - \frac{a_3}{x^3} + \dots + \frac{a_{2n}}{x^{2n}} \quad \text{... (ii)}$$

Now, $a_0^2 - a_1^2 + a_2^2 - a_3^2 + \dots + a_{2n}^2$ = coefficient of the term independent of x in

$$\left[a_0 + a_1 x + a_2 x^2 + \dots + a_{2n} x^{2n}\right]$$

$$\times \left[a_0 - \frac{a_1}{x} + \frac{a_2}{x^2} - \dots + \frac{a_{2n}}{x^{2n}}\right]$$

= Coefficient of the term independent of x in

$$\left(1+x+x^2\right)^n \left(1 - \frac{1}{x} + \frac{2}{x^2}\right)^n$$

Now, RHS $= \left(1+x+x^2\right)^n \left(1 - \frac{1}{x} + \frac{1}{x^2}\right)^n$

$$= \frac{\left(1+x+x^2\right)^n \left(x^2 - x + 1\right)^n}{x^{2n}}$$

$$= \frac{\left[\left(x^2+1\right)^2 - x^2\right]^n}{x^{2n}} = \frac{(1 + 2x^2 + x^4 - x^2)^n}{x^{2n}}$$

$$= \frac{\left(1 + x^2 + x^4\right)^n}{x^{2n}}$$

Thus, $a_0^2 - a_1^2 + a_2^2 - a_3^2 + \dots + a_{2n}^2$

= Coefficient of the term independent of x in

$$\frac{1}{x^{2n}}\left(1 + x^2 + x^4\right)^n$$

= Coefficient of x^{2n} in $\left(1 + x^2 + x^4\right)^n$

= Coefficient of t^n in $\left(1 + t + t^2\right)^n = a_n$

Sol 6: To show that

$$2^k \cdot {}^nC_0 \cdot {}^nC_k - 2^{k-1} \cdot {}^nC_1 \cdot {}^{n-1}C_{k-1} + 2^{k-2} \cdot {}^nC_2 \cdot {}^{n-2}C_{k-2} - \dots +$$

$$(-1)^k \, {}^nC_k \, {}^{n-k}C_0 = {}^nC_k$$

Taking LHS

$$2^k \cdot {}^nC_0 \cdot {}^nC_k - 2^{k-1} \cdot {}^nC_1 \cdot {}^{n-1}C_{k-1} + \dots + (-1)^k \cdot {}^nC_k = {}^nC_k$$

$$= \sum_{r=0}^{k} (-1)^r \cdot 2^{k-r} \cdot {}^nC_r \cdot {}^{n-r}C_{k-r}$$

$$= \sum_{r=0}^{k} (-1)^r \cdot 2^{k-r} \cdot \frac{n!}{r!(n-r)!} \cdot \frac{(n-r)!}{(k-r)!(n-k)!}$$

$$= \sum_{r=0}^{k} (-1)^r \cdot 2^{k-r} \cdot \frac{n!}{(n-k)!k!} \cdot \frac{k!}{r!(k-r)!}$$

$$= \sum_{r=0}^{k} (-1)^r \cdot 2^{k-r} \cdot {}^nC_k \cdot {}^kC_r$$

$$= 2^k \cdot {}^nC_k \left\{ \sum_{r=0}^{k} (-1)^r \cdot \frac{1}{2^r} \cdot {}^kC_r \right\}$$

$$= 2^k \cdot {}^nC_k \left(1 - \frac{1}{2}\right)^k = {}^nC_k = \text{RHS}$$

Sol 7: Let $y = \sum_{r=1}^{10} A_r (B_{10} B_r - C_{10} A_r)$

$$\sum_{r=1}^{10} A_r B_r = \text{coefficient of } x^{20} \text{ in } ((1+x)^{10} (x+1)^{20}) - 1$$

$= C_{20} - 1 = C_{10} - 1$ and $\sum_{r=1}^{10} (A_r)^2$ = coefficient of x^{10} in $((1+x)^{10} (x+1)^{10}) - 1 = B_{10} - 1$

$$\Rightarrow y = B_{10}(C_{10} - 1) - C_{10}(B_{10} - 1) = C_{10} - B_{10}.$$

Sol 8: Let T_{r-1}, T_r, T_{r+1} are three consecutive terms of $(1 + x)^{n+5}$

$T_{r-1} = {}^{n+5}C_{r-2}\,(x)^{r-2}$, $T_r = {}^{n+5}C_{r-1}x^{r-1}$, $T_{r+1} = {}^{n+5}C_r x^r$

Where, ${}^{n+5}C_{r-2} : {}^{n+5}C_{r-1} : {}^{n+5}C_r = 5 : 10 : 14$.

So $\dfrac{{}^{n+5}C_{r-2}}{5} = \dfrac{{}^{n+5}C_{r-1}}{10} \Rightarrow n - 3r = -3$ ... (i)

$\dfrac{{}^{n+5}C_{r-1}}{10} = \dfrac{{}^{n+5}C_r}{14} \Rightarrow 5n - 12r = -30$ ... (ii)

From equation (i) and (ii) $n = 6$

Sol 9: $2x_1 + 3x_2 + 4x_3 = 11$

Possibilities are $(0, 1, 2)$; $(1, 3, 0)$; $(2, 1, 1)$; $(4, 1, 0)$.

$\therefore$ Required coefficients

$= ({}^4C_0 \times {}^7C_1 \times {}^{12}C_2) + ({}^4C_1 \times {}^7C_3 \times {}^{12}C_0) + ({}^4C_2 \times {}^7C_1 \times {}^{12}C_1)$
$+ ({}^4C_4 \times {}^7C_1 \times 1)$

$= (1 \times 7 \times 66) + (4 \times 35 \times 1) + (6 \times 7 \times 12) + (1 \times 7)$

$= 462 + 140 + 504 + 7 = 1113.$

Sol 10: x^9 can be formed in 8 ways

i.e. x^9, x^{1+8}, x^{2+7}, x^{3+6}, x^{4+5}, $x^{1+2} + 6$, x^{1+3+5}, x^{2+3+4} and coefficient in each case is 1.

$\Rightarrow$ Coefficient of $x^9 = 1 + 1 + 1 + \underset{\text{8 times}}{\ldots\ldots\ldots} + 1 = 8$

Sol 11: $Z = \dfrac{-1 + i\sqrt{3}}{2} = \omega$

$P = \begin{bmatrix} (-\omega)^r & \omega^{2s} \\ \omega^{2s} & \omega^r \end{bmatrix}$

$P^2 = \begin{bmatrix} (-\omega)^r & \omega^{2s} \\ \omega^{2s} & \omega^r \end{bmatrix}\begin{bmatrix} (-\omega)^r & \omega^{2s} \\ \omega^{2s} & \omega^r \end{bmatrix}$

$= \begin{bmatrix} (-\omega)^{2r} + (\omega^{2s})^2 & \omega^{2s}(-\omega)^r + \omega^r\omega^{2s} \\ \omega^{2s}(-\omega) + \omega^r\omega^{2s} & \omega^{4s} + \omega^{2r} \end{bmatrix}$

$= \begin{bmatrix} \omega^{4s} + \omega^{2r} & \omega^{2s}(\omega^r + (-\omega)^r) \\ \omega^{2s}(\omega^r + (-\omega)^r) & \omega^{4s} + \omega^{2r} \end{bmatrix}$

$= -I$ (Given)

$\omega^{4s} + \omega^{2r} = -1$ and $\omega^{2s}(\omega^r + (-\omega)^r) = 0$

$\omega^r + (-\omega)^r = 0$

r	s
1	1
2	2

r	s
1	1
3	3

Total no. pairs = 1

4. PERMUTATIONS AND COMBINATIONS

1. INTRODUCTION

The main subject of this chapter is counting. Given a set of objects the problem is to arrange some or all of them according to some order or to select some or all of them according to some specification.

2. FUNDAMENTAL PRINCIPLE OF COUNTING

The rule of sum: If a first task can be performed in m ways, while a second task can be performed in n ways, and the tasks cannot be performed simultaneously, then performing either one of these tasks can be accomplished in any one of total m+n ways.

The rule of product: If a procedure can be broken down into first and second stages, and if there are m possible outcomes for the first stage and if, for each of these outcomes, there are n possible outcomes for the second stage, then the total procedure can be carried out, in the designated order, in a total of mn ways.

Illustration 1: There are three stations, A. B and C. Five routes for going from station A to station B and four routes for going from station B to station C. Find the number of different ways through which a person can go from A to C via B. **(JEE MAIN)**

Sol: This problem is an application of the Fundamental Principle of Counting. The rule of product can be used to solve this question easily.

Given there are five routes for going from A to B and four routes for going from B to C.

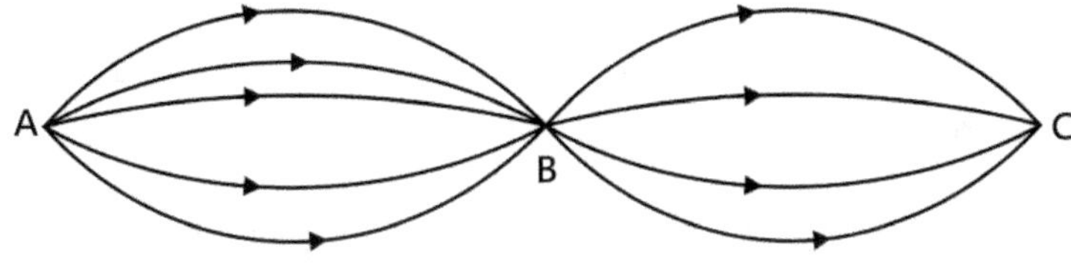

Figure 5.1

Hence, by the fundamental principle of multiplication the total number of different ways

= 5 x 4   (i.e., A to B and then B to C) = 20 ways

Illustration 2: A hall has 12 gates. In how many ways, can a man enter the hall through one gate and come out through a different gate. **(JEE MAIN)**

Sol: The rule of product can be used to solve this problem.

There are 12 ways of entering the hall. After entering the hall the man can come out through any of 11 different gates.

Hence, by the fundamental principle of multiplication, the total number of ways are 12 x 11 = 132 ways.

Illustration 3: How many numbers between 10 and 10,000 can be formed by using the digits 1, 2, 3, 4, 5 if

(i)   no digit is repeated in any number.   **(ii)**   digits can be repeated.   **(JEE MAIN)**

Sol: The numbers between 10 and 10,000 can be either two digit, three digit or four digit numbers. We consider each of these cases and try to find the number of possibilities using 1, 2, 3, 4 and 5. Finally, we add them up to get the desired result.

(i)   Number of two digit numbers = 5 x 4 = 20

   Number of three digit numbers = 5 x 4 x 3 = 60

   Number of four digit numbers = 5 x 4 x 3 x 2 = 120

   Total number of numbers = 20 + 60 +120 = 200

(ii)   Number of two digit numbers = 5 x 5 = 25

   Number of three digit numbers = 5 x 5 x 5 = 125

   Number of four digit numbers = $5 \times 5 \times 5 \times 5 = 625$

   Total number of numbers = 25 +125 +625 = 775

3. FACTORIAL NOTATION

An efficient way of writing a product of several consecutive integers is the factorial notation. The number n! (read as "n-factorial") is defined as follows :

For any positive integer n; $n! = n(n-1)(n-2) \ldots\ldots (3)(2)(1)$; For instance, $4! = 4.3.2.1 = 24$

Note: (i) $n! = n(n-1)(n-2) \ldots\ldots 3.2.1$; $n! = n.(n-1)!$; $0! = 1! = 1$; $(2n)! = 2^n.n![1.3.5.7 \ldots\ldots (2n-1)]$

(ii) $n! = n(n-1)! = n(n-1)(n-2)! = n(n-1)(n-2)(n-3)!$

(iii) $n(n-1) \ldots\ldots (n-r+1) = \dfrac{n!}{(n-r)!}$

Illustration 4: Find the sum of n terms of the series whose n^{th} term is n!×n.   **(JEE ADVANCED)**

Sol: Represent the general term in this question as a difference of two terms and then add them up to find the answer.

The required sum = $(1)! + 2(2)! + 3(3)! + \ldots\ldots + n(n!) = (2-1)! + (3-1)2! + (4-1)3! + \ldots\ldots + [(n+1)-1]n!$

$= (2! - 1!) + (3! - 2!) + (4! - 3!) + \ldots\ldots + [(n+1)! - n!] = (n+1)! - 1$

4. PERMUTATION

Each of the different arrangements, which can be made by taking some or all of a number of objects is called permutation. The number of permutations of n different objects taken r at a time is represented as

$${}^{n}P_{r} = \dfrac{n!}{(n-r)!} = n(n-1)(n-2) \ldots\ldots (n-r+1) \text{ (where, } 0 \le r \le n)$$

Note: (i) In permutation, the order of the items plays an important role.

(ii)   The number of all permutations of n distinct objects taken all at a time is n!

Illustration 5: If $^{56}P_{r+6} : {}^{54}P_{r+3} = 30800 : 1$, find $^{r}P_2$ **(JEE MAIN)**

Sol: Use the formula for $^{n}P_r$.

We have , $\dfrac{^{56}P_{r+6}}{^{54}P_{r+3}} = \dfrac{30800}{1} = \dfrac{56!}{(50-r)!} \times \dfrac{(51-r)!}{54!} = \dfrac{30800}{1} \Rightarrow 56 \times 55 (51-r) = 30800 \Rightarrow r = 41$

$\therefore \quad ^{41}P_2 = 41 \times 40 = 1640$

Illustration 6: Three men have 4 coats, 5 waist coats and 6 caps. In how many ways can they wear them?
          (JEE ADVANCED)

Sol: Use the concept and understanding of Permutation, i.e. arrangement to find the answer.

The total number of ways in which three men can wear 4 coats is the number of arrangements of 4 different coats taken 3 at a time. So, three men can wear 4 coats in $^{4}P_3$ ways. Similarly, 5 waist coats and 6 caps can be worn by three men in $^{5}P_3$ and $^{6}P_3$ ways respectively. Hence, the required no. of ways $= {}^{4}P_3 \times {}^{5}P_3 \times {}^{6}P_3 = (4!) \times (5 \times 4 \times 3) \times (6 \times 5 \times 4) = 172800$.

Illustration 7: Suppose 8 people enter an event in a swim meet. In how many ways could the gold, silver, and bronze prizes be awarded?   **(JEE ADVANCED)**

Sol: Use the formula for $^{n}P_r$. The required number of ways is an arrangement of 3 people out of 8 i.e.

$^{8}P_3 = \dfrac{8!}{5!} = \dfrac{8.7.6.5.4.3.2.1}{5.4.3.2.1} = 8 \cdot 7 \cdot 6 = 336.$

NOMORECLASS CONCEPTS

The following two steps are involved in the solution of a permutation problem:

Step 1: Recognizing the objects and the places involved in the problem.

Step 2: Checking whether the repetition of the objects is allowed or not.

4.1 Permutation with Repetition

These kinds of problems occur with permutations of different objects in which some of the objects can be repeated. The no. of permutations of n different objects taken r at a time when each object may be repeated any number of times is n^r.

Illustration 8: A student appears in an objective test which contains 10 multiple choice questions. Each question has four choices in which one is the correct option. What maximum number of different answers can the student give? How will the answer change if each question may have more than one correct answers?   **(JEE ADVANCED)**

Sol: Use the concept of Permutation with Repetition.

For the first part each question has four possible answers. So, the total possible answers $= 4 \times 4 \times \ldots 10$ times $= 4^{10}$.

For the second part. Suppose the choices for each question are denoted by A, B, C and D. Now the choice A is either correct or incorrect (two ways) similarly the other choices are either correct or incorrect. Thus, this particular question can have $2 \times 2 \times 2 \times 2 = 16$ possible answers. But this includes the case when all the four choices are incorrect. Thus the total number of answers $= 15$. Now, as there are 10 questions and each question has 15 possible answers. Therefore the total number of answers $= 15^{10}$.

The m^n or n^m dilemma

Let us start with an example.

Q. There are 7 letters and 5 letter-boxes. In how many ways can you put the letters in the boxes?

Sol: This is the typically confusing question asked frequently from the P & C area. Let's see how you can solve this type of question.

The Exhaustive Approach: One way to solve this question is through (what we will call) The Exhaustive Approach. While solving such problems, first decide which of the items (letters and letter-boxes here) is exhaustive. Exhaustive here means the entity which is sure to be used up completely.

In this example, all the "letters" are sure to be placed in the boxes, whereas there is no such constraint as regards the "letter-boxes". Some boxes could go empty. Having decided this, just go by the options we have for all the instances of the exhaustive item and you have your answer.

As you can see here, every letter has 5 boxes to choose from. Thus the total would

be $(5 \times 5 \, x........7 \text{ times}) = (5^7)$

A similar question could be: In how many ways can 10 rings be worn on 5 fingers? Try it yourself.

4.2 Permutation of Alike Objects

This kind of problems involve permutations of different objects in which some of them are similar.

The number of permutations of n objects taken all at a time in which, p are alike objects of one kind, q are alike objects of second kind & r are alike objects of a third kind and the rest $(n - (p + q + r))$ are all different is

$$\frac{n!}{p!q!r!}$$

Illustration 9: Determine the number of permutations of the letters of the word 'SIMPLETION' taken all at a time.

(JEE MAIN)

Sol: In the given word the letter I occurs twice and the remaining letters occur only once. So, the concept of Permutation of Alike Objects is used to find out the answer.

There are 10 letters in the word 'SIMPLETION' and out of these 10 letters two are identical. So, just selecting all 10 objects at a time will give twice the actual result. Hence, the number of permutations of taking all the letters at a time = $^{10}P_{10} / 2! = 10! / 2! = 181440$.

4.3 Permutation under Restriction

(a) The number of permutations of n different objects, taken r at a time, when a particular object is to be always included in each arrangement, is r. $^{n-1}P_{r-1}$

The number of permutations of n different objects, taken r at a time, when a particular object is never taken in each arrangement is $^{n-1}P_r$.

(b) **String method:** The number of permutations of n different objects, taken all at a time, when m specified objects always come together is $m! \times (n - m + 1)!$.

(c) **Gap Method:** The number of permutations when no two given objects occur together.

In order to find the number of permutations when no two given objects occur together.

(a) First of all, put the m objects for which there is no restriction, in a line. These m objects can be arranged in m! ways.

(b) Then count the number of gaps between every two m objects for which there is no restriction, including the end positions. Number of such gaps will be (m + 1).

(c) If m is the number of objects for which there is no restriction and n is the number of objects, two of which are not allowed to occur together, then the required number of ways = $m! \times {}^{m+1}C_n \times n!$

The number of permutations when two types of objects are to be arranged alternately

(a) If their numbers differ by 1 put the object whose number is greater in the first, third, fifth.... places, etc. and the other object in the second, fourth, sixth.... places.

(b) If the number of two types of objects is same, consider two cases separately keeping the first type of object in the first, third, fifth places, etc. and the second type of object in the first, third, fifth places.... and then add.

4.4 Non-Consecutive Selection

The number of selections of r consecutive objects out of n objects in a row $= n - r + 1$

The number of selections of r consecutive objects out of n objects along a circle $= \begin{cases} n, & \text{when } r < n \\ 1, & \text{when } r = n \end{cases}$

Illustration 10: Find the numbers between 300 and 3000 that can be formed with the digits 0, 1, 2, 3, 4 and 5, where no digit is repeated in any number. **(JEE MAIN)**

Sol: The numbers between 300 and 3000 can either be a three digit number or a four digit number. The solution is divided into these two different cases and their sum will give us the desired result.

Any number between 300 and 3000 must be of three or four digits.

Case I: When number is of three digits: The hundreds place can be filled by any one of the three digits 3, 4 and 5 in 3 ways. The remaining two places can be filled by the remaining five digits in ${}^5P_2$, ways.

$\therefore$ The number of numbers formed in this case $= 3 \times {}^5P_2 = 3 \times \dfrac{5!}{3!} = 60$

Case II: When number is of four digits: The thousands place can be filled by any one of the two digits 1 and 2 in 2 ways and the remaining three places can be filled by the remaining five digits in ${}^5P_3$ ways.

$\therefore$ The number of numbers formed in this case $= 2 \times {}^5P_3 = 2 \times \dfrac{5!}{2!} = 120$

$\therefore$ Total numbers = 60 + 120 = 180

Illustration 11: How many words can be formed from the letters of the word ARTICLE, so that vowels occupy the even places? **(JEE MAIN)**

Sol: Clearly, this is an example of Permutation under Restriction. We identify the even places and the odd places and try to find the number of ways in which the vowels and consonants can fill the spaces.

There are seven places: 3 even and 4 odd in which we have to fill 3 vowels and 4 consonants.

$\therefore$ The number of words $= {}^3P_3 \cdot {}^4P_4 = 3! \times 4! = 6 \times 24 = 144$.

Illustration 12: How many different words can be formed with the letters of the word ORDINATE so that

(a) Four vowels occupy the odd places (b) Beginning with O (c) Beginning with O and ending with E. **(JEE MAIN)**

Sol: The concept of Permutation under Restriction can be used to solve this problem.

There are 4 vowels and 4 consonants. Total 8 letters.

(a) No. of words = 4! × 4! = 24 × 24 = 576. Because 4 vowels are to be adjusted in 4 odd place and the 4 consonants in the remaining 4 even places.

(b) 7! ways, O being fixed.

(c) 6! ways, O fixed in first and E fixed in last.

Illustration 13: Find the number of ways in which 5 boys and 5 girls can be seated in a row so that

(a) No two girls may sit together.

(b) All the girls sit together and all the boys sit together

(c) All the girls are never together. **(JEE ADVANCED)**

Sol: Since the number of girls and the number of boys are equal they have to sit alternately. This can be used to solve (a). For (b), we keep the girls together and arrange the boys in five places. Also, the girls can be arranged amongst themselves in 5! ways. This gives us the number of arrangements. Use the answer of the second part to find (c).

(a) 5 boys can be seated in a row in $^5P_5$ = 5! ways. Now, in the 6 gaps between 5 boys, the 5 girls can be arranged in $^6P_5$ ways. Hence, the number of ways in which no two girls sit together = 5! $\times$ $^5P_5$ = 5! $\times$ 6!

(b) The two groups of girls and boys can be arranged in 2! ways. 5 girls can be arranged among themselves in 5! ways. Similarly, 5 boys can be arranged among themselves in 5! ways. Hence, by the fundamental principle of counting, the total number of requisite seating arrangements = 2!(5! $\times$ 5!) = 2(5!)^2.

(c) The total number of ways in which all the girls are never together = Total number of arrangements – Total number of arrangements in which all the girls are always together = 10! – 5! $\times$ 6!

Illustration 14: The letters of the word OUGHT are written in all possible orders and these words are written out as in a dictionary. Find the rank of the word TOUGH in this dictionary. **(JEE MAIN)**

Sol: The word TOUGH will appear after all the words that start with G, H and O. Then we look at the second letter of the words starting with T and then third. Hence, the rank of the word TOUGH will be one more than the sum of all the possibilities just mentioned.

Total number of letters in the word OUGHT is 5 and all the five letters are different, the alphabetical order of these letters is G, H, O, T, U.

Number of words beginning with G = 4! = 24

Number of words beginning with H = 4! = 24

Number of words beginning with O = 4! = 24

Number of words beginning with TG = 3! = 6

Number of words beginning with TH = 3! = 6

Number of words beginning with TOG = 2! = 2

Number of words beginning with TOH = 2! = 2

Next come the words beginning with TOU and TOUH is the first word beginning with TOU.

$\therefore$ Rank of 'TOUGH' in the dictionary = 24 + 24 + 24 + 6 + 6 + 2 + 2 + 1 = 89

Illustration 15: There are 21 balls which are either white or black and the balls of same color are alike. Find the number of white balls so that, the number of arrangements of these balls in a row is maximum. **(JEE ADVANCED)**

Sol: The property of a binomial coefficient can be used to solve this question.

Let there be r white balls so that the number of arrangements of these balls in a row be maximum. Number of arrangements of these balls is

$A = \dfrac{21!}{r!(21-r)!}$ A will be maximum when r = $\dfrac{21+1}{2}$ or $\dfrac{21-1}{2}$ i.e. 10 or 11

Illustration 16: The number plates of cars must contain 3 letters of the alphabet denoting the place and area to which its owner belongs. This is to be followed by a three-digit number. How many different number plates can be formed if:

(i) Repetition of letters and digits is not allowed.   (ii) Repetition of letters and digits is allowed. **(JEE ADVANCED)**

Sol: This is a simple application of Permutation with and without repetition.

There are 26 letters of alphabet and 10 digits from 0 to 9.

(i) When repetition is not allowed

3 letters selected in $26 \times 25 \times 24$ ways

3 digit numbers are $= 9 \times 9 \times 8$ (as zero can't be in the hundreds place)

$\therefore$ The Number of plates $= 26 \times 25 \times 24 \times 9 \times 9 \times 8 = 10108800$.

(ii) When repetition is allowed

3 letters are selected $26 \times 26 \times 26$ ways

3 digit numbers are $= 9 \times 10 \times 10 = 900$

$\therefore$ The number of plates $= 26 \times 26 \times 26 \times 900 = 15818400$.

Constraint based arrangement

Let us start with an example first:

Q. In how many ways can the word VARIETY be arranged so that exactly 2 vowels are together?

The problem could be easier if none or all vowels were to be kept together. Isn't it? Well, we will do exactly that!

Whenever question involving "constraints that has choices (2 vowels could be IE or AI or AE)" are asked, "go for the backward approach." Rather than finding the favorable cases, subtract the unfavorable ones from the total possible cases. This method is more reliable.

So, the solution to the above question would be:

(Total arrangements of VARIETY) – (Arrangements with no vowels together + Arrangements with all the vowels together)

5. COMBINATION

Each of the different groups or selection which can be made by some or all of a number of given objects without reference to the order of the objects in each group is called a combination.

The number of all combinations of n objects, taken r at a time is generally denoted by C(n, r) or $^{n}C_{r} = \dfrac{n!}{r!(n-r)!}$

$(0 \le r \le n) = \dfrac{^{n}p_{r}}{r!}$

Note:

(a) The number of ways of selecting r objects out of n objects, is the same as the number of ways in which the remaining (n - r) can be selected and rejected.

(b) The combination notation also represents the binomial coefficient. That is, the binomial coefficient $^{n}C_{r}$ is the combination of n elements chosen r at a time.

(c) (a) $^nC_r = {^nC_{n-r}}$

(b) $^nC_r + {^nC_{r-1}} = {^{n+1}C_r}$

(c) $^nC_x = {^nC_y} \Rightarrow x = y$ or $x + y = n$

(d) If n is even, then the greatest value of ^nC_r is $^nC_{n/2}$

(e) If n is odd, then the greatest value of ^nC_r is $^nC_{\frac{n+1}{2}}$

(f) $^nC_0 + {^nC_r} + \ldots\ldots\ldots + {^nC_n} = 2^n$

(g) $^nC_n + {^{n+1}C_n} + {^{n+2}C_n} + \ldots\ldots\ldots + {^{2n-1}C_n} = {^{2n}C_{n+1}}$

(d) Comparison of permutation and combination

Permutations	Combinations
Different orderings or arrangement of the r objects are different permutations $$^nP_r = \frac{n!}{(n-r)!}$$	Each choice or subset of r object give one combination. Order within the group of r objects does not matter. $$^nC_r = \frac{n!}{(n-r)!r!}$$
Clue words: arrangement, schedule, order	**Clue words:** group, committee, sample, selection, subset.

Illustration 17: A basketball coach must select two attackers and two defenders from among three attackers and five defenders. How many different combinations of attackers and defenders can he select? **(JEE MAIN)**

Sol: The number of ways two attackers and two defenders can be selected is $^3C_2$ and $^5C_2$ respectively.

$\therefore$ He can select in $^3C_2 \times {^5C_2} = \dfrac{5.4.3.2.1}{2.2} = 30$ different combinations.

Illustration 18: A soccer team of 11 players is to be chosen from 30 boys, of whom 4 can play only in goal, 12 can play only as forwards and the remaining 14 in any of the other positions. If the team is to include five forwards and of course, one goalkeeper, in how many ways can it be made up? **(JEE MAIN)**

Sol: Proceed according to the previous question.

There are: $^4C_1$ ways of choosing the goalkeeper. $^{12}C_5$ ways of choosing the forwards and $^{14}C_5$ ways of choosing the other 5 players. That is, $^4C_1 \times {^{12}C_5} \times {^{14}C_5}$ combinations altogether = 4 x 792 x 2002 = 6342336.

5.1 Combinations under Restrictions

(a) Number of ways of choosing r objects out of n different objects if p particular objects must be excluded.

Consider the objects $A_1, A_2, A_3,\ldots,A_p, A_{p+1}, \ldots\ldots, A_n$. If the p objects $A_1, A_2\ldots A_p$ are to be excluded then we will have to select r objects from the remaining n – p objects ($A_{p+1}, A_{p+2}, \ldots\ldots A_n$).

Hence the required number of ways = $^{(n-p)}C_r$

(b) Number of ways of choosing r objects out of n different objects if p particular objects must be included ($p \le r$).

Consider the objects $A_1, A_2, A_3 \ldots\ldots A_p, A_{p+1} \ldots\ldots A_n$. If the p particular objects $A_1, A_2, \ldots\ldots A_p$ (say) must be included in the selection then to complete the selection, we must select (r – p) more objects to complete the selection. These objects are to be selected from the remaining n – p objects.

Hence, the required number of ways = $^{n-p}C_{r-p}$

(c) The total number of combinations of n different objects taken one or more at a time = $2^n - 1$.

5.2 Combinations of Alike Objects

(a) The number of combinations of n identical objects taking ($r \leq n$) at a time is 1.

(b) The number of ways of selecting r objects out of n identical objects is n + 1.

(c) If out of (p + q + r + s) objects, p are alike of one kind, q are alike of a second kind, r are alike of the third kind and s are different, then the total number of combinations is $(p + 1)(q + 1)(r + 1)2^s - 1$

Note: The list of alike objects can be extended further

4. The number of ways in which r objects can be selected from a group of n objects of which p are identical, is

$$\sum_{0}^{t} {}^{n-p}C_r, \text{ if } r \leq p \text{ and } \sum_{r=p}^{t} {}^{n-p}C_r \text{ if } r > p$$

Illustration 19: There are 4 oranges, 5 apples and 6 mangoes in a fruit basket. In how many ways can a person select fruits from among the fruits in the basket? **(JEE MAIN)**

Sol: Use the concept of Combination of Alike objects described above.

Here, we consider all fruits of the same type as identical.

Zero or more oranges can be selected out of 4 identical oranges in 4 + 1 = 5 ways.

Zero or more apples can be selected out of 5 identical apples in 5 + 1 = 6 ways.

Zero or more mangoes can be selected out of 6 identical mangoes in 6 + 1 = 7 ways

$\therefore$ The total number of selections when all the three types of fruits are selected (the number of any type of fruit may be zero) = $5 \times 6 \times 7 = 210$.

But in one of these selections number of each type of fruit is zero and hence there is no selection, this must be excluded.

$\therefore$ The required number = 210 – 1 = 209.

Caution: When all fruits of same type are different, the number of selections

$= ({}^4C_0 + {}^4C_1 + \ldots\ldots + {}^4C_4)({}^5C_0 + {}^5C_1 + \ldots\ldots + {}^5C_5) ({}^6C_0 + \ldots\ldots\ldots + {}^6C_6) -1 = 2^4 \times 2^5 \times 2^6 - 1 = 2^{15} - 1$

Illustration 20: How many four digit numbers are there whose decimal notation contains not more than two distinct digits? **(JEE MAIN)**

Sol: A four digit number can consist of either only one digit or two digits as per the question. Clearly, there are nine four digit numbers with the same digit. Similarly, calculate the number of four digit numbers with two distinct digits and hence the sum gives us the desired result.

Evidently any number so formed of four digits contains

(i) Only one digit (like 1111, 2222,...) and there are 9 numbers.   **(ii)** Two digits

(a) if zero is one of the two, then the one more can be anyone of the nine, and these two digits can be arranged in ${}^9C_1 [{}^3C_1 + {}^3C_2 + {}^3C_2 + {}^3C_3] = 63$.

(b) if zero is not one of them, then two of the digits have to be selected from 9, and these   two can be arranged in ${}^9C_2[{}^4C_1 + {}^4C_2 + {}^4C_3] = 504$

Hence, the total number of required numbers = 567.

Illustration 21: In how many ways can a cricket team of eleven players be chosen out of a batch of 15 players, if

(a) there is no restriction on the selection

(b) a particular player is always chosen

(c) a particular player is never chosen   **(JEE MAIN)**

Sol: Using the concept of combination of alike objects we can get the answer.

(a) The total number of ways of selecting 11 players out of 15 is $= {}^{15}C_{11} = {}^{15}C_{15-11} = {}^{15}C_4 = \dfrac{15\times14\times13\times12}{4\times3\times2\times1} = 1365$

(b) A particular player is always chosen. This means that 10 players are selected out of the remaining 14 players.

$\therefore$ The required number of ways $= {}^{14}C_{10} = {}^{14}C_{14-10} = {}^{14}C_4 = 1001$

(c) The number of ways $= {}^{14}C_{11} = {}^{14}C_3 = \dfrac{14\times13\times12}{3\times2} = 364$

Illustration 22: In a plane there are 37 straight lines, of which 13 pass through the point A and 11 pass through the point B. Moreover, no three lines pass through one point, no line passes both points A and B. and no two are parallel. Find the number of points of intersection of the straight lines. **(JEE MAIN)**

Sol: Two non parallel straight lines give us a point of intersection. Using this idea we find the total number of points of intersection. Care should be taken that a point is not counted more than once.

The number of points of intersection of 37 straight lines is ${}^{37}C_2$. But 13 straight lines out of the given 37 straight lines pass through the same point A. Therefore, instead of getting ${}^{13}C_2$ points, we get merely one point A. Similarly, 11 straight lines out of the given 37 straight lines intersect at point B. Therefore, instead of getting ${}^{11}C_2$ points, we get only point B. Hence, the number of intersection points of the lines in ${}^{37}C_2 - {}^{13}C_2 - {}^{11}C_2 + 2 = 535$.

Illustration 23: How many five-digit numbers can be made having exactly two identical digits? **(JEE MAIN)**

Sol: Note that zero cannot occupy the first place. So we divide the solution into two cases when the common digit is 0 and otherwise. Calculate the number of possibilities in these two cases and their sum gives us the desired result.

Case I: Two identical digits are 0, 0.

The number of ways to select three more digits is ${}^9C_3$. The number of arrangements of these five digits is $(5!/2!) - 4! = 60 - 24 = 36$.

Hence, the number of such numbers is ${}^9C_3 \times 36 = 3024$ ... (i)

Case II: Two identical digits are (1, 1) or (2, 2) or.... or (9, 9).

If 0 is included, then number of ways of selection of two more digits is ${}^8C_2$.

The number of ways of arrangements of these five digits is $5! / 2! - 4! / 2! = 48$.

Therefore, the number of such numbers is ${}^8C_2 \times 48$.

If 0 is not included, then selection of three more digits is ${}^8C_3$.

Therefore, the number of such numbers is ${}^8C_3 \times 5! / 2! = {}^8C_3 \times 60$.

Hence, the total number of five-digit numbers with identical digits (1.1)......(9.9) is

$9 \times ({}^8C_2 \times 48 + {}^8C_3 \times 60) = 42336$ ... (ii)

From Eqs. (i) and (ii), the required number of numbers is $3024 + 42336 = 45360$.

Illustration 24: How many words can be made with letters of the word "INTERMEDIATE" if

(i) The words neither begin with I nor end with E,

(ii) The vowels and consonants alternate in the words,

(iii) The vowels are always consecutive,

(iv) The relative order of vowels and consonants does not change,

(v) No vowel is in between two consonants,

(vi) The order of vowels does not change? **(JEE MAIN)**

Sol: This is an application of Permutation under restriction and Permutation of Alike objects. Proceed according to the given conditions.

(i) The required number of words = (the number of words without restriction) – (the number of words beginning with I) – (the number of word ending with E) + (the number of words beginning with I and ending with E) (because words beginning with I as well as words ending with E contain some words beginning with I and ending with E).

The number of words without restriction $= \dfrac{12!}{2!\,3!\,2!}$

($\because$ There are 12 letters in which there are two I's, three E's and two T's).

The number of words beginning with I $= \dfrac{11!}{2!3!}$

($\because$ With E in the extreme left place we are left to arrange 11 letters INTERMEDIATE in which there are two T's and three E's).

The number of words ending with E $= \dfrac{11!}{2!2!2!}$

($\because$ With E in the extreme right place we are left to arrange 11 letters INTERMEDIATE in which there are two I's. two E's and two T's.)

The number of words beginning with I and ending with E $= \dfrac{10!}{2!2!}$

($\because$ With I in the extreme left and E in the extreme right places we are left to arrange 10 letters INTERMEDIATE in the which there are two T's and Two E's).

$\therefore$ the required number or words $= \dfrac{12!}{2!3!2!} - \dfrac{11!}{2!3!} - \dfrac{11!}{2!2!2!} + \dfrac{10!}{2!2!} = \dfrac{10!}{2!3!2!}(12 \times 11 - 11 \times 2 - 11 \times 3 + 6) = \dfrac{83 \times 10!}{24}$

(ii) There are 6 vowels and 6 consonants. So, the number of words in which vowels and consonants alternate = (the number of words in which vowels occupy odd places and consonants occupy even places) + (the number of words in which consonants occupy odd places and vowels occupy even places)

$$= \dfrac{6!}{2!3!} \times \dfrac{6!}{2!} + \dfrac{6!}{2!} \times \dfrac{6!}{2!3!} = 2.\, \dfrac{6!}{2!3!}\cdot\dfrac{6!}{2!} = 43200$$

(iii) Considering the 6 vowels IEEIAE as one object, the number of arrangements of this with 6 consonants $= \dfrac{7!}{2!}$
($\because$ there are two T's in the consonants).

For each of these arrangements, the 6 consecutive vowels can be arranged among themselves in $\dfrac{6!}{2!3!}$.

$\therefore$ The required number of words $= \dfrac{7!}{2!} \times \dfrac{6!}{2!3!}$ (as above) $= 151200$

(iv) The relative order of vowels and consonants will not change if in the arrangements of letters, the vowels occupy places of vowels, i.e.. 1st, 4th, 7th, 9th, 10th, 12th places and consonants occupy their places, i.e., 2nd. 3rd. 5th. 6th. 8th. 11th places, the required number of words

$$\dfrac{6!}{2!3!} \times \dfrac{6!}{2!} = 21600$$

(v) No vowel will be between two consonants if all the consonants become consecutive

$\therefore$ the required number of words = the number of arrangements when all the consonants are consecutive

$$= \dfrac{7!}{2!3!} \times \dfrac{6!}{2!} \text{ (as above) } = 151200$$

(vi) The order of vowels will not change if no two vowels interchange places, i.e., in the arrangement all the vowels are treated as identical.

(For example LATE. ATLE. TLAE. etc.. have the same order of vowels A. E. But LETA, ETLA.

TLEA. etc.. have changed order of vowels A. E. So. LATE is counted but LETA is not.

If A, E, are taken as identical say V then LVTV does not give a new arrangement by interchanging V, V.

The required number of words,

= The number of arrangements of 12 letters in which 6 vowels are treated as identical

$= \dfrac{12!}{6!2!}$ ($\because$ there are two T's also).

Illustration 25: India and South Africa play a one day international series until one team wins 4 matches. No match ends in a draw. Find, in how many ways can the series can be won. **(JEE ADVANCED)**

Sol: The team who wins the series is the team with more number of wins. The losing team wins either 0 or 1 or 2 or 3 matches. Using this we find the number of ways in which a team can win.

Let I for India and S for South Africa. We can arrange I and S to show the wins for India and South Africa respectively

Suppose India wins the series, then the last match is always won by India.

	Wins of S	Wins of I	No. of ways
(i)	0	4	1
(ii)	1	4	$4! / 3! = 4$
(iii)	2	4	$\dfrac{5!}{2!3!} = 10$
(iv)	3	4	$\dfrac{6!}{3!3!} = 20$

$\therefore$ Total no. of ways = 35

In the same number of ways South Africa can win the series

$\therefore$ Total no. of ways in which the series can be won = 35 × 2 = 70

Illustration 26: There are p intermediate railway stations on a railway line from one terminal to another. In how many ways can a train stop at three of these intermediate stations, if no two of these stations (where it stops) are to be consecutive? **(JEE ADVANCED)**

Sol: The train stops only at three intermediate stations implies that the train does not stop at $(p - 3)$ stations. Using this idea we proceed further to get the answer.

The problem then reduces to the following:

In how many ways can three objects be placed among $(p - 3)$ objects in a row such that no two of them are next to each other (at most 1 object is to be placed between any two of these $(p - 3)$ objects). Since there are $(p - 2)$ positions to place the three objects, the required number of ways = $^{p-2}C_3$.

Illustration 27: Five balls are to be placed in three boxes. Each can hold all the five balls. In how many different ways can we place the balls so that no box remains empty if

(i) balls and boxes are all different

(ii) balls are identical but boxes are different

(iii) balls are different but boxes are identical

(iv) balls as well as boxes are identical

(v) balls as well as boxes are identical but boxes are kept in a row? **(JEE MAIN)**

Sol: Use the different cases of combination to solve the question according to the given conditions

As no box is to remain empty, boxes can have balls in the following numbers:

Possibilities 1, 1, 3 or 1, 2, 2

(i) The number of ways to distribute the balls in groups of 1, 1, 3 = $^5C_1 \times {}^4C_1 \times {}^3C_3$.

But the boxes can interchange their content, no exchange gives a new way when boxes containing balls in equal number interchange.

$\therefore$ the total number of ways to distribute 1, 1, 3 balls to the boxes = $^5C_1 \times {}^4C_1 \times {}^3C_3 \times \dfrac{3!}{2!}$

Similarly, the total number of ways to distribute 1, 2, 2 balls to the boxes = $^5C_1 \times {}^4C_2 \times {}^2C_2 \times \dfrac{3!}{2!}$

$\therefore$ the required number of ways = $^5C_1 \times {}^4C_1 \times {}^3C_3 \times \dfrac{3!}{2!} + {}^5C_1 \times {}^4C_1 \times {}^2C_2 \times \dfrac{3!}{2!} = 5 \times 4 \times 3 + 5 \times 6 \times 3 = 60 + 90 = 150$

Note: Writing the whole answer in tabular form.

Possibilities	Combinations	Permutations
1, 1, 3	$^5C_1 \times {}^4C_1 \times {}^3C_3$	$^5C_1 \times {}^4C_1 \times {}^3C_3 \times \dfrac{3!}{2!} = 5 \times 4 \times 3 = 60$
1, 2, 2	$^5C_1 \times {}^4C_2 \times {}^2C_2$	$^5C_1 \times {}^4C_2 \times {}^2C_2 \times \dfrac{3!}{2!} = 5 \times 6 \times 3 = 90$

$\therefore$ the required number of ways = 60 + 90 = 150.

(ii) When balls are identical but boxes are different the number of combinations will be 1 in each case.

$\therefore$ the required number of ways = $1 \times \dfrac{3!}{2!} + 1 \times \dfrac{3!}{2!} = 3 + 3 = 6$

(iii) When the balls are different and boxes are identical, the number of arrangements will be 1 in each case.

$\therefore$ the required number of ways = $^5C_1 \times {}^4C_1 \times {}^3C_3 + {}^5C_1 \times {}^4C_2 \times {}^2C_2 = 5 \times 4 + 5 \times 6 = 20 + 30 = 50$

(iv) When balls as well as boxes are identical, the number of combinations and arrangements will be 1 each in both cases.

$\therefore$ the required number of ways = $1 \times 1 + 1 \times 1 = 2$

(v) When boxes are kept in a row, they will be treated as different. So, in this case the number of ways will be the same as in (ii).

Illustration 28: There are m points on one straight line AB and n points on another straight line AC, none of them being A. How many triangles can be formed with these points as vertices? How many can be formed if point A is also included? **(JEE MAIN)**

Sol: A triangle has three vertices, so we select two points on one line and one on the other and vice versa. Also, consider the case when one point of the triangle is the intersection of the two lines.

To get a triangle, we either take two points on AB and one point on AC. or one point on AB and two points on AC. Therefore, the number of triangles, we obtain

$$= (^mC_2)(^nC_1) + (^mC_1)(^nC_2) = \frac{m(m-1)}{2}n + m\frac{n(n-1)}{2} = \frac{1}{2}\,mn(m-1+n-1) = \frac{1}{2}mn(m+n-2)$$

If the point A is included, we get m n additional triangles. Thus, in this case we get

$$= \frac{mn}{2}\,(m+n-2) + mn = \frac{mn(m+n)}{2}\ \text{triangles.}$$

5.3 Division into Groups

(a) The number of ways in which (m + n) different objects can be divided into two unequal groups containing m and n objects respectively is $\dfrac{(m+n)!}{m!n!}$.

If m = n, the groups are equal and in this case the number of division is $\dfrac{(2n)!}{n!n!2!}$; as it is possible to interchange the two groups without obtaining a new distribution.

(b) However, if 2n objects are to be divided equally between two persons then the number of ways

$$= \frac{(2n)!}{n!n!2!} \cdot 2! = \frac{(2n)!}{n!n!}$$

(c) The number of ways in which $(m + n + p)$ different objects can be divided into three unequal groups containing

m, n and p objects respectively is $= \dfrac{(m+n+p)}{m!n!p!}$, $m \neq n \neq p$

If $m = n = P$ then the number of groups $= \dfrac{(3n)!}{n!n!n!3!}$. However, if 3n objects are to be divided equally among three

persons then the number of ways $= \dfrac{(3n)!}{n!n!n!3!} \cdot 3! = \dfrac{(3n)!}{(n!)^3}$

For example, the number of ways in which 15 recruits can be divided into three equal groups is $\dfrac{15!}{5!5!5!3!}$ and the

number of ways in which they can be drafted into three different regiments, five in each, is $\dfrac{15!}{5!5!5!}$

(d) The number of ways in which mn different objects can be divided equally into m groups if order of groups is

not important is $\dfrac{mn!}{(n!)^m m!}$

(e) The number of ways in which mn different objects can be divided equally into m groups if the order of groups

is important is $\dfrac{mn!}{(n!)^m m!} \times m! = \dfrac{(mn)!}{(n!)^m}$

Illustration 29: In how many ways can 12 balls be divided between 2 boys, one receiving 5 and the other 7 balls?

(JEE MAIN)

Sol: Simple application of division of objects into groups. Since the order is important, the number of ways in which 12 different balls can be divided between two boys who each get 5 and 7 balls respectively, is

$$\frac{12!}{5!7!} \times 2! = \frac{12.11.10.9.8.7!}{(5.4.3.2.1)7!} \times 2 = 1584$$

Alternative:

The first boy can be given 5 balls out of 12 balls in $^{12}C_5$ ways. The second boy can be given the remaining 7 balls in one way. But the order is important (the boys can interchange 2 ways).

Thus, the required number of ways $= {}^{12}C_5 \times 1 \times 2! = \dfrac{12!}{5!7!} \times 2 = \dfrac{12.11.10.9.8.7!.2}{5.4.3.2.1.7!} = 1584$

Illustration 30: Find the number of ways in which 9 different toys can be distributed among 4 children belonging to different age groups in such a way that the distribution among the 3 elder children is even and the youngest one is to receive one toy more.

(JEE ADVANCED)

Sol: Using the concept of division of objects into groups we can solve this problem very easily.

The distribution should be 2, 2, 2 and 3 to the youngest. Now, 3 toys for the youngest can be selected in $^9C_3$ ways, the remaining 6 toys can be divided into three equal groups in

$\dfrac{6!}{(2!)^3 . 3!}$ ways and can be distributed in 3 ! ways.

Thus, the required number of ways $= {}^9C_3 \cdot \dfrac{6!}{(2!)^3 3!} \cdot 3! = \dfrac{9!}{3!(2!)^3}$

Illustration 31: Divide 50 objects in 5 groups of size 10, 10, 10, 15 and 5 objects. Also find the number of distributions? **(JEE MAIN)**

Sol: Same as the above question. Number of ways of dividing 50 objects into 5 groups as given $= \dfrac{50!}{(10!)^3 (15)!(5)!(3)!}$

Number of ways of distributing 50 objects into above formed groups $= \dfrac{50!}{(10)!^3 .(15)!(5)!3!} \times 5!$

NOMORECLASS CONCEPTS

Identical Objects and Distinct Choices

Questions involving identical objects tend to be tricky, especially when the choices that they have are distinct.

Many choices image 1

An example would be :

Q. In how many ways can we place 10 identical oranges in 3 distinct baskets, such that every basket has at least 2 oranges each?

One method is to place 2 oranges in every basket and make cases for the rest of them. However, in such questions, the other approach results in fewer cases, and hence, simpler calculations and a more efficient solution

For this question: Divide 10 into groups of 3 rather than placing 2 in each and dividing the remaining four.

6. CIRCULAR PERMUTATION

Let us consider that persons A, B, C and D are sitting around a round table. If all of them (A, B, C, D) are shifted one place in an anticlockwise order, then we will get fig.(b) from fig.(a). Now, if we shift A, B, C, D in anticlockwise order again, we will get fig. (c). We shift them once more and we will get fig.(d); and in the next time fig(a).

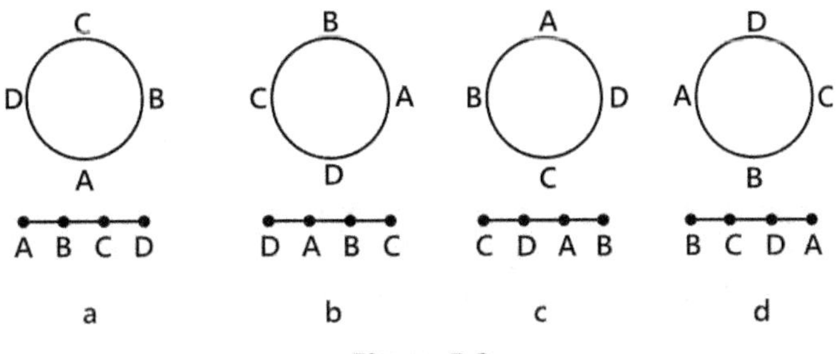

Figure 5.2

Thus, we see that if 4 persons are sitting at a round table, they can be shifted four times and the four different arrangements thus obtained will be same, because the anticlockwise order of A, B, C, D does not change. But if A, B, C and D are sitting in a row and they are shifted in such an order that the last occupies the place of first, then the four arrangements will be different.

Thus, if there are 4 objects, then for each circular arrangement number of linear arrangements is 4.

Similarly, if n different objects are arranged along a circle, for each circular arrangement the number of linear arrangements is n.

Therefore, the number of circular arrangements of n different objects = the number of linear arrangements of n different objects / n = n!/(n) = (n – 1)!

Clockwise and Anticlockwise Arrangements

Let the four persons A, B, C and D sit at a round table in anticlockwise as well as clockwise directions. These two arrangements are different. But if four flowers R (red), G (green), Y (yellow) and B (blue) are arranged to form a garland in anticlockwise and in clockwise order, then the two arrangements are same because if we view the garland from one side the four flowers R, G, Y, B will appear in anticlockwise direction and if seen from the other side the four flowers will appear in the clockwise direction. Here the two arrangements will be considered as one arrangement because the order of flowers does not change, rather only the side of observation changes. Here, two permutations will be counted as one.

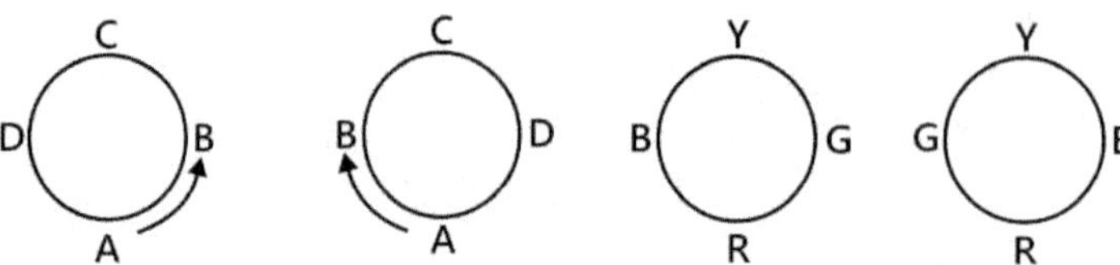

Figure 5.3

Therefore, when clockwise and anticlockwise arrangements are not different, i.e. when observations can be made from both sides, the number of circular arrangements of n different objects is (n – 1)!/2

Consider five persons A, B, C, D, E on the circumference of a circular table in an order which has no head. Now, shifting A, B, C, D, E one position in anticlockwise direction we will get arrangements as follows:

We see, that arrangements in all figures are different.

∴ The number of circular permutation of n different objects taken all at a time is (n – 1)!, if clockwise and anticlockwise orders are taken as different.

Note:

(a) The number of circular permutations of n different objects taken r at a time

$^nP_r/r$, when clockwise and anticlockwise orders are treated as different.

$^nP_r/2r$, when clockwise and anticlockwise orders are treated as same.

(b) The number of circular permutations of n different objects altogether

$^nP_n/n = (n – 1)!$, when clockwise and anticlockwise order are treated as different,

$^nP_n/2n = 1/2(n – 1)!$, when the above two orders are treated as same.

Illustration 32: In how many ways can 5 Indians and 4 Englishmen be seated at a round table if

(a) There is no restriction,   **(b)** All the four Englishmen sit together,

(c) All four Englishmen don't sit together,   **(d)** No two Englishmen sit together.   **(JEE MAIN)**

Sol: Clearly, this is a case of Circular Permutation. Using the formula (n – 1)!, we can find the answer according to the given cases.

(a) Total number of persons = 5 + 4 = 9. These 9 persons can be seated at the round table in 8! Ways.

∴ Required number of ways = 8!

(b) Regarding 4 Englishmen as one person, we have only 5 + 1 i.e. 6 persons.

These 6 persons can be seated at the round table in 5! ways. Also, the 4 Englishmen can be arranged among themselves in 4! ways.

∴ the required number of ways = 5! 4!

(c) The total number of arrangements when there is no restriction = 8!; the number of arrangements when all the four English men sit together = 5! 4!

∴ The number of arrangements when all the four Englishmen don't sit together = 8! – 5! 4!

(d) As there is no restriction on Indians, we first arrange the 5 Indians.

Now, 5 Indians can be seated around a table in 4! ways. If an Englishman sits between two Indians, then no two Englishmen will sit together. Now, there are 5 places for 4 English men, therefore, 4 Englishmen can be seated in $^5P_4$ ways.

$\therefore$ The required number of ways = $4! \times {}^5P_4 = 4 \times 5!$

Illustration 33: Consider 21 different pearls on a necklace. How many ways can the pearls be placed in this necklace such that 3 specific pearls always remains together? **(JEE MAIN)**

Sol: This is the case of circular permutation when there is no distinction between clockwise and anticlockwise arrangements.

After fixing the places of three pearls. Treating 3 specific pearls = 1 unit. So we have now 18 pearls + 1 unit = 19 and the number of arrangements will be $(19 - 1)! = 18!$. Also the number of ways 3 pearls can be arranged between themselves is $3! = 6$. As there is no distinction between the clockwise and anticlockwise arrangements, the required number of arrangements = $\dfrac{1}{2} 18!.\, 6 = 3\,(18!)$.

Illustration 34: Six persons A, B, C, D, E and F are to be seated at a circular table. Find the number of ways this can be done if A must have either B or C on his right and B must have either C or D on his right. **(JEE MAIN)**

Sol: Fix the position of some of the persons relative to each other as per the question and arrange the remaining in the seats available.

When A has B or C to his right we have either AB or AC

When B has C or D to his right we have BC or BD.

Thus, we must have ABC or ABD or AC and BD.

For ABC, D, E, F in a circular number of ways = $3! = 6$

For ABD, C, E, F in a circular number of ways = $3! = 6$

For AC, BD E, F the number of ways = $3! = 6$

Hence, the required number of ways = 18

7. LINEAR EQUATIONS WITH UNIT COEFFICIENTS

Consider the equation $x_1 + x_2 + x_3 + ... + x_k = m$, in k variables whose sum must always be m,

The number of non-negative solutions to the above equation is given by the fictitious partition method which is stated as:

Method of fictitious partition: Number of ways in which n identical objects may be distributed among p persons if each person may receive none, one or more objects is = $^{n+p-1}C_n$.

Coefficient Method

(a) The number of non-negative integral solutions of equation $x_1 + x_2 + ... + x_r = n$

= The number of ways of distributing n identical objects among r persons when each person can get zero, one or more objects = coeff. of x^n in $[(1 + x + x^2 + ... + x^n)(1 + x + x^2 + ... + x^n)\,(1 + x + x^2 + ... + x^n)...$upto r factors]

= coeff. of x^n in $(1 + x + x^2 + ... + x^n)^r$

= coeff. of x^n in $\left(\dfrac{1 - x^{n+1}}{1 - x}\right)^r$ = coeff. of x^n in $(1 - x^{n+1})^r (1 - x)^{-r}$ = coeff. of x^n in $(1 - x)^{-r}$

[leaving terms containing powers of x greater than n] = $^{n+r-1}C_{r-1}$

Note: If n is a positive integer, then

$$(1-x)^{-n} = 1 + \frac{(-n)}{1!}(x) + \frac{(-n)(-n-1)}{2!}(-x)^2 + \frac{(-n)(-n-1)(-n-2)}{3!}(-x)^3 + \dots \text{ to } \infty$$

$$= 1 + \frac{n}{1!}x + \frac{n(n+1)}{2!}x^2 + \frac{n(n+1)(n+2)}{3!}x^3 + \dots \text{ To } \infty = 1 + {}^nC_1x + {}^{n+1}C_2x^2 + {}^2C_3x^3 + \dots \text{ to } \infty$$

Coeff. X^r in $(1-x)^{-n} = {}^{n+r-1}C \Rightarrow$ coeff. of x^n in $(1-x)^{-r} = {}^{n+r-1}C_n = {}^{n+r-1}C_{r-1}$

(b) The number of positive integral solutions for the equation $x_1 + x_2 + \dots + x_r = n$

= The number of ways of distributing n identical objects among r persons when each person can get at least one object

= coeff. of x^n in $[x + x^2 + \dots + x^n)(x + x^2 + \dots + x^n)(x + x^2 + \dots + x^n) \dots$ Upto r factors]

$$= \text{coeff. of } x^n \text{ in } (x + x^2 + \dots + x^n)^r = \text{coeff. of } x^n \text{ in } x^r\left(\frac{1-x^n}{1-x}\right)^r$$

= coeff. Of x^{n-r} in $(1-x^n)^r(1-x)^{-r}$ = coeff. Of x^{n-r} in $(1-x)^{-r}$

[Leaving terms containing powers of x greater than $n - r$]

$$= {}^{n-r+r-1}C_{r-1} = {}^{n-1}C_{r-1}.$$

Illustration 35: How many integral solutions are there to $x + y + z + w = 29$, when $x \geq 1, y \geq 2, z \geq 3$ and $w \geq 0$?

(JEE ADVANCED)

Sol: Application of multinomial theorem.

$$x + y + z + w = 29 \qquad \dots(i)$$

$x \geq 1, y \geq 2, z \geq 3, w \geq 0 \quad \Rightarrow \quad x - 1 \geq 0, y - 2 \geq 0, z - 3 \geq 0, w \geq 0$

Let $x_1 = x - 1, x_2 = y - 2, x_3 = z - 3$

$\Rightarrow \quad x = x_1 + 1, y = x_2 + 2, z = x_3 + 3$ and then $x_1 \geq 0, x_2 \geq 0, x_3 \geq 0, w \geq 0$

From (i), $x_1 + 1 + x_2 + 2 + x_3 + 3 + w = 29$

$\Rightarrow \quad x_1 + x_2 + x_3 + w = 23$

Hence, the total number of solutions $= {}^{23+4-1}C_{4-1} = {}^{26}C_3 = 2600$

Illustration 36: Find the number of non-negative integral solution $3x + y + z = 24$. **(JEE MAIN)**

Sol: Application of multinomial theorem.

$3x + y + z = 24, x \geq 0, y \geq 0, z \geq 0$

Let $x = k$ $\therefore y + z = 24 - 3k$ ....(i)

Here, $0 \leq 24 - 3k \leq 24$. Hence, $0 \leq k \leq 8$

The total number of integral solutions of (1) is ${}^{24-3k+2-1}C_{2-1} = {}^{25-3k}C_1 = 25 - 3k$

Hence, the total number of solutions of the original equation

$$= \sum_{k=0}^{8}(25 - 3k) = 25\sum_{k=0}^{8}1 - 3\sum_{k=0}^{8}k \Rightarrow 25.9 - 3.\frac{8.9}{2} = 225 - 108 = 117.$$

Illustration 37: Find the number of solutions of the equation $x + y + z = 6$, where $x, y, z \in W$. **(JEE MAIN)**

Sol: The number of solutions $= {}^{6+3-1}C_{3-1} = {}^8C_2 = 280.$

Illustration 38: How many integers are there between 1 and 1000000 having the sum of the digits as 18?

(JEE ADVANCED)

Sol: Let the digits be $a_1, ..., a_6$ and use multinomial theorem we get the answer.

Any number between 1 and 1000000 must be of less than seven digits. Therefore, it must be of the form $a_1 \, a_2 \, a_3 \, a_4 \, a_5 \, a_6$

Where $a_1, a_2, a_3, a_4, a_5, a_6 \in \{0, 1, 2, ... 9\}$

Thus $a_1 + a_2 + a_3 + a_4 + a_5 + a_6 = 18$, where $0 \le a_i \le 9$, i = 1, 2, ... 9

The required number of ways = coeff. of x^{18} in $(1 + x + x^2 + ... x^9)^6$

$$= \text{coeff. of } x^{18} \text{ in } \left(\frac{1 - x^{10}}{1 - x}\right)^6$$

$= $ coeff. of x^{18} in $[(1 - x^{10})^6 (1 - x)^{-6}]$; $= $ coeff. of x^{18} in $[(1 - {}^6C_1 \, x^{10} ...) (1 - x)^{-6}]$

[leaving terms containing powers of x greater than 18]

$= $ coeff. of x^{18} in $(1 - x)^{-6} - {}^6C_1 .$ coeff. of x^8 in $(1 - x)^{-6} = {}^{6+18-1}C_5 - 6. \, {}^{6+8-1}C_5 = {}^{23}C_5 - 6.{}^{13}C_5$

$$= \frac{23.22.21.20.19}{120} - 6. \, \frac{13.12.11.10.9}{120} = 33649 - 7722 = 25927$$

NOMORECLASS CONCEPTS

- m different white balls and n different red balls are to be arranged in a line such that the balls of the same colour are always together = (m! n! 2!)

- m different white balls and n different red balls are to be arranged in a line such that all the red balls are together = ((m + 1)! n!)

- m different white balls and n different red balls are to be arranged in a line such that no two red balls are together (m ≥ n − 1) = (${}^{m+1}C_n$ m! n!)

- m different white balls and m different red balls are to be arranged in a line such that colour of the balls is alternating = (2 × (m!)^2)

- m identical white balls and n different red balls are to be arranged in a line such that no two red balls are together (m ≥ n − 1) = (${}^{m+1}C_n$)

- m identical white balls and n different red balls are to be arranged in a line such that no two red balls are together (m ≥ n − 1) = (${}^{m+1}C_n$ n!)

- If n objects are arranged in a line the number of selections of r objects (n ≥ 2r − 1) such that no two objects are adjacent is same number of ways of arranging n − r identical white balls and r identical red ball in a line such that no two balls are together = (${}^{n-r+1}C_r$). e.g. suppose there are n stations on trains's route and a train has to stop at r stations such that no two stations are adjacent. The number of ways must be ${}^{n-r+1}C_r$.

- suppose there are N seats in a particular row of a theatre. The number of ways of making n people sit (N ≥ 2n − 1) such that no two people sit side by side is same as number of ways of arranging N − n identical white balls (empty seats) and n different red balls (n people) such that no two red balls are together. The required number of ways are ${}^{N-n+1}C_n \times n!$.

8. DIVISIBILITY OF NUMBERS

The following table shows the conditions of divisibility of some numbers

Divisible by	Condition
2	Whose last digit is even
3	sum of whose digits is divisible by 3
4	whose last two digits number is divisible by 4
5	whose last digit is either 0 or 5
6	which is divisible by both 2 and 3
7	If you double the last digit and subtract it from the rest of the number, answer is a multiple of 7
8	whose last three digits number is divisible by 8
9	sum of whose digits is divisible by 9
10	Whose last digit is 0
11	If you sum every second digit and then subtract sum of all other digits, answer is a multiple of 11
25	whose last two digits are divisible by 25

Illustration 39: How many four digit numbers can be made with the digits 0, 1, 2, 3, 4, 5 which are divisible by 3 (digits being unrepeated in the same number)? How many of these will be divisible by 6?   **(JEE ADVANCED)**

Sol: A number is divisible by 3 if the sum of the digits is divisible by 3. This reduces the problem to the number of non-negative integral solutions of equation $x_1 + x_2 + ... + x_r = n$.

Here, $0 + 1 + 2 + 3 + 4 + 5 = 15$; so two digits are to be omitted whose sum is 3 or 6 or 9.

Hence, the number of four digits can be made by either

1, 2, 4, 5   or   0, 3, 4, 5   (omitting two digits whose sum is 3)

0, 1, 3, 5   or   0, 2, 3, 4   (omitting two digits whose sum 6)

0, 1, 2, 3   (omitting two digits whose sum is 9)

The number of 4-digit numbers that can be made with 1, 2, 4, 5 = $^4P_4$ = 4!

The number of 4-digit numbers that can be made by the digits in any one of remaining four groups (each containing 0) = 4! − 3!

∴ The required number of 4-digit numbers divisible by 3 = 4! + 4(4! − 3!) = 24 + 4(24 − 6) = 96

Now, a number is divisible by 6 if it is even as well as divisible by 3.

So, the number of 4-digit numbers divisible by 6 that can be made with 1, 2, 4, 5 = 2 × 3! (∵ the number should have an even digit in the units places).

The number of numbers of 4 digits, divisible by 6, that can be made with 0, 3, 4, 5 = (3! − 2!) + 3!

(∵ The number should have 4 or 0 in units place and 0 should not come in thousands place).

Similarly, the number of numbers of 4 digits, divisible by 6, that can be made with 0, 1, 2, 3 = (3! − 2!) + 3!

The number of 4-digit numbers divisible by 6 that can be made with the digits 0, 1, 3 = 3!

The number of numbers of 4 digits, divisible by 6, that can be made with 0, 2, 3, 4 = (3! − 2!) + (3! − 2!) + 3!

(∵ The number should have 4 or 2 or 0 in units place and 0 should not come in thousands place)

∴ the required 4-digit numbers divisible by 6

$= 2 \times 3! + (3! - 2!) + (3! - 2!) + 3! + 3! + (3! - 2!) + (3! - 2!) + (3! - 2!) + 3! = 12 + 4 + 6 + 4 + 6 + 6 + 4 + 4 + 6 = 52.$

9. SUM OF NUMBERS

(a) For given n different digits a_1, a_2, a_3 ... a_n the sum of the digits in the units place of all numbers formed (if numbers are not repeated) is

$(a_1 + a_2 + a_3 + ... + a_n)$ (n − 1)! i.e. (sum of the digits) (n − 1)!

(b) Sum of the total numbers which can be formed with given different digits a_1, a_2, a_3...... a_n is

$(a + a_2 + a_3 + ... + a_n)$ (n −1)! (111. .... n times)

Illustration 40: Find the sum of all 4 digit numbers formed using the digits 1, 2, 4 and 6. (JEE MAIN)

Sol: Use formula, Sum = $(a_1 + a_2 + a_3 + ... + a_n)$ (n− 1)! (111 .... N times)

Using formula, Sum = (1 + 2 + 4 + 6) 3! (1111) = 13 × 6 × 1111 = 86658

Alternate:

Here, the total 4-digit numbers will be 4! = 24. So, every digit will occur 6 times at every one of the four places. Since the sum of the given digits = 1 + 2 + 4 + 6 = 13. So, the sum of all the digits at every place of all the 24 numbers = 13 × 6 = 78.

The sum of the values of all the digits

At first place = 78

At the tens place = 780

At the hundreds place = 7800

At the thousands place = 78000

∴ The required sum 78 + 780 + 7800 + 78000 = 86658

10. FACTORS OF NATURAL NUMBERS

Let $N = p^a . q^b . r^c$ ... where p, q, r ... are distinct primes & a, b, c ... are natural numbers, then:

(a) The total number of divisors of N including 1 and N are = (a + 1) (b + 1) (c + 1) ...

(b) The sum of these divisors is = $(p^0 + p^1 + p^2 +...+ p^a)$ $(q^0 + q^1 + q^2 +...+ q^b)$ $(r^0 + r^1 + r^2 +...+ r^c)$...

(c) The number of ways in which N can be resolved as a product of two factors is

$$
= \begin{cases} \dfrac{1}{2}(a_1+1)(a_2+1)(a_3+1)... & \text{if N is not a perfect square} \\[2mm] \dfrac{1}{2}\left[(a_1+1)(a_2+1)(a_3+1)...+1\right] & \text{if N is a perfect square} \end{cases}
$$

(d) The number of ways in which a composite number N can be resolved into two factors which are relatively prime (or coprime) to each other is equal to 2^{n-1} where n is the number of different prime factors in N

Illustration 41: Find the number of factors of the number 38808 (excluding 1 and the number itself). Find also the sum of these divisors. (JEE MAIN)

Sol: Factorise 38808 into its product of primes and then use the concept of combination to find the answer.

$38808 = 2^3 . 3^2 . 7^2 . 11$

Hence, the total number of divisors (excluding 1 and itself) = (3 + 1) (2 + 1) (2 + 1) (1 + 1) – 2 = 70

The sum of these divisors = $(2^0 + 2^1 + 2^2 + 2^3) (3^0 + 3^1 + 3^2) (7^0 + 7^1 + 7^2) (11^0 + 11^1)$

$$= (15) (13) (57) (12) – 1 – 38808 = 94571$$

Illustration 42: In how many ways can the number 10800 be resolved as a product of two factors?   **(JEE MAIN)**

Sol: Check whether the number is a perfect square or not and accordingly use the formula to find the desired result.

$10800 = 2^4 . 3^3 . 5^2$

Here 10800 is not a perfect square ($\because$ power of 3 is odd).

Hence, the number of ways = $\dfrac{1}{2}$ (4 + 1) (3 + 1) (2 + 1) = 30.

Illustration 43: Find the number of positive integral solutions of $x_1.x_2.x_3 = 30$.   **(JEE ADVANCED)**

Sol: Factorise 30 into primes and then use combination to get the desired result.

$x_1x_2x_3 = 2 \times 3 \times 5$. If we treat 2, 3, 5 as objects and x_1, x_2, x_3 as distinct boxes then finding the number of positive integer solution is the same as finding the number of ways of distributing 3 distinct objects in 3 distinct boxes. Thus, the required number of solutions is $3^3 = 27$

(For example, if all the objects are held by x_1 the corresponding solution is $x_1 = 30$. $x_2 = 1$, $x_3 = 1$, if 2 and 3 are held by x_1 and 5 by x_3 then $x_1 = 6$, $x_2 = 1$, $x_3 = 5$ etc)

11. EXPONENT OF A PRIME P IN N!

Consider a prime p and we want to know its exponent in n!.

The number of multiples of p in n! is given by $\left[\dfrac{n}{p}\right]$. Even if a number k between 1 and n has two factors of p, this formula counts it as only one. Hence we need to evaluate $\left[\dfrac{n}{p^2}\right]$ also. Similarly, for three and four to infinity. Hence, the exponent of prime p in n! is given by $e_p(n) = \sum\limits_{i=1}^{\infty}\left[\dfrac{n}{p^i}\right]$ e_p is called Legendre's function.

Even though this is an infinite sum result, it is finite since for all p^i greater than n, step function becomes zero.

Let's understand this method by an example of finding exponent of 2 in 100!

Sol: $\left[\dfrac{100}{2}\right] = 50;\ \left[\dfrac{100}{2^2}\right] = 25;\ \left[\dfrac{100}{2^3}\right] = 12;\ \left[\dfrac{100}{2^4}\right] = 6;\ \left[\dfrac{100}{2^5}\right] = 3;\ \left[\dfrac{100}{2^6}\right] = 1;\ \left[\dfrac{100}{2^7}\right] = 0$

From this result we can infer that there are 50 numbers between 1 and 100 which have a factor of 2.

Out of these 50, there are 25 numbers that have a factor of 2^2.

Out of these 25, there are 12 numbers that have a factor of 2^3.

Out of these 12, there are 6 numbers that have a factor of 2^4.

Out of these 6, there are 3 numbers that have a factor of 2^5.

Out of these 3, there is 1 number that has a factor of 2^6.

But there is no number that has a factor of 2^7 or higher.

$\Rightarrow$ 100! can be written as $r.2^n$ where $2^n = (2^6)^1 (2^5)^{(3-1)} (2^4)^{(6-3)} (2^3)^{(12-6)} (2^2)^{(25-12)}(2)^{(50-25)}$ and r being a natural number which doesn't have 2 as a factor.

$\Rightarrow 2^n = (2^5.2)^1 \ (2^4.2)^{(3-1)}(2^3.2)^{(6-3)} \ (2^2.2)^{(12-6)} \ (2.2)^{(25-12)} \ (2)^{(50-25)}$

$= (2^5)^1(2^4)^{(3-1)} \ (2^3)^{(n-3)}(2^2)^{(12-6)} \ (2)^{(25-12)} \ (2)^{50}$

$= (2^4.2)^1 \ (2^3.2)^{(3-1)} \ (2^2.2)^{(6-3)} \ (2.2)^{(12-6)} \ (2)^{(25-12)} \ (2)^{50}$

$= (2^3.2)^1 \ (2^2.2)^{(3-1)} \ (2.2)^{(6-3)} \ (2)^{(12-6)} \ (2)^{25} \ (2)^{50}$

$= (2^2.2)^1 \ (2.2)^{(3-1)} \ (2)^{(6-3)} \ (2)^{12} \ (2)^{25} \ (2)^{50}$

$= (2.2)^1 \ (2)^{(3-1)} \ (2)^6 \ (2)^{12} \ (2)^{25} \ (2)^{50}$

$= (2)^1 \ (2)^3 \ (2)^6 \ (2)^{12} \ (2)^{25} \ (2)^{50}$

Hence, the exponent of 2 in 100! is $= 50 + 25 + 12 + 6 + 3 + 1 = 97$

NOMORECLASS CONCEPTS

Following method involves part of an advanced topic in mathematics called Modular Arithmetic. Legendre's function also has another result, which is

$$e_p(n) = \sum_{i=1}^{\infty}\left[\frac{n}{p^i}\right] = \frac{n - S_p(n)}{p-1}$$

Where $S_p(n)$ is the sum of digits of n when written in base p.

Converting n to base p is done by repeated division of n by p and by noting the remainders to form a number starting with units place.

This procedure is similar to converting a decimal number to binary. In binary, the base is equal to 2. Let us solve an example using this method.

Example: Determine the exponent of 3 in ((3!)!)!

Sol: $((3!)!)! = (6!)! = 720!$

Let us convert 720 to base 3

```
3|720
3|240 R 0
3|80  R 0
3|26  R 2
3|8   R 2
3|2   R 2
3|0   R 2
```

Hence $720 = (222200)_3$

$S_3(720) = 2 + 2 + 2 + 0 + 0 = 8$

$\Rightarrow e_3(720) = \dfrac{720-8}{3-1} = \dfrac{712}{2} = 356$

You can verify this answer using previous method.

If you are confused by base conversion then do not use this method.

Illustration 44: Find the exponent of 7 in 400!. **(JEE MAIN)**

Sol: Apply Legendre's formula.

$$e_7(400) = \left[\frac{400}{7}\right] + \left[\frac{400}{7^2}\right] + \left[\frac{400}{7^3}\right] = 57 + 8 + 1 = 66.$$

Illustration 45: Find all positive integers of n such that n! ends in exactly 1000 zeros. **(JEE ADVANCED)**

Sol: 10 is a multiple of 2 and 5. In order to get 1000 zeroes we must have 1000 as the exponent of 5 in n!. Now use the definition of the GIF to find the range of numbers satisfying the given condition.

There are clearly more 2's than 5's in the prime factorization of n!, hence it suffices to solve the equation $\left[\dfrac{n}{5}\right] + \left[\dfrac{n}{5^2}\right]$ + ... = 1000.

But $\left[\dfrac{n}{5}\right] + \left[\dfrac{n}{5^2}\right]$ + ... < $\dfrac{n}{5} + \dfrac{n}{5^2}$ + ... = $\dfrac{n}{5}\left(1 + \dfrac{1}{5} + ...\right)$ (as [x] < x) = $\dfrac{n}{5} \cdot \dfrac{1}{1-\dfrac{1}{5}} = \dfrac{n}{4}.$

Hence, n > 4000.

On the other hand, using the inequality [x] > x – 1, we have

$$1000 > \left(\frac{n}{5}-1\right) + \left(\frac{n}{5^2}-1\right) + \left(\frac{n}{5^3}-1\right) + \left(\frac{n}{5^4}-1\right) + \left(\frac{n}{5^5}-1\right) = \frac{n}{5}\left(1 + \frac{1}{5} + \frac{1}{5^2} + \frac{1}{5^3} + \frac{1}{5^4}\right) - 5 = \frac{n}{5} \cdot \frac{1-\left(\frac{1}{5}\right)^5}{1-\frac{1}{5}} - 5.$$

So, $\quad N < \dfrac{1005 \cdot 4 \cdot 3125}{3124} < 4022.$

We narrowed n down to {4001, 4002, ... , 4021}. Using Legendre's formula we find that 4005 is the first positive integer with the desired property and that 4009 is the last. Hence, n = 4005, 4006, 4007, 4008, 4009.

Second solution: It suffices to solve the equation $e_5(n) = 1000$. Using the second form of Legendre's formula, this becomes $n - s_5(n) = 4000$. Hence n > 4000. We work our way upward from 4000 looking for a solution. Since $e_5(n)$ can change only at multiples of 5 (why?), we step up 5 each time:

$$e_5(4000) = \frac{4000 - 4}{5 - 1} = 999.$$

$$e_5(4005) = \frac{4005 - 5}{5 - 1} = 1000.$$

$$e_5(4010) = \frac{4010 - 6}{5 - 1} = 1001.$$

Any n > 4010 will clearly have $e_5(n) \geq e_5(4010) = 1001$. Hence the only solutions are n = 4005, 4006, 4007, 4008, 4009.

12. INCLUSION-EXCLUSION PRINCIPLE

In its general form, the principle of inclusion-exclusion states that for finite sets $A_1, ... A_n$. One has the identity

$$\left|\bigcup_{i=1}^{n} A_i\right| = \sum_{i=1}^{n}|A_i| - \sum_{1 \leq i < j \leq n}|A_i \cap A_j| + \sum_{1 \leq i < j < k \leq n}|A_i \cap A_j \cap A_k| - ... + (-1)^{n-1}|A_1 \cap ... \cap A_n|.$$

This can be compactly written as $\left|\bigcup_{i=1}^{n} A_i\right| = \sum_{k=1}^{n}(-1)^{k+1}\left(\sum_{1 \leq i_1 < .... < i_k \leq n}|A_{i_1} \cap ... \cap A_{i_k}|\right)$

In words, to count the number of elements in a finite union of finite sets, first sum the cardinalities of the individual sets, then subtract the number of elements which appear in more than one set, then add back the number of elements which appear in more than two sets, then subtract the number of elements which appear in more than three sets, and so on. This process naturally ends since there can be no elements which appear in more than the number of sets in the union.

In applications it is common to see the principle expressed in its complementary form. That is, taking S to be a finite universal set containing all of the A_i and letting $\overline{A_i}$ denote the complement of A_i in S. By De Morgan's laws.

We have, $\left| \bigcap_{i=1}^{n} \overline{A_i} \right| = \left| S - \bigcup_{i=1}^{n} A_i \right| = |S| = \sum_{i=1}^{n} |A_i| + \sum_{1 \le i < j \le n} |A_i \cap A_j| - \dots + (-1)^n |A_1 \cap \dots \cap A_n|.$

Illustration 46: 105 students take an examination of whom 80 students pass in English. 75 students pass in Mathematics and 60 students pass in both subjects. How many students fail in both subjects? **(JEE MAIN)**

Sol: A simple application of Inclusion-Exclusion Principle.

Let X = the set of students who take the examination.

A = the set of students who pass in English

B = the set of students who pass in Mathematics

We are given that $n(X) = 105$, $n(A) = 80$, $n(B) = 75$, $n(A \cap B) = 60$

Since, $n(A \cup B) = n(A) + n(B) - n(A \cup B)$.

Therefore, $n(A \cup B) = 80 + 75 - 60 = 95$.

The required number $= n(X) - n(A \cup B) = 105 - 95 = 10$.

Thus, 10 students fail in both subjects.

Illustration 47: Find the number of permutations of the 8 letters AABBCCDD, taken all at a time, such that no two adjacent letters are alike. **(JEE ADVANCED)**

Sol: Divide the question into cases when A's are adjacent, B's are adjacent and so on. Similarly proceed to find the number of ways in which two alike objects are adjacent and so on. Then use Inclusion-Exclusion Principle to find the result.

First disregard the restriction that no two adjacent letters be alike.

The total number of permutation is then $N = \dfrac{8!}{2!2!2!2!} = 2250$

Now, apply the inclusion exclusion principle. Where a permutation has property α in case the A's are adjacent, property β in case the B's are adjacent, etc. It can be calculated that

$N(\alpha) = \dfrac{7!}{2!2!2!} = 630.$ $N(\alpha, \beta) = \dfrac{6!}{2!2!} = 180$

$N(\alpha, \beta, \gamma) = 60.$ $N(\alpha, \beta, \gamma, \delta) = 24.$

Hence, the answer is $N - 4N(\alpha) + 6N(\alpha, \beta) - 4N(\alpha, \beta, \gamma) + N(\alpha, \beta, \gamma, \delta) = 864$.

13. DERANGEMENTS THEOREM

Derangements theorem is an important application of inclusion exclusion principle.

Suppose, there are n letters and n corresponding envelopes. The number of ways in which letters can be placed in the

envelopes (one letter in each envelope) so that no letter is placed in correct envelope is $\quad n!\left[1 - \dfrac{1}{1!} + \dfrac{1}{2!} + \dots + \dfrac{(-1)^n}{n!}\right]$

Proof: n letters are denoted by 1, 2, 3, ... , n. Let, A_i denote the set of distribution of letters in envelopes (one letter in each envelope) so that the ith letter is placed in the corresponding envelope. Then, $n(A_i) = (n-1)!$ [since the remaining $n-1$ letters can be placed in $n-1$ envelopes in $(n-1)!$ Ways] Then, $n(A_i \cap A_j)$ represents the number of ways where letters i and j can be placed in their corresponding envelopes. Then $N(A_i \cap A_j) = (n-2)!$

Also, $n(A_i \cap A_j \cap A_k) = (n-3)!$

Hence, the required number is $n(A'_1 \cup A'_2 \cup \ldots \cup A'_n) = n! - n\,(A_i \cup A_2 \cup \ldots \cup A_n)$

$= n! - \left[\Sigma n(A_i) - \Sigma n(A_i \cap A_j) + \Sigma n(A_i \cap A_j \cap A_k) + \ldots + (-1)^n \Sigma n(A_1 \cap A_2 \ldots \cap A_n)\right]$

$= n! - [(^nC_1)(n-1)! - {}^nC_2\,(n-2)! + {}^nC_3\,(n-3)! + \ldots + (-1)^{n-1} \times {}^nC_n 1]$

$= n! - \left[\dfrac{n!}{1!(n-1)}(n-1)!\,\dfrac{n!}{2!(n-2)!}(n-2)! + \ldots + (-1)^{n-1}\right] = n!\left[1 - \dfrac{1}{1!} + \dfrac{1}{2!} + \ldots + \dfrac{(-1)^n}{n!}\right]$

Remark: If r objects go to wrong place out of n object then (n – r) objects goes to original place.

$A'_r = r!\left(1 - \dfrac{1}{1!} + \dfrac{1}{2!} - \dfrac{1}{3!} + \ldots + (-1)^r \dfrac{1}{r!}\right)$

NOMORECLASS CONCEPTS

Number of derangements $D_n = n! \displaystyle\sum_{i=0}^{n} \dfrac{(-1)^i}{i!}$; Interestingly, as $n \to \infty$, $D_n = \dfrac{1}{e}$

This results in an interesting relating $D_n = \left[\dfrac{n!}{e}\right]$. Where [x] is the nearest integer function.

Use this formula only if the given options are wide apart from one another.

Example You have 6 ball in 6 different colors, and for every ball you have a box of the same color. How many derangements do you have, if no ball is in a box of the same color?

Sol: We know that e = 2.71828. To make division simple let's round it to 2.7. You have to keep in mind that we have reduced the value of e, so the result which we get is greater than the actual result.

$\therefore\ D_n = \left[\dfrac{6!}{e}\right] = \left[\dfrac{720}{2.7}\right] = \left[\dfrac{800}{3}\right] = [266.66] = 267$

Hence, the result will be close to 266.

This is a pretty good approximation as the actual answer is 265.

But, if the given options are all close to 266, then it is advised to calculate using the original formula or by rounding the value of e to the number of significant digits equal to that of n! (numerator).

So if we use e = 2.72 we get 34

$D_n = \left[\dfrac{6!}{e}\right] = \left[\dfrac{720}{2.72}\right] = \left[\dfrac{4500}{17}\right] = [264.70] = 265$

Illustration 48: A person writes letters to six friends and addresses the corresponding envelopes. In how many ways can the letters be placed in the envelopes so that (i) atleast two of them are in the wrong envelopes. (ii) All the letters are in the wrong envelopes. **(JEE MAIN)**

Sol: Application of Derangement theorem.

(i) The number of ways in which at least two of them in the wrong envelopes $= \displaystyle\sum_{r=2}^{6} {}^6C_{6-r}D_r$

$= {}^6C_{6-2}\,D_2 + {}^6C_{6-3}D_3 + {}^6C_{6-4}\,D_4 + {}^6C_{6-5}\,D_5 + {}^6C_{6-6}\,D_6$

$= {}^6C_4.2!\left(1 - \dfrac{1}{1!} + \dfrac{1}{2!}\right) + {}^6C_3.3!\left(1 - \dfrac{1}{1!} + \dfrac{1}{2!} + \dfrac{1}{3!}\right) + {}^6C_2.4!\left(1 - \dfrac{1}{1!} + \dfrac{1}{2!} - \dfrac{1}{3!} + \dfrac{1}{4!}\right)$

$$+\,^6C_1.5!\left(1-\frac{1}{1!}+\frac{1}{2!}-\frac{1}{3!}+\frac{1}{4!}+\frac{1}{5!}\right)+\,^6C_0.6!\left(1-\frac{1}{1!}+\frac{1}{2!}-\frac{1}{3!}+\frac{1}{4!}-\frac{1}{5!}+\frac{1}{6!}\right)$$

(ii) The number of ways in which all letters be placed in wrong envelopes

$$=\,6!\left(1-\frac{1}{1!}+\frac{1}{2!}-\frac{1}{3!}+\frac{1}{4!}-\frac{1}{5!}+\frac{1}{6!}\right);\ =720\left(\frac{1}{2}-\frac{1}{6}+\frac{1}{24}-\frac{1}{120}+\frac{1}{720}\right);\ =360-120+30-6+1=265$$

14. MULTINOMIAL THEOREM

(a) If there are l objects of one kind, m objects of a second kind, n objects of a third kind and so on; then the number of ways of choosing r objects out these (i.e., $l + m + n$ ...) is the coefficient of x^r in the expansion of

$$(1 + x + x^2 + x^3 + ... + x^l)\,(1 + x + x^2 + x^3 + ... + x^m)\,(1 + x + x^2 + x^3 + ... + x^n)$$

Further, if one object of each kind is to be included, then the number of ways of choosing r objects out of these objects (i.e., $l + m + n + ...$) is the coefficient of x^r in the expansion of

$$(x + x^2 + x^3 + ... + x^l)\,(x + x^2 + x^3 + ... + x^m)\,(x + x^2 + x^3 + ... + x^n)...$$

(b) If there are objects of one kind, m objects of a second kind, n objects of a third kind and so on; then the number of possible arrangements/permutations of r objects out of these object (i.e., $l + m + m + ...$) is the coefficient of x^r in the expansion of

$$r!\left(1+\frac{x}{1!}+\frac{x^2}{2!}+.....\frac{x^l}{l!}\right)\left(1+\frac{x}{1!}+\frac{x^2}{2!}+...\frac{x^m}{m!}\right)\left(1+\frac{x}{1!}+\frac{x^2}{2!}+...\frac{x^n}{n!}\right)$$

Illustration 49: In an examination, the maximum marks for each of three papers is n and that for the fourth paper is 2n. Prove that the number of ways in which candidate can get 3n marks is

$\dfrac{1}{6}$ (n + 1) $(5n^2 + 10n + 6)$. **(JEE ADVANCED)**

Sol: The maximum marks in the four papers are n, n, n and 2n. Consider a polynomial $(1 + x + x^2 + ... + x^n)^3 (1 + x + ... + x^{2n})$. The number of ways of securing a total of 3n is equal to the co-efficient of the term containing x^{3n}.

The number of ways of getting 3n marks

$= $ coefficient of x^{3n} in $(1 + x + x^2 + ... + x^n)^3 (1 + x + ... + x^{2n})$

$= $ coefficient of x^{3n} in $(1 - x^{n+1})^3 (1 - x^{2n+1}) (1 - x)^{-4}$

$= $ coefficient of x^{3n} in $(1 - 3x^{n+1} + 3x^{2n+2} - x^{3n+3})(1 - x^{2n+1}) \times (1 + \,^4C_1 x + \,^5C_2 x^2 + \,^6C_3 x^3 + ...)$

$= $ coefficient of x^{3n} in $(1 - 3x^{n+1} - x^{2n+1} + 3x^{2n+2})(1 + \,^4C_1 x + \,^5C_2 x^2 ...)$

$= \,^{3n+3}C_{3n} - 3.^{2n+2}C_{2n-1} + 3 . \,^{n+1}C_{n-2} - \,^{n+2}C_{n-1}$

$$= \frac{(3n+3)!}{3!(3n)!} - 3 . \frac{(2n+2)!}{3!(2n-1)!} + 3\frac{(n-1)!}{3!(n-2)!} - \frac{(n+2)!}{3!(n-1)!}$$

$= 1/6\ (n + 1)\ (27n^2 + 27n + 6 - 24n^2 - 12n + 3n^2 - 3n - n^2 - 2n) = 1/6\ (n + 1)\ (5n^2 + 10n + 6)$

PROBLEM-SOLVING TACTICS

In any given problem, first try to understand whether it is a problem of permutations or combinations. Now, think if repetition is allowed and then try solving problem.

A simple method to solve these problems where repetition is not allowed is as follows -

First draw series of dashes representing the number of places you want to fill or number of items you want to select.

Now start filling dashes by the number of objects available to choose from and multiply the numbers. This is the final answer for a permutations problem.

If it is a combination problem then divide the answer with the factorial or number of items.

This calculation becomes complex if repetition is allowed.

FORMULAE SHEET

(a) **Permutation (Arrangement of Objects):** Each of the different arrangement, which can be made by taking some or all of a number of objects is called permutation.

(i) The number of permutations of n different objects taken r at a time is $^{n}P_{r} = \dfrac{n!}{(n-r)!}$.

(ii) The number of all permutations of n distinct objects taken all at a time is n!.

Permutation with Repetition: The number of permutations of n different objects taken r at a time when each object may be repeated any number of times is n^{r}.

Permutation of Alike Objects: The number of permutations of n objects taken all at a time in which, p are alike objects of one kind, q are alike objects of second kind & r are alike objects of a third kind and the rest $(n - (p + q + r))$ are all different, is $\dfrac{n!}{p!q!r!}$.

Permutation under Restriction: The number of permutations of n different objects, taken all at a time, when m specified objects always come together is $m! \times (n - m + 1)!$.

(b) **Combination (Selection of Objects):** Each of the different groups or selection which can be made by some or all of a number of given objects without reference to the order of the objects in each group is called a combination.

The number of all combinations of n objects, taken r at a time is generally denoted by C(n, r) or $^{n}C_{r} = \dfrac{n!}{r!(n-r)!}$

$(0 \le r \le n) = \dfrac{^{n}P_{r}}{r!}$

Note:

(a) The number of ways of selecting r objects out of n objects, is the same as the number of ways in which the remaining (n - r) can be selected and rejected.

(b) The combination notation also represents the binomial coefficient. That is, the binomial coefficient $^{n}C_{r}$ is the combination of n elements chosen r at a time.

(c) (i) $^{n}C_{r} = {}^{n}C_{n-r}$

(ii) $^{n}C_{r} + {}^{n}C_{r-1} = {}^{n+1}C_{r}$

(iii) $^{n}C_{x} = {}^{n}C_{y} \Rightarrow x = y \text{ or } x + y = n$

(iv) If n is even, then the greatest value of $^{n}C_{r}$ is $^{n}C_{n/2}$

(v) If n is odd, then the greatest value of $^{n}C_{r}$ is $^{n}C_{(n+1)/2}$

(vi) $^{n}C_{0} + {}^{n}C_{r} + \ldots\ldots\ldots + {}^{n}C_{n} = 2^{n}$

(vii) $^{n}C_{n} + {}^{n+1}C_{n} + {}^{n+2}C_{n} + \ldots\ldots\ldots + {}^{2n-1}C_{n} = {}^{2n}C_{n+1}$

Combinations under Restrictions

(a) The number of ways of choosing r objects out of n different objects if p particular objects must be excluded
$= {}^{(n-p)}C_r$

(b) The number of ways of choosing r objects out of n different objects if p particular objects must be included
$(p \le r) = {}^{n-p}C_{r-p}$

(c) The total number of combinations of n different objects taken one or more at a time $= 2^n - 1$.

Combinations of Alike Objects

(a) The number of combinations of n identical objects taking $(r \le n)$ at a time is 1.

(b) The number of ways of selecting r objects out of n identical objects is n + 1.

(c) If out of (p + q + r + s) objects, p are alike of one kind, q are alike of a second kind, r are alike of the third kind and s are different, then total number of combinations is $(p + 1)(q + 1)(r + 1)2^s - 1$

(d) The number of ways in which r objects can be selected from a group of n objects of which p are identical, is

$$\sum_0^t {}^{n-p}C_r \text{ , if } r \le p \text{ and } \sum_{r=p}^t {}^{n-p}C_r \text{ if } r > p$$

Division into Groups

(a) The number of ways in which (m + n) different objects can be divided into two unequal groups containing m and n objects respectively is $\dfrac{(m+n)!}{m!n!}$.

If m = n, the groups are equal and in this case the number of divisions is $\dfrac{(2n)!}{n!n!2!}$; as it is possible to interchange the two groups without obtaining a new distribution.

(b) However, if 2n objects are to be divided equally between two persons then the number of ways

$$= \frac{(2n)!}{n!n!2!}2! = \frac{(2n)!}{n!n!}$$

(c) The number of ways in which (m + n + p) different objects can be divided into three unequal groups containing

m , n and p objects respectively is $= \dfrac{(m+n+p)}{m!n!p!}$, $m \ne n \ne p$

If m = n = P then the number of groups $= \dfrac{(3n)!}{n!n!n!3!}$. However, if 3n objects are to be divided equally among

three persons then the number of ways $= \dfrac{(3n)!}{n!n!n!3!} \, 3! = \dfrac{(3n)!}{(n!)^3}$

(d) The number of ways in which mn different objects can be divided equally into m groups if the order of groups is not important is $\dfrac{mn!}{(n!)^m m!}$

(e) The number of ways in which mn different objects can be divided equally into m groups if the order of groups is important is $\dfrac{mn!}{(n!)^m m!} \times m! = \dfrac{(mn)!}{(n!)^m}$

Circular Permutation

(a) The number of circular permutations of n different objects taken r at a time

${}^nP_r/r$, when clockwise and anticlockwise orders are treated as different.

${}^nP_r/2r$, when clockwise and anticlockwise orders are treated as same.

(b) The number of circular permutations of n different objects altogether

${}^nP_n/n = (n - 1)!$, when clockwise and anticlockwise order are treated as different

$^nP_n/2n = 1/2(n-1)!$, when above two orders are treated as same

The number of non-negative integral solutions of equation $x_1 + x_2 + ... + x_r = n$

= The number of ways of distributing n identical objects among r persons when each person can get zero or one or more objects = $^{n+r-1}C_{r-1}$

The number of positive integral solutions for the equation $x_1 + x_2 + ... + x_r = n$

= The number of ways of distributing n identical objects among r persons when each person can get at least one object = $^{n-r+r-1}C_{r-1} = {}^{n-1}C_{r-1}$.

(c) For given n different digits $a_1, a_2, a_3 ... a_n$ the sum of the digits in the units place of all the numbers formed (if numbers are not repeated) is

$(a_1 + a_2 + a_3 + ... + a_n)(n-1)!$ i.e. (sum of the digits) $(n-1)!$

(d) The sum of the total numbers which can be formed with given different digits $a_1, a_2, a_3...... a_n$ is

$(a + a_2 + a_3 + ... + a_n)(n-1)!$ (111. .... n times)

Factors of Natural Numbers

Let $N = p^a . q^b . r^c ...$ where p, q, r ... are distinct primes & a, b, c ... are natural numbers, then:

(a) The total number of divisors of N including 1 and N are = $(a+1)(b+1)(c+1)...$

(b) The sum of these divisors is = $(p^0 + p^1 + p^2 + ... + p^a)(q^0 + q^1 + q^2 + ... + q^b)(r^0 + r^1 + r^2 + ... + r^c)...$

(c) The number of ways in which N can be resolved as a product of two factors is

$$= \begin{cases} \dfrac{1}{2}(a+1)(b+1)(c+1) & \text{if N is not a perfect square} \\ \dfrac{1}{2}\left[(a+1)(b+1)(c+1)... +1\right] & \text{if N is a perfect square} \end{cases}$$

(d) The number of ways in which a composite number N can be resolved into two factors which are relatively prime (or coprime) to each other is equal to 2^{n-1} where n is the number of different prime factors in N

Exponent of a Prime P in N! $= \sum\limits_{i=1}^{\infty}\left[\dfrac{n}{p^i}\right]$

Inclusion-Exclusion Principle: The principle of inclusion-exclusion states that for finite sets $A_1,...A_n$. One has the identity

$$\left|\bigcup_{i=1}^{n} A_i\right| = \sum_{i=1}^{n}|A_i| - \sum_{1\le i<j\le n}|A_i \cap Aj| + \sum_{1\le i<j<k\le n}|A_i \cap Aj \cap A_k| - ... + (-1)^{n-1}|A_1 \cap ... \cap A_n|.$$

This can be compactly written as $\left|\bigcup_{i=1}^{n} A_i\right| = \sum_{k=1}^{n}(-1)^{k+1}\left(\sum_{1\le i_1 < .... < i_k \le n}|A_{i_1} \cap ... \cap A_{i_k}|\right)$

Derangements Theorem: The number of ways in which letters n can be placed in n envelopes (one letter in each envelope) so that no letter is placed in the correct envelope is $n!\left[1 - \dfrac{1}{1!} + \dfrac{1}{2!} + ... + \dfrac{(-1)^n}{n!}\right]$

If n objects are arranged at n places then the number of ways to rearrange exactly r objects at right places is =

$$\dfrac{n!}{r}\left[1 - \dfrac{1}{1!} + \dfrac{1}{2!} - \dfrac{1}{3!} + \dfrac{1}{4!} - ... + (-1)^{n-r}\dfrac{1}{(n-r)!}\right]$$

Some Important results

(a) The number of totally different straight lines formed by joining n points on a plane of which m($<$n) are collinear is $^nC_2 - {}^mC_2 + 1$.

(b) The number of total triangles formed by joining n points on a plane of which m($<$n) are collinear is $^nC_3 - {}^mC_3$.

(c) The number of diagonals in a polygon of n sides is $^nC_2 - n$.

(d) If m parallel lines in a plane are intersected by a family of other n parallel lines. Then total number of parallelograms so formed are $^mC_2 \times {}^nC_2$.

(e) Given n points on the circumference of a circle, then

the number of straight lines between these points are $^nC_2$

the number of triangles between these points are $^nC_3$

the number of quadrilaterals between these points are $^nC_4$

(f) If n straight lines are drawn in the plane such that no two lines are parallel and no three lines are concurrent. Then, the number of parts into which these lines divide the plane is $= 1 + Sn$

Solved Examples

JEE Main/Boards

Example 1: Find the number of ways in which 5 identical balls can be distributed among 10 different boxes, if exactly one ball goes into a box.

Sol: It is same as selecting 5 boxes from 10 boxes and distributing the balls in those 5 boxes.

Number of boxes = 10 and Number of balls = 5.

$\therefore$ Possible number of ways $= {}^{10}C_5$

Example 2: There are n intermediate stations on a railway line from one terminal to another. In how many ways can the train stop at 3 of these intermediate stations if

(i) all the three stations are consecutive.

(ii) at least two of the stations are consecutive.

Sol: The first part is very trivial. For the second part consider a pair of consecutive stations and then select a station such that it is not consecutive. Check for multiple counting.

Let the intermediate stations be $S_1, S_2, \ldots, S_n$

(i) The number of triplets of consecutive stations, as $S_1S_2S_3, S_2S_3S_4, S_3S_4S_5, \ldots S_{n-2}S_{n-1}S_n$, is $(n-2)$.

(ii) The total number of consecutive pairs of stations, as $S_1S_2, S_2S_3, \ldots, S_{n-1}S_n$ is $(n-1)$.

Each of the above pairs can be associated with a third station in $(n-2)$ ways. Thus, choosing a pair of stations and any third station can be done in $(n-1)(n-2)$ ways.

The above count also includes the case of three consecutive stations. However, we can see that each such case has been counted twice. For example, the pair S_4S_5 combined with S_6 and the pair S_5S_6 combined with S_4 are identical.

Hence, subtracting the excess counting, the number of ways in which three stations can be chosen so that at least two of them are consecutive

$= (n-1)(n-2) - (n-2) = (n-2)^2$

Example 3: How many ways are there to invite 1 of 3 friends for dinner on 6 successive nights such that no friend is invited more than 3 times?

Sol: Divide the solution in different possible cases. 6 can be partitioned in the following ways

$1 + 2 + 3$

$0 + 3 + 3$

$2 + 2 + 2$

Using this we can form different possibilities and calculate the number of ways the friends can be invited.

Let x, y, z be the friends and let (a, b, c) denote the case where x is invited a times, y, b times and z, c times. For example, one possible arrangement corresponding to the triplet (3, 2, 1) is x, x, y, x, y, z

Then we have the following possibilities:

(i) (a, b, c) = (1, 2, 3); (1, 3, 2); (2, 3, 1); (2, 1, 3):

(ii) (a, b, c) = (3, 3,0); (3,0,3) ; (0, 3, 3).

(iii) (a, b, c) = (2, 2, 2). So the total number of ways is 6 × 6!/1! 2! 3! + 3 × 6!/3!3! + 6!/2!2!2!

Note: We can also solve this problem using linear equations.

Example 4: There are 2n guests at a dinner party. Supposing that the master and mistress of the house have fixed seats opposite one another. And that there are two specified guests who must not be placed next to one another. Show that the number of ways in which the company can be placed is $(2n - 2)! (4n^2 - 6n + 4)$

Sol: This is an application of division into groups. Find the total number of ways of arrangement of the guests and then subtract the number of ways in which the two mentioned guests are together.

Excluding the two specified guests, $2n - 2$ persons can be divided into two groups, one containing n and the other $(n - 2)$ in $\dfrac{(2n-2)!}{n!(n-2)!}$ and can sit on either side of mister and mistress in 2! ways and can arrange themselves in $n!(n - 2)!$

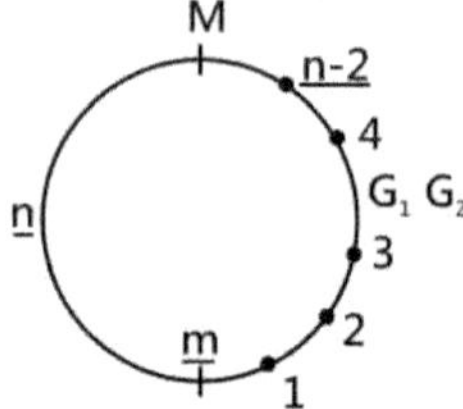

Now, the two specified guests where $(n - 2)$ guests are seated will have $(n - 1)$ gaps and can arrange themselves in 2! Ways. Number of ways when G_1 G_2 will always be together

$$= \dfrac{(2n-2)!}{n!(n-2)!} \; 2! \; n! \; (n-2)! \; (n-1) \; 2! = (2n-2)! \; 4(n-1)$$

Hence, the number of ways when G_1 G_2 are never together

$$= \dfrac{2!}{n! \; n! \; 2!} \; 2! \; n! \; n! - 4(n-1)(2n-2)!$$

$$= (2n-2)! \; [2n(2n-1) - 4(n-1)] = (2n-2)! \; [4n^2 - 6n + 4]$$

Example 5: Find the number of words of 5 letters that can be formed with the letters of the word Proposition.

Sol: Divide the cases into words having 5 distinct letters,

2 alike of one kind and 3 alike of different kind and so on. Count the number of words in these cases and their sum gives us the answer.

Proposition contains 11 letters PP, R, OOO, S, II, T, N.

Following table given the number of words.

Repeated letters: O(2), P(2), I(2)

Different letters: R, S, T, N

	Letters	No. of Words	Total
A	5 Distinct	$^7C_5.5!$	2520
B	3 Alike 2 Alike	$1.2C1.(51/3!2!)$	20
C	3 Alike 2 Different	$1.^6C_2.(5!/3!)$	300
D	2 Alike 2Other Alike 1 Different	$^3C_2.^5C_1.(5!/2!2!)$	450
E	2 Alike 3 Different	$^3C_1.^6C_3.(5!/2!)$	3600

Total no. of words = 6890

Example 6: There are 10 points in a plane where no three points are collinear except for 4 points which are collinear. Find the number of triangles formed by the points as vertices.

Sol: A triangle is formed from three non-collinear points. Select 3 points from 10 points in $^{10}C_3$ ways and subtract the cases when the points are collinear, as they would not form a triangle.

Let us suppose that the 10 points are such that no three of them are collinear. Now, a triangle will be formed by any three of these ten points. Thus forming a triangle amount to selecting any three of the 10 points.

Now 3 points can be selected out of 10 point in $^{10}C_3$ ways.

∴ Number of triangles formed by 10 points when no three of them are collinear = $^{10}C_3$.

Similarly, the number of triangles formed by 4 points then no 3 of them are collinear = $^4C_3$

∴ Required number of triangle formed = $^{10}C_3 - ^4C_3 = 120 - 4 = 116$.

Example 7: From 6 gentlemen and 4 ladies, a committee of 5 is to be formed. In how many ways can this be done if the committee is to include at least one lady?

Sol: According to the question, the committee should include atleast one lady. Consider cases when the committee consists of 1, 2, 3 or 4 ladies and find the number of ways for all these cases.

Different combinations are listed below:

No. of Ladies	No. of Gentlemen	No. of Committees
1	4	$^4C_1 \, ^6C_4$
2	3	$^4C_2 \, ^6C_3$
3	2	$^4C_3 \, ^6C_2$
4	1	$^4C_4 \, ^6C_1$

Total number of committees

$= {}^4C_1 \, ^6C_4 + {}^4C_2 \, ^6C_3 + {}^4C_3 \, ^6C_2 + {}^4C_4 \, ^6C_1 = 246$

Example 8: (a) In how many ways can the following diagram be coloured, subject to two conditions: Each of the smaller triangle is to be painted with one of three colours: red, blue, green and no two adjacent regions should have the same color?

(b) How many numbers of four digits can be formed with the digits 1, 2, 3, 4 and 5?

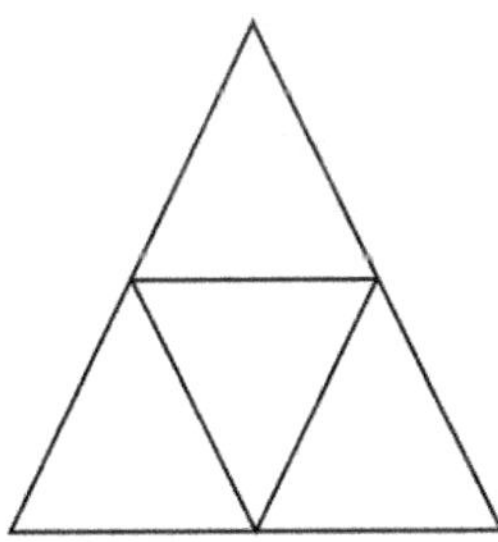

(c) A gentleman has 6 friends to invite. In how many ways can he send invitation cards to them if he has 3 servants to carry the cards?

(d) Find the number of arrangements of the letters of the word 'BENEVOLENT'.

Sol: For each of the parts (a), (b), (c) and (d), identify the number of ways a particular cell can be coloured or filled in and then use permutation / combination to get the result.

(a) These conditions are satisfied if we proceed as follows: Just color the central triangle by one color, this can be done in three ways. Next paint other three triangles with remaining 2 colors. By the fundamental principle of counting. This can be done in $3 \times 2 \times 2 \times 2 = 24$ ways.

(b) 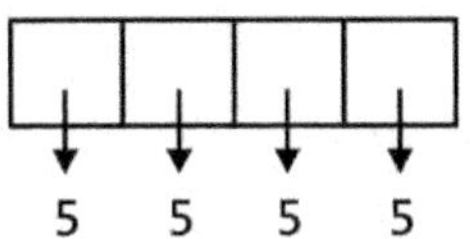

Each place can be filled by any of the 5 numbers. Therefore, the total number of arrangements is 5^4.

(c) Each card can be given to any of the 3 servants.

$\therefore$ No. of ways $= 3 \times 3 \times 3 \times 3 \times 3 \times 3 = 3^6 = 729$.

(d) There are ten letters in the word BENEVOLENT of which three are E and two are N, and the rest five are different.

$\therefore$ Total number of arrangements $= \dfrac{10!}{3! \, 2!}$

Example 9: n_1 and n_2 are five-digit numbers. Find the total number of ways of forming n_1 and n_2, so that n_2 can be subtracted from n_1 without borrowing at any stage.

Sol: Two numbers can be subtracted without borrowing if all the digits in n_1 is greater than all the corresponding digits in the number n_2. Using this information, find the number of ways for different possible cases and add them up to get the answer.

Let $n_1 = x_1 x_2 x_3 x_4 x_5$ and $n_2 = y_1 y_2 y_3 y_4 y_5$ be two numbers. n_1 and n_2 can be subtracted without borrowing at any stage if $x_i \geq y_i$.

Here, x_i and y_i denotes the digits at various places in the number n_1 and n_2 respectively.

Value of x_5	Value of y_5
9	0,1,2,...9
8	0,1,2,...8
7	0,1,2,...7
6	0,1,2,3,4,5,6
5	0,1,2,3,4,5
4	0,1,2,3,4
3	0,1,2,3
2	0,1
1	0
0	

Thus, x_5 and y_5 can be selected collectively by $10 + 9 + 8 + \dots 1 = 55$ ways. Similarly, each pair (x_4, y_4), (x_3, y_3), (x_2, y_2) can be selected in 55 ways. But, pair (x_1, y_1) can be selected in $1 + 2 + 3 + \dots + 9 = 45$ ways as in this pair we cannot have 0.

Therefore total number of ways $= 45(55)^4$.

Example 10: Prove that the product of r consecutive positive integers is divisible by r!.

Sol: Simple application of the definition of ^nP_r.

Let P be the product of r consecutive positive integers ending with n; then

$P = n(n-1) \dots (n-r+1)$

$$\frac{P}{r!} = \frac{n(n-1)\dots(n-r+1)}{r!}$$

$$\frac{[n(n-1)(n-2)\dots(n-r+1)][(n-r)\dots3.2.1]}{r!(n-r)\dots3.2.1}$$

$$= \frac{n!}{r!\,n-r!} = \,^nC_r = \text{an integer}$$

$\therefore$ P is divisible by r!.

JEE Advanced/Boards

Example 1: How many numbers of n digits can be made with the non-zero digits in which no two consecutive digits are the same?

Sol: Using Permutation under Restriction we can easily find the answer.

There are nine non-zero digits, namely 1, 2, 3, ... and 9.

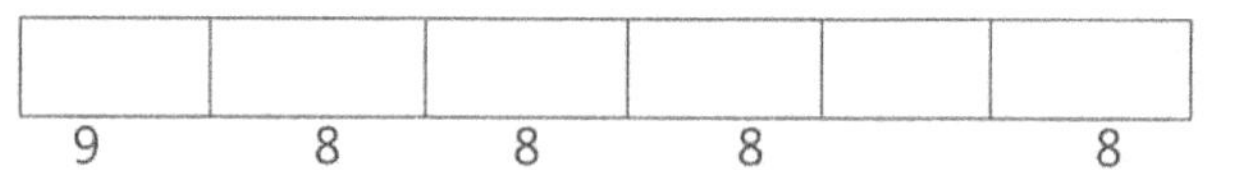

9	8	8	8	8

In order the make an n-digit number we have to fill n places by using the nine digits. As no two consecutive digits are to be the same, a digit used in a place cannot be used in the next place but it can be used again in the place coming after the next place.

So, the first place can be filled in 9 ways;

the second place can be filled in 8 ways (rejecting the digit used in the first place)

the third place can be filled in 7 + 1, i.e., 8 ways (rejecting the digit used in the second place but including the digit used in the first place) and so on.

$\therefore$ The total number of desired numbers

$= 9 \times 8 \times 8 \times 8 \times \dots$ to n factors $= 9 \times 8^{n-1}$.

Example 2: A dice is a six-faced cube, with the faces reading 1, 2, 3, 4, 5 and 6. When two dice are thrown, we add the digits they show on top and take that sum as the result of the throw. In how many different ways the first throw of the 2 dice shows a total of 5, and second throw of the 2 dice shows a total of 4?

Sol: List down different ways in which we get the sum of 5 and 4 and get the answer.

Event E (the first throw resulting in 5) can happen in one of four ways as:

$3 + 2; 4 + 1; 2 + 3; 1 + 4.$

Event F (the second throw resulting in 4) can happen in one of three ways as:

$2 + 2; 1 + 3; 3 + 1.$

The two events can together happen in $4 \times 3 = 12$ ways.

Example 3: An eight-oared boat is to be manned by a crew chosen from 11 men of whom 3 can steer but cannot row and the rest cannot steer. In how many ways can the crew be arranged if two of the men can only row in bow side?

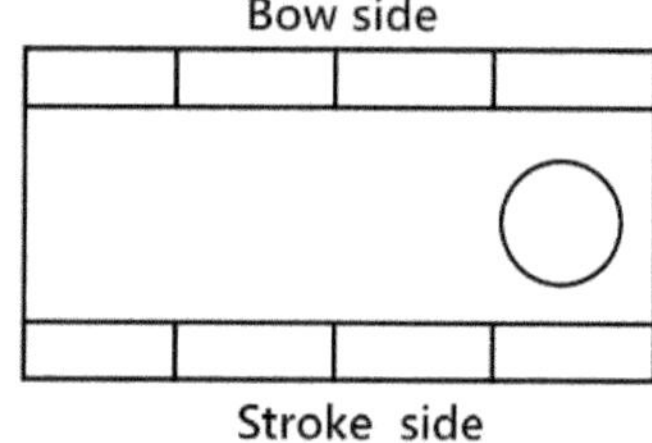

Sol: Find the number of ways we can select for steering, rowing and arranging the remaining men. Their product gives us the required result.

The total number of men = 11

The number of men who can only steer = 3

The number of other men = 8

The number of ways of selecting one man for steering out of 3 = $^3C_1$.

The number of ways in which the two particular men who only row on bow side

Can be arranged on bow side = $^4P_2$

The number of ways in which remaining 6 men can be arranged in remaining 6 places = 6!

$\therefore$ The required number = $^3C_1 . \,^4P_2 . 6!$

$$= 3 \times 4 \times 3 \times 6! = 25920$$

Example 4: The members of a chess club took part in a round robin competition in which each plays every one else once. All members scored the same number of points, except four juniors whose total score were 17.5. How many members were there in the club? (Assume that for each win a player scores 1 point, for draw $\frac{1}{2}$ point and zero for losing.)

Sol: Form an equation of the total number of points scored using the given information. Solve the equation to find the answer.

Let the number of members be n.

Total number of point $= {}^nC_2$.

$\therefore {}^nC_2 - 17.5 = (n - 4) x$ (where x is the number of point scored by each player)

$n (n - 1) - 35 = 2 (n - 4)x$

$2x = \dfrac{n(n-1) - 35}{n - 4}$ (where x takes the values 0.5, 1, 1.5 etc.)

$= \dfrac{n^2 - n - 35}{n - 4}$ (must be an integer)

$= \dfrac{n(n - 4) + 3(n - 4) - 23}{n - 4} = (n + 3) - \dfrac{23}{n - 4}$

$\Rightarrow \dfrac{23}{n - 4}$ must be an integer

$\Rightarrow n = 27$ is the only possibility.

Example 5: If p, q, r, s, t are prime numbers. Find the number of ways in which the product, pq^2r^3st can be expressed as product of two factors, excluding 1 as a factor.

Sol: Use the standard result to find the answer.

Total factors $= 2 \times 3 \times 4 \times 2 \times 2 = 96$

Hence, the total ways $= \dfrac{96}{2} = 48$. but this includes 1 and the number itself also. Hence, the required number of ways $= 48 - 1 = 47$

Example 6: In the given figure you have the road plan of a city. A man standing at X wants to reach the cinema hall at Y by the shortest path. What is the number of different paths that he can take?

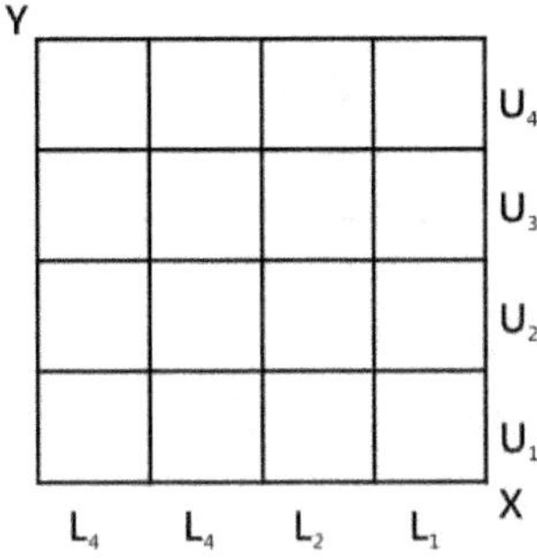

Sol: If the man moves only in the upward and the leftward direction, then the path will be the shortest. Use this idea to calculate total number of shortest paths.

As the man wants to travel by one of the many possible shortest paths, he will never turn to the right or turn downward. So a travel by one of the shortest paths is to take 4 horizontal pieces and 4 vertical pieces of roads.

$\therefore$ A shortest path is an arrangement of eight objects L_1, L_2, L_3, L_4, U_1, U_2, U_3, U_4 so that the order of L's and U's do not change.

($\because$ Clearly L_2 cannot be taken without taking L_1, U_2 cannot be taken without taking U_1, etc.)

Hence, the number of shortest paths

= The number of arrangements of L_1, L_2, L_3, L_4, U_1, U_2, U_3, U_4 where the order of Ls as well as the order of Us do not change

= The number of arrangement treating Ls as identical and Us as identical

$= \dfrac{8!}{4!4!} = \dfrac{8.7.6.5}{24} = 2.7.5 = 70.$

Example 7: A condolence meeting being held in a hall which has 7 doors, by which mourners enter the hall. One can use any of the 7 doors to enter and can come at any time during the meeting. At each door, a register is kept in which mourner has to affix his signature while entering the hall. If 200 people attend the meeting, how many different sequences of 7 lists of signatures can arise?

Sol: Clearly, the total number of people is 200, hence the sum of the entries is 200. Apply Multinomial theorem to find the total number of ways list can be made and hence the answer.

There are 7 lists, say 1, 2,.... 7. Suppose, that list i has x_i names; then,

$x_1 + ... + x_7 = 200$ where $x_i \geq 0$ is an integer.

We need to first find the number of solutions of this equation.

(Note that this does not complete the solution to the questions as list 1 may contain 7 names which would remain the same in 7!, arrangements of the names)

The number of solutions are $= {}^{200+7-1}C_{7-1} = {}^{206}C_6$

But corresponding to any one solution $(x_1...x_7)$ (i.e. list f contains x_f names) we can have 200! arrangements consistent with distribution of x_j names to j^{th} list

$\therefore$ The number of different sequences of 7 lists

$= {}^{206}C_6 \times 200! = \dfrac{206}{6!}$.

Exercise 1

Q.1 How many odd numbers less than 1000 can be formed using the digits 0, 1, 4 and 7 if repetition of digits is allowed?

Q.2 In how many ways can five people be seated in a car with two people in the front seat and three in the rear, if two particular persons out of the five cannot drive?

Q.3 A team consisting of 7 boys and 3 girls play singles matches against another team consisting of 5 boys and 5 girls. How many matches can be scheduled between the two teams if a boy plays against a girl and a girl plays against a boy?

Q.4 Prove that $\dfrac{(2n+1)}{n!} = 2^n[1.3.5...(2n-1)(2n+1)]$

Q.5 If $^nP_4 = 360$, find n.

Q.6 Find the number of numbers between 300 and 3000 which can be formed with the digits 0, 1, 2, 3, 4 and 5, with no digit being repeated in any number.

Q.7 How many even numbers are there with three digits such that if 5 is one of the digits in a number then 7 is the next digit in that number?

Q.8 Find the sum of 3 digit numbers formed by digits 1, 2, 3 is

Q.9 A telegraph has 5 arms and each arm is capable of 4 distinct positions, including the position of rest. What is the total number of signals that can be made?

Q.10 In telegraph communication, the Morse code is used in which all the letters of the English alphabet, digits 0 to 9 and even the punctuation marks, all usually referred as characters, are represented by 'dots' and 'dashes'

For example, E is represented by a dot (.), T by a dash (–), O by three dashes (- - -), S by three dots (. . .) and so on. Thus, SOS is represented by (. . . – – – . . .).

(i) How many characters can be transmitted using one symbol (dot or dash), two symbols, three symbols, four symbols? Also find the total number of characters which can be transmitted using at most four symbols.

(ii) How many characters can be transmitted by using (a) exactly five symbols? (b) at most five symbols?

Q.11 In how many of the distinct permutation of the letter in MISSISSIPPI do the four I's not come together?

Q.12 In how many ways 4 boys and 3 girls can be seated in a row so that they are alternate?

Q.13 A biologist studying the genetic code is interested to know the number of possible arrangements of 12 molecules in a chain. The chain contains 4 different molecules represented by the initials A (for adenine), C (for Cytosine), G (for Guanine) and T (for Thymine) and 3 molecules of each kind. How many different such arrangements are possible in all?

Q.14 Find the number of rearrangement of the letters of the word 'BENEVOLENT'. How many of them end in L?

Q.15 How many words can be formed with the letters of the word PATALIPUTRA' without changing the relative order of the vowels and consonants?

Q.16 A person is to walk from A to B. However, he is restricted to walk only to the right of A or upwards of A, but not necessarily in this order. One such path is shown in the given figure Determine the total number of paths available to the person from A to B.

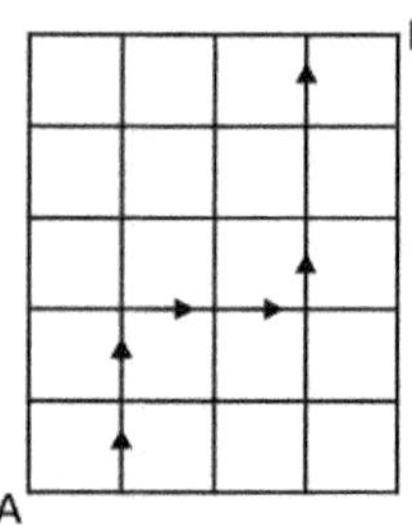

Q.17 In how many ways can three jobs I, II and III be assigned to three persons A, B and C, if one person is assigned only one job and all are capable of doing each job? Which assignment of jobs will take the least time to complete the jobs, if time taken (in hours) by an individual on each job as follows?

Job persons	I	II	III
A	5	4	4
B	$4\dfrac{1}{4}$	$3\dfrac{1}{2}$	4
C	5	3	5

Q.18 If $^{15}C_{3r} = {}^{15}C_{r+3}$, find r.

Q.19 Prove that $^{n}C_{r} \times {}^{r}C_{s} = {}^{n}C_{s} \times {}^{n-s}C_{r-s}$.

Q.20 Find the value of the expression

$$^{47}C_{4} + \sum_{j=1}^{5} {}^{52-j}C_{3}.$$

Q.21 Prove that the product of r consecutive integers is divisible by r!.

Q.22 From a class of 25 students, 10 are to be chosen for a field trip. There are 3 students who decide that either all of them will join or none of them will join. In how many ways can the field trip members be chosen?

Q.23 There are ten points in a plane. Of these ten points, four points are in a straight line and with the exception of these four points, no three points are in the same straight line. Find-

(i) The number of triangles formed.

(ii) The number of straight lines formed

(iii) The number of quadrilaterals formed, by joining these ten points.

Q.24 In an examination a minimum of is to be secured in each of 5 subjects for a pass. In how many ways can a student fail?

Q.25 In how many ways 50 different objects can be divided in 5 sets three of them having 12 objects each and two of them having 7 objects each.

Q.26 Six "X"s (crosses) have to be placed in the squares of the figure given below, such that each row contains at least one X. In how many different ways can this be done?

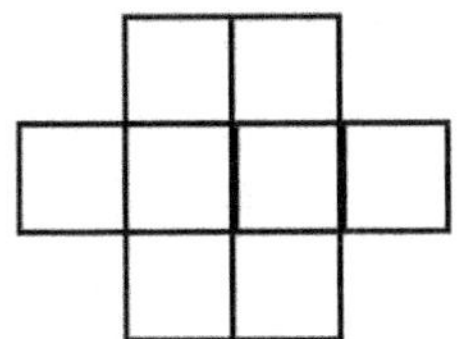

Q.27 Five balls of different colors are to be placed in three boxes of different sizes. Each box can hold all five balls. In how many different ways can we place the balls so that no box remains empty?

Q.28 How many different words of 4 letters can be formed with the letters of the word "EXAMINATION"?

Exercise 2

Single Correct Choice Type

Q.1 If the letters of the word "VARUN" are written in all possible ways and then are arranged as in a dictionary, then rank of the word VARUN is:

(A) 98   (B) 99   (C) 100   (D) 101

Q.2 Number of natural numbers between 100 and 1000 such that at least one of their digits is 7, is

(A) 225   (B) 243   (C) 252   (D) none

Q.3 The 120 permutations of MAHES are arranged in dictionary order, as if each were an ordinary five-letter word. The last letter of the 86th word in the list is

(A) A   (B) H   (C) S   (D) E

Q.4 A new flag is to be designed with six vertical strips using some or all of the colors yellow green, blue and red. Then the number of ways this can be done such that no two adjacent strips have the same color is

(A) 12×81 (B) 16×192 (C) 20×125 (D) 24×216

Q.5 The number of 10-digit numbers such that the product of any two consecutive digits in the number is a prime number, is

(A) 1024   (B) 2048   (C) 512   (D) 64

Q.6 Consider the five points comprising of the vertices of a square and the intersection point of its diagonals. How many triangles can be formed using these points?

(A) 4   (B) 6   (C) 8   (D) 10

Q.7 How many of the 900 three digit numbers have at least one even digit?

(A) 775   (B) 875   (C) 100   (D) 101

Q.8 A 5 digit number divisible by 3 is to be formed using the numbers 0, 1, 2, 3, 4 & 5 without repetition. The total number of ways in which this can be done is:

(A) 3125   (B) 600   (C) 240   (D) 216

Q.9 The number of different seven digit numbers that can be written using only three digits 1, 2 & 3 under the condition that the digit 2 occurs exactly twice in each number is

(A) 672   (B) 640   (C) 512   (D) none

Q.10 Out of seven consonants and four vowels, the number of words of six letters, formed by taking four consonants and two vowels is (Assume that each ordered group of letter is a word):

(A) 210   (B) 462   (C) 151200   (D) 332640

Q.11 All possible three digits even numbers which can be formed with the condition that if 5 is one of the digit, then 7 is the next digit is:

(A) 5   (B) 325   (C) 345   (D) 365

Q.12 Number of 5 digit numbers which are divisible by 5 and each number containing the digit 5, digits being all different is equal to k(4!), the value of k is

(A) 84   (B) 168   (C) 188   (D) 208

Q.13 The number of six digit numbers that can be formed from the digits 1, 2, 3, 4, 5, 6, & 7 so that digits do not repeat and the terminal digit are even is:

(A) 144   (B) 72   (C) 288   (D) 720

Q.14 The number of natural numbers from 1000 to 9999 (both inclusive) that do not have all 4 different digits is

(A) 4048   (B) 4464   (C) 4518   (D) 4536

Q.15 Number of positive integers which have no two digits having the same value with sum of their digits being 45, is

(A) 10!   (B) 9!   (C) 9.9!   (D) 17.8!

Q.16 Number of 3 digit number in which the digit at hundredth's place is greater than the other two digit is

(A) 285   (B) 281   (C) 240   (D) 204

Q.17 Number of permutation of 1, 2, 3, 4, 5, 6, 7, 8 and 9 taken all at a time, such that the digit 1 appearing somewhere to the left of 2, 3 appearing to somewhere the left of 4 and 5 somewhere to the left of 6, is (e.g. 815723946 would be one such permutation)

(A) 9.7!   (B) 8!   (C) 5!.4!   (D) 8!.4!

Q.18 Number of odd integers between 1000 and 8000 which have none of their digit repeated, is

(A) 1014   (B) 810   (C) 690   (D) 1736

Q.19 The number of ways in which 5 different books can be distributed among 10 people if each person can get at most one book is:

(A) 252   (B) 10^5 (C) 5^{10} (D) $^{10}C_5.5!$

Q.20 A students have to answer 10 out of 13 questions in an examination. The number of ways in which he can answer if he must answer at least 3 of the first five questions is

(A) 276   (B) 267   (C) 80   (D) 1200

Q.21 The number of three digit numbers having only two consecutive digits identical is:

(A) 153   (B) 162   (C) 180   (D) 161

Q.22 The interior angles of a regular polygon measure $150°$ each. The number of diagonals of the polygon is

(A) 35   (B) 44   (C) 54   (D) 78

Q.23 The number of n digit numbers which consists of the digits 1 & 2 only if each digit is to be used at least once, is equal to 510 then n is equal to:

(A) 7   (B) 8   (C) 9   (D) 10

Q.24 Number of four digit numbers with all digits different and containing the digit 7 is

(A) 2016   (B) 1828   (C) 1848   (D) 1884

Q.25 An English school and a Vernacular school are both under one superintendent. Suppose that the superintendentship, the four teachership of English and Vernacular school each, are vacant, if there be altogether 11 candidates for the appointments, 3 of whom apply exclusively for the superintendentship and 2 exclusively for the appointment in the English school, the number of ways in which the different appointment can be disposed of is :

(A) 4320   (B) 268   (C) 1080   (D) 25920

Q.26 A committee of 5 is to be chosen from a group of 9 people. Number of ways in which it can be formed if two particular persons either serve together or not at all and two other particular persons refuse to serve with each other, is

(A) 41   (B) 36   (C) 47   (D) 76

Q.27 A question paper on mathematics consists of twelve questions divided into three parts A, B and C, each containing four questions, in how many ways can an examinee answer five questions, selecting at least one from each part?

(A) 624   (B) 208   (C) 1248   (D) 2304

Q.28 Number of ways in which 7 green bottles and 8 blue bottles can be arranged in a row if exactly 1 pair of green bottles is side by side, is (Assume all bottles to be alike except for the color).

(A) 84   (B) 360   (C) 504   (D) None

Q.29 The kindergarden teacher has 25 kids in her class. She takes 5 of them at a time, to zoological garden as often as she can, without taking the same 5 kids more than once. Then the number of visits, the teacher makes to the garden exceeds that of a kid by:

(A) $^{25}C_5 - ^{24}C_5$ (B) $^{24}C_5$ (C) $^{24}C_4$ (D) None

Q.30 A rack has 5 different pairs of shoes. The number of ways in which 4 shoes can be chosen from it so that there will be no complete pair is:

(A) 1920   (B) 200   (C) 110   (D) 80

Q.31 Number of ways in which 9 different toys can be distributed among 4 children belonging to different age groups in such a way that distribution among the 3 elder children is even and the youngest one is to receive one toy more, is:

(A) $\dfrac{(5!)^2}{8}$ (B) $\dfrac{9!}{2}$ (C) $\dfrac{9!}{3!(2!)^3}$ (D) None

Q.32 There are 10 red balls of different shades & 9 green balls of identical shades. Then the number of such arrangements such that no two green balls are together in the row is:

(A) $(10!).^{11}P_9$ (B) $(10!).^{11}C_9$ (C) $10!$ (D) $10!\,9!$

Q.33 A shelf contains 20 different books of which 4 are in single volume and the others form sets of 8, 5 and 3 volumes respectively. Number of ways in which the books may be arranged on the shelf, if the volumes of each set are together and in their due order is

(A) $\dfrac{20!}{8!5!3!}$ (B) $7!$ (C) $8!$ (D) $7.8!$

Q.34 Number of ways in which 3 men and their wives can be arranged in a line such that none of the 3 men stand in a position that is ahead of his wife, is

(A) $3!.3!$ (B) $2.3!.3!$ (C) $3!$ (D) $\dfrac{6!}{2!2!2!}$

Q.35 The number of different ways in which five 'dashes' and eight 'dots' can be arranged, using only seven of these 13 'dashes' & 'dots' is

(A) 1287   (B) 119   (C) 120   (D) 1235520

Q.36 Number of different words that can be formed using all the letters of the word "DEEPMALA" if two vowels are together and the other two are also together but separated from the first two is

(A) 960   (B) 1200   (C) 2160   (D) 1440

Q.37 In a unique hockey series between India & Pakistan, they decide to play on till a team wins 5 matches. The number of ways in which the series can be won by India, if no match ends in a draw is:

(A) 126   (B) 252   (C) 225   (D) None

Q.38 Sameer has to make a telephone call to his friend Harish, Unfortunately he does not remember the 7 digit phone number. But he remembers that the first three digits are 635 or 674, the number is odd and there is exactly one 9 in the number. The maximum number of trials that Sameer has to make to be successful is

(A) 10,000   (B) 3402   (C) 3200   (D) 5000

Q.39 There are 12 guests at a dinner party. Supposing that the master and mistress of the house have fixed seats opposite one another, and that there are two specified guests who must always, be placed next to one another; the number of ways in which the company can be placed is :

(A) 20 . 10 ! (B) 22 . 10 !   (C) 44 . 10 !   (D) None

Q.40 In a conference 10 speakers are present. If S_1 wants to speak before S_2 and S_2 wants to speak after S_3, then the number of ways all the 10 speakers can give their speeches with the above restriction if the remaining seven speakers have no objection to speak at any number is

(A) $^{10}C_3$ (B) $^{10}P_8$ (C) $^{10}P_3$ (D) $\dfrac{10!}{3}$

Q.41 The number of all possible selection of one or more questions from 10 given questions, each question having an alternative is:

(A) 3^{10} (B) $2^{10}-1$ (C) $3^{10}-1$ (D) 2^{10}

Q.42 Number of 7 digit numbers the sum of whose digits is 61 is:

(A) 12   (B) 24   (C) 28   (D) None

Q.43 There are 2 identical white balls, 3 identical red balls and 4 green balls of different shades. The number of ways in which they can be arranged in a row so that at least one ball is separated from the balls of the same color, is:

(A) $6\,(7! - 4!)$ (B) $7\,(6! - 4!)$ (C) $8! - 5!$ (D) None

Q.44 Product of all the even divisors of N = 1000, is

(A) $32 \cdot 10^2$ (B) $64 \cdot 2^{14}$ (C) $64 \cdot 10^{18}$ (D) $128 \cdot 10^6$

Q.45 A lift with 7 people stops at 10 floors. People varying from zero to seven go out at each floor. The number of ways in which the lift can get emptied, assuming each way only differs by the number of people leaving at each floor, is

(A) $^{16}C_6$ (B) $^{17}C_7$ (C) $^{16}C_7$ (D) None

Q.46 You are given an unlimited supply of each of the digits 1, 2, 3 or 4. Using only these four digits, you construct n digit numbers. Such n digit numbers will be called LEGITIMATE if it contains the digit 1 either an even number times or not at all. Number of n digit legitimate numbers are

(A) $2^n + 1$ (B) $2^{n+1} + 2$ (C) $2^{n+2} + 4$ (D) $2^{n-1}(2^n + 1)$

Q.47 Distinct 3 digit numbers are formed using only the digits 1, 2, 3 and 4 with each digit used at most once in each number thus formed. The sum of all possible numbers so formed is

(A) 6660   (B) 3330   (C) 2220   (D) None

Q.48 An ice cream parlor has ice creams in eight different varieties. Number of ways of choosing 3 ice creams taking at least two ice creams of the same variety, is (Assume that ice creams of the same variety to be identical & available in unlimited supply)

(A) 56   (B) 64   (C) 100   (D) None

Q.49 There are 12 books on Algebra and Calculus in our library, the books of the same subject being different. If the number of selection each of which consists of 3 books on each topic is greatest then the number of books of Algebra and Calculus in the library are respectively:

(A) 3 and 9   (B) 4 and 8   (C) 5 and 7   (D) 6 and 6

Q.50 A person writes letters to his 5 friends and addresses the corresponding envelopes. Number of ways in which the letters can be placed in the envelope, so that at least two of them are in the wrong envelopes, is,

(A) 1   (B) 2   (C) 118   (D) 119

Q.51 For a game in which two partners oppose two other partners, 8 men are available. If every possible pair must play with every other pair, the number of games played is

(A) $^8C_2 \cdot {}^6C_2$ (B) $8C_2 \cdot {}^6C_2 \cdot 2$ (C) $^8C_4 \cdot 3$ (D) None

Q.52 The number 916238457 is an example of nine digit number which contains each of the digit 1 to 9 exactly once. It also has the property that the digits 1 to 5 occur in their natural order, while the digits 1 to 6 do not. Number of such numbers are

(A) 2268   (B) 2520   (C) 2975   (D) 1560

Q.53 Number of functions defined from f : {1, 2, 3, 4, 5, 6} → {7, 8, 9, 10} such that the sum f(1) + f(2) + f(3) + f(4) + f(5) + f(6) is odd, is

(A) 2^{10} (B) 2^{11} (C) 2^{12} (D) $2^{12} - 1$

Multiple Correct Choice Type

Q.54 The continued product, 2.6.10.14... to n factors is equal to:

(A) $^{2n}C_n$ (B) $^{2n}P_n$

(C) $^{2n+1}C_n$ (D) None

Q.55 The maximum number of permutations of 2n letters in which there are only a's & b's, taken all at a time is given by :

(A) $^{2n}C_n$

(B) $\dfrac{2}{1} \cdot \dfrac{6}{2} \cdot \dfrac{10}{3} \cdots \dfrac{4n-6}{n-1} \cdot \dfrac{4n-2}{n}$

(C) $\dfrac{n+1}{1} \cdot \dfrac{n+2}{2} \cdot \dfrac{n+3}{3} \cdot \dfrac{n+4}{4} \cdots \dfrac{2n-1}{n-1} \cdot \dfrac{2n}{n}$

(D) $\dfrac{2^n\left[1.3.5...(2n-3)(2n-1)\right]}{n!}$

(E) All of the above

Q.56 Number of ways in which 3 numbers in A.P. can be selected from 1, 2, 3,... n is :

(A) $\left(\dfrac{n-1}{2}\right)^2$ if n is even   (B) $\dfrac{n(n-2)}{4}$ if n is odd

(C) $\dfrac{(n-1)}{4}$ if n is odd   (D) $\dfrac{n(n-2)}{4}$ if n is even

Previous Years' Questions

Q.1 The value of the expression $^{47}C_4 + \sum\limits_{j=1}^{5} {}^{52-j}C_3$ is equal to   *(1982)*

(A) $^{47}C_5$ (B) $^{52}C_5$

(C) $^{52}C_4$ (D) None of these

Q.2 Eight chairs are numbered 1 to 8. Two women and three men wish to occupy one chair each. First the women choose the chairs from amongst the chairs marked 1 to 4, and then the men select the chairs from amongst the remaining. The number of possible arrangements is *(1982)*

(A) $^6C_3 \times {}^4C_2$ (B) $^4P_2 \times {}^4P_3$

(C) $^4C_2 + {}^4P_3$ (D) None

Q.3 A five digits number divisible by 3 is to be formed using the numbers 0, 1, 2, 3, 4 and 5, without repetition. The total number of ways this can be done, is *(1989)*

(A) 216   (B) 240   (C) 600   (D) 3125

Q.4 Number of divisors of the form $(4n + 2)$, $n \geq 0$ of integer 240 is *(1998)*

(A) 4   (B) 8   (C) 10   (D) 3

Q.5 If r, s, t are prime numbers and p, q are the positive integers such that LCM of p, q is $r^2 s^4 t^2$, then the number of ordered pairs (p, q) is *(2006)*

(A) 252   (B) 254   (C) 225   (D) 224

Q.6 The letters of the word COCHIN are permuted and all the permutations are arranged in an alphabetical order as in an English dictionary. The number of words that appear before the word COCHIN is *(2007)*

(A) 360   (B) 192   (C) 96   (D) 48

Q.7 The number of seven digit integers, with sum of the digits equal to 10 and formed by using the digits 1, 2 and 3 only, is *(2009)*

(A) 55   (B) 66   (C) 77   (D) 88

JEE Advanced/Boards

Exercise 1

Q.1 Consider all the six digit numbers that can be formed using the digits 1, 2, 3, 4, 5 and 6, each digit being used exactly once. Each of such six digit numbers have the property that for each digit, not more than two digits, smaller than that digit, appear to the right of that digit. Find the number of such six digit numbers having the desired property

Q.2 Find the number of five digit number that can be formed using the digits 1, 2, 3, 4, 5, 5, 7, 9 in which one digit appears once and two digits appear twice (e.g 41174 is one such number but 75355 is not.)

Q.3 Find the number of ways in which 3 distinct numbers can be selected from the set $\{3^1, 3^2, 3^3, \ldots 3^{100}, 3^{101}\}$ so that they form a G.P.

Q.4 Find the number of odd numbers between 3000 to 6300 that have all different digits.

Q.5 A man has 3 friend. In how many ways he can invite one friend every day for dinner on 6 successive nights so that no friend is invited more than 3 times.

Q.6 In an election for the managing committee of a reputed club, the number of candidates contesting elections exceeds the number of members to be elected by r(r > 0). If a voter can vote in 967 different ways to elect managing committee by voting at least 1 of them & can vote in 55 different ways to elect (r – 1) candidates by voting in the same manner. Find the number of candidates contesting the election & the number of candidates losing the elections.

Paragraph for question nos. 7 to 9:

2 American men; 2 British men; 2 Chinese men and one each of Dutch, Egyptian, French and German persons are to be seated for a round table conference.

Q.7 If the number of ways in which they can be seated if exactly to pairs of persons of same nationality are together is p(6!), then find p.

Q.8 If the number of ways in which only American pair is adjacent is equal to q(6!), then find q.

Q.9 If the number of ways in which no two people of the same nationality are together given by r (6!), find r.

Q.10 For each positive integer k, let S_k denote the increasing arithmetic sequence of integers whose first term is 1 and whose common difference is k. For example, S_3 is the sequence 1, 4, 7, 10 ..... Find the number of values of k for which S_k contain the term 36!

Q.11 A shop sells 6 different flavors of ice-cream. In how many ways can a customer choose 4 ice-cream cones if

(i) They are all of different flavors

(ii) They are not necessarily of different flavors

(iii) They contain only 3 different flavors

(iv) They contain only 2 or 3 different flavors?

Q.12 (a) How many divisors are there of the number 21600. Find also the sum of these divisors.

(b) In how many ways the number 7056 can be resolved as a product of 2 factors.

(c) Find the number of ways in which the number 300300 can be split into 2 factors which are relatively prime.

(d) Find the number of positive integers that are divisors of at least one of the number 10^{10}; 15^7; 18^{11}.

Q.13 How many 15 letter arrangement of 5A's, 5 B's and 5 C's have no A's in the first 5 letters, no B's in the next 5 letters, and to C's in the last 5 letters.

Q.14 Determine the number of paths from the origin to the point (9, 0) in the Cartesian plane which never pass through (5, 5) in paths consisting only of steps going 1 unit North and 1 unit East.

Q.15 There are n triangles of positive area that have one vertex A(0, 0) and the other two vertices whose coordinates are drawn independently with replacement from the set {0, 1, 2, 3, 4} e.g. (1. 2), (0, 1) (2, 2) etc. Find the value of n.

Q.16 How many different ways can 15 Candy bars be distributed between Ram, Shyam, Ghanshyam and Balram, if Ram cannot have more than 5 candy bars and Shyam must have at least two. Assume all Candy bars to be alike

Q.17 Find the number of three digits number from 100 to 999 inclusive which have any one digit that is the average of the other two.

Q.18 (a) Find the number of non-empty subsets S of {1, 2, 3, 4, 5, 6, 7, 8, 9, 10, 11, 12} such that if, S contains k elements, then S contains no number less than k.

(b) If the number of ordered pairs (S, T) of subsets of {1, 2, 3, 4, 5, 6} are such that $S \cup T$ contains exactly three elements 10λ, then find the value of λ.

Q.19 Find the number of permutation of the digits 1, 2, 3, 4 and 5 taken all at a time so that the sum of the digits at the first two places is smaller than the sum of the digit at the last two places.

Q.20 In a league of 8 teams, each team played every other team 10 times. The number of wins of the 8 teams formed an arithmetic sequence. Find the least possible number of games won by the champion.

Q.21 Find the sum of all numbers greater than 10000 formed by using the digits 0, 1, 2, 4, 5 no. digit being repeated in any number.

Q.22 There are 3 cars of different make available to transport 3 girls and 5 boys on a field trip. Each car can hold up to 3 children. Find

(a) the number of ways in which they can be accommodated.

(b) the numbers of ways in which they can be accommodated if 2 or 3 girls are assigned to one of the cars.

In both the cases internal arrangement of children inside the car is considered to be immaterial.

Q.23 Find the number of three elements sets of positive integers {a, b, c} such that a × b × c = 2310.

Q.24 Find the number of integer between 1 and 10000 with a least one 8 nd at least one 9 as digits

Q.25 Let N be the number of ordered pairs of non-empty sets A and B that have the following properties:

(a) $A \cup B$ = {1, 2, 3, 4, 5, 6, 7, 8, 9, 10}

(b) $A \cap B = \phi$

(c) The number of elements of A is not the element of B.

(d) The number of elements of B is not an element of A.

Find N.

Q.26 In how many other ways can be letters of the word MULTIPLE be arranged:

(i) Without changing the order of the vowels

(ii) Keeping the position of each vowel fixed and without changing the relative order/position or vowels & consonants.

Q.27 Let N denotes the number of all 9 digits numbers if

(a) The digit of each number are all from the set {5, 6, 7, 8, 9} and

(b) Any digit that appears in the number, repeats at least three times. Find the value of N/5.

Q.28 How many integers between 1000 and 9999 have exactly one pair of equal digit such as 4049 or 9902 but not 4449 or 4040?

Q.29 How many 6 digits odd numbers greater than 60,000 can be formed from the digits 5, 6, 7, 8, 9, 0 if

(i) Repetitions are not allowed

(ii) Repetitions are allowed.

Exercise 2

Single Correct Choice Type

Q.1 An eight digit number divisible by 9 is to be formed by using 8 digits out of the digits 0, 1, 2, 3, 4, 5, 6, 7 8, 9 without replacement. The number of ways in which this can be done is

(A) 9! (B) 2(7!) (C) 4(7!) (D) (36) (7!)

Q.2 Number of 4 digit numbers of the form N = abcd which satisfy following three conditions

(i) $4000 \le N < 6000$

(ii) N is a multiple of 5

(iii) $3 \le b < c \le 6$ is equal to

(A) 12 (B) 18 (C) 24 (D) 48

Q.3 5 Indian & 5 American couples meet at a party and shake hands. If no wife shakes hands with her own husband and no Indian wife shakes hands with a male then the number of handshakes that takes place in the party is

(A) 95 (B) 110 (C) 135 (D) 150

Q. 4 The 9 horizontal and 9 vertical lines on an 8 × 8 chessboard form 'r' rectangles and 's' squares, The ratio s/r in its lowest terms is

(A) $\dfrac{1}{6}$ (B) $\dfrac{17}{108}$ (C) $\dfrac{4}{27}$ (D) None

Q.5 Number of different natural numbers which are smaller than two hundred million and use only the digits 1 or 2 is

(A) $(3) \cdot 2^8 - 2$ (B) $(3) \cdot 2^8 - 1$

(C) $2(2^9 - 1)$ (D) None

Q.6 There are counters available in x different colors. The counters are all alike except for the color. The total number of arrangements consisting of y counters, assuming sufficient number of counters of each color, if no arrangement consists of all counters of the same color is:

(A) $x^y - x$ (B) $x^y - y$ (C) $y^x - x$ (D) $y^x - y$

Q.7 If m denotes the number of 5 digit numbers of each successive digits are in their descending order magnitude and n is the corresponding figure, when the digits are in their ascending order of magnitude then (m − n) has the value

(A) $^{10}C_4$ (B) $^{9}C_5$ (C) $^{10}C_3$ (D) $^{9}C_3$

Q.8 There are m points on straight line AB & n points on the line AC none of them being the point A. Triangles are formed with these points as vertices, when

(i) A is excluded

(ii) A is included. The ration of number of triangles in the two cases is:

(A) $\dfrac{m+n-2}{m+n}$ (B) $\dfrac{m+n-2}{m+n-1}$

(C) $\dfrac{m+n-2}{m+n+2}$ (D) $\dfrac{n(n-1)}{(m+1)(n+1)}$

Q.9 The number of 5 digit numbers such that the sum of their digits is even is

(A) 50000 (B) 45000 (C) 60000 (D) None

Q.10 Number of ways in which 8 people can be arranged in a line if A and B must be next each other and C must be somewhere behind D, is equal to

(A) 10080 (B) 5040 (C) 5050 (D) 10100

Q.11 Seven different coins are to be divided amongst three persons. If no two of the persons receive the same number of coins but each receives at least one coin & none is left over, then the number of ways in which the division may be made is

(A) 420 (B) 630 (C) 710 (D) None

Q.12 Let there be 9 fixed point on the circumference of a circle. Each of these points is joined to every one of the remaining 8 points by a straight line and the points are so positioned on the circumference that at most 2 straight lines meet in any interior point of the circle. The number of such interior intersection points is:

(A) 126   (B) 351   (C) 756   (D) None

Q.13 The number of ways in which 8 distinguishable apples can be distributed among 3 boys such that every boy should get at least 1 apple & at most 4 apples is K. $^7P_3$ where K has the value equal to

(A) 14   (B) 66   (C) 44   (D) 22

Q.14 There are five different peaches and three different apples. Number of ways they can be divided into two packs of four fruits if each pack must contain at least one apple, is

(A) 95   (B) 65   (C) 60   (D) 30

Q.15 Let P_n denote the number of ways in which three people can be selected out of 'n' people sitting in a row, if no two of them are consecutive. If $P_{n+1} - P_n = 15$ then the value of 'n' is

(A) 7   (B) 8   (C) 9   (D) 10

Q.16 The number of positive integers not greater than 100, which are not divisible by 2, 3 or 5 is

(A) 26   (B) 18   (C) 31   (D) None

Q.17 There are six periods in each working day of a school. Number of ways in which 5 subjects can be arranged if each subject is allotted at least one period and no period remains vacant is

(A) 210   (B) 1800   (C) 360   (D) 3600

Q.18 An old man while dialing a 7 digit telephone number remembers that the first four digits consists of one 1's, one 2's and two 3's. He also remembers that the fifth digit is either a 4 or 5 while has no memorizing of the sixth digit, he remembers that the seventh digit is 9 minus the sixth digit. Maximum number of distinct trials he has to try to make sure that he dials the correct telephone number, is

(A) 360   (B) 240   (C) 216   (D) None

Q.19 Number of rectangles in the grid shown which are not squares is

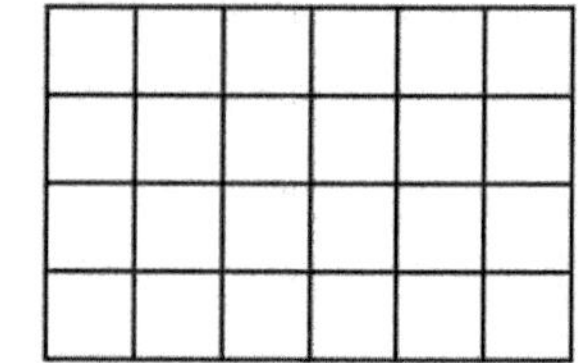

(A) 160   (B) 162   (C) 170   (D) 185

Q.20 All the five digit numbers in which each successive digit exceeds its predecessor are arranged in the increasing order of their magnitude. The 97th number in the list does not contains the digit

(A) 4   (B) 5   (C) 7   (D) 8

Q.21 There are n identical red balls & m identical green balls. The number of different linear arrangements consisting of n red ball but not necessarily all the green balls' is ^xC_y then

(A) $x = m + n,\ y = m$

(B) $x = m + n + 1,\ y = ,m$

(C) $x = m + n + 1,\ y = m + 1$

(D) $x = m + n,\ y = n$

Q.22 A gentleman invites a party of $m + n$ $(m \neq n)$ friends to a dinner & places m at one table T_1 and n at another table T_2, the table being round, if not all people shall have the same neighbor in any two arrangement, then the number of ways in which he can arrange the guests, is

(A) $\dfrac{(m+n)!}{4mn}$ (B) $\dfrac{1}{2}\dfrac{(m+n)!}{mn}$

(C) $2\dfrac{(m+n)!}{mn}$ (D) None

Q.23 Consider a determinant of order 3 all whose entries are either 0 or 1. Five of these entries are 1 and four of them are '0'. Also $a_{ij} = a_{ji}\ \forall\ 1 \leq i, j \leq 3$. The number of such determinants, is equal to

(A) 6   (B) 8   (C) 9   (D) 12

Q.24 A team of 8 students goes on an excursion, in two cars, of which one can seat 5 and the other only 4. If internal arrangement inside the car does not matter then the number of ways in which they can travel, is

(A) 91   (B) 182   (C) 126   (D) 3920

Q.25 One hundred management students who read at least one of the three business magazines are surveyed to study the readership pattern. It is found that 80 read Business India, 50 read Business world, and 30 read Business Today. Five students read all the three magazines. How many read exactly two magazines?

(A) 50 (B) 10 (C) 95 (D) 45

Q.26 Six people are going to sit in a row on a bench. A and B are adjacent, C does not want to sit adjacent to D. E and F can sit anywhere. Number of ways in which these six people can be seated, is

(A) 200 (B) 144 (C) 120 (D) 56

Q.27 Number of cyphers at the end of $^{2002}C_{1001}$ is

(A) 0 (B) 1 (C) 2 (D) 200

Q.28 Three vertices of a convex n sided polygon are selected. If the number of triangles that can be constructed such that none of the sides of the triangle is also the side of the polygon is 30, then the polygon is a

(A) Heptagon   (B) Octagon

(C) Nonagon   (D) Decagon

Q.29 Given 11 points, of which 5 lie on one circle, other than these 5, no 4 lie on one circle. Then the maximum number of circles that can be drawn so that each contains at least three of the given points is:

(A) 216 (B) 156 (C) 172 (D) None

Q.30 Number of 5 digit numbers divisible by 25 that can be formed using only the digits 1, 2, 3, 4, 5, & 0 taken five at a time is

(A) 2 (B) 32 (C) 42 (D) 52

Q.31 Let P_n denotes the number of ways of selecting 3 people out of 'n' sitting in a row, if no two of them are consecutive and Q_n is the corresponding figure when they are in a circle. If $P_n - Q_n = 6$, then 'n' is equal to

(A) 8 (B) 9 (C) 10 (D) 12

Q.32 Let m denote the number of ways in which 4 different books are distributed among 10 persons, each receiving none or one only and let n denote the number of ways of distribution if the books are all alike. Then:

(A) m = 4n (B) n = 4m (C) m = 24 n (D) none

Q.33 The number of ways of choosing a committee of 2 women & 6 men, if Mr. A refuses to serve on the committee if Mr. B is a member & Mr. B can only serve, if Miss C is the member of the committee, is

(A) 60 (B) 84 (C) 124 (D) None

Q.34 Six person A, B, C, D, E and F are to be seated at a circular table. The number of ways this can be done if A must have either B or C on his right and B must have either C or D on his right is:

(A) 36 (B) 12 (C) 24 (D) 18

Q.35 There are 100 different books in a shelf. Number of ways in which 3 books can be selected so that no two of which are adjacent is

(A) $^{100}C_3 - 98$ (B) $^{97}C_3$ (C) $^{96}C_3$ (D) $^{98}C_3$

Q.36 Number of ways in which four different toys and five indistinguishable marbles can be distributed between Amar, Akbar and Anthony, if each child receives at least one toy and one marble, is

(A) 42 (B) 100 (C) 150 (D) 216

Q.37 A 3 digit palindrome is a 3 digit number (not starting with zero) which reads the same backwards as forwards. For example 171. The sum of all even 3 digit palindromes, is

(A) 22380 (B) 25700 (C) 22000 (D) 22400

Q.38 Two classrooms A and B having capacity of 25 and (n–25) seats respectively A_n denotes the number of possible seating arrangements of room 'A', when 'n' students are to be seated in these rooms, starting from room 'A' which is to be filled up full to its capacity. If $A_n - A_{n-1} = 25! (^{49}C_{25})$ then 'n' equals

(A) 50 (B) 48 (C) 49 (D) 51

Q.39 Number of positive integral solution satisfying the equation $(x_1 + x_2 + x_3)(y_1 + y_2) = 77$, is

(A) 150 (B) 270 (C) 420 (D) 1024

Q.40 There are counters available in 3 different colors (at least four of each color). Counters are all alike except for the color. If 'm' denotes the number of arrangements of four counters if no arrangement consists of counters of same color and 'n' denotes the corresponding figure when every arrangement consists of counters of each color, then:

(A) m = 2 n   (B) 6 m = 13 n

(C) 3 m = 5 n   (D) 5 m = 3 n

Q.41 Three digit numbers in which the middle one is a perfect square are formed using the digits 1 to 9. Their sum is:

(A) 134055   (B) 270540

(C) 170055   (D) none

Q.42 A guardian with 6 wards wishes every one of them to study either Law of Medicine or Engineering. Number of ways in which he can make up his mind with regard to the education of his wards if every one of them be fit for any of the branches to study, and at least one child is to be sent in each discipline is:

(A) 120   (B) 216   (C) 729   (D) 540

Q.43 There are $(p + q)$ different books on different topics in Mathematics $(p \neq q)$

If L = The number of ways in which these books are distributed between two students X and Y such that X get p books and Y gets q books.

M = The number of ways in which these books are distributed between two students X and Y such that one of them gets p books and another gets q books.

N = The number of ways in which these books are divided into two groups of p books and q books then,

(A) L = M = N   (B) L = 2M = 2N

(C) 2L = M = 2N   (D) L = M = 2N

Q.44 Number of ways in which 5A' and 6B's can be arranged in a row which reads the same backwards and forwards, is

(A) 6   (B) 8   (C) 10   (D) 12

Q.45 Coefficient of $x^2 y^3 z^4$ in the expansion of $(x + y + z)^9$ is equal to

(A) The number of ways in which 9 objects of which 2 alike of one kind, 3 alike of 2nd kind, and 4 alike of 3rd kind can be arranged.

(B) The number of ways in which 9 identical objects can be distributed in 3 persons each receiving at least two objects.

(C) The number of ways in which 9 identical objects can be distributed in 3 persons each receiving none one or more.

(D) The number of ways in which 9 different books can be tied up in to three bundles one containing 2, other 3 and third containing 4 books.

Multiple Correct Choice Type

Q.46 The combinatorial coefficient C(n, r) is equal to

(A) number of possible subsets of r members from a set of n distinct members.

(B) number of possible binary messages of length n with exactly r l's.

(C) number of non-decreasing 2-D paths from the lattice point (0, 0) to (r, n).

(D) number of ways of selecting r objects out of n different objects when a particular object is always included plus the number of ways of selecting 'r' objects out of n, when a particular object out of n, when a particular object is always excluded.

Q.47 There are 10 questions, each question is either True or False. Number of different sequences of incorrect answers is also equal to

(A) Number of ways in which a normal coin tossed 10 times would fall in a definite order if both Heads and Tails are present.

(B) Number of ways in which a multiple choice question containing 10 alternatives with one or more than one correct alternatives, can be answered.

(C) Number of ways in which it is possible to draw a sum of money with 10 coins of different denominations taken some or all at a time.

(D) Number of different selection of 10 indistinguishable objects taken some or all at a time.

Q.48 The number of ways in which five different books can be distributed among 3 persons so that each person gets at least one book, is equal to the number of ways in which

(A) 5 persons are allotted 3 different residential flats so that and each person is allotted at most one flat and no two persons are allotted the same flat.

(B) Number of parallelograms (some of which may be overlapping) formed by one set of 6 parallel lines and other set of 5 parallel lines that goes in other direction.

(C) 5 different toys are to be distributed among 3 children, so that each child gets at least one toy.

(D) 3 mathematics professors are assigned five different lecturers to be delivered, so that each professor gets at least one lecturer.

Q.49 If k is odd then ^kC_r is maximum for r equal to

(A) $r = \dfrac{1}{2}(k-1)$ (B) $r = \dfrac{1}{2}(k+1)$

(C) $k-1$ (D) k

Q.50 Which of the following statements are correct?

(A) Number of words that can be formed with 6 only of the letters of the word "CENTRIFUGAL' if each word must contain all the vowels is $3 . 7!$

(B) There are 15 balls of which some are white and the rest black. If the number of ways in which the balls con be arranged in a row, is maximum then the number of white balls must be equal to 7 or 8. Assume balls of the same color to be alike.

(C) There are 12 objects, 4 alike of one kind, 5 alike and of another kind and the rest are all different. The total number of combinations in 240

(D) Number of selections that can be made of 6 letters from the word "COMMITTEE" is 35.

Q.51 Number of ways in which the letters of the word 'B U L B U L' can be arranged in a line is a definite order is also equal to the

(A) Number of ways in which 2 alike Apples and 4 alike Mangoes can be distributed in 3 children so that each child receives any number of fruits.

(B) Number of ways in which 6 different books can be tied up into 3 bundles, if each bundle is to have equal number of books.

(C) Coefficient of $x^2y^2z^2$ in the expansion of $(x + y + z)^6$.

(D) Number of ways in which 6 different prizes can be distributed equally in three children.

Comprehension Type

Paragraph 1: Consider the word W = MISSISSIPPI

Q.52 If N denotes the number of different selections of 5 letters from the word W = MISSISSIPPI then N belongs to the set

(A) {15, 16, 17, 18, 19}   (B) {20, 21, 22, 23, 24}

(C) {25, 26, 27, 28, 29}   (D) {30, 31, 32, 33, 34}

Q.53 Number of ways in which the letters of the word W can be arranged if at least one vowel is separated from rest of the vowels

(A) $\dfrac{8!.161}{4!.4!.2!}$ (B) $\dfrac{8!.161}{4.4!.2!}$ (C) $\dfrac{8!.161}{4!.2!}$ (D) $\dfrac{8!}{4!.2!}.\dfrac{165}{4!}$

Q.54 If the number of arrangements of the letters of the word W if all the S's and P's are separated is (k) $\left(\dfrac{10!}{4!.4!}\right)$ then k equals

(A) $\dfrac{6}{5}$ (B) 1   (C) $\dfrac{4}{3}$ (D) $\dfrac{3}{2}$

Paragraph 2: 16 players P_1, P_2, P_3, ... P_{16} take part in a tennis tournament. Lower suffix player is better than any higher suffix player. These players are to be divided into 4 groups each comprising of 4 players and the best from each group is selected for semifinals.

Q.55 Number of ways in which 16 players can be divided into four equal groups, is

(A) $\dfrac{35}{27}\prod_{r=1}^{8}(2r-1)$ (B) $\dfrac{35}{24}\prod_{r=1}^{8}(2r-1)$

(C) $\dfrac{35}{52}\prod_{r=1}^{8}(2r-1)$ (D) $\dfrac{35}{6}\prod_{r=1}^{8}(2r-1)$

Q.56 Number of ways in which they can be divided into 4 equal groups if the players P_1, P_2, P_3 and P_4 are in different groups, is:

(A) $\dfrac{(11)!}{36}$ (B) $\dfrac{(11)!}{72}$ (C) $\dfrac{(11)!}{108}$ (D) $\dfrac{(11)!}{216}$

Match the Columns

Q.57

Column-I	Column-II
(A) Number of increasing permutations of m symbols are there from the n set numbers $\{a_1, a_2, ..., a_n\}$ where the order among the number is given by $a_1 < a_2 < a_3 < ... a_{n-1} < a_n$ is	(p) n^m
(B) There are m men and n monkeys. Number of ways in which every monkey has a master, if a man can have any number of monkeys	(q) ^mC_n
(C) Number of ways in which n red balls are (m – 1) green balls can be arranged in a line, so that no two red balls are together, is (balls of the same color are alike)	(r) ^nC_m
(D) Number of ways in which 'm' different toys can be distributed in 'n' children if every child may receive any number of toys, is	(s) m^n

Q.58

Column-I	Column-II
(A) Four different movies are running in a town. Ten students go to watch these four movies. The number of ways in which every movie is watched by at least one student, is (Assume each way differs only by number of students watching a movies)	(p) 11
(B) Consider 8 vertices of a regular octagon and its center. If T denotes the number of triangles and S denotes the number of straight lines that can be formed with these 9 points then the value of $(T - S)$ equals	(q) 36
(C) In an examination, 5 children were found to have their mobiles in their pocket. The Invigilator fired them and took their mobiles in his possession. Towards the end of the test, Invigilator randomly returned their mobiles. The number of ways in which at most two children did not get their own mobiles is	(r) 52
(D) The product of the digits of 3214 is 24. The number of 4 digit natural numbers such that the product of their digits is 12, is	(s) 60
(E) The number of ways in which a mixed double tennis game can be arranged from amongst 5 married couple if no husband & wife plays in the same game, is	(t) 84

Previous Years' Questions

Q.1 Five balls of different colors are to be placed in three boxes of different sizes. Each box can hold all five. In how many different ways can we place the balls so that no box remains empty? *(1981)*

Q.2 7 relatives of a man comprises 4 ladies and 3 gentlemen, his wife has also 7 relatives; 3 of them are ladies and 4 gentlemen. In how many ways can they invite a dinner party of 3 ladies and 3 gentlemen so that there are 3 of man's relative and 3 of the wife's relatives? *(1985)*

Q.3 A box contains two white balls, three black balls and four red balls. In how many ways can three balls be drawn from the box, if at least one black ball is to be included in the draw? *(1986)*

Q.4 Eighteen guests have to be seated half on each side of a long table. Four particular guests desire to sit one particular side and three other on the other side. Determine the number of ways in which the sitting arrangements can be made. *(1991)*

Q.5 A committee of 12 is to be formed from 9 women and 8 men. In how many ways this can be done if at least five women have to be included in a committee? In how many of these committees

(a) the women are in majority?

(b) the men are in majority? *(1994)*

Q.6 Match the conditions/expressions in column I with statement in column II.

Consider all possible permutations of the letters of the word ENDEANOEL. *(2008)*

Column I	Column II
(A) The number of permutations containing the word ENDEA. is	(p) 5!
(B) The number of permutations in which the letter E occurs in the first and the last positions, is	(q) 2 × 5!
(C) The number of permutations in which none of the letters D, L, N occurs in the last five positions, is	(r) 7 × 5!
(D) The number of permutations in which the letters A, E, O occur only in odd positions, is	(s) 21×5!

Important Questions

JEE Main/Boards

Exercise 1

Q.6	Q.10	Q.13	Q.16	Q.22
Q.26	Q.27			

Exercise 2

Q.3	Q.13	Q.15	Q.18	Q.25
Q.33	Q.43	Q.46	Q.47	Q.50
Q.52				

Previous Years' Questions

Q.4	Q.5	Q.7

JEE Advanced/Boards

Exercise 1

Q.5	Q.6	Q.12	Q.13	Q.16
Q.20	Q.22	Q.28	Q.30	

Exercise 2

Q.1	Q.3	Q.8	Q.13	Q.20
Q.26	Q.32	Q.39	Q.42	Q.43
Q.49	Q.55	Q.58		

Previous Years' Questions

Q.1	Q.3	Q.4	Q.6

Answer Key

JEE Main/Boards

Exercise 1

Q.1 $2 + 6 + 24 = 32$

Q.2 $3 \times 4 \times 3 \times 2 \times 1 = 72$

Q.3 $35 + 15 = 50$

Q.5 $n = 6$

Q.6 180

Q.7 365

Q.8 1332

Q.9 1023

Q.10 (i) $2^1 + 2^2 + 2^3 + 2^4 = 30$ (ii) $= 2^1 + 2^2 + 2^3 + 2^4 + 2^5 = 2 + 4 + 8 + 16 + 32 = 62$

Q.11 $34650 - 840 = 33810$

Q.12 $4!\ 3!$

Q.13 369600

Q.14 302399, 30240

Q.15 3600

Q.16 126

Q.17 $3 + 4\frac{1}{4} + 4 = 11\frac{1}{2}$ Hours.

Q.18 $r = 3$

Q.20 $^{52}C_4 = 270725$

Q.22 817190

Q.23 (i) 116 (ii) 40 (iii) 185

Q.24 31

Q.25 $\dfrac{50!}{(12!)^3.(7!)^2 3!}$

Q.26 26

Q.27 150

Q.28 2454

Exercise 2

Single Correct Choice Type

Q.1 C	**Q.2** C	**Q.3** D	**Q.4** A	**Q.5** B	**Q.6** C
Q.7 A	**Q.8** D	**Q.9** A	**Q.10** C	**Q.11** D	**Q.12** B
Q.13 D	**Q.14** B	**Q.15** A	**Q.16** A	**Q.17** A	**Q.18** D
Q.19 D	**Q.20** A	**Q.21** B	**Q.22** C	**Q.23** C	**Q.24** C
Q.25 D	**Q.26** A	**Q.27** A	**Q.28** C	**Q.29** B	**Q.30** D
Q.31 C	**Q.32** B	**Q.33** C	**Q.34** D	**Q.35** C	**Q.36** D
Q.37 A	**Q.38** B	**Q.39** A	**Q.40** D	**Q.41** C	**Q.42** C
Q.43 A	**Q.44** C	**Q.45** C	**Q.46** D	**Q.47** A	**Q.48** B
Q.49 D	**Q.50** D	**Q.51** C	**Q.52** B	**Q.53** B	**Q.54** B
Q.55 E	**Q.56** D				

Previous Years' Questions

Q.1 C	**Q.2** D	**Q.3** A	**Q.4** A	**Q.5** C	**Q.6** C
Q.7 C					

JEE Advanced/Boards

Exercise 1

Q.1 162 **Q.2** 7560 **Q.3** 2500 **Q.4** 826 **Q.5** 510 **Q.6** 10, 3

Q.7 60 **Q.8** 64 **Q.9** 244 **Q.10** 24 **Q.11** (i)15, (ii) 126, (iii) 60 (iv) 105

Q.12 (a) 72; 78120; (b) 23 (c) 32; (d) 435 **Q.13** 2252 **Q.14** 30980 **Q.15** 276

Q.16 440 **Q.17** 121 **Q.18** (a) 128; (b) 54 **Q.19** 48 **Q.20** 42

Q.21 3119976 **Q.22** (a) 1680; (b) 1140 **Q.23** 40 **Q.24** 974 **Q.25** 186

Q.26 (i) 3359; (ii) 59; (iii) 359 **Q.27** 4201 **Q.28** 3888 **Q.29** (i) 240, (ii) 15552

Exercise 2

Single Correct Choice Type

Q.1 D	**Q.2** C	**Q.3** C	**Q.4** B	**Q.5** A	**Q.6** A
Q.7 B	**Q.8** A	**Q.9** B	**Q.10** B	**Q.11** B	**Q.12** A
Q.13 D	**Q.14** D	**Q.15** D	**Q.16** A	**Q.17** B	**Q.18** B
Q.19 A	**Q.20** B	**Q.21** B	**Q.22** A	**Q.23** D	**Q.24** C
Q.25 A	**Q.26** B	**Q.27** B	**Q.28** C	**Q.29** B	**Q.30** C
Q.31 C	**Q.32** C	**Q.33** C	**Q.34** D	**Q.35** D	**Q.36** D
Q.37 C	**Q.38** A	**Q.39** C	**Q.40** B	**Q.41** A	**Q.42** D
Q.43 C	**Q.44** C	**Q.45** D			

Multiple Correct Choice Type

Q.46 A, B, D   **Q.47** B, C   **Q.48** B, C, D   **Q.49** A, B   **Q.50** B, D   **Q.51** C, D

Comprehension Type

Q.52 C   **Q.53** B   **Q.54** B   **Q.55** A   **Q.56** C

Matric Match Type

Q.57 A → r; B → s; C → q; D → p   **Q.58** A → t; B → r; C → p; D → q; E → s

Previous Years' Questions

Q.1 300   **Q.2** 485   **Q.3** 64   **Q.4** $^9P_4 \times {}^9P_3 \times (11)!$

Q.5 6062, (a) 2702 (b) 1008   **Q.6** A → p; B → s; C → q; D → q.

Solutions

JEE Main/Boards

Exercise 1

Sol 1: 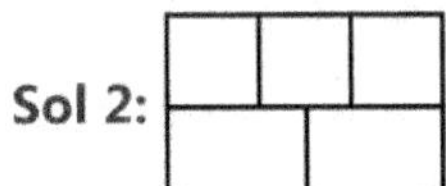

Total numbers = 4 × 4 × 2 = 32

Sol 2:

First arrange those 3 persons in rear seats

Then remaining in front.

Total ways to seat = 5!

Two particular people cannot seat on the driver self

So for this case total ways $\Rightarrow$ 4! + 4!

Therefore, required number of ways = 5! – 4! – 4! = 4! (5 – 2) = 3 × 4 × 3 × 2 × 1 = 72

Sol 3: $^7C_1 \times {}^5C_1 + {}^3C_1 \times {}^5C_1$ = 35 + 15 = 50

Sol 4: $(2n + 1)! = 1 \times 2 \times 3 \ldots\ldots (2n + 1)$

$= (2 \times 4 \times 6 \ldots 2n)(1 \times 3 \times 5 \ldots (2n + 1))$

$= 2^n (1 \times 2 \times 3 \ldots n)(1 \times 3 \times 5 \ldots (2n +1))$

$= 2^n.n! (1 \times 3 \times 5 \ldots (2n + 1))$

$\dfrac{(2n+1)}{n!} = 2^n. (1 \times 3 \times 5 \ldots (2n + 1))$

Sol 5: $^nP_4 = 360$; $\dfrac{n!}{(n-4)!} = 360$

$n(n - 1)(n - 2)(n - 3) = 360 \Rightarrow 6 \times 5 \times 4 \times 3 = 360$

$n = 6$

Sol 6: Case-I: 4 digits

$2 \times {}^5C_3 \times 3! = 120$

Case-II: 3 digits

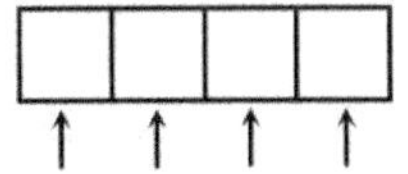

$3 \times {}^5C_2 \times 2! = 60$

Total number of numbers = 60 + 120 = 180

Sol 7: 5 can be there only in the thousand's digit

Case-I: 5 is there

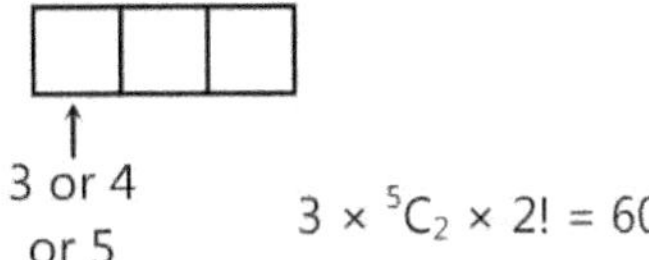

Therefore, total number for case I = 5

Case-II: 5 is not there

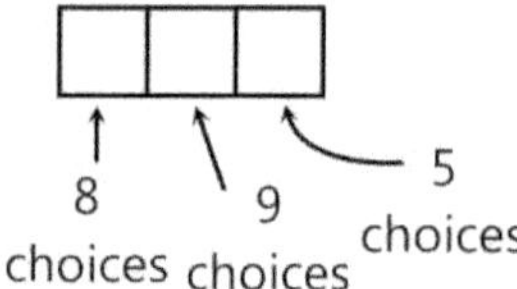

Therefore, total number for case II = 8 × 9 × 5 = 360

Total ways = 5 + 360 = 365

Sol 8: 123+132+213+231+312+321 = 1332

Sol 9: $4^5 - 1 = 1024 - 1 = 1023$

One case is excluded when all arms are at rest.

Sol 10: (i) $2^1 + 2^2 + 2^3 + 2^4 = 30$

(ii) (a) $2^5 = 32$ (b) $2^1 + 2^2 + 2^3 \ldots \ldots 2^5 = 62$

Sol 11: M I S S I S S I P P I

M (1)

S (4)

I (4)

P (2)

Ways = Total permutations – Permutations with 4 I's together

$$= \frac{11!}{4! \ 4! \ 2!} - \frac{8!}{4! \ 2!} = 33810$$

Sol 12: $\downarrow \downarrow \downarrow$ B B B B
| G | G | G |

× × ×

first arrange girls → 3!

4 seats remains in which 4 boys will be

seated → 4!

Total ways = 3! × 4! = 144

Sol 13: $\dfrac{12!}{(3!)^4} = 369600$

Sol 14: BENEVOLENT

No. of letters = 10

B (1), E (3), N (2), V (1), O (1), L (1), T (1)

Total rearrangements = $\dfrac{10!}{3! \ 2!} - 1 = 302400 - 1 = 302399$

Total rearrangements with L in the end = $\dfrac{9!}{3! \ 2!} = 30240$

Sol 15: PATALI PUTRA

Vowels = 5

Consonants = 6

Vowels: A(3) I(1) U(1)

Consonants: P(2) T(2)L(1)R(1)

Total no. of words = $\dfrac{5!}{3!} \times \dfrac{6!}{2! \ 2!} = 3600$

Sol 16: $\dfrac{9!}{4! \ 5!} = 126$

Sol 17: 3! ways

A → III

B → I

C → II

Time = $4 + 4\dfrac{1}{4} + 3 = 11\dfrac{1}{4}$ hours

Sol 18: $^{15}C_{3r} = {}^{15}C_{r+3}$

(1) $3r = r + 3 \Rightarrow r = \dfrac{3}{2}$

(2) $3r + r + 3 = 15 \ \Rightarrow 4r = 12 \ \Rightarrow r = 3$

$r = \dfrac{3}{2}$ Not possible

r = 3 Possible

Sol 19: $^nC_r \times {}^rC_s = {}^nC_s \times {}^{n-s}C_{r-s}$

LHS: $^nC_r + {}^rC_s = \dfrac{n!}{r!(n-r)!} , \dfrac{r!}{s!(r-s)!}$

$= \dfrac{n!}{s!} \times \dfrac{(n-s)!}{(n-s)!} \times \dfrac{1}{(n-r)!.(r-s)!} = \dfrac{n!}{s!(n-s)!} \times \dfrac{(n-s)!}{(n-r)!(r-s)!}$

$= {}^nC_s \times {}^{n-s}C_{r-s}$ = RHS

Sol 20: $^{47}C_4 + \sum\limits_{j=1}^{5} {}^{52-j}C_3$

$= {}^{47}C_4 + ({}^{47}C_3 + {}^{48}C_3 \ldots \ldots {}^{51}C_3 = {}^{48}C_4 + ({}^{48}C_3 + \ldots \ldots + {}^{51}C_3)$

$= {}^{51}C_4 + {}^{51}C_3 = {}^{52}C_4 = 270725$

Sol 21: Product = (n + 1)(n + 2) …… (n + r)

$$= \frac{(n+r)!}{n!} = r! \times \frac{(n+r)!}{n!\ r!} = r! \times {}^{n+r}C_r$$

${}^{n+r}C_r$ will be integer

Hence product is divisible by r!.

Sol 22: ${}^{22}C_{10} + {}^{3}C_3\,{}^{22}C_7 = 817190$

Sol 23:

(i) No. of triangles = ${}^{10}C_3 - {}^{4}C_3 = 116$

(ii) No. of straight lines = ${}^{6}C_2 + 6 \times 4 + 1$

$= 15 + 24 + 1\ = 40$

(iii) No. of quadrilaterals = ${}^{10}C_4 - {}^{4}C_3 \times 6 - {}^{4}C_4 = 185$

Sol 24: $2^5 - 1 = 31$

Sol 25: $\dfrac{50!}{(12!)^3 3!\ (7!)^2 2!}$

Sol 26: Cases not allowed

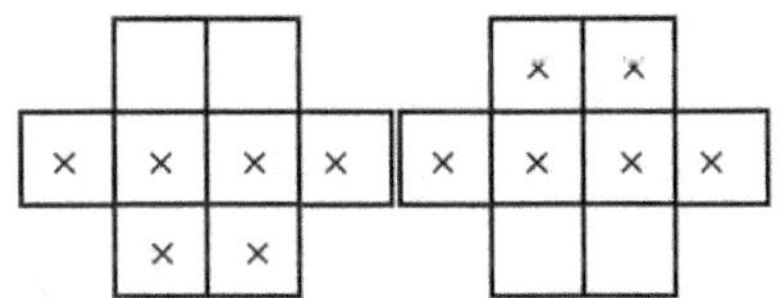

${}^{8}C_6 - 2 = 26$

Sol 27: Possible groups

1 1 3

1 2 2

No. of ways

$$= 3! \times \left(\frac{5!}{1!\ 1!\ 3!\ (2!)} + \frac{5!}{1!\ 2!\ 2!\ (2!)} \right) = 60 + 90 = 150$$

Sol 28: EXAMINATION

4 DIFF

${}^{8}C_4 \times 4! = 1680$

2 diff. 2 alike

$${}^{3}C_1 \times {}^{7}C_2 \times \frac{4!}{2!} = 756$$

2A, 2A

$${}^{3}C_2 \times \frac{4!}{2!\ 2!} = 18$$

Total = 2454

Exercise 2

Single Correct Choice Type

Sol 1: (C)

Sol 2: (C) Numbers = Total – Numbers with no digit 7

Total = 900

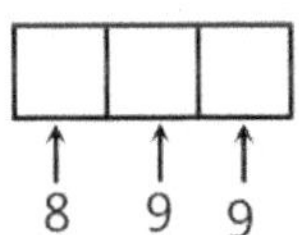

Number of numbers with at least

One digit 7 = $900 - 8 \times 9 \times 9 = 252$

Sol 3: (D)

Sol 4: (A) 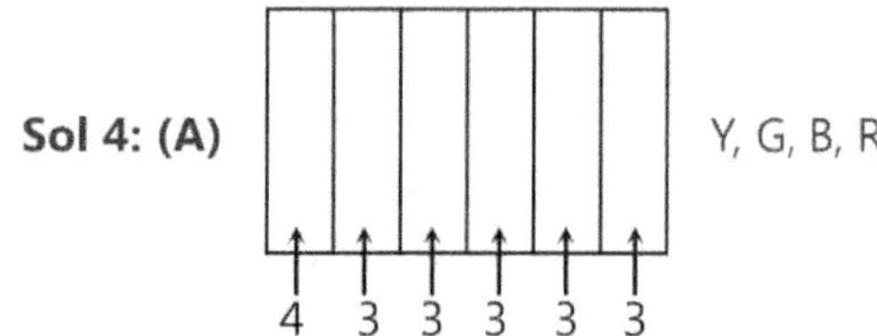 Y, G, B, R

$4 \times 3^5 = 12 \times 81$

Sol 5: (B) Prime number in 0 – 9

2, 3, 5, 7

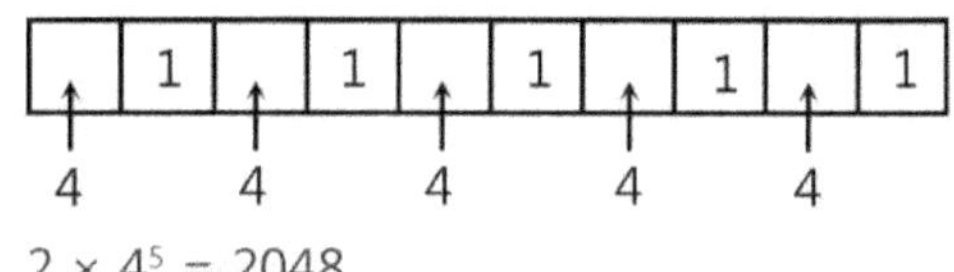

$2 \times 4^5 = 2048$

Sol 6: (C) 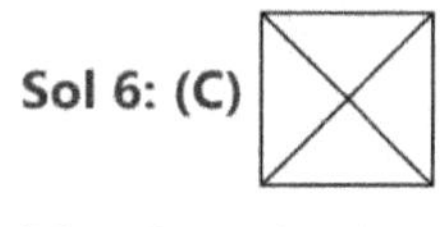

$^5C_3 - 2 = 10 - 2 = 8$

Sol 7: (A) 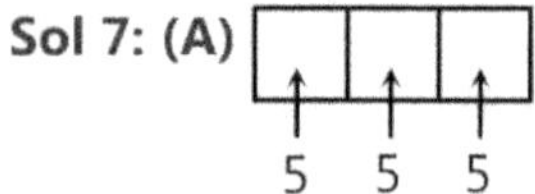

No. 5 with no even digit

$900 - 5^3 = 775$

Sol 8: (D) The sum of the 5-digits used must be divisible by 3.

Only possible combinations are:

1, 2, 3, 4, 5 & 0, 1, 2, 4, 5

  ↓↓

 5!   4.4!

Total = 9.4!

= 9 × 24 = 216

Sol 9: (A) $^7C_2 \times 2^5 = 672$

Sol 10: (C) $^7C_4 \times {}^4C_2 \times 6! = 151200$

Sol 11: (D) 5 can be there only in the thousand's digit

Case-I: 5 is there

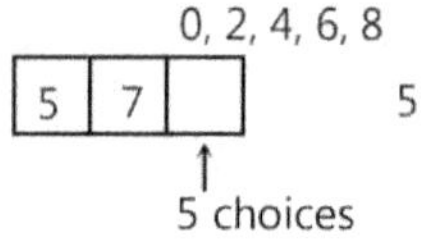

Case-II: 5 is not there

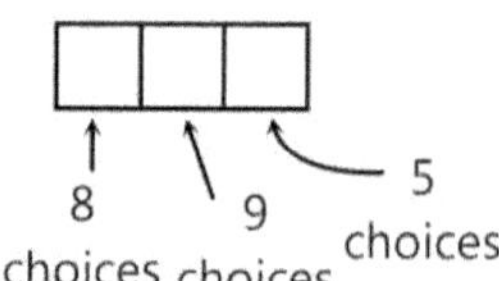

$8 \times 9 \times 5 = 360$

Total ways = 5 + 360 = 365

Sol 12: (B) 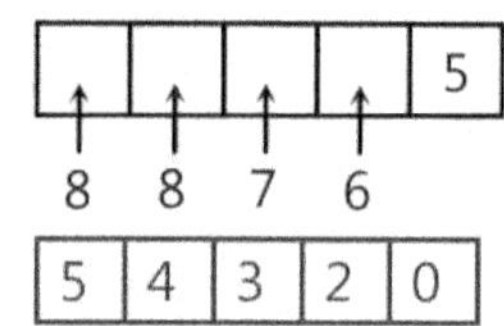

$^4C_1 \times 8 \times 7 \times 6$

Total ways = $8 \times 8 \times 7 \times 6 + {}^4C_1 \times 8 \times 7 \times 6 = 168 \times 4!$

Sol 13: (D) 4 odd, 3 even

Arrangements: 4 + 2 3 + 3

Total numbers = $^3C_2 \times 2 \times 4! + {}^4C_3 \times {}^3C_2 \times 2 \times 4! = 720$

Sol 14: (B) $9000 - 9 \times 9 \times 8 \times 7 = 4464$

Sol 15: (A) It can be a digit number with digits 1 to 9 or a 10 digit number with digits 0 to 9

$9! + 9.9! = 10!$

Sol 16: (A) 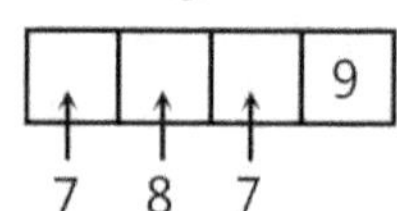

$$\sum_{n=1}^{9} n^2 = \frac{9 \times 10 \times 19}{6} = 285$$

Sol 17: (A) Consider 1 & 2, 3 & 4, 5 & 6 to be identical

Permutations = $\dfrac{9!}{(2!)^3} = 9.7!$

Sol 18: (D)

Last digit 9   Last digit not 9

Total = $7 \times 8 \times 7 + 6 \times 8 \times 7 \times 4 = 1736$

Sol 19: (D)

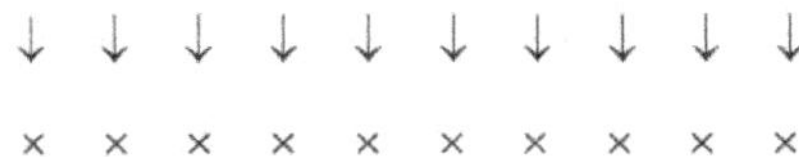

$\frac{10 \ \text{persons}}{^{10}C_5 \ . \ 5!}$

Choose 5 persons from 10 getting books & distribute books to then 5! ways .

Sol 20: (A) $^5C_3 \times {}^8C_7 + {}^5C_4 \times {}^8C_6 + {}^5C_5 \times {}^8C_5$

$= 80 + 5 \times 28 + 56 = 276$

Sol 21: (B)

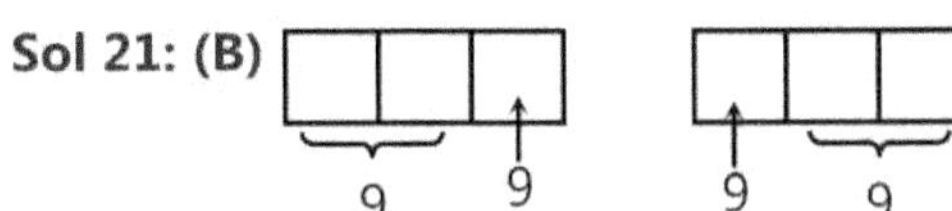

Total = 162

Sol 22: (C) $\frac{(n-2)}{n} \times 180 = 150$

$6n - 12 = 5n \Rightarrow n = 12$

Diagonals $= {}^{12}C_2 - 12 = 54$

Sol 23: (C) $2^n - 2 = 510$

$2^n = 512$

$n = 9$

Sol 24: (C)

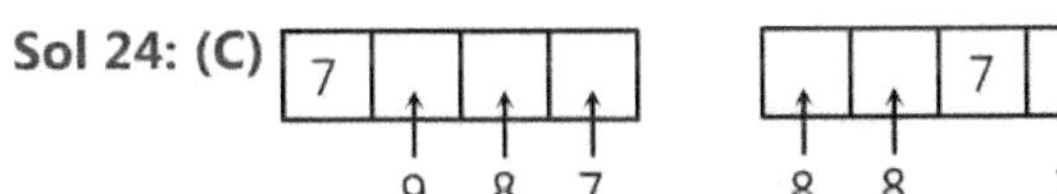

$9 \times 8 \times 7 + 8 \times 8 \times 7 \times 3 = 33 \times 8 \times 7 = 1848$

Sol 25: (D) Total appointment = 11

$11 \rightarrow \boxed{2} \ \boxed{3} \ \boxed{6}$

total ways to disposed $= (2! \ 3! \ 6!) \times 3 = 8640 \times 3 = 25920$

Sol 26: (A) Illegal ways $= 2 \times {}^7C_4 + {}^7C_3 - 2 \times {}^5C_2 = 85$

No. of possible ways $= {}^9C_5 - 85 = 126 - 85 = 41$

Sol 27: (A) 1 1 3

1 2 2

$3 \times ({}^4C_1 \times {}^4C_1 \times {}^4C_3 + {}^4C_1 \times {}^4C_2 \times {}^4C_2)$

$= 3 \times (64 + 144) = 624$

Sol 28: (C) Combine those 2 green bottles

× × × × × × × ×

9 spaces

$6 \times {}^9C_6 = 84 \times 6 = 504$

Sol 29: (B) $^{25}C_5 - {}^{24}C_4 = {}^{24}C_5$

Sol 30: (D) $^5C_4 \times 2^4 = 80$

Select 4 pairs out of 5 different pairs. Now in each pair you can choose 2 different shoes.

Sol 31: (C) 2 2 2 3

$\frac{3! \ 9!}{3! \ (2!)^3 \ 3!} = \frac{9!}{3! \ (2!)^3}$

Sol 32: (B) × × × × × × × × × ×

10 red balls

11 spaces

choose 9 spaces to fill green balls. 10! ways to arrange red balls.

$^{11}C_9 \times 10!$

Sol 33: (C) $7! \times 2^3$

7! ways to arrange

2 order (ascending/descending)

Sol 34: (D) $\frac{6!}{2! \ 2! \ 2!}$

Consider man and wife to be identical and arrange them.

Similar concept as used in Q. 18

Sol 35: (C) $\frac{7!}{7! \ 0!} + \frac{7!}{6! \ 1!} + \frac{7!}{5! \ 2!} \ \ldots\ldots \ \frac{7!}{2! \ 5!}$

$^7C_0 + {}^7C_1 + {}^7C_2 \ \ldots\ldots \ {}^7C_5 = 2^7 - {}^7C_6 - {}^7C_7 = 128 - 7 - 1 = 120$

Sol 36: (D)

EE AA

EAEA

AEAE

EAAE

$4! \times ({}^5C_2 \times (2! + 2! + 1 + 1)) = 1440$

Sol 37: (A) The last match has to be won by India

$1 + {}^5C_1 + {}^6C_2 + {}^7C_3 + {}^8C_4 = 126$

Sol 38: (B)

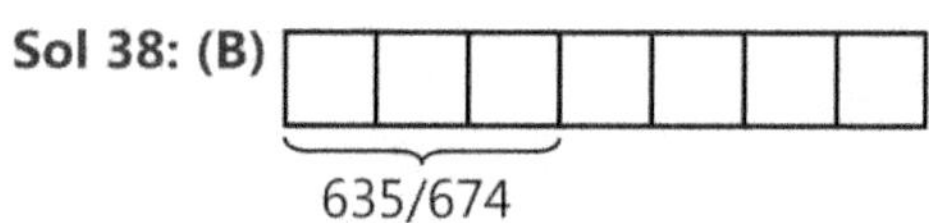

635/674

$2(9^3 \times 1 + 3 \times 4 \times 9^2) = 2(1701) = 3402$

Sol 39: (A)

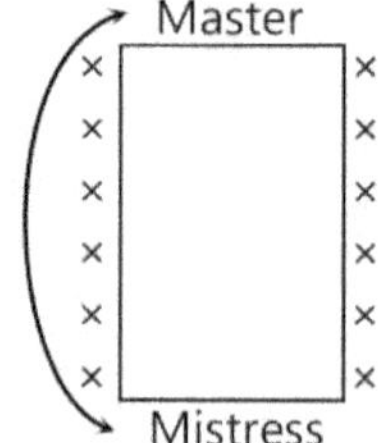

$2 \times 10 \times 10! = Sol.20\ 10!$

Sol 40: (D) $S_1\ S_3\ S_2$

$S_3 S_1 S_2$

Consider $S_1\ S_2\ S_3$ identical & arrange

$2 \times \dfrac{10!}{3!} = \dfrac{10!}{3}$

Sol 41: (C) $3^{10} - 1$

3 choices question its alternative & no question.

Sol 42: (C) 9 9 9 9 9 9 7

9 9 9 9 9 8 8

$\dfrac{7!}{6!} + \dfrac{7!}{5!\ 2!}\ f = 7 + 21 = 2s$

Sol 43: (A) $\dfrac{9!}{2!\ 3!} - 4!\ 3! = 6(7! - 4!)$

Sol 44: (C) $1000 = 2^3 \times 5^3$

The product of even divisors of 1000 will be

$= \left(2 \times 2^2 \times 2^3\right) \times \left(2 \times 5\right) \times \left(2 \times 5^2\right) \times \left(2 \times 5^3\right) \times \left(2^2 \times 5\right)$

$\times \left(2^2 \times 5^2\right) \times \left(2^2 \times 5^3\right) \times \left(2^3 \times 5\right) \times \left(2^3 \times 5^2\right) \times \left(2^3 \times 5^3\right)$

$= (2^6)^4 \times (5 \times 5^2 \times 5^3) = (2^6)^4 \times (5^6)^3 = 64 \times 10^{18}$

Sol 45: (C) $x_1 + x_2 + \ldots\ldots x_{10} = 7$

7 people are distributed to 10 floor

Total ways $= {}^{7 + 10 - 1}C_{10-1} = {}^{16}C_9 = {}^{16}C_7$

Sol 46: (D) Total ways $= 3^n + {}^nC_2 .3^{n-2} + {}^nC_4 .3^{n-4} + \ldots\ldots$

$= \dfrac{(1+3)^n}{2} + \dfrac{2^n}{2} = 2^{2n-1} + 2^{n-1} = 2^{n-1}(2^n + 1)$

Sol 47: (A) Each digit will be present at unit's ten's and hundred's place 6 times.

$(1 + 2 + 3 + 4) \times 6 = 60$

$$60$$
$$60 \times$$

Sum of digits $= \dfrac{60 \times \times}{6660}$

Sol 48: (B) ${}^8C_1 + {}^8C_1 \times {}^7C_1 = 64$

Sol 49: (D) No. of books of algebra = No of books of calculus

Sol 50: (D) No. of ways = Total ways − No letter is in wrong envelope $= 5! - 1 = 119$

Sol 51: (C) $P_1\ P_2\ P_3\ P_4$

$\dfrac{4!}{(2!)^2\ 2!} = 3$

Total matches $= {}^8C_4 \times 3$

Sol 52: (B) ${}^9C_6 \times 5 \times 3! = 2520$

Sol 53: (B) ${}^6C_1 \times 2^1 \times 2^5 + {}^6C_3 \times 2^3 \times 2^3 + {}_6C^5$

$\times 2^5 \times 2^1 = 2^6\ ({}^6C_1 + {}^6C_3 + {}^6C_5)$

$= 2^6 \times \dfrac{2^6}{2} = 2^{11}$

Sol 54: (B) $2.6.10 \ldots\ldots (4n - 2)$

$= 2^n(1.3.5 \ldots\ldots (2n - 1)) \times \dfrac{n!}{n!} = \dfrac{(2n)!}{n!} = {}^{2n}P_n$

Sol 55: (E) No. of permutations

$= {}^{2n}C_n = \dfrac{(2n)!}{n!\ n!} = \dfrac{2^n \ .\ n![1.3.5\ldots\ldots(2n-3)(2n-1)]}{n!}$

Sol 56: (D) n is odd

$1,\ 2,\ 3 \ldots\ldots \dfrac{n+1}{2} \ldots\ldots n$

Total A.P.S. $= 0 + 1..... + \dfrac{n-1}{2} + ... + 1 + 0 = \dfrac{(n-1)^2}{4}$

n is even

$1, 2, ... \dfrac{n}{2}, \dfrac{n}{2} + 1, ... n$

Total A.P.S

$= 2 \times \left(0 + 1 + ... \dfrac{n}{2} - 1 \right) = \dfrac{n(n-2)}{4}$

Previous Years' Questions

Sol 1: (C) Here, $^{47}C_4 + \displaystyle\sum_{j=1}^{5} {}^{52-j}C_3$

$= {}^{47}C_4 + {}^{51}C_3 + {}^{50}C_3 + {}^{49}C_3 + {}^{47}C_3$

$= ({}^{47}C_4 + {}^{47}C_3) + {}^{48}C_3 + {}^{49}C_3 + {}^{50}C_3 + {}^{51}C_3$

(using $^{n}C_r + {}^{n}C_{r-1} = {}^{n+1}C_r$)

$= ({}^{48}C_4 + {}^{48}C_3) + {}^{49}C_3 + {}^{50}C_3 + {}^{51}C_3$

$= ({}^{49}C_4 + {}^{49}C_3) + {}^{50}C_{43} + {}^{51}C_3$

$= ({}^{50}C_4 + {}^{50}C_3) + {}^{51}C_3 = {}^{51}C_4 + {}^{51}C_3 = {}^{52}C_4$

Sol 2: (D) Since, the first 2 women select the chairs amongst 1 to 4 in $^4P_2$ ways

Now, from the remaining 6 chairs, three men could be arranged in $^6P_3$.

$\therefore$ Total number of arrangements $= {}^4P_2 \times {}^6P_3$.

Sol 3: (A) Since, a five digits number is formed using the digits {0, 1, 2, 3, 4 and 5} divisible by 3 ie, only possible when sum of the digits is multiple of three.

Case I : Using digits 0, 1, 2, 4, 5

Number of ways = $4 \times 4 \times 3 \times 2 \times 1 = 96$

Case II : Using digits 1, 2, 3, 4, 5

Number of ways = $5 \times 4 \times 3 \times 2 \times 1 = 120$

$\therefore$ Total number formed $= 120 + 96 = 216$

Sol 4: (A) Since, $240 = 2^4 \times 3 \times 5$

$\therefore$ Total number of divisors = $(4 + 1)(2)(2) = 20$

Out of these 2, 6, 10, and 30 are of the fomr $4n + 2$.

Therefore, (a) is the answer.

Sol 5: (C) Since, r, s, t are prime numbers.

$\therefore$ Selection of p and q are as under

Pq Number of ways

$R^0 r^2$ 1 way

$R^1 r^2$ 1 way

$R^2 r^0, r^1, r^2$ 3 ways

$\therefore$ Total number of ways to select r = 5.

Selection of s as under

$s^0 s^4$ 1 way

$s^1 s^4$ 1 way

$s^2 s^4$ 1 way

$s^3 s^4$ 1 way

$s^4 s^4$ 1 way

$\therefore$ Total number of ways to select s = 9.

Similarly, the number of ways to select t = 5.

$\therefore$ Total number of ways = $5 \times 9 \times 5 = 2$Sol.25

Sol 6: (C) Arrange the letters of the word COCHIN as in the order of dictionary CCHINO.

Consider the words starting from C.

There are 5! Such words. Number of words with the two C' s occupying first and second place = 4!.

Sol 7: (C) There are two possible cases

Case-I Five 1's , one 2's, one 3's

Number of numbers = $\dfrac{7!}{5!} = 42$

Case-II Four 1's three 2's

Number of numbers = $\dfrac{7!}{4!\,3!} = 35$

Total number of numbers = $42 + 35 = 77$

JEE Advanced/Boards

Exercise 1

Sol 1:

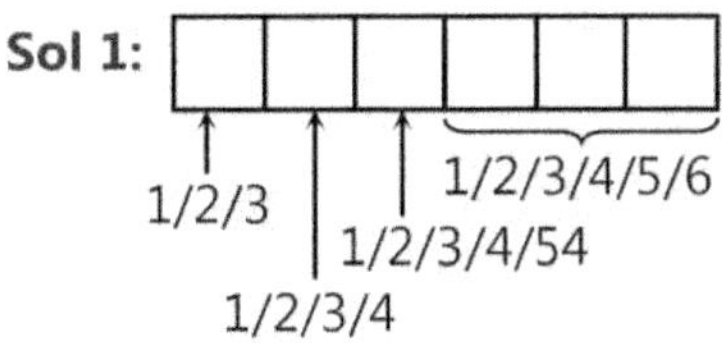

$\underbrace{3 \times 3 \times 3 \times 3}_{\substack{\text{Filling first} \\ \text{4 places}}} \times \underbrace{2}_{\text{last two places}} = 81 \times 2 = 162$

Sol 2: $^9C_1 \times {}^8C_2 \times \dfrac{5!}{2!\ 2!} = 9 \times 28 \times \dfrac{120}{84} = 7560$

Sol 3: $3^1\ 3^2\ 3^3\ \dots\dots\ 3^{51}\ \dots\dots\ 3^{98}\ 3^{99}\ 3^{100}\ 3^{101}$

Fix the middle element of G.P. the find the number of G.P.S possible

$0 + 1 + 2 + \dots + 49 + 50 + 49 + \dots + 1 + 0 = 2500$

Sol 4: 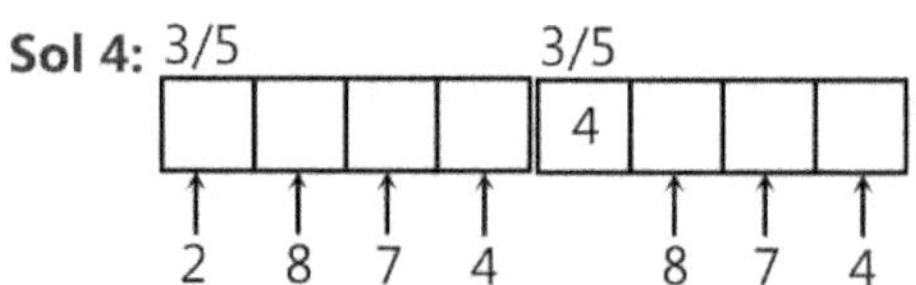

$2 \times 8 \times 7 \times 4 = 448$ $8 \times 7 \times 5 = 280$

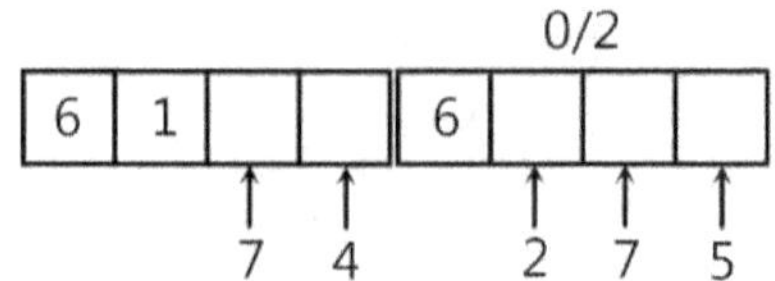

$7 \times 4 = 28$ $2 \times 7 \times 5 = 70$

Total = 826

Sol 5: Case-I: 2 2 2

$\dfrac{6!}{(2!)^3} = 90$

Case-II: 1 2 3

$\dfrac{6!}{1!\ 2!\ 3!} \times 3! = 360$

Case-III: 0 3 3

$\dfrac{6!}{(3!)^2\ 2!} \times 3! = 60$

Total = 90 + 360 + 60 = 510

Sol 6: 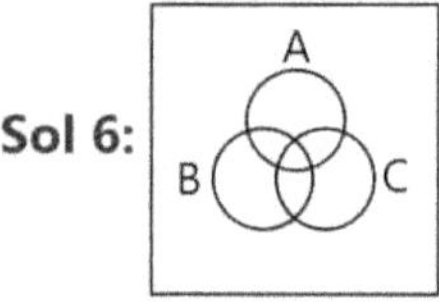

$n(A) = 8! = n(B) = n(C)$

$n(A \cap B) = n(B \cap C) = n(A \cap C) = 7!$

$n(A \cap B \cap C) = 6!$

Sol 7: $3 \times n(A \cap B) - 3n(A \cap B \cap C)$

$3 \times 7! - 3 \times 6!$

$= (63 - 3)6! = 60(6!)$

Sol 8: $n(A) - 2n(A \cap B) + 2n(A \cap B \cap C)$

$= 8! - 2 \times 7! + 6! = (56 + 14 + 1$

$6! = 43\ 6!$

Sol 9: $9! - n(A \cup B \cup C)$

Sol 10: $\dfrac{360}{k}$ Should be an integer.

$\therefore$ k is a factor of 360

$360 = 3^2 \times 2^3 \times 5$

Total factors = $4 \times 3 \times 2 = 24$

Sol 11: (i) $^6C_4 = 15$

(ii) All, 4 diff. $^6C_4 = 15$

2diff, 2Alike $^6C_2 \times {}^4C_1 = 60$

2 alike, 2alike $^6C_2 = 15$

1 diff, 3alike $2 \times {}^6C_2 = 30$

4 alike $^6C_1 = 6$

126

(iii) 3 different flavours = 60

(iv) 2 or 3 different flavours = 60 + 15 + 30 = 105

Sol 12: (a) x = 21600

$x = 6^3 \times 100 = 2^5\ 3^3\ 5^2$

No. of divisors = $6 \times 4 \times 3 = 72$

Sum = $(2^0 + 2^1 + \dots + 2^5)$

$(3^0 + 3^1 + \dots 3^3)(5^0 + 5^1 + 5^2) = 60 \times 40 \times 31 = 78120$

(b) x = 7056

$x = 2^4 \times 3^2 \times 7^2$

Total factors = $5 \times 3 \times 3 = 45$

Answer = $\dfrac{45 + 1}{2} = 23$

(c) $300300 = 7 \times 11 \times 13 \times 10^2 \times 3$

$= 2^2 \times 3 \times 5^2 \times 7 \times 11 \times 13$

$\dfrac{2^6}{2} = 32$

(d) $10^{10}\ 15^7\ 18^{11}$

$2^{10} \times 5^{10},\ 3^7 \times 5^7,\ 2^{11} \times 3^{22}$

HCF of 10^{10}, 15^7 & 18^{11} = 1

HCF of 10^{10} & 15^7 = 5^7

HCF of 10^{10} & 18^{11} = 2^{10}

HCF of 15^7 & $18^{11} = 3^7$

Total divisors $= (11 \times 11 + 8 \times 8 + 12 \times 23) - (8 + 11 + 8) + 1 = 435$

Sol 13: $\displaystyle\sum_{x=0}^{5} ({}^5C_x)^3 = 1 + 5^3 + 10^3 + 10^3 + 5^3 + 1 = 2252$

Sol 14: ${}^{18}C_9 - {}^{10}C_5 \times {}^8C_4 = 48620 - 252 \times 70 = 30980$

Sol 15: $5^2 - 1 = 24 \quad \therefore {}^{24}C_2$ triangles

Sol 16: $x + y + 2 + w = 13$

$x \le 5y \ge 2$

${}^{10}C_2 + {}^{11}C_2 + \ldots {}^{15}C_2 = 45 + 55 + 66 + 78 + 91 + 105 = 440$

Sol 17: 2 digits should be odd or

2 digits should be even for average to be integer

111 222 .... 999 each repeats 3 times

$${}^3C_2 \times \underbrace{{}^5C_1 \times {}^5C_1}_{\text{odd}} + {}^3C_2 \times \underbrace{{}^4C_1 \times {}^4C_1}_{\text{even/0}} + \underbrace{{}^2C_1 \times 4 \times 3}_{\text{zero}} - \underbrace{2 \times 9}_{\text{repetition}} = 121$$

Sol 18: (a) 1 element ${}^{12}C_1 = 12$

2 elements ${}^{10}C_2 = 45$

3 elements ${}^8C_3 = 56$

4 elements ${}^6C_4 = 15$

5 elements 0

Total $= 128$

(b) Select 3 elements

Each is this element is either present in S, T or both S and T

$\therefore$ Total $= {}^6C_3 \times 3^3 = 540$

$\lambda = 54$

Sol 19: Sum at first places can be 6 at max.

4 3 1 2 5 not allowed

	First 2 digits	Last 2 digits	Permutations
Sum 6:	15	43	4
	24	53	4
Sum5:	14	53/52	8
	23	54/51	8
Sum4:	13	54/52/42	12
Sum3:	12	54/53	12
		Total	48

Sol 20: Total wins = Total matches $= 10 \times {}^8C_2$

Wins of champion $= A + 7d$

$$\frac{8}{2}[2A + 7d] = 280$$

$$A = \frac{70 - 7d}{2}$$

Now $d \neq 0$ $d = 2$

$A = 28$

Wins of champion $= 28 + 7 \times 2 = 42$

Sol 21:

S_1	S_2	S_3	S_4	S_5

$S_1 = (1 + 2 + 4 + 5) \times 4! = 288$

$S_2 = S_3 = S_4 = S_5 = (1 + 2 + 4 + 5) \times 3 \times 3! = 216$

Sum $= S_1 \times 10^4 + S_2(10^3 + 10^2 + 10 + 1)$

$= 288 \times 10^4 + 216(1111) = 3119976$

Sol 22: (a) ${}^8C_3 \times {}^5C_3 \times 3! = 1680$

(b) ${}^3C_1 \dfrac{5!}{3!\,2!} 2 + {}^3C_2 \times 3 \times 2 \times \dfrac{5!}{1!\,2!\,2!} \times 2!$

$= 60 + 1080 = 1140$

a, b, c distinct

Sol 23: $\{1, x, x\}$ $({}^5C_1 + {}^5C_2) = 15$

$\{x, x, x\}$ $\dfrac{5!}{1!\,1!\,3!\,2!} + \dfrac{5!}{1!\,2!\,2!\,2!} = 25$

Total $= 15 + 25 = 40$

Sol 24: No. of integers

2 digit no. 89, 98 2

3 digit no.

Zero included 890, 809 4

980, 908

Zero excludes

$2 \times {}^3C_2 + {}^7C_1 \times 3!$ 48

4 digit no.

Tow zeroes ${}^3C_2 \times 26$

One zero ${}^3C_1 \times ({}^7C_1 \times 3! + 6)$ 144

No zero $2 \times {}^4C_2 \times 7 \times 7 + 2 \times {}^4C_2 \times 2 \times 7$ 770 $+ 2 \times {}^4C_3 + {}^4C_2 = 974$

Sol 25: A $-$ 1 B $-$ 9 1

A $-$ 2 B $-$ 8 ${}^8C_1$

A – 3B – 7$^8C_2$

A – 4B – 6$^8C_3$

A – 5B – 50

N = 2(1 + $^8C_1$ + $^8C_2$ + $^8C_3$)

N = 186

Sol 26: (i) M^U LT^I PL^E

$$\frac{8!}{3! \; 2!} - 1 = 3359$$

(ii) $\dfrac{5!}{2!} - 1 = 59$

(iii)$3! \times \dfrac{5!}{2!} - 1 = 359$

Sol 27: 3 digits $^5C_3 \times \dfrac{9!}{3! \; 3! \; 3!} = 16{,}800$

2 digits $2 \times {}^5C_2 \times \left(\dfrac{9!}{3! \; 6!} + \dfrac{9!}{4! \; 5!}\right) = 4{,}200$

1 digit $^5C_1 = 5$

Total = 21005

$\dfrac{N}{5} = 4201$

Sol 28: $^3C_2 \times 9 \times 8 + 9 \times {}^3C_2 \times 8 \times 8 + 9 \times 3 \times 9 \times 8 = 3888$

Sol 29: (i) $4 \times 4! + 2 \times 3 \times 4! = 240$

(ii) $4 \times 6^4 \times 3 = 15552$

Exercise 2

Single Correct Choice Type

Sol 1: (D) We have $0 + 1 + 2 + 3 \ldots + 8 + 9 = 45$

To obtain an eight digit number exactly divisible by 9, we must not use either (0, 9) or (2, 7) or (3, 6) or (4, 5). [Sum of the remaining eight digits is 36 which is exactly divisible by 9.]

When, we do not use (0, 9), then the number of required 8 digit number is 8!.

When, one of (1, 8) or (2, 7) or (3, 6) or (4, 5) is not used, the remaining digits can be arranged in 8! – 7! ways as 0 cannot be at extreme left.

Hence, there are 8! + 4(8! – 7!) = (36) (7!) numbers in the desired category.

Sol 2: (C) 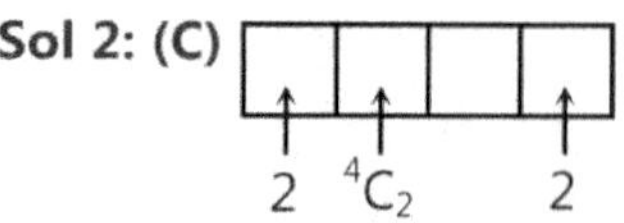

$2 \times {}^4C_2 \times 2 = 24$

Sol 3: (C) $^{20}C_2 - 10 - 5 \times 9 = 190 - 10 - 45 = 135$

Sol 4: (B) $r = {}^9C_2 \times {}^9C_2 = 36^2$

$s = 8_2 + 7^2 + 6^2 + \ldots\ldots 1^2 = \dfrac{8 \times 9 \times 17}{6}$

$\dfrac{s}{r} = \dfrac{17}{108}$

Sol 5: (A) 200,000,000

$2^8 + 2^8 + 2^7 + 2^6 + \ldots\ldots 2^1 = 2^8 + 2 . \left(\dfrac{2^8 - 1}{2 - 1}\right) = 3 . (2)^8 - 2$

Sol 6: (A) 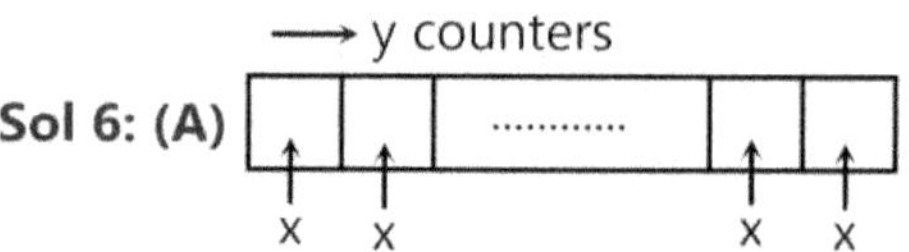

$x^y - y$

Sol 7: (B) $M - n = {}^{10}C_5 - {}^9C_5 = {}^9C_5$

Sol 8: (A) $N_1 = {}^{m+n}C_3 - {}^mC_3 - {}^nC_3$

$N_2 = {}^{m+n+1}C_3 - {}^{m+1}C_3 - {}^{n+1}C_3$

$\dfrac{N_1}{N_2} = \dfrac{m + n - 2}{m + n}$

Sol 9: (B) 10000 to 99999

Number of numbers $= \dfrac{90000}{2} = 45000$

Sol 10: (B) $\boxed{AB}$ D C E F G H

$2 \times \dfrac{7!}{2!} = 5040$

Sol 11: (B) $\dfrac{7!}{1! \; 2! \; 4!} \times 3! = 630$

Sol 12: (A) $^9C_4 = 126$

Sol 13: (D) Case-I: 1 3 4

$$\frac{8!}{1!\ 3!\ 4!} \times 3!$$

Case-II: 2 3 3

$$\frac{8!}{2!\ 3!\ 3!\ 2!} \times 3!$$

Case-III: 2 2 4

$$\frac{8!}{2!\ 2!\ 4!\ 2!} \times 3!$$

Total = 4620 = 22 $\times$ $^7P_3$

Sol 14: (D) Pack 1 Pack 2

Peaches Apples Peaches Apples

3 1 2 2

$^5C_3 \times {}^3C_1 = 30$

Sol 15: (D) $\downarrow\quad\downarrow\quad\downarrow\quad\downarrow\quad\downarrow$

$\underbrace{\times \quad \times \quad \times \quad \times}_{n-3}$

$(n-2)$ choices. select 3

$P_n = {}^{n-2}C_3$

$P_{n+1}\quad P_n = {}^{n-1}C_3 - {}^{n-2}C_3 - {}^{n-2}C_2 = 15$

$n - 2 = 6 \Rightarrow n = 8$

Sol 16: (A) $n(2) = 50\ n(2 \cap 3) = 16$

$n(3) = 33\ n(3 \cap 5) = 6\ n(2 \cap 3 \cap 5) = 5$

$n(5) = 20\ n(2 \cap 5) = 10$

$n(2 \cup 3 \cup 5) = (50 + 33 + 20) - (16 + 6 + 10) + 3$

$n(\overline{2 \cup 3 \cup 5}) = 100 - 74 = 26$

Sol 17: (B) $^5C_1 \times \dfrac{6!}{2!} = 1800$

Choose subject with two periods and then arrange.

Sol 18: (B)

•	•	•	•			

$^4C_2 \times 2 \quad \times \quad 2 \quad \times \quad 10$

240

Sol 19: (A) Rectangles = $^7C_2 \times {}^5C_2 = 21 \times 10 = 210$

Squares = $6 \times 4 + 5 \times 3 + 4 \times 2 + 3 \times 1$

= 24 + 15 + 8 + 3 = 50

Rectangles which are not squares = 210 - 50 = 160

Sol 20: (B)

$$1\text{-----}\left(\begin{array}{l}\{15+10+6+3+1\}+\\ \{10+6+3+1\}\\ +\{6+3+1\}+\{3+1\}+1\end{array}\right)=70$$

$2\underline{3}\text{-----}\ 10+6+3+1 = 20+70 = 90$

$2\ 4\ 5\text{----}\ 3+2+1 = 6+90 = 96$

$2\ 4\ 6\ 7\ 8 = 97^{\text{th}}$

Sol 21: (B) $^nC_n + {}^{n+1}C_n + {}^{n+2}C_n + \ldots\ldots {}^{n+m}C_n$

$= {}^nC_0 + {}^{n+1}C_1 + {}^{n+2}C_2 + \ldots {}^{n+m}C_m - {}^{n+1}C_0 + {}^{n+1}C_0 = {}^{n+m+1}C_m$

Sol 22: (A) $\dfrac{1}{2} \times \dfrac{1}{2} \times {}^{m+n}C_m(m-1)(n-1)!$

$$= \frac{1}{4} \times \frac{(m+n)!}{mn}$$

Sol 23: (D) $\begin{vmatrix} 1 & x & x \\ x & 1 & x \\ x & x & 1 \end{vmatrix}\ {}^3C_1 = 3$

$\begin{vmatrix} 1 & x & x \\ x & 0 & x \\ x & x & 0 \end{vmatrix}\ {}^3C_1 \times {}^3C_2 = 9$

Total = 12

Sol 24: (C) 5   3   or   4   4

$^8C_5 + {}^8C_4 = 126$

Sol 25: (A) $100 = (80 + 50 + 30) - n + 5$

$n = 65$

People reading exactly 2 magazines

$= 65 - 3 \times 5 = 50$

Sol 26: (B) $2(5! - 4! \times 2) = 144$

Sol 27: (B)

Degree of 5 in prime factorization of $^{2002}C_{1001} = 1$

$^{2002}C_{1001}$ is clearly divisible by 2

Sol 28: (C) No. of triangles = $^nC_3 - [(n(n-4)) + n]$

$$= \frac{n(n-1)(n-2)}{3!} - [(n^2 - 3n)] = 30 \Rightarrow n = 9$$

Sol 29: (B) $^{11}C_3 - {}^5C_3 + 1 = 156$

Sol 30: (C)

$3! + 3 \times 2 \times 2! \ {}^4C_3 \times 3! = 18 = 24$

Total no.s = 18 + 24 = 42

Sol 31: (C) $P_n - Q_n$

The difference in P_n & Q_n is the number of ways in which first and last person of the row is selected

$P_n - Q_n = n - 4 = 6$

$n = 10$

Sol 32: (C) $m = {}^{10}C_4 \times 4!$

$n = {}^{10}C_4; \ m = 24 \ n$

Sol 33: (C) No. of ways of selecting a committee of 2W and 3M from 5W and 6M

$= {}^4C_1 \times {}^4C_2 + {}^5C_2 \times {}^5C_2 = 124$

Sol 34: (D) ABC, ACD, ABD

$3! + 3! + 3! = 18$

Sol 35: (D) Ways in which 2 are neighbours

$= 2 \times 97 + 97 \times 96 = 98 \times 97$

ways in which all 3 are neighbours = 98

Required ways = $^{100}C_3 - 98^2 = 152096 = {}^{98}C_3$

Sol 36: (D) Make 3 groups of boys 1, 1, 2

$$\frac{4!}{2! \ (1!)^2 \ 2!} \times 3! \text{ ways to distribute}$$

Identical marbles distribution

1 2 2– 3 ways

1 1 3 – 3 ways

Total ways = $\dfrac{4!}{2! \ 2!} \times 3! \times 6 = 216$

Sol 37: (C) $S_1 \ S_2 \ S_3$

$S_1 = S_3 = (2 + 4 + 6 + 8) \times 10 = 200$

$S_2 = (0 + 1 + 2 + \ldots\ldots 9) \times 4 = 180$

Sum = $200 + 180 \times 10 + 200 \times 100 = 22000$

Sol 38: (A) $A_n = {}^nC_{25} \times 25!$

$A_n - A_{n-1} = 25! \ ({}^nC_{25} - {}^{n-1}C_{25}) = 25!({}^{n-1}C_{24})$

$n = 50$

Sol 39: (C) $(x_1 + x_2 + x_3)(y_1 + y_2) = 77 = 7 \times 11$

$x_1 + x_2 + x_3 = 4 \ \& \ x_1 + x_2 + x_3 = 8$

$y_1 + y_2 = 9y_1 + y_2 = 5$

No. of possible solutions

$= {}^6C_2 \times {}^{10}C_1 + {}^{10}C_2 \times {}^6C_1 = 15 \times 10 + 45 \times 6$

$= 150 + 270 = 420$

Sol 40: (B) $m = 34 - 3 = 78$

$n = 3^4 - [3(2^4 - 2) + 3] = 81 - 45 = 36$

Hence, $\dfrac{m}{n} = \dfrac{78}{36} = \dfrac{13}{6}$

Sol 41: (A) Sum of digits at units place

$= (1 + 2 + \ldots + 9) \times 3 \times 9 = 45 \times 27 = 1215$

Sum of digits at ten's place

$= (1 + 4 + 9) \times 9 \times 9 = 14 \times 81 = 1134$

Sum of digits at hundred's place

$= (1 + 2 + \ldots\ldots 9) \times 3 \times 9 = 45 \times 27 = 1215$

$$\text{Sum} = \frac{\begin{array}{r} 1215 \\ 1134\times \\ 1215\times\times \end{array}}{134055}$$

Sol 42: (D) 3 possible distribution of wards for each subject.

1 1 4

1 2 3

2 2 2

Total ways = $3! \left(\dfrac{6!}{4! \ (1!)^2 \ 2!} + \dfrac{6!}{1! \ 2! \ 3!} + \dfrac{6!}{(2!)^3 3!} \right) = 540$

Sol 43: (C) $M = L \times 2!$

$L = N \qquad \therefore 2L = M = 2N$

Sol 44: (C) There are 11 position

At the 6th position A should be present. In the 5 positions left to 6th positions 2positions will have A.

$^5C_2$ ways

Sol 45: (D) $^9C_2 \times {}^7C_3 \times {}^4C_4$

(A) $\dfrac{9!}{2! \ 3! \ 4!}$

(D) $\dfrac{9!}{2! \ 3! \ 4!}$

Multiple Correct Choice Type

Sol 46: (A, B, D) $^{n-1}C_{r-1} + {}^{n-1}C_r = {}^nC_r$

Sol 47: (B, C) $2^{10} - 1$

Sol 48: (B, C, D) 2 2 1

$\dfrac{5!}{(2!)^2 \ 1! \ 2!} \times 3! = \dfrac{120}{8} \times 6 = 90$

Sol 49: (A, B) Let $k = 2n + 1$, then $^{2n+1}C_r$ is maximum when $r = n$. Also $^{2n+1}C_n = {}^{2n+1}C_{n+1}$. Thus, ^kC_r is maximum

when $r = \dfrac{1}{2}(k-1) \quad$ or $\quad r = \dfrac{1}{2}(k+1)$

Sol 50: (B, D) (A) 4 vowels, 7 consonants

$^7C_2 \times 6! = 3.7!$

(B) $\dfrac{15}{n!(15-x)!}$ n = no of white balls

(C) $\dfrac{12!}{4! \ 5!}$

(D) 35

Sol 51: (C, D) BULBUL

$\dfrac{6!}{2! \ 2! \ 2!} = 90$

(A) $^4C_2 \times {}^6C_2 = 6 \times 15 = 90$

(B) $\dfrac{6!}{(2!)^3 \ 3!} = 15$

(C) $\dfrac{6!}{(2!)^3} = 90$

(D) $\dfrac{6!}{(2!)^3 \ 3!} \times 3! = 90$

Comprehension Type

Sol 52: (C) M I(4) S(4)P(2)

(i) $^2C_1 \times {}^3C_1 = 6$

(ii) $^2C_1 \times {}^2C_1 = 4$

(iii) $^2C_1 \times {}^3C_1 = 6$

(iv) $^3C_2 \times 2 = 6$

(v) $^3C_1 = 3$

Adding all these, we get $= 25$

Sol 53: (B) Total ways in which all vowels are together

$= \dfrac{11!}{4! \ 4! \ 2!} - \dfrac{8!}{4! \ 2!} = \dfrac{8!}{4! \ 2!} \left(\dfrac{165}{4} - 1\right) = \dfrac{8!.161}{4 \ 4! \ 2!}$

Sol 54: (B) $^6C_2 \times {}^8C_4 \times 5 = 1 \times \left(\dfrac{10!}{4! \ 4!}\right)$

Sol 55: (A) No of ways $= \dfrac{16!}{(4!)^5}$

$= \dfrac{(1 \times 3 \times 5 .... \times 15) \times 2^8 \times 8^1}{(4!)^5}$

$= \prod_{r=1}^{8}(2r-1) \times \dfrac{2^8 \times 8 \times 7 \times 6 \times 5 \times 4!}{(4!)^5}$

$= 35 \prod_{r=1}^{8}(2r-1) \times \dfrac{2^8 \times 48}{(24)^4} = \dfrac{35}{27} \prod_{r=1}^{8}(2r-1)$

Sol 56: (C) $\dfrac{12!}{(3!)^4 \ 4!} \times 4! = \dfrac{12 \times 11!}{6^4} = \dfrac{11!}{108}$

Match the Columns

Sol 57: A $\to$ r; B $\to$ s; C $\to$ q; D $\to$ p

(A) $^nC_m \qquad$ (B) $m^n \qquad$ (C) $^mC_n \qquad$ (D) n^m

Sol 58: A $\to$ t; B $\to$ r; C $\to$ p; D $\to$ q; E $\to$ s

(A) x_1, x_2, x_3 & x_4 are the students watching a particular movie

$x_2 + x_2 + x_3 + x_4 = 10$

$x_i \geq 1$

$^9C_3 = \dfrac{9 \times 8 \times 7}{6} \, 84$

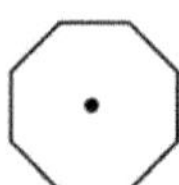

(B) $T = {}^9C_3 - 4 = 80$

$S = {}^8C_2 = 28$

$T - S = 52$

(C) $1 + {}^5C_2 = 11$

(D) $12 = 1 \times 2 \times 2 \times 3 = 1 \times 1 \times 4 \times 3 = 1 \times 1 \times 2 \times 6$

$\dfrac{4!}{2!} + \dfrac{4!}{2!} + \dfrac{4!}{2!} = 36$

(E) $^5C_2 \times {}^3C_2 \times 2 = 10 \times 3 \times 2 = 60$

$M_1 M_2 \, M_1 \, M_2$

$F_1 F_2 \, F_2 F_1$

Previous Years' Questions

Sol 1: Since, each box can hold five balls.

$\therefore$ Number of ways in which balls could be distributed so that none is empty are (2, 21) or (3, 1, 1).

ie, $({}^5C_2 \, {}^3C_2 \, {}^1C_1 + {}^5C_3 \, {}^2C_1 \, {}^1C_1) \times 3!$

$= (30 + 20) \times 6 = 300$

Sol 2: The possible cases are

Case-I: A man invites 3 ladies and women invites 3 gentlemen

Number of ways $= {}^4C_3 \cdot {}^4C_3 = 16$

Case-II: A man invites (2 ladies, 1 gentlemen) and women invites (2 gentlemen, 1 lady).

Number of ways

$= ({}^4C_2 \cdot {}^3C_1) \, ({}^3C_1 \cdot {}^4C_1) = 324$

Case-III: A man invites (1 lady, 2 gentlemen) and women invites (2 ladies, 1 gentleman).

Number of ways

$= ({}^4C_1 \cdot {}^3C_2) \cdot ({}^3C_2 \cdot {}^4C_1) = 144$

Case-IV: A man invites (3 gentlemen) and women invites (3 ladies).

Number of ways $= {}^3C_3 \cdot {}^3C_3 = 1$

$\therefore$ Total number of ways $= 16 + 324 + 144 + 1 = 485$

Sol 3: Case-I: When one black and two others ball S are drawn

$\Rightarrow$ number of ways $= {}^3C_1 \cdot {}^6C_2 = 45$

Case-II: When two black and one other balls are drawn

$\Rightarrow$ Number of ways $= {}^3C_2 \cdot {}^6C_1 = 18$

Case-III : When all three black balls are drawn

$\Rightarrow$ Number of ways $= {}^3C_3 = 1$

$\therefore$ Total number of ways $= 45 + 18 + 1 = 64$

Sol 4: Let the two sides be A and B. Assume that four particular guests wish to sit on side A. Four guests who wish to sit on side A can be accommodated on nine chairs in $^9P_4$ ways and there guests who wish to sit on side B and be accommodated in $^9P_3$ ways.

Now, the remaining guests are left who can sit on 11 chairs on both the sides of the table in (11!) ways.

Hence, the total number of ways in which 18 persons can be seated $= {}^9P_4 \times {}^9P_3 \times (11)!$.

Sol 5: There are 9 women and 8 men. A committee of 12, consisting of at least 5 women, can be formed by choosing:

(i) 5 women and 7 men

(ii) 6 women and 6 men

(iii) 7 women and 5 men

(iv) 8 women and 4 men

(v) 9 women and 3 men

$\therefore$ Total number of ways forming the committee

$= {}^9C_5 \times {}^8C_7 + {}^9C_6 \times {}^8C_6 + {}^9C_7 \times {}^8C_5 + {}^9C_8 \times {}^8C_4 + {}^9C_9 \times {}^8C_3$

$= 126 \times 8 + 84 \times 28 + 36 \times 56 + 9 \times 70 + 1 \times 56 = 6062$

(i) Clearly, women are in majority in (iii), (iv) and (v) cases as discussed above.

(ii) So, total number of committees in which women are in majority

$= {}^9C_7 \times {}^8C_5 + {}^9C_8 \times {}^8C_4 + {}^9C_9 \times {}^8C_3$

$= 36 \times 56 + 9 \times 70 + 1 \times 56 = 2702$

Clearly, men are in majority in only (i) case as discussed above.

So, total number of committees in which men are in majority

$= {}^9C_5 \times {}^8C_7 = 126 \times 8 = 1008$

Sol 6: A $\to$ p; B $\to$ s; C $\to$ q; D $\to$ q

(A) If ENDEA is fixed word, then assume this as a single letter.

Total number of letters = 5

Total number of arrangements = 5!

(B) If E is at first and last places, then total number of permutation of $\dfrac{7!}{2!} = 21 \times 5!$

(C) If D, L, N are not in last five positions

$\leftarrow$ D, I, N, N $\to$ $\leftarrow$ E, E, E, A, O $\to$

Total number of permutation

$= \dfrac{4!}{2!} \times \dfrac{5!}{3!} = 2 \times 5!$

(D) Total number of odd position = 5

Permutation of AEEEO are $\dfrac{5!}{3!}$

Total number of even positions = 4

Number of permutations of N, N, D, L = $\dfrac{4!}{2!}$

Hence, total number of permutation = $\dfrac{5!}{3!} \times \dfrac{4!}{2!} = 2 \times 5!$